Student Solutic

for

Finite Mathematics

Fourth Edition

Stefan Waner

Hofstra University

Steven R. Costenoble

Hofstra University

THOMSON

BROOKS/COLE

Australia • Brazil • Canada • Mexico • Singapore • Spain • United Kingdom • United States

Printed in the United States of America
1 2 3 4 5 6 7 10 09 08 07 06

Printer: Thomson/West

0-495-01689-6

Cover image: Pebbles. Graeme Harris

Thomson Higher Education
10 Davis Drive
Belmont, CA 94002-3098
USA

For more information about our products, contact us at:
Thomson Learning Academic Resource Center
1-800-423-0563

For permission to use material from this text or product, submit a request online at
http://www.thomsonrights.com.
Any additional questions about permissions can be submitted by email to **thomsonrights@thomson.com.**

Table of Contents

Chapter 0

0.1

1. $2(4 + (-1))(2 \cdot -4)$
$= 2(3)(-8) = (6)(-8) = -48$

3. `20/(3*4)-1`
$= \dfrac{20}{12} - 1 = \dfrac{5}{3} - 1 = \dfrac{2}{3}$

5. $\dfrac{3 + ([3 + (-5)])}{3 - 2 \times 2} = \dfrac{3 + (-2)}{3 - 4}$
$= \dfrac{1}{-1} = -1$

7. `(2-5*(-1))/1-2*(-1)`
$= \dfrac{2 - 5 \cdot (-1)}{1} - 2 \cdot (-1)$
$= \dfrac{2+5}{1} + 2 = 7 + 2 = 9$

9. $2 \cdot (-1)^2 / 2 = \dfrac{2 \times (-1)^2}{2} = \dfrac{2 \times 1}{2} = \dfrac{2}{2} = 1$

11. $2 \cdot 4^2 + 1 = 2 \times 16 + 1 = 32 + 1 = 33$

13. `3^2+2^2+1`
$= 3^2 + 2^2 + 1 = 9 + 4 + 1 = 14$

15. $\dfrac{3 - 2(-3)^2}{-6(4 - 1)^2} = \dfrac{3 - 2 \times 9}{-6(3)^2} = \dfrac{3 - 18}{6 \times 9}$
$= \dfrac{-15}{54} = \dfrac{5}{18}$

17. `10*(1+1/10)^3`
$= 10\left(1 + \dfrac{1}{10}\right)^3 = 10(1.1)^3$
$= 10 \times 1.331 = 13.31$

19. $3\left[\dfrac{-2 \cdot 3^2}{-(4 - 1)^2}\right] = 3\left[\dfrac{-2 \times 9}{-3^2}\right] = 3\left[\dfrac{-18}{-9}\right]$
$= 3 \times 2 = 6$

21. $3\left[1 - \left(-\dfrac{1}{2}\right)^2\right]^2 + 1 = 3\left[1 - \dfrac{1}{4}\right]^2 + 1$
$= 3\left[\dfrac{3}{4}\right]^2 = 3\left[\dfrac{9}{16}\right] + 1 = \dfrac{27}{16} + 1 = \dfrac{43}{16}$

23. `(1/2)^2-1/2^2`
$= \left[\dfrac{1}{2}\right]^2 - \dfrac{1}{2^2} = \dfrac{1}{4} - \dfrac{1}{4} = 0$

25. $3 \times (2-5) =$ `3*(2-5)`

27. $\dfrac{3}{2-5} =$ `3/(2-5)`
Note `3/2-5` is wrong, since it corresponds to $\dfrac{3}{2} - 5$.

29. $\dfrac{3-1}{8+6} =$ `(3-1)/(8+6)`
Note `3-1/8-6` is wrong, since it corresponds to $3 - \dfrac{1}{8} - 6$.

31. $3 - \dfrac{4+7}{8} =$ `3-(4+7)/8`

33. $\dfrac{2}{3+x} - xy^2 =$ `2/(3+x)-x*y^2`

35. $3.1x^3 - 4x^{-2} - \dfrac{60}{x^2-1}$
$=$ `3.1x^3-4x^(-2)-60/(x^2-1)`

37. $\dfrac{\left[\dfrac{2}{3}\right]}{5} =$ `(2/3)/5`
Note that we use only (round) parentheses in technology formulas, and not brackets.

39. $3^{4-5} \times 6 =$ `3^(4-5)*6`
Note that the entire exponent is in parentheses.

41. $3\left[1 + \dfrac{4}{100}\right]^{-3} =$ `3*(1+4/100)^(-3)`
Note that we use only (round) parentheses in technology formulas, and not brackets.

43. $3^{2x-1} + 4^x - 1 = $ `3^(2*x-1)+4^x-1`
Note that the entire exponent of 3 is in parentheses.

45. $2^{2x^2-x+1} = $ `2^(2x^2-x+1)`
Note that the entire exponent is in parentheses.

47. $\dfrac{4e^{-2x}}{2-3e^{-2x}}$

= `4*e^(-2*x)/(2-3e^(-2*x))`
or `4(*e^(-2*x))/(2-3e^(-2*x))`
or `(4*e^(-2*x))/(2-3e^(-2*x))`

49. $3\left[1 - \left(-\frac{1}{2}\right)^2\right]^2 + 1$

= `3(1-(-1/2)^2)^2+1`
Note that we use only (round) parentheses in technology formulas, and not brackets.

0.2

1. $3^3 = 27$

3. $-(2 \cdot 3)^2 = -(2^2 \cdot 3^2) = -(4 \cdot 9) = -36$ or
$-(2 \cdot 3)^2 = -(6^2) = -36$

5. $\left(\dfrac{-2}{3}\right)^2 = \dfrac{(-2)^2}{3^2} = \dfrac{4}{9}$

7. $(-2)^{-3} = \dfrac{1}{(-2)^3} = \dfrac{1}{-8} = -\dfrac{1}{8}$

9. $\left(\dfrac{1}{4}\right)^{-2} = \dfrac{1}{(1/4)^2} = \dfrac{1}{1/4^2} = \dfrac{1}{1/16} = 16$

11. $2 \cdot 3^0 = 2 \cdot 1 = 2$

13. $2^3 \, 2^2 = 2^{3+2} = 2^5 = 32$ or $2^3 \, 2^2 = 8 \cdot 4 = 32$

15. $2^2 \, 2^{-1} \, 2^4 \, 2^{-4} = 2^{2-1+4-4} = 2^1 = 2$

17. $x^3 x^2 = x^{3+2} = x^5$

19. $-x^2 x^{-3} y = -x^{2-3} y = -x^{-1} y = -\dfrac{y}{x}$

21. $\dfrac{x^3}{x^4} = x^{3-4} = x^{-1} = \dfrac{1}{x}$

23. $\dfrac{x^2 y^2}{x^{-1} y} = x^{2-(-1)} y^{2-1} = x^3 y$

25. $\dfrac{(xy^{-1}z^3)^2}{x^2 yz^2} = \dfrac{x^2 (y^{-1})^2 (z^3)^2}{x^2 yz^2} = x^{2-2} y^{-2-1} z^{6-2} =$
$y^{-3} z^4 = \dfrac{z^4}{y^3}$

27. $\left(\dfrac{xy^{-2}z}{x^{-1}z}\right)^3 = \dfrac{(xy^{-2}z)^3}{(x^{-1}z)^3} = \dfrac{x^3 y^{-6} z^3}{x^{-3} z^3} = x^{3-(-3)} y^{-6} z^{3-3}$
$= x^6 y^{-6} = \dfrac{x^6}{y^6}$

29. $\left(\dfrac{x^{-1}y^{-2}z^2}{xy}\right)^{-2} = (x^{-1-1}y^{-2-1}z^2)^{-2} = (x^{-2}y^{-3}z^2)^{-2}$
$= x^4 y^6 z^{-4} = \dfrac{x^4 y^6}{z^4}$

31. $3x^{-4} = \dfrac{3}{x^4}$

33. $\dfrac{3}{4} x^{-2/3} = \dfrac{3}{4x^{2/3}}$

35. $1 - \dfrac{0.3}{x^{-2}} - \dfrac{6}{5} x^{-1} = 1 - 0.3x^2 - \dfrac{6}{5x}$

37. $\sqrt{4} = 2$

39. $\sqrt{\dfrac{1}{4}} = \dfrac{\sqrt{1}}{\sqrt{4}} = \dfrac{1}{2}$

41. $\sqrt{\dfrac{16}{9}} = \dfrac{\sqrt{16}}{\sqrt{9}} = \dfrac{4}{3}$

43. $\dfrac{\sqrt{4}}{5} = \dfrac{2}{5}$

45. $\sqrt{9} + \sqrt{16} = 3 + 4 = 7$

47. $\sqrt{9 + 16} = \sqrt{25} = 5$

49. $\sqrt[3]{8 - 27} = \sqrt[3]{-19} \approx -2.668$

51. $\sqrt[3]{\dfrac{27}{8}} = \dfrac{\sqrt[3]{27}}{\sqrt[3]{8}} = \dfrac{3}{2}$

53. $\sqrt{(-2)^2} = \sqrt{4} = 2$

Section 0.2

55. $\sqrt{\frac{1}{4}(1+15)} = \sqrt{\frac{16}{4}} = \frac{\sqrt{16}}{\sqrt{4}} = \frac{4}{2} = 2$

57. $\sqrt{a^2 b^2} = \sqrt{a}\sqrt{b} = ab$

59. $\sqrt{(x+9)^2} = x+9$ ($x + 9 > 0$ because x is positive)

61. $\sqrt[3]{x^3(a^3+b^3)} = \sqrt[3]{x^3}\sqrt[3]{a^3+b^3} = x\sqrt[3]{a^3+b^3}$

(Notice: *Not* $x(a + b)$.)

63. $\sqrt{\frac{4xy^3}{x^2 y}} = \sqrt{\frac{4y^2}{x}} = \frac{\sqrt{4}\sqrt{y^2}}{\sqrt{x}} = \frac{2y}{\sqrt{x}}$

65. $\sqrt{3} = 3^{1/2}$

67. $\sqrt{x^3} = x^{3/2}$

69. $\sqrt[3]{xy^2} = (xy^2)^{1/3}$

71. $\frac{x^2}{\sqrt{x}} = \frac{x^2}{x^{1/2}} = x^{2-1/2} = x^{3/2}$

73. $\frac{3}{5x^2} = \frac{3}{5}x^{-2}$

75. $\frac{3x^{-1.2}}{2} - \frac{1}{3x^{2.1}} = \frac{3}{2}x^{-1.2} - \frac{1}{3}x^{-2.1}$

77. $\frac{2x}{3} - \frac{x^{0.1}}{2} + \frac{4}{3x^{1.1}} = \frac{2}{3}x - \frac{1}{2}x^{0.1} + \frac{4}{3}x^{-1.1}$

79. $\frac{1}{(x^2+1)^3} - \frac{3}{4\sqrt[3]{(x^2+1)}} =$

$\frac{1}{(x^2+1)^3} - \frac{3}{4(x^2+1)^{1/3}} =$

$(x^2+1)^{-3} - \frac{3}{4}(x^2+1)^{-1/3}$

81. $2^{2/3} = \sqrt[3]{2^2}$

83. $x^{4/3} = \sqrt[3]{x^4}$

85. $(x^{1/2}y^{1/3})^{1/5} = \sqrt[5]{\sqrt{x}\,\sqrt[3]{y}}$

87. $-\frac{3}{2}x^{-1/4} = -\frac{3}{2x^{1/4}} = -\frac{3}{2\sqrt[4]{x}}$

89. $0.2x^{-2/3} + \frac{3}{7x^{-1/2}} = \frac{0.2}{x^{2/3}} + \frac{3x^{1/2}}{7} = \frac{0.2}{\sqrt[3]{x^2}} + \frac{3\sqrt{x}}{7}$

91. $\frac{3}{4(1-x)^{5/2}} = \frac{3}{4\sqrt{(1-x)^5}}$

93. $4^{-1/2}4^{7/2} = 4^{-1/2+7/2} = 4^3 = 64$

95. $3^{2/3}3^{-1/6} = 3^{2/3-1/6} = 3^{1/2} = \sqrt{3}$

97. $\frac{x^{3/2}}{x^{5/2}} = x^{3/2-5/2} = x^{-1} = \frac{1}{x}$

99. $\frac{x^{1/2}y^2}{x^{-1/2}y} = x^{1/2+1/2}y^{2-1} = xy$

101. $\left(\frac{x}{y}\right)^{1/3}\left(\frac{y}{x}\right)^{2/3} = \left(\frac{y}{x}\right)^{-1/3}\left(\frac{y}{x}\right)^{2/3} = \left(\frac{y}{x}\right)^{1/3}$

103. $x^2 - 16 = 0,\ x^2 = 16,\ x = \pm\sqrt{16} = \pm 4$

105. $x^2 - \frac{4}{9} = 0,\ x^2 = \frac{4}{9},\ x = \pm\sqrt{\frac{4}{9}} = \pm\frac{2}{3}$

4

107. $x^2 - (1 + 2x)^2 = 0$, $x^2 = (1 + 2x)^2$, $x = \pm(1 + 2x)$; if $x = 1 + 2x$ then $-x = 1$, $x = -1$; if $x = -(1 + 2x)$ then $3x = -1$, $x = -1/3$. So, $x = -1$ or $-1/3$.

109. $x^5 + 32 = 0$, $x^5 = -32$, $x = \sqrt[5]{-32} = -2$

111. $x^{1/2} - 4 = 0$, $x^{1/2} = 4$, $x = 4^2 = 16$

113. $1 - \dfrac{1}{x^2} = 0$, $1 = \dfrac{1}{x^2}$, $x^2 = 1$, $x = \pm\sqrt{1} = \pm 1$

115. $(x - 4)^{-1/3} = 2$, $x - 4 = 2^{-3} = \dfrac{1}{8}$, $x = 4 + \dfrac{1}{8} = \dfrac{33}{8}$

0.3

1. $x(4x + 6) = 4x^2 + 6x$

3. $(2x - y)y = 2xy - y^2$

5. $(x + 1)(x - 3) = x^2 + x - 3x - 3 = x^2 - 2x - 3$

7. $(2y + 3)(y + 5) = 2y^2 + 3y + 10y + 15 = 2y^2 + 13y + 15$

9. $(2x - 3)^2 = 4x^2 - 12x + 9$

11. $\left(x + \dfrac{1}{x}\right)^2 = x^2 + 2 + \dfrac{1}{x^2}$

13. $(2x - 3)(2x + 3) = (2x)^2 - 3^2 = 4x^2 - 9$

15. $\left(y - \dfrac{1}{y}\right)\left(y + \dfrac{1}{y}\right) = y^2 - \left(\dfrac{1}{y}\right)^2 = y^2 - \dfrac{1}{y^2}$

17. $(x^2 + x - 1)(2x + 4) = (x^2 + x - 1)2x + (x^2 + x - 1)4 = 2x^3 + 2x^2 - 2x + 4x^2 + 4x - 4 = 2x^3 + 6x^2 + 2x - 4$

19. $(x^2 - 2x + 1)^2 = (x^2 - 2x + 1)(x^2 - 2x + 1) = x^2(x^2 - 2x + 1) - 2x(x^2 - 2x + 1) + (x^2 - 2x + 1) = x^4 - 2x^3 + x^2 - 2x^3 + 4x^2 - 2x + x^2 - 2x + 1 = x^4 - 4x^3 + 6x^2 - 4x + 1$

21. $(y^3 + 2y^2 + y)(y^2 + 2y - 1) = y^3(y^2 + 2y - 1) + 2y^2(y^2 + 2y - 1) + y(y^2 + 2y - 1) = y^5 + 2y^4 - y^3 + 2y^4 + 4y^3 - 2y^2 + y^3 + 2y^2 - y = y^5 + 4y^4 + 4y^3 - y$

23. $(x + 1)(x + 2) + (x + 1)(x + 3) = (x + 1)(x + 2 + x + 3) = (x + 1)(2x + 5)$

25. $(x^2 + 1)^5(x + 3)^4 + (x^2 + 1)^6(x + 3)^3 = (x^2 + 1)^5(x + 3)^3(x + 3 + x^2 + 1) =$

$(x^2 + 1)^5(x + 3)^3(x^2 + x + 4)$

27. $(x^3 + 1)\sqrt{x + 1} - (x^3 + 1)^2\sqrt{x + 1} = (x^3 + 1)\sqrt{x + 1}\,[1 - (x^3 + 1)] = -x^3(x^3 + 1)\sqrt{x + 1}$

29. $\sqrt{(x + 1)^3} + \sqrt{(x + 1)^5} = \sqrt{(x + 1)^3} \cdot [1 + \sqrt{(x + 1)^2}\,] = \sqrt{(x + 1)^3}\,(1 + x + 1) = (x + 2)\sqrt{(x + 1)^3}$

31. (a) $2x + 3x^2 = x(2 + 3x)$ **(b)** $x(2 + 3x) = 0$; $x = 0$ or $2 + 3x = 0$; $x = 0$ or $-2/3$

33. (a) $6x^3 - 2x^2 = 2x^2(3x - 1)$
(b) $2x^2(3x - 1) = 0$; $x = 0$ or $3x - 1 = 0$; $x = 0$ or $1/3$

35. (a) $x^2 - 8x + 7 = (x - 1)(x - 7)$
(b) $(x - 1)(x - 7) = 0$; $x - 1 = 0$ or $x - 7 = 0$; $x = 1$ or 7

37. (a) $x^2 + x - 12 = (x - 3)(x + 4)$
(b) $(x - 3)(x + 4) = 0$; $x - 3 = 0$ or $x + 4 = 0$; $x = 3$ or -4

39. (a) $2x^2 - 3x - 2 = (2x + 1)(x - 2)$
(b) $(2x + 1)(x - 2) = 0$; $2x + 1 = 0$ or $x - 2 = 0$; $x = -1/2$ or 2

41. (a) $6x^2 + 13x + 6 = (2x + 3)(3x + 2)$
(b) $(2x + 3)(3x + 2) = 0$; $2x + 3 = 0$ or $3x + 2 = 0$; $x = -3/2$ or $-2/3$

43. (a) $12x^2 + x - 6 = (3x - 2)(4x + 3)$
(b) $(3x - 2)(4x + 3) = 0$; $3x - 2 = 0$ or $4x + 3 = 0$; $x = 2/3$ or $-3/4$

45. (a) $x^2 + 4xy + 4y^2 = (x + 2y)^2$
(b) $(x + 2y)^2 = 0$; $x + 2y = 0$; $x = -2y$

47. (a) $x^4 - 5x^2 + 4 = (x^2 - 1)(x^2 - 4) =$
$(x - 1)(x + 1)(x - 2)(x + 2)$
(b) $(x - 1)(x + 1)(x - 2)(x + 2) = 0$; $x - 1 = 0$
or $x + 1 = 0$ or $x - 2 = 0$ or $x + 2 = 0$; $x = \pm 1$
or ± 2

0.4

1. $\dfrac{x-4}{x+1} \cdot \dfrac{2x+1}{x-1} = \dfrac{(x-4)(2x+1)}{(x+1)(x-1)} =$

$\dfrac{2x^2 - 7x - 4}{x^2 - 1}$

3. $\dfrac{x-4}{x+1} + \dfrac{2x+1}{x-1} =$

$\dfrac{(x-4)(x-1) + (x+1)(2x+1)}{(x+1)(x-1)} =$

$\dfrac{3x^2 - 2x + 5}{x^2 - 1}$

5. $\dfrac{x^2}{x+1} - \dfrac{x-1}{x+1} = \dfrac{x^2 - (x-1)}{x+1} = \dfrac{x^2 - x + 1}{x+1}$

7. $\dfrac{1}{\left(\dfrac{x}{x-1}\right)} + x - 1 = \dfrac{x-1}{x} + x - 1 =$

$\dfrac{x - 1 + x(x-1)}{x} = \dfrac{x^2 - 1}{x}$

9. $\dfrac{1}{x}\left(\dfrac{x-3}{xy} + \dfrac{1}{y}\right) = \dfrac{1}{x}\left(\dfrac{x-3+x}{xy}\right) = \dfrac{2x-3}{x^2 y}$

11. $\dfrac{(x+1)^2(x+2)^3 - (x+1)^3(x+2)^2}{(x+2)^6} =$

$\dfrac{(x+1)^2(x+2)^2[(x+2) - (x+1)]}{(x+2)^6} = \dfrac{(x+1)^2}{(x+2)^4}$

13. $\dfrac{(x^2-1)\sqrt{x^2+1} - \dfrac{x^4}{\sqrt{x^2+1}}}{x^2+1} =$

$\dfrac{(x^2-1)(x^2+1) - x^4}{(x^2+1)\sqrt{x^2+1}} = \dfrac{-1}{\sqrt{(x^2+1)^3}}$

15. $\dfrac{\dfrac{1}{(x+y)^2} - \dfrac{1}{x^2}}{y} = \dfrac{x^2 - (x+y)^2}{yx^2(x+y)^2} =$

$\dfrac{x^2 - x^2 - 2xy - y^2}{yx^2(x+y)^2} = \dfrac{-y(2x+y)}{yx^2(x+y)^2} = \dfrac{-(2x+y)}{x^2(x+y)^2}$

0.5

1. $x + 1 = 0$, $x = -1$

3. $-x + 5 = 0$, $x = 5$

5. $4x - 5 = 8$, $4x = 13$, $x = 13/4$

7. $7x + 55 = 98$, $7x = 43$, $x = 43/7$

9. $x + 1 = 2x + 2$, $-x = 1$, $x = -1$

11. $ax + b = c$, $ax = c - b$, $x = (c - b)/a$

13. $2x^2 + 7x - 4 = 0$, $(2x - 1)(x + 4) = 0$, $x = -4, \dfrac{1}{2}$

15. $x^2 - x + 1 = 0$, $\Delta = -3 < 0$, so this equation has no real solutions

17. $2x^2 - 5 = 0$, $x^2 = \dfrac{5}{2}$, $x = \pm\sqrt{\dfrac{5}{2}}$

19. $-x^2 - 2x - 1 = 0$, $-(x + 1)^2 = 0$, $x = -1$

21. $\dfrac{1}{2}x^2 - x - \dfrac{3}{2} = 0$, $x^2 - 2x - 3 = 0$, $(x + 1)(x - 3) = 0$, $x = -1, 3$

23. $x^2 - x = 1$, $x^2 - x - 1 = 0$, $x = \dfrac{1 \pm \sqrt{5}}{2}$ by the quadratic formula

25. $x = 2 - \dfrac{1}{x}$, $x^2 = 2x - 1$, $x^2 - 2x + 1 = 0$, $(x - 1)^2 = 0$, $x = 1$

27. $x^4 - 10x^2 + 9 = 0$, $(x^2 - 1)(x^2 - 9) = 0$, $x^2 = 1$ or $x^2 = 0$, $x = \pm1, \pm3$

29. $x^4 + x^2 - 1 = 0$, $x^2 = \dfrac{-1 \pm \sqrt{5}}{2}$ by the quadratic formula, $x = \pm\sqrt{\dfrac{-1 \pm \sqrt{5}}{2}}$

31. $x^3 + 6x^2 + 11x + 6 = 0$, $(x + 1)(x + 2)(x + 3) = 0$, $x = -1, -2, -3$

33. $x^3 + 4x^2 + 4x + 3 = 0$, $(x + 3)(x^2 + x + 1) = 0$, $x = -3$ (For $x^2 + x + 1 = 0$, $\Delta = -3 < 0$, so there are no real solutions to this quadratic equation.)

35. $x^3 - 1 = 0$, $x^3 = 1$, $x = \sqrt[3]{1} = 1$

37. $y^3 + 3y^2 + 3y + 2 = 0$, $(y + 2)(y^2 + y + 1) = 0$, $y = -2$ (For $y^2 + y + 1 = 0$, $\Delta = -3 < 0$, so there are no real solutions to this quadratic equation.)

39. $x^3 - x^2 - 5x + 5 = 0$, $(x - 1)(x^2 - 5) = 0$, $x = 1, \pm\sqrt{5}$

41. $2x^6 - x^4 - 2x^2 + 1 = 0$, $(2x^2 - 1)(x^4 - 1) = 0$, [or $(2x^2 - 1)(x^2 - 1)(x^2 + 1) = 0$; in any case, think of the cubic you get by substituting y for x^2], $x = \pm1, \pm\dfrac{1}{\sqrt{2}}$

43. $(x^2 + 3x + 2)(x^2 - 5x + 6) = 0$, $(x + 2)(x + 1)(x - 2)(x - 3) = 0$, $x = -2, -1, 2, 3$

0.6

1. $x^4 - 3x^3 = 0$, $x^3(x - 3) = 0$, $x = 0, 3$

3. $x^4 - 4x^2 = -4$, $x^4 - 4x^2 + 4 = 0$, $(x^2 - 2)^2 = 0$, $x = \pm\sqrt{2}$

5. $(x + 1)(x + 2) + (x + 1)(x + 3) = 0$, $(x + 1)(x + 2 + x + 3) = 0$, $(x + 1)(2x + 5) = 0$, $x = -1, -5/2$

7. $(x^2 + 1)^5(x + 3)^4 + (x^2 + 1)^6(x + 3)^3 = 0$, $(x^2 + 1)^5(x + 3)^3(x + 3 + x^2 + 1) = 0$, $(x^2 + 1)^5(x + 3)^3(x^2 + x + 4) = 0$, $x = -3$ (Neither $x^2 + 1 = 0$ nor $x^2 + x + 4 = 0$ has a real solution.)

9. $(x^3 + 1)\sqrt{x + 1} - (x^3 + 1)^2\sqrt{x + 1} = 0$, $(x^3 + 1)\sqrt{x + 1}\,[1 - (x^3 + 1)] = 0$, $-x^3(x^3 + 1)\sqrt{x + 1} = 0$, $x = 0, -1$

11. $\sqrt{(x + 1)^3} + \sqrt{(x + 1)^5} = 0$, $\sqrt{(x + 1)^3}\,(1 + x + 1) = 0$, $(x + 2)\sqrt{(x + 1)^3} = 0$, $x = -1$ ($x = -2$ is not a solution because $\sqrt{(x + 1)^3}$ is not defined for $x = -2$.)

13. $(x + 1)^2(2x + 3) - (x + 1)(2x + 3)^2 = 0$, $(x + 1)(2x + 3)(x + 1 - 2x - 3) = 0$, $(x + 1)(2x + 3)(-x - 2) = 0$, $x = -2, -3/2, -1$

15. $\dfrac{(x + 1)^2(x + 2)^3 - (x + 1)^3(x + 2)^2}{(x + 2)^6} = 0$,

$\dfrac{(x + 1)^2(x + 2)^2[(x + 2) - (x + 1)]}{(x + 2)^6} = 0$, $\dfrac{(x + 1)^2}{(x + 2)^4}$

$= 0$, $(x + 1)^2 = 0$, $x = -1$

17. $\dfrac{2(x^2 - 1)\sqrt{x^2 + 1} - \dfrac{x^4}{\sqrt{x^2 + 1}}}{x^2 + 1} = 0$,

$\dfrac{2(x^2 - 1)(x^2 + 1) - x^4}{(x^2 + 1)\sqrt{x^2 + 1}} = 0$, $\dfrac{x^4 - 2}{(x^2 + 1)\sqrt{x^2 + 1}} = 0$,

$x^4 - 2 = 0$, $x = \pm\sqrt[4]{2}$

19. $x - \dfrac{1}{x} = 0$, $x^2 - 1 = 0$, $x = \pm 1$

21. $\dfrac{1}{x} - \dfrac{9}{x^3} = 0$, $x^2 - 9 = 0$, $x = \pm 3$

23. $\dfrac{x - 4}{x + 1} - \dfrac{x}{x - 1} = 0$,

$\dfrac{(x - 4)(x - 1) - x(x + 1)}{(x + 1)(x - 1)} = 0$,

$\dfrac{-6x + 4}{(x + 1)(x - 1)} = 0$, $-6x + 4 = 0$, $x = 2/3$

25. $\dfrac{x + 4}{x + 1} + \dfrac{x + 4}{3x} = 0$,

$\dfrac{3x(x + 4) + (x + 1)(x + 4)}{3x(x + 1)} = 0$,

$\dfrac{(x + 4)(3x + x + 1)}{3x(x + 1)} = 0$, $\dfrac{(x + 4)(4x + 1)}{3x(x + 1)} = 0$,

$(x + 4)(4x + 1) = 0$, $x = -4, -1/4$

Chapter 1
1.1

1. Using the table,

(a) $f(0) = 2$ **(b)** $f(2) = 0.5$.

3. Using the table,

(a) $f(2) - f(-2) = 0.5 - 2 = -1.5$

(b) $f(-1)f(-2) = (4)(2) = 8$

(c) $-2f(-1) = -2(4) = -8$

5. $f(x) = 4x - 3$

(a) $f(-1) = 4(-1) - 3 = -4 - 3 = -7$

(b) $f(0) = 4(0) - 3 = 0 - 3 = -3$

(c) $f(1) = 4(1) - 3 = 4 - 3 = 1$

(d) Substitute y for x to obtain

$$f(y) = 4y - 3$$

(e) Substitute $(a+b)$ for x to obtain

$$f(a+b) = 4(a+b) - 3$$

7. $f(x) = x^2 + 2x + 3$

(a) $f(0) = (0)^2 + 2(0) + 3$

$= 0 + 0 + 3 = 3$

(b) $f(1) = 1^2 + 2(1) + 3$

$= 1 + 2 + 3 = 6$

(c) $f(-1) = (-1)^2 + 2(-1) + 3$

$= 1 - 2 + 3 = 2$

(d) $f(-3) = (-3)^2 + 2(-3) + 3$

$= 9 - 6 + 3 = 6$

(e) Substitute a for x to obtain

$$f(a) = a^2 + 2a + 3$$

(f) Substitute $(x+h)$ for x to obtain

$$f(x+h) = (x+h)^2 + 2(x+h) + 3$$

9. $g(s) = s^2 + \dfrac{1}{s}$

(a) $g(1) = 1^2 + \dfrac{1}{1} = 1 + 1 = 2$

(b) $g(-1) = (-1)^2 + \dfrac{1}{(-1)} = 1 - 1 = 0$

(c) $g(4) = 4^2 + \dfrac{1}{4} = 16 + \dfrac{1}{4}$

$= 16\tfrac{1}{4}$ or $\dfrac{65}{4}$ or 16.25

(d) Substitute x for s to obtain

$$g(x) = x^2 + \dfrac{1}{x}$$

(e) Substitute $(s+h)$ for s to obtain

$$g(s+h) = (s+h)^2 + \dfrac{1}{s+h}$$

(f) $g(s+h) - g(s)$

$= $ Answer to part (e) $-$ Original function

$$= \left((s+h)^2 + \dfrac{1}{s+h}\right) - \left(s^2 + \dfrac{1}{s}\right)$$

11. $f(t) = \begin{cases} -t & \text{if } t < 0 \\ t^2 & \text{if } 0 \le t < 4 \\ t & \text{if } t \ge 4 \end{cases}$

(a) $f(-1) = -(-1) = 1$ (using the first formula, since $-1 < 0$).

(b) $f(1) = 1^2 = 1$ (using the second formula, since $0 \le 1 < 4$).

(c) $f(4) = 4$ (using the third formula, since $4 \ge 4$).

$f(2) = 2^2 = 4$ (using the second formula, since $0 \le 2 < 4$).

Therefore,

$f(4) - f(2) = 4 - 4 = 0$

(d) $f(3) = 3^2 = 9$ (using the second formula, since $0 \le 3 < 4$).

$f(-3) = -(-3) = 3$ (using the first formula, since $-3 < 0$).

Therefore,

$f(3)f(-3) = (9)(3) = 27$.

13. $f(x) = x - \dfrac{1}{x^2}$, with domain $(0, +\infty)$

(a) Since 4 is in $(0, +\infty)$, $f(4)$ is defined.

$$f(4) = 4 - \dfrac{1}{4^2} = 4 - \dfrac{1}{16} = \dfrac{63}{16}$$

(b) Since 0 is not in $(0, +\infty)$, $f(0)$ is not defined.

11

(c) Since -1 is not in $(0, +\infty)$, $f(-1)$ is not defined.

15. $f(x) = \sqrt{x+10}$, with domain $[-10, 0)$
(a) Since 0 is not in $[-10, 0)$, $f(0)$ is not defined.
(b) Since 9 is not in $[-10, 0)$, $f(9)$ is not defined.
(c) Since -10 is in $[-10, 0)$, $f(-10)$ is defined.
$f(-10) = \sqrt{-10+10} = \sqrt{0} = 0$

17. $f(x) = x^2$
(a) $f(x+h) = (x+h)^2$
Therefore,
$$\begin{aligned} f(x+h) - f(x) &= (x+h)^2 - x^2 \\ &= x^2+2xh+h^2 - x^2 \\ &= 2xh + h^2 \\ &= h(2x + h) \end{aligned}$$
(b) Using the answer to part (a)
$$\begin{aligned} \frac{f(x+h) - f(x)}{h} &= \frac{h(2x + h)}{h} \\ &= 2x + h \end{aligned}$$

19. $f(x) = 2-x^2$
(a) $f(x+h) = 2-(x+h)^2$
Therefore,
$$\begin{aligned} &f(x+h) - f(x) \\ &= 2-(x+h)^2 - (2-x^2) \\ &= 2-x^2-2xh-h^2 - 2+x^2 \\ &= -2xh-h^2 \\ &= -h(2x + h) \end{aligned}$$

(b) Using the answer to part (a)
$$\begin{aligned} \frac{f(x+h) - f(x)}{h} &= -\frac{h(2x + h)}{h} \\ &= -(2x + h) \end{aligned}$$

21. $f(x) = 0.1x^2 - 4x + 5$
Technology formula: `0.1*x^2-4*x+5`

Table of Values:

x	0	1	2	3
$f(x)$	5	1.1	-2.6	-6.1
x	4	5	6	7
$f(x)$	-9.4	-12.5	-15.4	-18.1
x	8	9	10	
$f(x)$	-20.6	-22.9	-25	

23. $h(x) = \dfrac{x^2-1}{x^2+1}$

Technology formula: `(x^2-1)/(x^2+1)`
Table of Values (rounded to four decimal places):

x	0.5	1.5	2.5	3.5
$h(x)$	-0.6000	0.3846	0.7241	0.8491
x	4.5	5.5	6.5	7.5
$h(x)$	0.9059	0.9360	0.9538	0.9651
x	8.5	9.5	10.5	
$h(x)$	0.9727	0.9781	0.9820	

25.
(a) Using the table,
$P(5) = 117 \qquad P(10) = 132$
Since $P(9) = 130$ and $P(10) = 132$,
we estimate $P(9.5)$ to be midway between these 130 and 132:
$$P(9.5) \approx \frac{130+132}{2} = 131.$$

Interpretation:
Since $P(t)$ gives the number of million people employed in July 1 of year t, we have:
$P(5) = 117$: Approximately 117 million people were employed in the US on July 1, 1995.
$P(10) = 132$: Approximately 132 million people were employed in the US on July 1, 2000.
$P(9.5) \approx 131$: Approximately 131 million people were employed in the US on January 1, 2000. (Note that January 1, 2000 is midway between July 1, 1999 and July 1, 2000.)
(b) The table gives $P(t)$ for $5 \le t \le 11$. Hence, the domain of P is the set of numbers t with $5 \le t \le 11$, that is, $[5, 11]$.

27. (a) The model is valid for the range 1994 ($t = 0$) through 2004 ($t = 10$). Thus, an appropriate domain is [0, 10]. $t \geq 0$ is not an appropriate domain because it would predict U.S. trade with China into the indefinite future with no basis.

(b) $C(t) = 3t^2 - 7t + 50$

$C(10) = 3(10)^2 - 7(10) + 50 = 300 - 70 + 50 = \$280$ billion.

Since $t = 10$ represents 2004, the answer tells us that U.S. trade with China in 2004 was valued at approximately $280 billion.

29. The technology formulas are:

(1) `-0.2*t^2+t+16`

(2) `0.2*t^2+t+16`

(3) `t+16`

The following table shows the values predicted by the three models.

t	0	2	4	6	7
S	16	18	22	28	30
(1)	16	17.2	16.8	14.8	13.2
(2)	16	18.8	23.2	29.2	32.8
(3)	16	18	20	22	23

As shown in the table, the values predicted by model (2) are much closer to the observed values S than those predicted by the other models.

(b) Since 1998 corresponds to $t = 8$,

$$S(t) = 0.2\,t^2 + t + 16$$
$$S(8) = 0.2(8)^2 + 8 + 16 = 36.8$$

So the spending on corrections in 1998 was predicted to be $36.8 billion.

31. $q(p) = 361{,}201 - (p+1)^2$

(a) Note p is expressed in cents, not dollars, so 50¢ is represented by $p = 50$, not 0.50.

$$q(50) = 361{,}201 - (50+1)^2$$
$$= 361{,}201 - 2601$$
$$= 358{,}600 \text{ brownie dishes}$$

(b) If they give them away, $p = 0$, so

$$q(0) = 361{,}201 - (0+1)^2$$
$$= 361{,}201 - 1$$
$$= 361{,}200 \text{ brownie dishes}$$

(c) If they sell no dishes, $q = 0$, and so

$$0 = 361{,}201 - (p+1)^2$$
$$(p+1)^2 = 361{,}201$$
$$p+1 = \sqrt{361{,}201} = 601$$
$$p = 601 - 1 = 600¢, \text{ or } \$6.00.$$

33. $P(t) = \begin{cases} 75t + 200 & if\ 0 \leq t \leq 4 \\ 600t - 1900 & if\ 4 < t \leq 9 \end{cases}$

(a) $P(0) = 75(0) + 200 = 200$. We used the first formula, since 0 is in [0, 4].

$P(4) = 75(4) + 200 = 300 + 200 = 500$. We used the first formula, since 4 is in [0, 4].

$P(5) = 600(5) - 1900 = 3000 - 1900 = 1100$. We used the second formula, since 5 is in (4, 9].

Interpretation:

Since $P(t)$ gives the speed of Intel processors at the start of year 1995 + t, we have:

$P(0) = 200$: At the start of 1995, the processor speed was 200 megahertz.

$P(4) = 500$: At the start of 1999, the processor speed was 500 megahertz.

$P(5) = 1100$: At the start of 2000, the processor speed was 1100 megahertz.

(b) Since 2.0 gigahertz = 2000 megahertz, we want $P(t) = 2000$. The first formula gives speeds no higher than 500 megahertz (see part (a)), so we use the second formula instead:

$$P(t) = 2000$$
$$600t - 1900 = 2000$$
$$600t = 3900$$
$$t = 3900/600 = 6.5,$$

or midway through 2001

(c) A technology formula for *P* is

```
(75*t+200)*(t<=4)+
     (600*t-1900)*(t>4)
```

(For a graphing calculator, use x instead of t, and ≤ instead of <=)

Table of Values:

t	P(t)
0	200
1	275
2	350
3	425
4	500
5	1100
6	1700
7	2300
8	2900
9	3500

35. $C(t) = \begin{cases} 0.08t + 0.6 & \text{if } 0 \le t < 8 \\ 0.355t - 1.6 & \text{if } 8 \le t \le 11 \end{cases}$

(a) A technology formula for *C* is

```
(0.08*t+0.6)*(t<8)+(0.355*t-
1.6)*(t>=8)
```

(For a graphing calculator, use x instead of t, and ≥ instead of >=)

Table of Values:

t	0	1	2	3	4	5
C(t)	0.6	0.68	0.76	0.84	0.92	1
t	6	7	8	9	10	11
C(t)	1.08	1.16	1.24	1.595	1.95	2.305

(b) The costs of a Superbowl ad in 1998, 1999, and 2000 were:

1998: $C(8) = \$1.24$ million

1999: $C(9) = \$1.595$ million

2000: $C(10) = \$1.95$ million

From 1998 to 1999, the cost increased by $1.595 − $1.24 = \$0.355$ million dollars.

From 1999 to 2000, the cost increased by $1.95 − $1.595 = \$0.355$ million dollars.

Thus, the cost increased at a rate of $0.355 million dollars (or $355,000) per year between 1998 and 2000.

37. A taxable income of $26,000 falls in the category "Over $7,300 but not over $29,700" and so we use the formula "$730.00 + 15% of the amount over $7,300."

$$T(26,000) = \$730 + 0.15(26,000 - 7300)$$
$$= \$3535.00$$

A taxable income of $65,000 falls in the category "Over $29,700 but not over $71,950" and so we use the formula "$4,090.00 + 25% of the amount over $29,700."

$$T(65,000) = \$4090 + 0.25(65,000 - 29,700)$$
$$= \$12,915$$

39. $C(q) = 2000 + 100q^2$

(a) $C(10) = 2000 + 100(10)^2 = 2000 + 10,000$
$$= \$12,000$$

(b) Net Cost = Total Cost − Subsidy

$$N(q) = C(q) - S(q)$$
$$= 2000 + 100q^2 - 500q$$
$$N(20) = 2000 + 100(20)^2 - 500(20)$$
$$= 2000 + 40,000 - 10,000$$
$$= \$32,000$$

41. $p(t) = 100\left(1 - \dfrac{12,200}{t^{4.48}}\right)$ $\qquad (t \ge 8.5)$

(a) Technology formula:

```
100*(1-12200/t^4.48)
```

(b) Table of values:

t	9	10	11	12	13	14
p(t)	35.2	59.6	73.6	82.2	87.5	91.1
t	15	16	17	18	19	20
p(t)	93.4	95.1	96.3	97.1	97.7	98.2

(c) From the table, $p(12) = 82.2$, so that 82.2% of children are able to speak in at least single words by the age of 12 months.

(d) We seek the first value of t such that $p(t)$ is at least 90. Since $t = 14$ has this property ($p(14) = 91.1$) we conclude that, at 14 months, 90% or more children are able to speak in at least single words.

43. The dependent variable is a function of the independent variable. Here, the market price of gold m is a function of time t. Thus, the independent variable is t and the dependent variable is m.

45. To obtain the function notation, write the dependent as a function of the independent variable. Thus $y = 4x^2 - 2$ can be written as
$$f(x) = 4x^2 - 2 \quad \text{or} \quad y(x) = 4x^2 - 2$$

47. Number of sound files
= Starting number + New files
= $200 + 10 \times$ Number of days
So, $\quad N(t) = 200 + 10t$
(N = number of sound files, t = time in days)

49. As the text reminds us: to evaluate f of a quantity (such as $x+h$) replace x everywhere by the *whole quantity* $x+h$:
$$f(x) = x^2 - 1$$
$$f(x+h) = (x+h)^2 - 1.$$

51. Functions with infinitely many points in their domain (such as $f(x) = x^2$) cannot be specified numerically. So, the assertion is false.

1.2

1. From the graph, we find
(a) $f(1) = 20$ **(b)** $f(2) = 30$

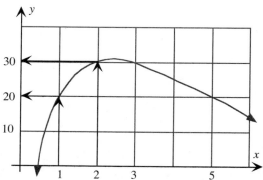

In a similar way, we find

(c) $f(3) = 30$ **(d)** $f(5) = 20$

(e) $f(3) - f(2) = 30 - 30 = 0$

3. From the graph,

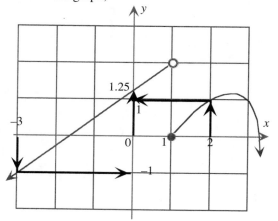

(a) $f(-3) = -1$ **(b)** $f(0) = 1.25$

(c) $f(1) = 0$ since the solid dot is on $(1, 0)$
    ($f(1) \neq 2$ since the hollow dot at $(1, 2)$
    indicates that $(1, 2)$ is not a point of the graph.)

(d) $f(2) = 1$
(e) Since $f(3) = 1$ and $f(2) - 1$,
$$\frac{f(3) - f(2)}{3 - 2} = \frac{1 - 1}{3 - 2} = 0$$

5.
(a) $f(x) = x \ (-1 \leq x \leq 1)$
Since the graph of $f(x) = x$ is a diagonal $45°$ line through the origin inclining up from left to right, the correct graph is (I)
(b) $f(x) = -x \ (-1 \leq x \leq 1)$
Since the graph of $f(x) = -x$ is a diagonal $45°$ line through the origin inclining down from left to right, the correct graph is (IV)
(c) $f(x) = \sqrt{x} \ (0 < x < 4)$
Since the graph of $f(x) = \sqrt{x}$ is the top half of a sideways parabola, the correct graph is (V)
(d) $f(x) = x + \dfrac{1}{x} - 2 \ (0 < x < 4)$

If we plot a few points like $x = 0.1, 1, 2,$ and 3 we find that the correct graph is (VI).
(e) $f(x) = |x| \ (-1 \leq x \leq 1)$
Since the graph of $f(x) = |x|$ is a "V"-shape with its vertex at the origin, the correct graph is (III).
(f) $f(x) = x - 1 \ (-1 \leq x \leq 1)$
Since the graph of $f(x) = x-1$ is a straight line through $(0, -1)$ and $(1, 0)$, the correct graph is (II).

7. $f(x) = -x^3$ (domain $(-\infty, +\infty)$)
Technology formula: $-(x^3)$

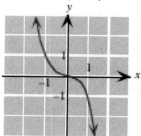

9. $f(x) = x^4$ (domain $(-\infty, +\infty)$)

Technology formula: **x^4**

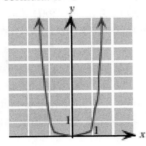

11. $f(x) = \dfrac{1}{x^2}$ $(x \neq 0)$

Technology formula: **1/x^2**

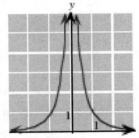

13. $f(x) = \begin{cases} x & \text{if } -4 \leq x < 0 \\ 2 & \text{if } 0 \leq x \leq 4 \end{cases}$

Technology formula: **x*(x<0)+2*(x>=0)**

(For a graphing calculator, use $\geq$ instead of >=.)

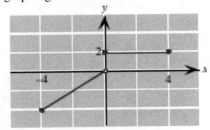

(a) $f(-1) = -1$. We used the first formula, since -1 is in $[-4, 0)$.

(b) $f(0) = 2$. We used the second formula, since 0 is in $[0, 4]$.

(c) $f(1) = 2$. We used the second formula, since 1 is in $[0, 4]$.

15. $f(x) = \begin{cases} x^2 & \text{if } -2 < x \leq 0 \\ 1/x & \text{if } 0 < x \leq 4 \end{cases}$

Technology formula:
(x^2)*(x<=0)+(1/x)*(0<x)

(For a graphing calculator, use $\leq$ instead of <=.)

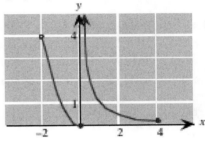

(a) $f(-1) = 1^2 = 1$. We used the first formula, since -1 is in $(-2, 0]$.

(b) $f(0) = 0^2 = 0$. We used the first formula, since 0 is in $(-2, 0]$.

(c) $f(1) = 1/1 = 1$. We used the second formula, since 1 is in $(0, 4]$.

17. $f(x) = \begin{cases} x & \text{if } -1 < x \leq 0 \\ x+1 & \text{if } 0 < x \leq 2 \\ x & \text{if } 2 < x \leq 4 \end{cases}$

Technology formula:

$$x*(x<=0)+(x+1)*(0<x)*(x<=2)+$$
$$x*(2<x)$$

(For a graphing calculator, use $\leq$ instead of <=.)

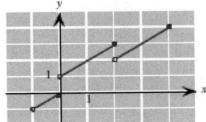

(a) $f(0) = 0$. We used the first formula, since 0 is in $(-1, 0]$.

(b) $f(1) = 1+1 = 2$. We used the second formula, since 1 is in $(0, 2]$.

(c) $f(2) = 2+1 = 3$. We used the second formula, since 2 is in (0, 2].

(d) $f(3) = 3$. We used the third formula, since 3 is in (2, 4].

19. Reading from the graph, $f(6) \approx 2000$, $f(9) \approx 2800$, $f(7.5) \approx 2500$. Interpretation: Since $f(t)$ is the is the number of SUVs (in thousands) sold in the U.S. in the year starting $t+1990$, we interpret the results as follows:

$f(6) \approx 2000$: In 1996 (1990+6), 2,000,000 SUVs were sold.

$f(9) \approx 2800$: In 1999 (1990+9), 2,800,000 were sold.

$f(7.5) \approx 2500$: Since 1990+7.5 = 1997.5, or half way through 1997, we interpret this by saying that, in the year beginning July, 1997, 2,500,000 were sold.

21. From the graph,

$f(6)-f(5) \approx 2000-1750 = 250$

$f(10)-f(9) \approx 2900-2750 = 150$

Therefore, $f(6)-f(5)$ is larger.

Interpretation: In general, the difference $f(b)-f(b)$ measures the change in sales from year a to year b. So, SUV sales increased more from 1995 to 1996 than from 1999 to 2000.

23.

(a) From the graph, we see that $N(t)$ is defined for $-1.5 \leq t \leq 1.5$. Thus, the domain of N is $[-1.5, 1.5]$.

(b) From the graph, $N(-0.5) \approx 131$, $N(0) \approx 132$, $N(1) \approx 132$. Since $N(t)$ gives the number of million people employed in the U.S. in year Jan. 2000+t, we interpret the results as follows:

$N(-0.5) \approx 131$: In July, 1999 (Jan. 2000 $-$ 0.5), approximately 131 million people were employed.

$N(0) \approx 132$: In January 2000, approximately 132 million people were employed.

$N(1) \approx 132$: In January 2001 Jan. 2000 + 1), approximately 132 million people were employed.

(c) The graph is descending from around $t = 0.5$ to $t = 1.5$. Thus, $N(t)$ is falling on the interval [0.5, 1.5], showing that employment was falling during the period July, 2000–July, 2001.

25.

(a) The technology formulas are:

(A): `0.005*x+2.75`

(B): `0.01*x+20+25/x`

    (C): `0.0005*x^2-0.07*x+23.25`

    (D): `25.5*1.08^(x-5)`

The following table shows the values predicted by the four models.

x	5	25	40	100	125
$A(x)$	22.91	21.81	21.25	21.25	22.31
(A)	2.775	2.875	2.95	3.25	3.375
(B)	25.05	21.25	21.025	21.25	21.45
(C)	22.913	21.813	21.25	21.25	22.313
(D)	25.5	118.85	377.03	38177	261451

Model (C) fits the data almost perfectly—more closely that any of the other models.

(b) Graph of model (C):

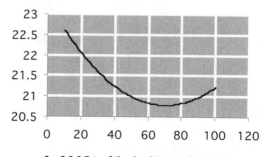

`0.0005*x^2-0.07*x+23.25`

The lowest point on the graph occurs at $x = 70$ with a y-coordinate of 20.8. Thus, the lowest cost per shirt is $20.80, which the team can obtain by buying 70 shirts.

27. A plot of the given data suggests a curve with a low point somewhere between $t = 0$ and $t = 5$. A linear model would predict perpetually increasing or decreasing value of the euro (depending on whether the slope is positive or negative) and an exponential model $n(t) = Ab^t$ would also be perpetually increasing or decreasing (depending whether b is larger than 1 or less than 1). This leaves a quadratic model as the only possible choice. In fact, a quadratic can always be found that passes through any three points not on the same straight line with different x-coordinates. Therefore, a quadratic model would give an exact fit.

29. $p(t) = 100\left(1 - \dfrac{12{,}200}{t^{4.48}}\right)$ $\qquad$ $(t \geq 8.5)$

(a) Technology formula:
$$100*(1-12200/t\char`^4.48)$$

(b) Graph:

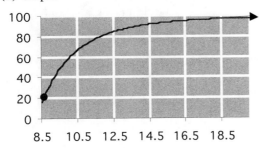

(c) If we zoom in to the graph near $x = 12$ we find that $p(12) \approx 82$. Thus, 82% of children are able to speak in at least single words by the age of 12 months.

(d) We need to find the approximate value of t so that $p(t) = 90$. From the graph above, it occurs somewhere between $t = 12.5$ and 14.5. Here is a close-up of the graph near that point:

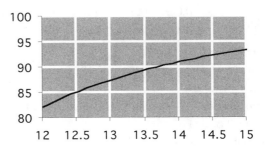

This close-up shows that $f(13.7) \approx 90$. Therefore, 90% of children speaking in at least single words at approximately 13.7 months of age, or 14 months if we round to the nearest month.

31. (a) Technology formula:
$$(75*t+200)*(t<=4)+$$
$$(600*t-1900)*(t>4)$$
(For a graphing calculator, use x instead of t, and $\geq$ instead of >=)
Graph:

(b) To answer the question, we need to find the value of t such that $P(t) = 2000$ (because 2.0 gigahertz = 2000 megahertz). From the graph, $P(6.5) \approx 2000$, and 6.5 years since the start of 1995 was midway through 2001. Thus, processor speeds first reached 2.0 gigahertz midway through 2001.

33. $C(t) = \begin{cases} 0.08t + 0.6 & \text{if } 0 \le t < 8 \\ 0.355t - 1.6 & \text{if } 8 \le t \le 11 \end{cases}$

(a) Technology formula:

`(0.08*t+0.6)*(t<8)+(0.355*t-`
`1.6)*(t>=8)`

(For a graphing calculator, use x instead of t, and $\ge$ instead of >=)

Graph:

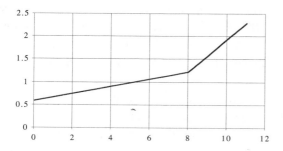

(b) To answer the question, we need to find the first integer value of t such that $C(t)$ exceeds 2. Although $C(10) \approx 2$, it is a little less than 2. On the other hand, $C(11) \ge 2$, and $t = 11$ represents the first year such that $C(t) \ge 2$. Thus, a Superbowl ad first exceeded \$2 million in 1990+11 = 2001.

35. True. A graphically specified function is specified by a graph. Given a graph, we can read off a set of values to construct a table, and hence specify the function numerically. (The more accurate the graph is, the more accurate the numerical values are.)

37. False. A numerically specified function with domain [0, 10] is specified by a table of some values between 0 and 10. Since only certain values of the function are specified, we can obtain only certain points on the graph.

39. If two functions are specified by the same formula $f(x)$ say, their graphs must follow the same

curve $y = f(x)$. However, it is the domain of the function that specifies what portion of the curve appears on the graph. Thus, if the functions have different domains, their graphs will be different portions of the curve $y = f(x)$.

41. Suppose we already have the graph of f and want to construct the graph of g. We can plot a point of the graph of g as follows: Choose a value for x ($x = 7$, say) and then "look back" 5 units to read off $f(x-5)$ ($f(2)$ in this instance). This value gives the y-coordinate we want. In other words, points on the graph of g are obtained by "looking back 5 units" to the graph of f and then copying that portion of the curve. Put another way, the graph of g is the same as the graph of f, but shifted 5 units to the right.

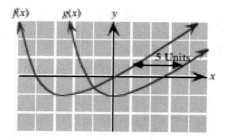

1.3

1.

x	-1	0	1
y	5	8	

We calculate the slope m first. The first two points shown give a changes in x and y of

$$\Delta x = 0 - (-1) = 1$$
$$\Delta y = 8 - 5 = 3$$

This gives a slope of

$$m = \frac{\Delta y}{\Delta x} = \frac{3}{1} = 3.$$

Now look at the second and third points: The change in x is again

$$\Delta x = 1 - 0 = 1$$

and so Δy must be given by the formula

$$\Delta y = m\Delta x$$
$$\Delta y = 3(1) = 3.$$

This means that the missing value of y is

$$8 + \Delta y = 8 + 3 = 11.$$

3.

x	2	3	5
y	-1	-2	

We calculate the slope m first. The first two points shown give a changes in x and y of

$$\Delta x = 3 - 2 = 1$$
$$\Delta y = -2 - (-1) = -1$$

This gives a slope of

$$m = \frac{\Delta y}{\Delta x} = \frac{-1}{1} = -1.$$

Now look at the second and third points: The change in x is

$$\Delta x = 5 - 3 = 2$$

and so Δy must be given by the formula

$$\Delta y = m\Delta x$$
$$\Delta y = (-1)(2) = -2.$$

This means that the missing value of y is

$$-2 + \Delta y = -2 + (-2) = -4.$$

5.

x	-2	0	2
y	4		10

We calculate the slope m first. The first and third points shown give a changes in x and y of

$$\Delta x = 2 - (-2) = 4$$
$$\Delta y = 10 - 4 = 6$$

This gives a slope of

$$m = \frac{\Delta y}{\Delta x} = \frac{6}{4} = \frac{3}{2}.$$

Now look at the first and second points: The change in x is

$$\Delta x = 0 - (-2) = 2$$

and so Δy must be given by the formula

$$\Delta y = m\Delta x$$
$$\Delta y = (\frac{3}{2})(2) = 3.$$

This means that the missing value of y is

$$4 + \Delta y = 4 + 3 = 7.$$

7. From the table,

$$b = f(0) = -2.$$

The slope (using the first two points) is

$$m = \frac{y_2 - y_2}{x_2 - x_1} = \frac{-2 - (-1)}{0 - (-2)} = \frac{-1}{2} = -\frac{1}{2}.$$

Thus, the linear equation is

$$f(x) = mx + b = -\frac{1}{2}x - 2,$$

or $f(x) = -\frac{x}{2} - 2.$

9. The slope (using the first two points) is

$$m = \frac{y_2 - y_2}{x_2 - x_1} = \frac{-2 - (-1)}{-3 - (-4)} = \frac{-1}{1} = -1.$$

To obtain $f(0) = b$, notice that, since the slope is -1, y decreases by 1 for every one-unit increase in x. Thus,

$$f(0) = f(-1) + m = -4 - 1 = -5.$$

This gives

$$f(x) = mx + b = -x - 5.$$

11. In the table, x increases in steps of 1 and f increases in steps of 4, showing that f is linear with slope

$$m = \frac{\Delta y}{\Delta x} = \frac{4}{1} = 4$$

and intercept

$$b = f(0) = 6$$

giving

$$f(x) = mx + b = 4x + 6.$$

The function g does not increase in equal steps, so g is not linear.

13. In the first three points listed in the table, x increases in steps of 3, but f does not increase in equal steps, whereas g increases in steps of 6. Thus, based on the first three points, only g could possibly be linear, with slope

$$m = \frac{\Delta y}{\Delta x} = \frac{6}{3} = 2$$

and intercept

$$b = g(0) = -1$$

giving

$$g(x) = mx + b = 2x - 1.$$

We can now check that the remaining points in the table fit the formula $g(x) = 2x - 1$, showing that g is indeed linear.

15. Slope = coefficient of $x = -\dfrac{3}{2}$

17. Write the equation as

$$y = \frac{x}{6} + \frac{1}{6}$$

Slope = coefficient of $x = \dfrac{1}{6}$

19. If we solve for x we find that the given equation represents the vertical line $x = -1/3$, and so its slope is infinite (undefined).

21. $3y + 1 = 0$

Solving for y:

$$3y = -1$$
$$y = -\frac{1}{3}$$

Slope = coefficient of $x = 0$

23. $4x + 3y = 7$

Solve for y:

$$3y = -4x + 7$$
$$y = -\frac{4}{3}x + \frac{7}{3}$$

Slope = coefficient of $x = -\dfrac{4}{3}$

25. $y = 2x - 1$

y–intercept $= -1$, slope $= 2$

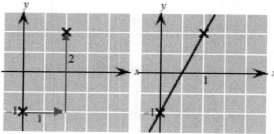

27. $y = -\frac{2}{3}x + 2$

y–intercept $= 2$, slope $= -\dfrac{2}{3}$

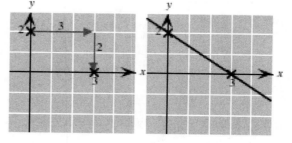

29. $y + \frac{1}{4}x = -4$

Solve for y to obtain $y = -\frac{1}{4}x - 4$

y−intercept = −4, slope = $-\frac{1}{4}$

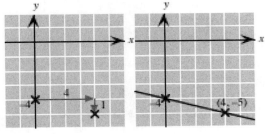

31. $7x - 2y = 7$

Solve for y:

$$-2y = -7x + 7$$
$$y = \frac{7}{2}x - \frac{7}{2}$$

y−intercept = $-\frac{7}{2}$ = −3.5, slope = $\frac{7}{2}$ = 3.5

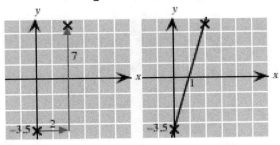

33. $3x = 8$

Solve for x to obtain $x = \frac{8}{3}$.

The graph is a vertical line:

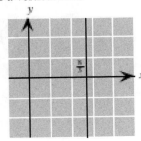

35. $6y = 9$

Solve for y to obtain $y = \frac{9}{6} = \frac{3}{2} = 1.5$

y−intercept = $\frac{3}{2}$ = 1.5, slope = 0

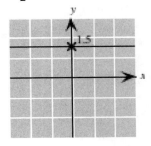

37. $2x = 3y$

Solve for y to obtain $y = \frac{2}{3}x$

y−intercept = 0, slope = $\frac{2}{3}$

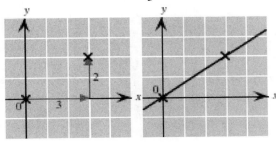

39. (0, 0) and (1, 2)

$$m = \frac{y_2 - y_1}{x_2 - x_1} = \frac{2 - 0}{1 - 0} = 2$$

41. (−1, −2) and (0, 0)

$$m = \frac{y_2 - y_1}{x_2 - x_1} = \frac{0 - (-2)}{0 - (-1)} = \frac{2}{1} = 2$$

43. (4, 3) and (5, 1)

$$m = \frac{y_2 - y_1}{x_2 - x_1} = \frac{1 - 3}{5 - 4} = \frac{-2}{1} = -2$$

45. $(1, -1)$ and $(1, -2)$

$$m = \frac{y_2 - y_1}{x_2 - x_1} = \frac{-2 - (-1)}{1 - 1} \text{ Undefined}$$

47. $(2, 3.5)$ and $(4, 6.5)$

$$m = \frac{y_2 - y_1}{x_2 - x_1} = \frac{6.5 - 3.5}{4 - 2} = \frac{3}{2} = 1.5$$

49. $(300, 20.2)$ and $(400, 11.2)$

$$m = \frac{y_2 - y_1}{x_2 - x_1} = \frac{11.2 - 20.2}{400 - 300}$$
$$= \frac{-9}{100} = -0.09$$

51. $(0, 1)$ and $(-\frac{1}{2}, \frac{3}{4})$

$$m = \frac{y_2 - y_1}{x_2 - x_1} = \frac{\frac{3}{4} - 1}{-\frac{1}{2} - 0}$$
$$= \frac{-\frac{1}{4}}{-\frac{1}{2}} = \frac{1}{4} \cdot 2 = \frac{1}{2}$$

53. (a, b) and (c, d) $(a \neq c)$

$$m = \frac{y_2 - y_1}{x_2 - x_1} = \frac{d - b}{c - a}$$

55.

(a)

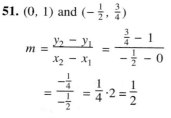

$$m = \frac{\Delta y}{\Delta x} = \frac{1}{1}$$
$$= 1$$

(b)

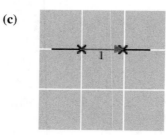

$$m = \frac{\Delta y}{\Delta x} = \frac{1}{2}$$

(c)

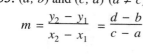

$$m = \frac{\Delta y}{\Delta x} = \frac{0}{1}$$
$$= 0$$

(d)

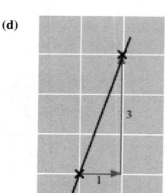

$$m = \frac{\Delta y}{\Delta x} = \frac{3}{1}$$
$$= 3$$

(e)

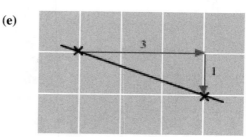

$$m = \frac{\Delta y}{\Delta x} = \frac{-1}{3} = -\frac{1}{3}$$

(f)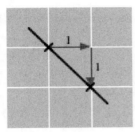

$$m = \frac{\Delta y}{\Delta x} = \frac{-1}{1}$$
$$= -1$$

$$y = 3x$$

(g)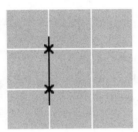

Vertical line; undefined slope

59. Through $(1, -\frac{3}{4})$ with slope $\frac{1}{4}$

Point: $(1, -\frac{3}{4})$ **Slope:** $m = \frac{1}{4}$

$$b = y_1 - mx_1$$
$$= -\frac{3}{4} - \frac{1}{4}\cdot 1 = -1$$

Thus, the equation is

$$y = mx + b$$
$$y = \frac{1}{4}x - 1$$

(h)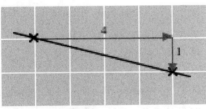

$$m = \frac{\Delta y}{\Delta x} = \frac{-1}{4} = -\frac{1}{4}$$

61. Through $(20, -3.5)$ and increasing at a rate of 10 units of y per unit of x

Point: $(20, -3.5)$

Slope: $m = \frac{\Delta y}{\Delta x} = \frac{10}{1} = 10$

$$b = y_1 - mx_1$$
$$= -3.5 - (10)(20)$$
$$= -3.5 - 200 = -203.5$$

Thus, the equation is

$$y = mx + b$$
$$y = 10x - 203.5$$

(i)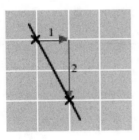

$$m = \frac{\Delta y}{\Delta x} = \frac{-2}{1}$$
$$= -2$$

63. Through $(2, -4)$ and $(1, 1)$

Point: $(2, -4)$

Slope: $m = \frac{y_2 - y_1}{x_2 - x_1} = \frac{1 - (-4)}{1 - 2} = \frac{5}{-1} = -5$

$$b = y_1 - mx_1$$
$$= -4 - (-5)(2) = 6$$

Thus, the equation is

$$y = mx + b$$
$$y = -5x + 6$$

57. Through $(1, 3)$ with slope 3

Point: $(1, 3)$ **Slope:** $m = 3$

$$b = y_1 - mx_1$$
$$= 3 - 3\cdot 1 = 0$$

Thus, the equation is

$$y = mx + b$$
$$y = 3x + 0$$

65. Through $(1, -0.75)$ and $(0.5, 0.75)$

Point: $(1, -0.75)$

Slope: $m = \frac{y_2 - y_1}{x_2 - x_1} = \frac{0.75 - (-0.75)}{0.5 - 1}$
$$= \frac{1.5}{-0.5} = -3$$

25

$b = y_1 - mx_1$

$= -0.75 - (-3)(1) = -0.75+3 = 2.25$

Thus, the equation is

$y = mx + b$

$y = -3x + 2.25$

67. Through (6, 6) and parallel to the line $x + y = 4$

Point: (6, 6)

Slope: Same as slope of $x + y = 4$. To find the slope, solve for y, getting

$y = -x + 4$

Thus, $m = -1$.

$b = y_1 - mx_1$

$= 6 - (-1)(6) = 6+6 = 12$

Thus, the equation is

$y = mx + b$

$y = -x + 12$

69. Through (0.5, 5) and parallel to the line $4x - 2y = 11$

Point: (0.5, 5)

Slope: Same as slope of $4x - 2y = 11$. To find the slope, solve for y, getting

$2y = 4x - 11$

$y = 2x - \dfrac{11}{2}$

Thus, $m = 2$.

$b = y_1 - mx_1$

$= 5 - (2)(0.5) = 5 - 1 = 4$

Thus, the equation is

$y = mx + b$

$y = 2x + 4$

71. A table of values of x and y comes from a linear equation precisely when the successive ratios $\Delta y/\Delta x$ are all the same. Thus, to test such a table of values, compute the corresponding successive changes Δx in x and Δy in y, and compute the ratios $\Delta y/\Delta x$. If the answer is always the same number,

then the values in the table come from a linear function.

73. To find the linear function, solve the equation $ax + by = c$ for y:

$by = -ax + c$

$y = -\dfrac{a}{b}x + \dfrac{c}{b}$

Thus, the desired function is $f(x) = -\dfrac{a}{b}x + \dfrac{c}{b}$.

If $b = 0$, then $\dfrac{a}{b}$ and $\dfrac{c}{b}$ are undefined, and y cannot be specified as a function of x. (The graph of the resulting equation would be a vertical line.)

75. The slope of the line is $m = \dfrac{\Delta y}{\Delta x} = \dfrac{3}{1} = 3$.

Therefore, if, in a straight line, y is increasing three times as fast as x, then its <u>slope</u> is <u>3</u>.

77. If m is positive then y will increase as x increases; if m is negative then y will decrease as x increases; if m is zero then y will not change as x changes.

79.

	A	B	C	D
1	x	y	m	b
2	1		2	=(B3-B2)/(A3-A2)	=B2-C2*A2
3	3	-1	Slope	Intercept

The slope computed in cell C2 is given by

$$m = \dfrac{y_2 - y_1}{x_2 - x_1} = \dfrac{-1 - 2}{3 - 1} = -1.5$$

If we increase the y-coordinate in cell B3, this increases y_2, and thus increases the numerator $\Delta y = y_2 - y_1$ without effecting the denominator Δx. Thus the slope will increase.

1.4

1. For a linear cost function,

$$C(x) = mx + b$$

m = marginal cost = \$1500 per piano

b = fixed cost = \$1200

Thus, the daily cost function is

$$C(x) = 1500x + 1200.$$

(a) The cost of manufacturing 3 pianos is

$$C(3) = 1500(3) + 1200$$
$$= 4500 + 1200 = \$5700$$

(b) The cost of manufacturing each additional piano (such as the third one or the 11th one) is the marginal cost, m = \$1500.

(c) Same answer as (b).

3. We are given two points on the graph of the linear cost function: (100, 10,500) and (120, 11,0000) (x is the number of items, and y is the cost C).

Marginal cost:

$$m = \frac{C_2 - C_1}{x_2 - x_1} = \frac{11{,}000 - 10{,}500}{120 - 100}$$
$$= \frac{500}{20} = \$25 \text{ per bicycle.}$$

Fixed cost:

$$b = C_1 - mx_1 = 10{,}500 - (25)(100)$$
$$= 10{,}500 - 2500 = \$8000$$

5.

(a) For a linear cost function,

$$C(x) = mx + b.$$

m = marginal cost = \$0.40 per copy

b = fixed cost = \$70

Thus, the cost function is

$$C(x) = 0.4x + 70.$$

The revenue function is

$$R(x) = 0.50x \quad (x \text{ copies @ } 50\text{¢ per copy})$$

The profit function is

$$P(x) = R(x) - C(x)$$

$$= 0.5x - (0.4x + 70)$$
$$= 0.5x - 0.4x - 70$$
$$= 0.1x - 70$$

(b) $P(500) = 0.1(500) - 70$
$$= 50 - 70 = -20$$

Since P is negative, this represents a loss of \$20.

(c) For break-even,

$$P(x) = 0$$
$$0.1x - 70 = 0$$
$$0.1x = 70$$
$$x = \frac{70}{0.1} = 700 \text{ copies}$$

7. A linear demand function has the form

$$q = mp + b.$$

(x is the price p, and y is the demand q). We are given two points on its graph: (1, 1960) and (5, 1800).

Slope:

$$m = \frac{q_2 - q_1}{p_2 - p_1} = \frac{1800 - 1960}{5 - 1} = \frac{-160}{4} = -40$$

Intercept:

$$b = q_1 - mp_1 = 1960 - (-40)(1)$$
$$= 1960 + 40 = 2000$$

Thus, the demand equation is

$$q = mp + b$$
$$q = -40p + 2000$$

9. (a) A linear demand function has the form

$$q = mp + b.$$

(x is the price p, and y is the demand q). We are given two points on its graph:

2004 second quarter data: (111, 45.4)

2004 fourth quarter data: (105, 51.4)

Slope:

$$m = \frac{q_2 - q_1}{p_2 - p_1} = \frac{51.4 - 45.4}{105 - 111} = \frac{6}{-6} = -1$$

Intercept:

$$b = q_1 - mp_1 = 45.4 - (-1)(111) = 156.4$$

Thus, the demand equation is

$$q = mp + b$$
$$q = -p + 156.4$$
If $p = \$103$, then
$$q = -p + 156.4 = -103 + 156.4$$
$$= 53.4 \text{ million phones}$$
(b) Since the slope is -1 million phones per unit increase in price, we interpret of the slope as follows: For every __$1__ increase in price, sales of cellphones decrease by __1 million__ units.

11.

(a) Demand Function: The given points are
$$(p, q) = (1, 90) \text{ and } (2, 30)$$
Slope:
$$m = \frac{q_2 - q_1}{p_2 - p_1} = \frac{30 - 90}{2 - 1} = -60$$
Intercept:
$$b = q_1 - mp_1 = 90 - (-60(1) = 150$$
Thus, the demand equation is
$$q = mp + b$$
$$q = -60p + 150$$
Supply Function: The given points are
$$(p, q) = (1, 20) \text{ and } (2, 100)$$
Slope:
$$m = \frac{q_2 - q_1}{p_2 - p_1} = \frac{100 - 20}{2 - 1} = 80$$
Intercept:
$$b = q_1 - mp_1 = 20 - (80)1 = -60$$
Thus, the supply equation is
$$q = mp + b$$
$$q = 80p - 60$$
(b) For equilibrium,
$$\text{Supply} = \text{Demand}$$
$$80p - 60 = -60p + 150$$
$$140p = 210$$
$$p = \frac{210}{140} = 1.5$$
Thus, the chias should be marked at \$1.50 each.

13.

(a) Here is the given graph with successive points joined by line segments.

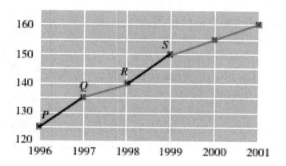

The segments PQ and RS are the steepest (they have the largest Δy ($\Delta y = 10$) for the fixed $\Delta x = 1$), and hence have the largest slope ($m = \Delta y/\Delta x = 10$). (You can check that all line segments joining non-successive pairs of points—for example P and R—have smaller slopes.) Thus, we have two possible answers:
PQ: Points $P(1996, 125)$ and $Q(1997, 135)$
RS: Points $R(1998, 140)$ and $(1999, 150)$.
(b) Consulting the textbook, we find that the slope m measures the rate of change of the number of in-ground swimming pools, and is measured in thousands of pools per year (units of y per unit of x). Thus, the number of new in-ground pools increased most rapidly during the periods 1996–1997 and 1998–1999, when it rose by 10,000 new pools in a year.

15. We are asked for a linear model of N as a function of t. That is, $N = mt + b$. (N is playing the role of y and t is playing the role of x.) The two points we are given are
$$(t, N) = (-1, 350) \quad t = -1 \text{ represents } 1999$$
and
$$(t, N) = (1, 450) \quad t = 1 \text{ represents } 2001$$

(Notice that we are also thinking of N in millions of transactions.)

Slope:

$$m = \frac{N_2 - N_1}{t_2 - t_1} = \frac{450 - 350}{1 - (-1)} = \frac{100}{2} = 50$$

Intercept:

$$b = N_1 - mt_1 = 350 - (50)(-1) = 400$$

Thus, the linear model is

$$N = mt + b$$

$$N = 50t + 400 \text{ million transactions}$$

Consulting the textbook, we find that the slope m measures the rate of change of the number of online shopping transactions, and is measured in millions of transactions per year (units of N per unit of t).

17.

(a) We are asked for a linear model of s as a function of t. That is, $s = mt + b$. The two points we are given are

$\quad (t, s) = (0, 240) \qquad t = 0$ represents 2000

and

$\quad (t, s) = (25, 600) \qquad t = 1$ represents 2001

(Notice that we are also thinking of s in billions of dollars.)

Slope:

$$m = \frac{s_2 - s_1}{t_2 - t_1} = \frac{600 - 240}{25 - 0} = \frac{360}{25} = 14.4$$

Intercept:

$\quad b = 240$ (The point $(0, 240)$ gives us the y-intercept.)

Thus, the linear model is

$$s = mt + b$$

$$s = 14.4t + 240 \text{ billion dollars.}$$

The quantity s increases by $m = 14.4$ for every one-unit increase in t. Thus, Medicare spending is predicted to rise at a rate of \$14.4 billion per year.

(b) In 2040, $t = 40$, so

$$s(40) = 14.4(40) + 240 = 576 + 240 = 816.$$

Thus, Medicare spending in 2040 will be \$816 billion.

19. $s(t) = 2.5t + 10$

(a) Velocity = slope = 2.5 feet/sec.

(b) After 4 seconds, $t = 4$, so

$$s(4) = 2.5(4) + 10 = 10 + 10 = 20$$

Thus the model train has moved 20 feet along the track.

(c) The train will be 25 feet along the track when $s = 25$. Substituting gives

$$25 = 2.5t + 10$$

Solving for time t gives

$$2.5t = 25 - 10 = 15$$

$$t = \frac{15}{2.5} = 6 \text{ seconds}$$

21.

(a) Take s to be displacement from Jones Beach, and t to be time in hours. We are given two points

$\quad (t, s) = (10, 0) \quad s = 0$ for Jones Beach.

$\quad (t, s) = (10.1, 13) \qquad 6$ minutes $= 0.1$ hours

We are asked for the speed, which equals the magnitude of the slope.

$$m = \frac{s_2 - s_1}{t_2 - t_1} = \frac{13 - 0}{10.1 - 10} = \frac{13}{0.1} = 130$$

$$\text{Units of slope} \quad = \text{units of } s \text{ per unit of } t$$

$$= \text{miles per hour}$$

Thus, the police car was traveling at 130 mph.

(b) For the displacement from Jones Beach at time t, we want to express s as a linear function of t; namely, $s = mt + b$. We already know $m = 130$ from part (a). For the intercept, use

$$b = s_1 - mt_1 = 0 - 130(10) = -1300$$

Therefore, the displacement at time t is

$$s = mt + b$$

$$s = 130t - 1300$$

23. F = Fahrenheit temperature,

$\quad C$ = Celsius temperature,

and we want F as a linear function of C. That is,

$\quad F = mC + b$

(F plays the role of y and C plays the role of x.)

We are given two points:

$\quad (C, F) = (0. 32)$ Freezing point

$\quad (C, F) = (100. 212)$ Boiling point

Slope:

$$m = \frac{F_2 - F_1}{C_2 - C_1} = \frac{212 - 32}{100 - 0} = \frac{180}{100} = 1.8$$

Intercept:

$\quad b = F_1 - mC_1 = 32 - 1.8(0) = 32.$

Thus, the linear relation is

$\quad F = mC + b$

$\quad F = 1.8C + 32$

When $C = 30°$

$\quad F = 1.8(30) + 32 = 54 + 32 = 86°$

When $C = 22°$

$\quad F = 1.8(22) + 32 = 39.6 + 32 = 71.6°$

Rounding to the nearest degree gives 72°F.

When $C = -10°$

$\quad F = 1.8(-10) + 32 = -18 + 32 = 14°$

When $C = -14°$

$\quad F = 1.8(-14) + 32 = -25.2 + 32 = 6.8°$

Rounding to the nearest degree gives 7°F.

25. Income = royalties + screen rights

$\quad I$ = 5% of net profits + 50,000

$\quad I = 0.05N + 50,000$ Equation notation

$\quad I(N) = 0.05N + 50,000$ Function notation

For an income of \$100,000,

$\quad 100,000 = 0.05N + 50,000$

$\quad 0.05N = 50,000$

$\quad N = \dfrac{50,000}{0.05} = \$1,000,000$

Her marginal income is her increase in income per \$1 increase in net profit. This is the slope, $m = 0.05$ dollars of income per dollar of net profit, or 5¢ per dollar of net profit.

27. We want w as a linear function of n:

$\quad w = mn + b$

Thus, w plays the role of y and n plays the role of x. Here is the (milk) data listed in the customary way ($x = n$ first, and $y = w$ second).

n	57	59
w	56	60

Slope:

$$m = \frac{w_2 - w_1}{n_2 - n_1} = \frac{60-56}{59-57} = \frac{4}{2} = 2$$

Intercept:

$\quad b = w_1 - mn_1 = 56-(2)(57)$

$\quad\quad = 56-114 = -58$

Thus, the linear function is

$\quad w = 2n - 58.$

For the second part of the question, we are told that $n = 50$. Thus,

$\quad w = 2(50) - 58 = 100-58$

$\quad\quad = 42$ billion pounds of milk

29. We want c (cheese production) as a linear function of m (milk production) in the western states.

$\quad c = km + b$ We are using k for the slope

Thus, c plays the role of y and m plays the role of x. Here is the (western states) data listed in the customary way ($x = m$ first, and $y = c$ second).

m	56	60
c	2.7	3.0

Slope:

$$k = \frac{c_2 - c_1}{m_2 - m_1} = \frac{3.0 - 2.7}{60 - 56} = \frac{0.3}{4} = 0.075$$

$\quad b = c_1 - km_1 = 2.7-(0.075)(56)$

$\quad\quad = 2.7-4.2 = -1.5$

Thus, the linear equation is

$\quad c = km + b$

$\quad c = 0.075m - 1.5$

For the second part of the question, we want the number of pounds of cheese produced for every

10 pounds of milk. The slope $k = 0.075$ gives us the number of pounds of cheese produced per (additional) *one* pound of milk. Multiplying this by 10 gives 0.75 pounds of cheese per 10 pounds of milk.

31. We want the temperature T as a linear function of the rate r of chirping. That is,
$$T(r) = mr + b.$$
Thus, T plays the role of y and r plays the role of x. We are given two points:
$$(r, T) = (140, 80) \text{ and } (120, 75)$$
Slope:
$$m = \frac{T_2 - T_1}{r_2 - r_1} = \frac{75-80}{120-140} = \frac{-5}{-20} = \frac{1}{4}$$
Intercept:
$$b = T_1 - mr_1 = 80 - \tfrac{1}{4}(140) = 80-35 = 45$$
Thus, the linear function is
$$T(r) = mr + b$$
$$T(r) = \tfrac{1}{4}r + 45$$

When the chirping rate is 100 chirps per minute, $r = 100$, and so the temperature is
$$T(100) = \tfrac{1}{4}(100) + 45 = 25+45 = 70°\text{F}$$

33. The hourly profit function is given by
$$\text{Profit = Revenue – Cost}$$
$$P(x) = R(x) - C(x)$$
(Hourly) cost function: This is a fixed cost of \$5132 only:
$$C(x) = 5132$$
(Hourly) revenue function: This is a variable of \$100 per passenger cost only:
$$R(x) = 100x$$
Thus, the profit function is
$$P(x) = R(x) - C(x)$$
$$P(x) = 100x - 5132$$
For the domain of $P(x)$, the number of passengers x cannot exceed the capacity: 405.

Also, x cannot be negative. Thus, the domain is given by $0 \le x \le 405$, or $[0. 405]$.

For break-even, $P(x) = 0$
$$100x - 5132 = 0$$
$$100x = 5132, \text{ or } x = \frac{5132}{100} = 51.32$$
If x is larger than this, then the profit function is positive, and so there should be at least 52 passengers; $x \ge 52$, for a profit.

35. To compute the break-even point, we use the profit function:
$$\text{Profit = Revenue – Cost}$$
$$P(x) = R(x) - C(x)$$
$$R(x) = 2x \quad \text{\$2 per unit}$$
$$\begin{aligned} C(x) &= \text{Variable Cost + Fixed Cost} \\ &= 40\% \text{ of Revenue} + 6000 \\ &= 0.4(2x) + 6000 \\ &= 0.8x + 6000 \end{aligned}$$
Thus,
$$P(x) = R(x) - C(x)$$
$$P(x) = 2x - (0.8x + 6000)$$
$$P(x) = 1.2x - 6000$$
For break-even, $P(x) = 0$
$$1.2x - 6000 = 0$$
$$1.2x = 6000$$
$$x = \frac{6000}{1.2} = 5000$$
Therefore, 5000 units should be made to break even.

37. To compute the break-even point, we use the revenue and cost functions:
$$\begin{aligned} R(x) &= \text{Selling price} \times \text{Number of units} \\ &= SPx \end{aligned}$$
$$\begin{aligned} C(x) &= \text{Variable Cost + Fixed Cost} \\ &= VCx + FC \end{aligned}$$
(Note that "variable cost per unit" is marginal cost.) For break-even

$R(x) = C(x)$

$SPx = VCx + FC$

$SPx - VCx = FC$

$x(SP - VC) = FC$

$x = \dfrac{FC}{SP - VC}$

39. Take x to be the number of grams of perfume he buys and sells. The profit function is given by

Profit = Revenue – Cost

$P(x) = R(x) - C(x)$

Cost function $C(x)$:

Fixed costs:	20,000
Cheap perfume @ $20 per g:	$20x$
Transportation @ $30 per 100 g:	$0.3x$

Thus the cost function is

$C(x) = 20x + 0.3x + 20,000$

$C(x) = 20.3x + 20,000$

Revenue function $R(x)$

$R(x) = 600x$ $600 per gram

Thus, the profit function is

$P(x) = R(x) - C(x)$

$P(x) = 600x - (20.3x + 20,000)$

$P(x) = 579.7x - 20,000,$

with domain $x \geq 0$.

For break-even, $P(x) = 0$

$579.7x - 20,000 = 0$

$579.7x = 20,000$

$x = \dfrac{20,000}{579.7} \approx 34.50$

Thus, he should buy and sell 34.50 grams of perfume per day to break even.

41. $C(t) = \begin{cases} 0.08t + 0.6 & \text{if } 0 \leq t < 8 \\ 0.355t - 1.6 & \text{if } 8 \leq t \leq 11 \end{cases}$

million dollars.

Since 1999 is represented by $t = 9$, we use the second formula,

$C(t) = 0.355t - 1.6.$

Since $m = 0.355$ million dollars per year, or $355,000 per year, the cost of an ad was increasing by $355,000 per year.

43. The data is

t	0	5	9
y	200	50	250

(a) 1995–2000 (first two data points):

Slope: $m = \dfrac{y_2 - y_1}{t_2 - t_1} = \dfrac{50 - 200}{5 - 0} = -30$

Intercept: $b = 200$ Specified in first data point

Thus, the linear model is

$y = mt + b$

$y = -30t + 200$

(b) 2000–2004 (second and third data points):

Slope: $m = \dfrac{y_2 - y_1}{t_2 - t_1} = \dfrac{250 - 50}{9 - 5} = 50$

Intercept: $b = y_1 - mt_1 = 50 - 50(5) = -200$

Thus, the linear model is

$y = mt + b$

$y = 50t - 200$

(c) Since the first model is valid for $0 \leq t \leq 5$ and the second one for $5 \leq t \leq 9$, we put them together as

$y = \begin{cases} -30t + 200 & \text{if } 0 \leq t \leq 5 \\ 50t - 200 & \text{if } 5 < t \leq 9 \end{cases}$

Notice that, since both formulas agree at $t = 5$, we can also say

$y = \begin{cases} -30t + 200 & \text{if } 0 \leq t < 5 \\ 50t - 200 & \text{if } 5 \leq t \leq 9 \end{cases}$.

(d) Since 2002 is represented by $t = 7$, we use the second formula to obtain

$y = 50(7) - 200 = 150$

45.

1989–1994 ($0 \leq t \leq 5$)

Points: $(t, C) = (0, 30,000)$ 1989 data

      $(t, C) = (5, 23,000)$ 1994 data

Slope: $m = \dfrac{C_2 - C_1}{t_2 - t_1} = \dfrac{23{,}000 - 30{,}000}{5 - 0}$

$\qquad = \dfrac{-7000}{5} = -1400$

Intercept: $b = 30{,}000$

Thus, the linear model is

$\qquad C = mt + b$

$\qquad C = -1400t + 30{,}000$

1994–1999 $(5 \le t \le 10)$

Points: $(t, C) = (5, 23{,}000)$ 1994 data

$\qquad\quad (t, C) = (10, 60{,}000)$ 1999 data

Slope: $m = \dfrac{C_2 - C_1}{t_2 - t_1} = \dfrac{60{,}000 - 23{,}000}{10 - 5}$

$\qquad = \dfrac{37{,}000}{5} = 7400$

Intercept: $b = C_1 - mt_1 = 23{,}000 - 7400(5)$

$\qquad\qquad = 23{,}000 - 37{,}000 = 14{,}000$

Thus, the linear model is

$\qquad C = mt + b$

$\qquad C = 7400t + 14{,}000$

Putting them together gives

$C(t) = \begin{cases} -1400t + 30{,}000 & \text{if } 0 \le t \le 5 \\ 7400t - 14{,}000 & \text{if } 5 < t \le 10 \end{cases}$

Since 1992 corresponds to $t = 3$, we use the first formula to obtain

$\qquad C(3) = -1400(3) + 30{,}000$

$\qquad\qquad = -4200 + 30{,}000 = 25{,}800$ students

47. We want d as a piecewise-linear function of r, using the three points (r, d) given:

$\quad (r, d) = (1.3, 22), (1.6, 35),$ and $(1.1, 30)$

Caution: These points are not given in ascending order of r. We rearrange them in increasing order of r:

$\quad (r, d) = (1.1, 30), (1.3, 22),$ and $(1.6, 35)$

First pair of points:

$\quad (r, d) = (1.1, 30)$ and $(1.3, 22)$

Slope $m = \dfrac{d_2 - d_1}{r_2 - r_1} = \dfrac{22 - 30}{1.3 - 1.1} = \dfrac{-8}{0.2} = -40$

Intercept: $b = d_1 - mr_1 = 30 - (-40)1.1$

$\qquad\qquad = 30 + 44 = 74$

Thus, when $1.1 \le r \le 1.3$, the linear model is

$\qquad d = mr + b$

$\qquad d = -40 + 74$

Second pair of points:

$\quad (r, d) = (1.3, 22)$ and $(1.6, 35)$

Slope $m = \dfrac{d_2 - d_1}{r_2 - r_1} = \dfrac{35 - 22}{1.6 - 1.3} = \dfrac{13}{0.3} = \dfrac{130}{3}$

Intercept: $b = d_1 - mr_1 = 22 - \dfrac{130}{3}(1.3)$

$\qquad\qquad = 22 - \dfrac{169}{3} = -\dfrac{103}{3}$

Thus, when $1.3 \le r \le 1.6$, the linear model is

$\qquad d = mr + b$

$\qquad d = \dfrac{130}{3}r - \dfrac{103}{3}$

Putting the two linear models together gives

$d(r) = \begin{cases} -40r + 74 & \text{if } 1.1 \le r \le 1.3 \\ \dfrac{130r}{3} - \dfrac{103}{3} & \text{if } 1.3 < r \le 1.6 \end{cases}$

When there are the same number of available men as women, the ratio r of available men to women is 1, so we are asked to calculate $d(1)$. Since $1 \le 1.3$, we extrapolate the first formula:

$\qquad d(1) = -40(1) + 74 = 34\%$

Thus, the divorce rate is 34%.

49. The units of the slope m are units of y (bootlags) per unit of x (zonars). The intercept b is on the y-axis, and is thus measured in units of y (bootlags). Thus, m is measured in <u>bootlags per zonar</u> and b is measured in <u>bootlags</u>.

51. If a quantity changes linearly with time, it must change by the same amount for every unit change in time. Thus, since it increases by 10

units in the first day, it must increase by 10 units each day, including the third.

53. $v = 0.1t + 20$ m/sec

Since the slope is 0.1, the velocity is increasing at a rate of 0.1 m/sec every second. Since the velocity is increasing, the object is accelerating (choice B).

55. Increasing the number of items from the breakeven results in a profit: Because the slope of the revenue graph is larger than the slope of the cost graph, it is higher than the cost graph to the right of the point of intersection, and hence corresponds to a profit.

1.5

1. (1, 1), (2, 2), (3, 4) ; $y = x-1$

x	y	Predicted $\hat{y} = x-1$	Residual $y - \hat{y}$	Residual^2 $(y - \hat{y})^2$
1	1	0	1	1
2	2	1	1	1
3	4	2	2	4

SSE = Sum of squares of residuals
$$= 4+1+1 = 6$$

3. (0,−1), (1,3), (4,6), (5,0); $y = -x+2$

x	y	Predicted $\hat{y} = -x+2$	Residual $y - \hat{y}$	Residual^2 $(y - \hat{y})^2$
0	−1	2	−3	9
1	3	1	2	4
4	6	−2	8	64
5	0	−3	3	9

SSE = Sum of squares of residuals
$$= 9 + 4 + 64 + 9 = 86$$

5. (1, 1), (2, 2), (3, 4)

(a) $y = 1.5x-1$

x	y	$\hat{y}$	$y - \hat{y}$	$(y - \hat{y})^2$
1	1	0.5	0.5	0.25
2	2	2	0	0
3	4	3.5	0.5	0.25

SSE = Sum of squares of residuals = 0.5

(b) $y = 2x - 1.5$

x	y	$\hat{y}$	$y - \hat{y}$	$(y - \hat{y})^2$
1	1	0.5	0.5	0.25
2	2	2.5	−0.5	0.25
3	4	4.5	−0.5	0.25

SSE = Sum of squares of residuals = 0.75
The model that gives the better fit is (a) because it gives the smaller value of SSE.

7. (0, −1), (1, 3), (4, 6), (5, 0)

(a) $y = 0.3x + 1.1$

x	y	$\hat{y}$	$y - \hat{y}$	$(y - \hat{y})^2$
0	−1	1.1	−2.1	4.41
1	3	1.4	1.6	2.56
4	6	2.3	3.7	13.69
5	0	2.6	−2.6	6.76

SSE = Sum of squares of residuals = 27.42

(b) $y = 0.4x+0.9$

x	y	$\hat{y}$	$y - \hat{y}$	$(y - \hat{y})^2$
0	−1	0.9	−1.9	3.61
1	3	1.3	1.7	2.89
4	6	2.5	3.5	12.25
5	0	2.9	−2.9	8.41

SSE = Sum of squares of residuals = 27.16
The model that gives the better fit is (b) because it gives the smaller value of SSE.

9. (1,1), (2,2), (3,4)

x	y	xy	x^2
1	1	1	1
2	2	4	4
3	4	12	9
Σ (Sum)	6	7	17	14

n = 3 (number of data points)

Slope: $m = \dfrac{n(\Sigma xy) - (\Sigma x)(\Sigma y)}{n(\Sigma x^2) - (\Sigma x)^2}$

$$= \dfrac{3(17) - (6)(7)}{3(14) - 6^2} = \dfrac{9}{6} = 1.5$$

Intercept: $b = \dfrac{\Sigma y - m(\Sigma x)}{n}$

$$= \dfrac{7 - 1.5(6)}{3} = -\dfrac{2}{3} \approx -0.6667$$

Thus, the regression line is
$$y = mx + b$$
$$y = 1.5x - 0.6667$$

Graph:

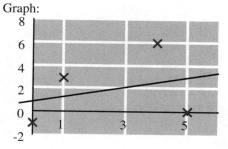

11. $(0, -1)$, $(1, 3)$, $(4, 6)$, $(5, 0)$

x	y	xy	x^2
0	-1	0	0
1	3	3	1
4	6	24	16
5	0	0	25
Σ (Sum) 10	8	27	42

n = 4 (number of data points)

Slope: $m = \dfrac{n(\Sigma xy) - (\Sigma x)(\Sigma y)}{n(\Sigma x^2) - (\Sigma x)^2}$

$= \dfrac{4(27) - (10)(8)}{4(42) - 10^2} = \dfrac{28}{68} \approx 0.4118$

Intercept: $b = \dfrac{\Sigma y - m(\Sigma x)}{n}$

$= \dfrac{8 - \left(\dfrac{28}{68}\right)(10)}{4} \approx 0.9706$

Thus, the regression line is

$y = mx + b$

$y = 0.4118x + 0.9706$

Graph:

13. (a) $\{(1, 3), (2, 4), (5, 6)\}$

x	y	xy	x^2	y^2
1	3	3	1	9
2	4	8	4	16
5	6	30	25	36
Σ 8	13	41	30	61

n = 3 (number of data points)

$r = \dfrac{n(\Sigma xy) - (\Sigma x)(\Sigma y)}{\sqrt{n(\Sigma x^2) - (\Sigma x)^2} \cdot \sqrt{n(\Sigma y^2) - (\Sigma y)^2}}$

$= \dfrac{3(41) - (8)(13)}{\sqrt{3(30)-(8)^2} \sqrt{3(61)-(13)^2}}$

$\approx \dfrac{19}{19.078784} \approx 0.9959$

(b) $\{(0, -1), (2, 1), (3, 4)\}$

x	y	xy	x^2	y^2
0	-1	0	0	1
2	1	2	4	1
3	4	12	9	16
Σ 5	4	14	13	18

n = 3 (number of data points)

$r = \dfrac{n(\Sigma xy) - (\Sigma x)(\Sigma y)}{\sqrt{n(\Sigma x^2) - (\Sigma x)^2} \cdot \sqrt{n(\Sigma y^2) - (\Sigma y)^2}}$

$= \dfrac{3(14) - (5)(4)}{\sqrt{3(13)-(5)^2} \sqrt{3(18)-(4)^2}}$

$\approx \dfrac{22}{23.0651252} \approx 0.9538$

(c) $\{(4, -3), (5, 5), (0, 0)\}$

x	y	xy	x^2	y^2
4	-3	-12	16	9
5	5	25	25	25
0	0	0	0	0
Σ 9	2	13	41	34

n = 3 (number of data points)

$$r = \frac{n(\Sigma xy) - (\Sigma x)(\Sigma y)}{\sqrt{n(\Sigma x^2) - (\Sigma x)^2} \cdot \sqrt{n(\Sigma y^2) - (\Sigma y)^2}}$$

$$= \frac{3(13) - (9)(2)}{\sqrt{3(41)-(9)^2} \sqrt{3(34)-(2)^2}}$$

$$\approx \frac{21}{64.1560597} \approx 0.3273$$

The value of r in part (a) has the largest absolute value. Therefore, the regression line for the data in part (a) is the best fit.

The value of r in part (c) has the smallest absolute value. Therefore, the regression line for the data in part (c) is the worst fit.

Since r is not ± 1 for any of these lines, none of them is a perfect fit.

15. The entries in the xy column are obtained by multiplying the entries in the x column by the corresponding entries in the y column. The entries in the x^2 column are the squares of the entries in the x column.

	x	y	xy	x^2
	3	500	1500	9
	5	600	3000	25
	7	800	5600	49
Σ (Sum)	15	1900	10100	83

$n = 3$ (number of data points)

Slope: $m = \dfrac{n(\Sigma xy) - (\Sigma x)(\Sigma y)}{n(\Sigma x^2) - (\Sigma x)^2}$

$$= \frac{3(10100) - (15)(1900)}{3(83)-(15)^2}$$

$$= \frac{1800}{24} = 75$$

Intercept: $b = \dfrac{\Sigma y - m(\Sigma x)}{n}$

$$= \frac{1900 - (75)(15)}{3}$$

$$= \frac{775}{3} \approx 258.33 \quad \text{(to 2 decimal places)}$$

Thus, the regression line is

$$y = mx + b$$
$$y = 75x + 258.33$$

To estimate the 2008 sales we put $x = 8$:
$$y = 75(8) + 258.33 \approx 858.33 \text{ million}$$

17. Calculation of the regression line:

x	y	xy	x^2
0	6	0	0
2	10	20	4
4	16	64	16
Σ (Sum) 6	32	84	20

$n = 3$ (number of data points)

Slope: $m = \dfrac{n(\Sigma xy) - (\Sigma x)(\Sigma y)}{n(\Sigma x^2) - (\Sigma x)^2}$

$$= \frac{3(84) - (6)(32)}{3(20) - (6)^2}$$

$$= \frac{60}{24} = 2.5$$

Intercept: $b = \dfrac{\Sigma y - m(\Sigma x)}{n}$

$$= \frac{32 - (2.5)(6)}{3}$$

$$= \frac{17}{3} \approx 5.67 \quad \text{(to 2 decimal places)}$$

Thus, the regression line is

$$y = mt + b \quad \text{Independent variable is called } t$$

$$y = 2.5t + 5.67$$

$$y(3) \approx 2.5(3) + 5.67 = \$13.17 \text{ billion}$$

19. Calculation of the regression line:

x	y	xy	x^2
20	3	60	400
40	6	240	1,600
80	9	720	6,400
100	15	1500	10,000
Σ (Sum)	240	33	2520	18,400

$n = 4$ (number of data points)

Slope: $m = \dfrac{n(\Sigma xy) - (\Sigma x)(\Sigma y)}{n(\Sigma x^2) - (\Sigma x)^2}$

$= \dfrac{4(2520) - (240)(33)}{4(18400) - (240)^2}$

$= \dfrac{2160}{16,000} = 0.135$

Intercept: $b = \dfrac{\Sigma y - m(\Sigma x)}{n}$

$= \dfrac{33 - (0.135)(240)}{4}$

$= \dfrac{0.6}{4} = 0.15$

Thus, the regression line is

$y = mx + b$

$y = 0.135x + 0.15$

$y(50) = 0.135(50) + 0.15$

$\quad\quad = 6.9$ million jobs

21. (a) Since production is a function of cultivated area, we take x as cultivated area, and y as production:

x	25	30	32	40	52
y	15	25	30	40	60

Using the method of Example 3, we obtain the following regression line and plot (coefficients rounded to two decimal places):

$y = 1.62x - 23.87$

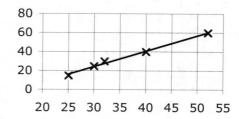

(b) To interpret the slope $m = 1.62$, recall that units of m are units of y per unit of x; That is, millions of tons of production of soybeans per million acres of cultivated land. Thus, production increases by 1.62 million tons of soybeans per million acres of cultivated land. More simply, each acre of cultivated land produces about 1.62 tons of soybeans.

23. (a) Using x = median household income and y = poverty rate gives the following table of values:

x	y
39	13
38	14
39	14
41	13
44	12
42	11
41	12
40	12

Using the method of Example 3, we obtain the following regression line and plot (coefficients rounded to two significant digits):

$y = -0.40x + 29$

Graph:

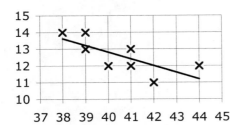

Notice from the graph that y decreases and x increases. Therefore, the graph suggests a relationship between x and y.

(b) To interpret the slope, recall that units of the slope are units of y (percentage points of the poverty rate) per unit of x (median house income in thousands of dollars). Thus,

$m = -0.40$ percentage points per $1000 dollars,

indicating that the poverty rate declines by 0.40% for each $1000 increase in the median household income.

(c) Using the method of Example 4, we can use technology to show the value of r^2:

$$r^2 \approx 0.5385$$

so $r = \sqrt{r^2} \approx -\sqrt{0.5385} \approx -0.7338$.

(We used the negative square root because the slope of the regression equation is negative.)

Since r is not close to 1, the correlation between x and y is not a "strong" one.

25. (a) Using the method of Example 3, we obtain the following regression line and plot (coefficients rounded to two significant digits):

$p = 0.13t + 0.22$

Graph:

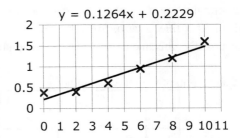

$$y = 0.1264x + 0.2229$$

(b) The first and last points lie above the regression line, while the central points lie below it, suggesting a curve.

(c) Here is an Excel worksheet showing the computation of the residuals (based on Example 1 in the text):

◇	A	B	C	D
1	t	p (Observed)	p (predicted)	Residual
2	0	0.38	=0.13*A2+0.22	=B2-C2
3	2	0.4		
4	4	0.6		
5	6	0.95		
6	8	1.2		
7	10	1.6		

◇	A	B	C	D
1	t	p (Observed)	p (predicted)	Residual
2	0	0.38	0.22	0.16
3	2	0.4	0.48	-0.08
4	4	0.6	0.74	-0.14
5	6	0.95	1	-0.05
6	8	1.2	1.26	-0.06
7	10	1.6	1.52	0.08

Notice that the residuals are positive at first, become negative, and then become positive, confirming the impression from the graph.

27. The regression line is defined to be the line that gives the lowest sum-of-squares error, SSE. If we are given two points, (a, b) and (c, d) with $a \neq c$ then there is a line that passes through these two points, giving SSE = 0. Since 0 is the smallest value possible, this line must be the regression line.

29. If the points (x_1, y_1), (x_2, y_2), . . . , (x_n, y_n) lie on a straight line, then the sum-of-squares error, SSE, for this line is zero. Since 0 is the smallest value possible, this line must be the regression line.

31. Calculation of the regression line:

	x	y	xy	x^2
	0	0	0	0
	$-a$	a	$-a^2$	a^2
	a	a	a^2	a^2
Σ	0	$2a$	0	$2a^2$

n = 3 (number of data points)

Slope: $m = \dfrac{n(\Sigma xy) - (\Sigma x)(\Sigma y)}{n(\Sigma x^2) - (\Sigma x)^2}$

$= \dfrac{3(0) - (0)(2a)}{3(2a^2) - 0^2} = 0$

Regression Coefficient:

$r = \dfrac{n(\Sigma xy) - (\Sigma x)(\Sigma y)}{\sqrt{n(\Sigma x^2) - (\Sigma x)^2} \cdot \sqrt{n(\Sigma y^2) - (\Sigma y)^2}}$

has the same numerator as m, and we have just seen that this numerator is zero, Hence, $r = 0$.

33. No. The regression line through $(-1, 1)$, $(0, 0)$, and $(1, 1)$ passes through none of these points.

Chapter 1 Review Exercises

1.

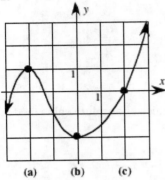

(a) 1 **(b)** −2 **(c)** 0

(d) $f(2) - f(-2) = 0 - 1 = -1$

2.

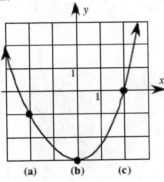

(a) −1 **(b)** −3 **(c)** 0

(d) $f(2) - f(-2) = 0 - (-1) = 1$

3.

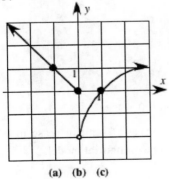

(a) **(b) (c)**

(a) 1 **(b)** 0 **(c)** 0

(d) $f(1) - f(-1) = 0 - 1 = -1$

4.

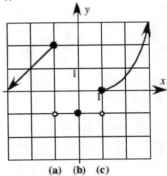

(a) **(b) (c)**

(a) 2 **(b)** −1 **(c)** 0

(d) $f(1) - f(-1) = 0 - 2 = -2$

5. $y = -2x + 5$

y−intercept = 5 slope = −2

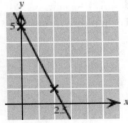

6. $2x - 3y = 12$

Solving for y gives

$$-3y = -2x + 12$$

$$y = \frac{2}{3}x - 4$$

y−intercept = −4 slope = $\frac{2}{3}$

41

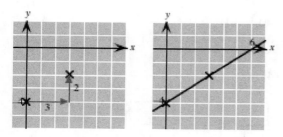

7. $y = \begin{cases} \frac{1}{2}x & \text{if } -1 \le x \le 1 \\ x - 1 & \text{if } 1 < x \le 3 \end{cases}$

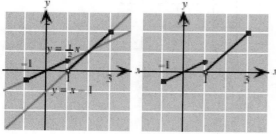

8. $f(x) = 4x - x^2$ with domain [0, 4]
Technology formula: `4*x-x^2`

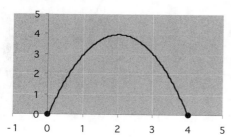

x	-2	0	1	2	4
$f(x)$	4	2	1	0	2
$g(x)$	-5	-3	-2	-1	1
$h(x)$	1.5	1	0.75	0.5	0
$k(x)$	0.25	1	2	4	16
$u(x)$	0	4	3	0	-12

Here are the graphs of the five functions, with the points connected:

9.

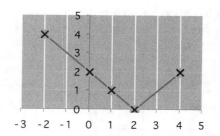

$f(x)$

V-shape indicates **absolute value** model.

10.

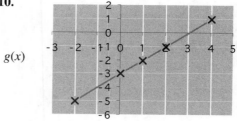

$g(x)$

Linear

11.

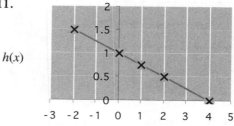

$h(x)$

Linear

12.

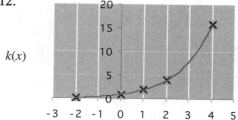

$k(x)$

y doubles for each 1-unit increase in x, indicating an **exponential** model.

13.

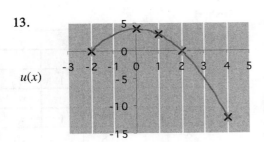

$u(x)$

Parabolic shape of graph indicates a **quadratic** model.

14. Line through $(3, 2)$ with slope -3

Point: $(3, 2)$

Slope: $m = -3$

Intercept: $b = y_1 - mx_1$
$$= 2 - (-3)(3) = 2 + 9 = 11$$

Thus, the equation is
$$y = mx + b$$
$$y = -3x + 11$$

15. Through $(-1, 2)$ and $(1, 0)$

Point: $(1, 0)$

Slope: $m = \dfrac{y_2 - y_1}{x_2 - x_1} = \dfrac{0 - 2}{1 - (-1)} = \dfrac{-2}{2} = -1$

Intercept: $b = y_1 - mx_1 = 0 - (-1)(1) = 1$

Thus, the equation is
$$y = mx + b$$
$$y = -x + 1$$

16. Through $(1, 2)$ parallel to $x - 2y = 2$

Point: $(1, 2)$

Slope: Parallel to the line $x - 2y = 2$, and so has the same slope. Solve for y to obtain
$$-2y = -x - 2$$
$$y = \tfrac{1}{2}x + 1$$

Thus the slope is $m = \tfrac{1}{2}$

Intercept: $b = y_1 - mx_1 = 2 - \tfrac{1}{2}(1) = \dfrac{3}{2}$

Thus, the equation is
$$y = mx + b$$

$$y = \tfrac{1}{2}x + \tfrac{3}{2}$$

17. With slope $1/2$ crossing $3x + y = 6$ at its x-intercept

Point: We need the x-intercept of $3x + y = 6$. This is given by setting $y = 0$ and solving for x:
$$3x + 0 = 6$$
$$x = 2.$$

Thus, the point is $(2, 0)$ On the x-axis, so $y = 0$

Slope: $m = \tfrac{1}{2}$

Intercept: $b = y_1 - mx_1 = 0 - \tfrac{1}{2}(2) = -1$

Thus, the equation is
$$y = mx + b$$
$$y = \tfrac{1}{2}x - 1$$

18. $y = -0.5x + 1$:

x	observed y	predicted y	residual^2
-1	1	1.5	0.25
1	1	0.5	0.25
2	0	0	0
		SSE:	0.5

$y = -x/4 + 1$:

x	observed y	predicted y	residual^2
-1	1	1.25	0.0625
1	1	0.75	0.0625
2	0	0.5	0.25
		SSE:	0.375

The second line, $y = -x/4 + 1$, is a better fit.

19. $y = x + 1$:

x	observed y	predicted y	residual^2
-2	-1	-1	0
-1	1	0	1
0	1	1	0
1	2	2	0
2	4	3	1
3	3	4	1
		SSE:	3

$y = x/2 + 1$:

x	observed y	predicted y	residual^2
−2	−1	0	1
−1	1	0.5	0.25
0	1	1	0
1	2	1.5	0.25
2	4	2	4
3	3	2.5	0.25
		SSE:	5.75

The first line, $y = x + 1$, is the better fit.

20.

x	y	xy	x^2	y^2
−1	1	−1	1	1
1	2	2	1	4
2	0	0	4	0
Σ (Sum)	2	3	1	6	5

n = 3 (number of data points)

Slope: $m = \dfrac{n(\Sigma xy) - (\Sigma x)(\Sigma y)}{n(\Sigma x^2) - (\Sigma x)^2}$

$= \dfrac{3(1) - (2)(3)}{3(6) - 2^2} = \dfrac{-3}{14} \approx -0.214$

Intercept: $b = \dfrac{\Sigma y - m(\Sigma x)}{n}$

$= \dfrac{3 - (-0.214)(2)}{3} \approx 1.14$

Thus, the regression line is

$y = mx + b$

$y = -0.214x + 1.14$

The correlation coefficient is

$r = \dfrac{n(\Sigma xy) - (\Sigma x)(\Sigma y)}{\sqrt{n(\Sigma x^2) - (\Sigma x)^2} \cdot \sqrt{n(\Sigma y^2) - (\Sigma y)^2}}$

$= \dfrac{3(1) - (2)(3)}{\sqrt{3(6)-(2)^2} \sqrt{3(5)-(3)^2}}$

≈ -0.33

21.

x	y	xy	x^2	y^2
−2	−1	2	4	1
−1	1	−1	1	1
0	1	0	0	1
1	2	2	1	4
2	4	8	4	16
3	3	9	9	9
Σ (Sum)	3	10	20	19	32

n = 6 (number of data points)

Slope: $m = \dfrac{n(\Sigma xy) - (\Sigma x)(\Sigma y)}{n(\Sigma x^2) - (\Sigma x)^2}$

$= \dfrac{6(20) - (3)(10)}{6(19) - 3^2} = \dfrac{90}{105} \approx 0.857$

Intercept: $b = \dfrac{\Sigma y - m(\Sigma x)}{n}$

$= \dfrac{10 - (0.857)(3)}{6} \approx 1.24$

Thus, the regression line is

$y = mx + b$

$y = 0.857x + 1.24$

The correlation coefficient is

$r = \dfrac{n(\Sigma xy) - (\Sigma x)(\Sigma y)}{\sqrt{n(\Sigma x^2) - (\Sigma x)^2} \cdot \sqrt{n(\Sigma y^2) - (\Sigma y)^2}}$

$= \dfrac{6(20) - (3)(10)}{\sqrt{6(19)-(3)^2} \sqrt{6(32)-(10)^2}}$

≈ 0.92

22. $n(x) = \begin{cases} 0.02x & \text{if } 0 \le x \le 1000 \\ 0.025x - 5 & \text{if } 1000 < x \le 2000 \end{cases}$

(a) $n(500) = 0.02(500) = 10$ books per day

We used the first formula, since 500 is in [0, 1000].

$n(1000) = 0.02(1000) = 20$ books per day

We used the first formula, since 1000 is in [0, 1000].

$n(1500) = 0.025(1500) - 5$

$= 37.6 - 5 = 32.5$ books per day

We used the second formula, since 1500 is in (1000, 2000).

(b) The coefficient 0.025 is the slope of the second formula, indicating that, for web site traffic of more than 1000 hits per day and up to 2000 hits per day ($1000 < x \leq 2000$) book sales are increasing by 0.025 books per additional hit.

(c) To sell an average of 30 books per day, we desire $n(x) = 30$. Of we try the first formula, we get

$$0.02x = 30$$

giving $x = 1500$, which is not in the domain of the first formula. So, we try the second formula:

$$0.025x - 5 = 30$$
$$0.025x = 35$$
$$x = \frac{35}{0.025} = 1400 \text{ hits,}$$

which is in the domain of the second formula. Thus, 1400 hits per day will result in average sales of 30 books per day.

23.

t	1	2	3	4	5	6
$S(t)$	12.5	37.5	62.5	72.0	74.5	75.0

(a) Technology formulas:

    (A): `300/(4+100*5^(-t))`

    (B): `13.3*t+8.0`

    (C): `-2.3*t^2+30.0*t-3.3`

    (D): `7*3^(0.5*t)`

Here are the values for the three given models (rounded to 1 decimal place):

t	1	2	3	4	5	6
(A)	12.5	37.5	62.5	72.1	74.4	74.9
(B)	21.3	34.6	47.9	61.2	74.5	87.8
(C)	24.4	47.5	66.0	79.9	89.2	93.9
(D)	12.1	21.0	36.4	63.0	109.1	189.0

Model (A) gives an almost perfect fit, whereas the other models are not even close.

(b) Looking at the table, we see the following behavior as t increases:

(A) Leveling off; (B) Rising (C) Rising (they begin to fall after 7 months, however) (D) Rising

24.

c	$2000	$5000
h	1900	2050

Point: (2000, 1900)

Slope: $m = \dfrac{h_2 - h_1}{c_2 - c_1} = \dfrac{2050 - 1900}{5000 - 2000}$

$$= \frac{150}{3000} = 0.05$$

Intercept: $b = y_1 - mx_1$

$$= 1900 - (0.05)(2000)$$
$$= 1900 - 100 = 1800$$

Thus, the equation is

$$h = mc + b$$
$$h = 0.05c + 1800$$

(b) A budget of $6000 per month for banner ads corresponds to $c = 6000$

$$h(6000) = 0.05(6000) + 1800$$
$$= 300 + 1800 = 2100 \text{ hits per day}$$

(c) We are given $h = 2500$ and want c.

$$2500 = 0.05c + 1800$$
$$0.05c - 1500 - 1800 = 700$$
$$c = \frac{700}{0.05} = \$14000 \text{ per month}$$

25. $h = -0.000005c^2 + 0.085c + 1750$

(a) Currently, $c = \$6000$, so

$$h = -0.000005(6000)^2 + 0.085(6000) + 1750$$
$$= -180 + 510 + 1750 = 2080 \text{ hits per day}$$

(b) Here is a portion of the graph of h:

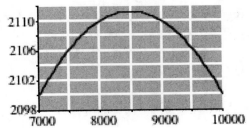

Technology formula:

```
-0.000005*x^2+0.085*x+1750
```

For $c = 8500$ or larger, we see that Web site traffic is projected to decrease as advertising increases, and then drop toward zero. Thus, the model does not appear to give a reasonable prediction of traffic at expenditures larger than \$8,500 per month.

26.

Cost function: $C = mx + b$

b = fixed cost = \$900 per month

m = marginal cost = \$4 per novel

Thus, the linear cost function is

$C = 4x + 900$

Revenue function: $R = mx + b$

b = fixed revenue = 0

m = marginal revenue = \$5.50 per novel

Thus, the linear revenue function is

$R = 5.50x$

Profit function: $P = R - C$

$P = 5.50x - (4x + 900)$

$\quad = 5.50x - 4x - 900$

$\quad = 1.50x - 900$

(b) For break-even, $P = 0$

$1.50x - 900 = 0$

$1.50x = 900$

$x = \dfrac{900}{1.50} = 600$ novels per month

(c) $R = 5.00x$

$P = 5.00x - (4x + 900)$

$\quad = 5x - 4x - 900 = x - 900$

For break-even,

$x - 900 = 0,$

$x = 900$ novels per month

27.

(a) Demand: We are given two points:

$\quad (p, q) = (10, 350)$ and $(5.5, 620)$ Slope:

$$m = \frac{q_2 - q_1}{p_2 - p_1} = \frac{620-350}{5.5-10} = \frac{270}{-4.5} = -60$$

Intercept:

$b = q_1 - mp_1 = 350-(-60)(10)$

$\quad = 350 + 600 = 950$

Thus, the demand equation is

$\quad q = mp + b$

$\quad q = -60p + 950$

(b) When $p = \$15$, the demand is

$\quad q = -60(15) + 950 = -900 + 950$

$\quad = 50$ novels per month

(c) From Question 8, the cost function is

$\quad C = 4q + 900$

We use q for the monthly sales rather than x

$\quad = 4(-60p + 950) + 900$

We want everything expressed in terms of p, so we used the demand equation.

$\quad = -240p + 3800 + 900$

$C = -240p + 4700$

To compute the profit in terms of price, we need the revenue as well:

$\quad R = pq = p(-60p + 950)$

$\quad = -60p^2 + 950p$

Profit: $P = R - C$

$P = -60p^2 + 950p - (-240p + 4700)$

$\quad = -60p^2 + 1190p - 4700$

Now we compare profits for the three prices:

$P(5.50) = -60(5.5)^2 + 1190(5.5) - 4700$

$\quad = \$30$

$P(10) = -60(10)^2 + 1190(10) - 4700$

$\quad = \$1200$

$P(15) = -60(15)^2 + 1190(15) - 4700$

$\quad = -\$350$ (loss)

Thus, charging \$10 will result in the largest monthly profit of \$1200.

28.

p	5.50	10	12	15
q	620	350	300	100

(a) Calculation of the regression line:

x	y	xy	x^2
5.5	620	3410	30.25
10	350	3500	100
12	300	3600	144
15	100	1500	225
Σ	42.5	1370	12010	499.25

n = 4 (number of data points)

Slope: $m = \dfrac{n(\Sigma xy) - (\Sigma x)(\Sigma y)}{n(\Sigma x^2) - (\Sigma x)^2}$

$= \dfrac{4(12010) - (42.5)(1370)}{4(499.25) - (42.5)^2}$

$= \dfrac{-10185}{190.75} \approx 53.3945$

Intercept: $b = \dfrac{\Sigma y - m(\Sigma x)}{n}$

$\approx \dfrac{1370 - (-53.394495)(42.5)}{4}$

$\approx \dfrac{3639.26606}{4} \approx 909.8165$

Thus, the regression equation is

$q = -53.3945p + 909.8165$

(q plays the role of y and p plays the role of x)

(b) We are given $p = \$8$, so the demand is

$q = -53.3945(8) + 909.8165$

≈ 483 novels per month (rounded)

Chapter 2
2.1

1. $x - y = 0$

$\quad x + y = 4$

Adding gives

$\quad 2x = 4$

$\quad x = 2$

Substituting $x=2$ in first equation:

$\quad 2 - y = 0$

$\quad y = 2$

Solution: $(2, 2)$

Graph: $y = x$; $y = 4 - x$

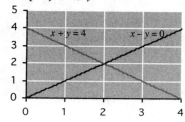

3. $x + y = 4$

$\quad x - y = 2$

Adding gives

$\quad 2x = 6$

$\quad x = 3$

Substituting $x = 3$ in first equation:

$\quad 3 + y = 4$

$\quad y = 4 - 3 = 1$

Solution: $(3, 1)$

Graph: $y = 4 - x$; $y = x - 2$

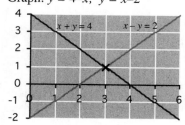

5. $3x - 2y = 6$

$\quad 2x - 3y = -6$

Multiply first equation by 2 and second by –3:

$\quad 6x - 4y = 12$

$\quad -6x + 9y = 18$

Adding gives

$\quad 5y = 30$

$\quad y = 6$

Substituting $y=6$ in first equation gives

$\quad 3x - 12 = 6$

$\quad 3x = 18$

$\quad x = 6$

Solution: $(6, 6)$

Graph: $y = \dfrac{3}{2}x - 3$; $y = \dfrac{2}{3}x + 2$

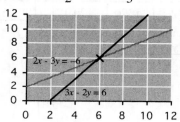

7. $0.5x + 0.1y = 0.7$

$\quad 0.2x - 0.2y = 0.6$

Multiply both equations by 10:

$\quad 5x + y = 7$

$\quad 2x - 2y = 6$

Divide second equation by 2:

$\quad 5x + y = 7$

$\quad x - y = 3$

Adding gives

$\quad 6x = 10$

$\quad x = \dfrac{10}{6} = \dfrac{5}{3}$

Substituting $x = \dfrac{5}{3}$ in $x - y = 3$ gives

$\quad \dfrac{5}{3} - y = 3$

$\quad y = \dfrac{5}{3} - 3 = -\dfrac{4}{3}$

Solution: $\left(\dfrac{5}{3}, -\dfrac{4}{3}\right)$

Graph: $y = -5x+7$; $y = x-3$

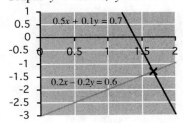

9. $\dfrac{x}{3} - \dfrac{y}{2} = 1$

$\dfrac{x}{4} + y = -2$

Multiply first equation by 6 and second by 4:

$2x - 3y = 6$

$x + 4y = -8$

Multiply the second equation by -2:

$2x - 3y = 6$

$-2x - 8y = 16$

Adding gives

$-11y = 22$

$y = \dfrac{22}{-11} = -2$

Substituting $y=-2$ into $x+4y = -8$ gives

$x + 4(-2) = -8$

$x - 8 = -8$

$x = 0$

Solution: $(0, -2)$

Graph: $y = -x/4-2$; $y = 2(x/3-1)$

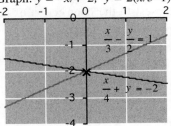

11. $2x + 3y = 1$

$-x - \dfrac{3y}{2} = -\dfrac{1}{2}$

Multiply the second equation by 2

$2x + 3y = 1$

$-2x - 3y = -1$

Adding gives $0 = 0$, indicating that the system is redundant: The graphs are the same, so there are infinitely many solutions. We find the solutions by solving either equation for y:

$2x + 3y = 1$

$3y = -2x+1$

$y = (-2x+1)/3$

Solution: $(x, [-2x+1]3)$; x arbitrary

Graph: $y = (-2x+1)/3$

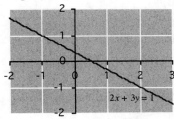

Every point on the line represents a solution.

13. $2x + 3y = 2$

$-x - \dfrac{3y}{2} = -\dfrac{1}{2}$

Multiply the second equation by 2:

$2x + 3y = 2$

$-2x - 3y = -1$

Adding gives $0 = 1$, indicating that the system is inconsistent. There is no solution; the lines are parallel.

Graph: $y = (-2x+2)/3$; $y = (-2x+1)/3$

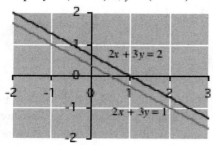

Parallel lines; no solution

15. $2x + 8y = 10$

$x + y = 5$

To graph these, solve for y:

$y = (-2x + 10)/8$

$y = -x + 5$

Graph, with vertical and horizontal scales of 0.1:

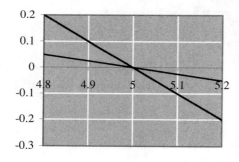

The grid point closest to the intersection of the lines gives the approximate solution (in this case the exact solution)

Solution: (5, 0)

17. $3.1x - 4.5y = 6$

$4.5x + 1.1y = 0$

To graph these, solve for y:

$y = (3.1x - 6)/4.5$

$y = -4.5x/1.1$

Graph, with vertical and horizontal scales of 0.1:

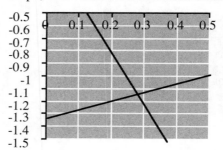

The grid point closest to the intersection of the lines gives the approximate solution.

Approximate solution: $(0.3, -1.1)$

19. $10.2x + 14y = 213$

$4.5x + 1.1y = 448$

To graph these, solve for y:

$y = (-10.2x + 213)/14$

$y = (-4.5x + 448)/1.1$

Graph, with vertical and horizontal scales of 0.1:

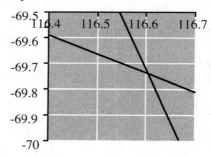

The grid point closest to the intersection of the lines gives the approximate solution.

Approximate solution: $(116.6, -69.7)$

21. Line through (0, 1) and (4.2, 2)

Point: (0, 1)

Slope: $m = \dfrac{y_2 - y_1}{x_2 - x_1} = \dfrac{2 - 1}{4.2 - 0} = \dfrac{1}{4.2}$

Intercept: $b = 1$ Given

Thus, the equation is

$$y = mx + b$$
$$y = \frac{1}{4.2}x + 1$$

Line through $(2.1, 3)$ and $(5.2, 0)$

Point: $(5.2, 0)$

Slope: $m = \frac{y_2 - y_1}{x_2 - x_1} = \frac{0 - 3}{5.2 - 2.1} = -\frac{3}{3.1}$

Intercept:
$$b = y_1 - mx_1$$
$$= 0 - \left(-\frac{3}{3.1}\right)5.2 = \frac{15.6}{3.1}$$

Thus, the equation is
$$y = mx + b$$
$$y = -\frac{3}{3.1}x + \frac{15.6}{3.1}$$

Graph, with vertical and horizontal scales of 0.1:

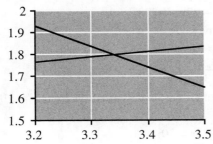

The grid point closest to the intersection of the lines gives the approximate solution.

Approximate solution: $(3.3, 1.8)$

23. Line through $(0, 0)$ and $(5.5, 3)$

Point: $(0, 0)$

Slope: $m = \frac{y_2 - y_1}{x_2 - x_1} = \frac{3 - 0}{5.5 - 0} = \frac{3}{5.5}$

Intercept: 0   Given

Thus, the equation is
$$y = mx + b$$
$$y = \frac{3}{5.5}x$$

Line through $(5, 0)$ and $(0, 6)$.

Point: $(0, 6)$

Slope: $m = \frac{y_2 - y_1}{x_2 - x_1} = \frac{6 - 0}{0 - 5} = -\frac{6}{5}$

Intercept: 6   Given

Thus, the equation is
$$y = mx + b$$
$$y = -\frac{6}{5}x + 6$$

Graph, with vertical and horizontal scales of 0.1:

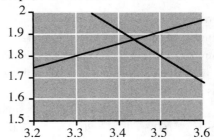

The grid point closest to the intersection of the lines gives the approximate solution.

Approximate solution: $(3.4, 1.9)$

25. Unknowns:

x = the number of quarts of Creamy Vanilla

y = the number of quarts of Continental Mocha

Arrange the given information in a table with unknowns across the top:

	Vanilla (x)	Mocha (y)	Available
Eggs	2	1	500
Cream	3	3	900

We can now set up an equation for each of the items listed on the left:

Eggs: $2x + y = 500$

Cream: $3x + 3y = 900$

Multiply the first equation by -1 and divide the second by 3:
$$-2x - y = -500$$
$$x + y = 300$$

Adding gives

$$-x = -200$$
$$x = 200 \text{ quarts of vanilla}$$

Substituting $x = 200$ in the first equation gives

$$2(200) + y = 500$$
$$400 + y = 500$$
$$y = 500 - 400 = 100 \text{ quarts of mocha}$$

Solution: Make 200 quarts of vanilla and 100 quarts of mocha.

27. Unknowns:

x = the number of servings of Mixed Cereal

y = the number of servings of Mango Tropical Fruit

Arrange the given information in a table with unknowns across the top:

	Cereal (x)	Mango (y)	Desired
Calories	60	80	200
Carbs.	11	21	43

We can now set up an equation for each of the items listed on the left:

Calories: $60x + 80y = 200$

Carbs: $11x + 21y = 43$

Divide first equation by 20:

$$3x + 4y = 10$$
$$11x + 21y = 43$$

Multiply first equation by -11 and second by 3:

$$-33x - 44y = -110$$
$$33x + 63y = 129$$

Adding gives

$$19y = 19$$
$$y = 1 \text{ serving of Mango Tropical Fruit}$$

Substituting $y=1$ in the equation $3x+4y = 10$ gives

$$3x + 4(1) = 10$$
$$3x = 6$$
$$x = 2 \text{ servings of Mixed Cereal}$$

Solution: Use 1 serving of Mango Tropical Fruit and 2 servings of Mixed Cereal.

29. (a) one half of the US RDA for protein, is 30g.

Unknowns:

x = the number of servings of Pork & Beans

y = the number of slices of white bread

Arrange the given information in a table with unknowns across the top:

	Pork & Beans (x)	Bread (y)	Desired
Protein	5	2	30
Carbs.	21	11	139

We can now set up an equation for each of the items listed on the left:

Protein $5x + 2y = 30$

Carbs: $21x + 11y = 139$

Multiply first equation by 11 and second by -2:

$$55x + 22y = 330$$
$$-42x - 22y = -278$$

Adding gives

$$13x = 52$$
$$x = \frac{52}{13} = 4 \text{ servings of Pork \& Beans}$$

Substituting $x=4$ in the equation $5x+2y = 30$ gives

$$5(4) + 2y = 30$$
$$20 + 2y = 30$$
$$2y = 10$$
$$y = 5 \text{ slices of bread}$$

Solution: Prepare 4 servings of Pork & Beans and 5 slices of bread.

(b) Decreasing the Carbohydrate total to 100 gives:

Protein $5x + 2y = 30$

Carbs: $21x + 11y = 100$

Multiply first equation by 11 and second by -2:

$$55x + 22y = 330$$
$$-42x - 22y = -200$$

Adding gives

$$13x = 130$$
$$x = \frac{130}{13} = 10 \text{ servings of Pork \& Beans}$$

Substituting $x=10$ in the equation $5x+2y = 30$ gives

$5(10) + 2y = 30$

$50 + 2y = 30$

$2y = 30-50 = -20$

$y = -10$ slices of bread

Since it is impossible to prepare a negative number of slices of bread, we conclude that it is not possible to prepare such a meal.

31. Unknowns:

x = number of servings of Cell-Tech

y = number of servings of Ribo-Force HP

Arrange the given information in a table with unknowns across the top (ALA stands for Alpha Lipoic Acid):

	Cell-Tech (x)	Ribo-Force (y)	Desired
Creatine	10	5	80
ALA	200	0	1000
Cost	$2.20	$1.60	

We can now set up an equation for the two nutrients listed on the left:

Creatine: $\qquad 10x + 5y = 80$

Alpha Lipoic Acid: $\qquad 200x = 1000$

The second equation gives

$$x = \frac{1000}{200} = 5 \text{ servings of Cell-Tech.}$$

Substituting $x = 5$ into the first equation gives

$50 + 5y = 80$

$5y = 30$

$y = \frac{30}{5} = 6$ servings of Ribo-Force HP.

To compute the cost, we use the information on the third row of the above table:

$$\begin{aligned} \text{Cost} &= 2.20x + 1.60y \\ &= 2.20(5) + 1.60(6) \\ &= \$20.60 \end{aligned}$$

Solution: Mix 5 servings of Cell-Tech and 6 servings of Ribo-Force HP for a cost of $20.60.

33. Unknowns:

x = number of shares of CSCO

y = number of shares of NOK

Arrange the given information in a table with unknowns across the top:

	CSCO (x)	NOK (y)	Total
November 1	20	17	4550
November 30	22.50	18	4950

We can now set up equations for the start and end of March:

March 1: $\qquad 20x + 17y = 4550$

March 31: $\qquad 22.5x + 18y = 4950$

Multiply the second equation by 2:

$20x + 17y = 4550$

$45x + 36y = 9900$

Divide the second equation by 9:

$20x + 17y = 4550$

$5x + 4y = 1100$

Multiply the second by -4:

$20x + 17y = 4550$

$-20x - 16y = -4400$

Adding gives

$y = 150$ shares of NOK

Substituting $y = 150$ into $5x + 4y = 1100$ gives

$5x + 4(150) = 1100$

$5x + 600 = 1100$

$5x = 1100 - 600 = 500$

$x = \frac{500}{5} = 100$ shares of CSCO

Solution: You bought 100 shares of CSCO and 150 shares of NOK.

35. Unknowns:

x = number of shares of ED

y = number of shares of KSE

The total investment was $11,200:

Investment in ED = x shares @ $40 = $40x$

Investment in KSE = y shares @ $36 = 36$y
Thus,
$$40x + 36y = 11{,}200$$
You earned $145 in dividends:
    ED dividend = 5.5% of $40x$ invested
      = $0.055(40x) = 2.2x$
    KSE dividend = 5% of $36y$ invested
      = $0.05(36y) = 1.8y$
Thus,
$$2.2x + 1.8y = 580$$
We therefore have the following system:
$$40x + 36y = 11{,}200$$
$$2.2x + 1.8y = 580$$
Eliminate decimals: Multiply the second equation by 10:
$$40x + 36y = 11{,}200$$
$$22x + 18y = 5800$$
Simplify: Divide the first equation by 4 and the second by 2:
$$10x + 9y = 2800$$
$$11x + 9y = 2900$$
Subtracting gives
$$x = 100 \text{ shares of ED}$$
Substituting $x = 100$ in $10x + 9y = 2800$ gives
$$1000 + 9y = 2800$$
$$9y = 1800$$
$$y = 200 \text{ shares of KSE}$$
Solution: you purchased 100 shares of ED and 200 shares of KSE.

37. Unknowns:
x = number of members voting in favor
y = number of members voting against
We are given two pieces of information:
(1) Total number of votes is 435
    $x + y = 435$
(2) 49 more members voted in favor than against.
    Rephrasing this gives:

The number of members voting in favor exceeded the number of members voting against by 49.
    $x - y = 49$
Thus we have two equations:
    $x + y = 435$
    $x - y = 49$
Adding gives
    $2x = 484$
    $x = 242$
Substituting $x = 242$ in the first equation gives
    $242 + y = 435$
    $y = 435 - 242 = 193$
Solution: 242 voted in favor and 193 voted against.

39. Unknowns:
x = number of soccer games won
y = number of football games won
We are given two pieces of information:
(1) The total number of games was 12
    $x + y = 12$
(2) Total number of points earned was 38
    $2x + 4y = 38$
(Two points per soccer game and 4 per football game)
Thus we have two equations:
    $x + y = 12$
    $2x + 4y = 38$
Dividing the second by -2:
    $x + y = 12$
    $-x - 2y = -19$
Adding gives
    $-y = -7$
    $y = 7 \text{ football games}$
Substituting $y = 7$ in the first equation gives
    $x + 7 = 12$
    $x = 5 \text{ soccer games}$

Solution: Lombardi House won 5 soccer games and 7 football games.

41. Unknowns:

x = number of brand X pens

y = number of brand Y pens

We are given two pieces of information:

(1) The total number of pens is 12

$x + y = 12$

(2) The total amount spent was \$42.

$4x + 2.8y = 42$

(\$4 per band X pen and \$2.80 per brand Y pen)

Thus we have two equations:

$x + y = 12$

$4x + 2.8y = 42$

Multiply the second by 10:

$x + y = 12$

$40x + 28y = 420$

Multiply the first by 10 and divide the second by –4:

$10x + 10y = 120$

$-10x - 7y = -105$

Adding,

$3y = 15$

$y = 5$ brand Y pens

Substituting this value in the first equation $x+y = 12$ gives

$x + 5 = 12$

$x = 7$ brand X pens

Solution: Elena purchased 7 brand X pens and 5 brand Y pens

43. Rewrite the demand and supply equations in standard form:

Demand: $q + 60p = 150$

Supply: $q - 80p = -60$

For neither a shortage nor surplus, both equations must be satisfied. Multiplying the second equation by –1 and adding gives

$140p = 210$

$p = \dfrac{210}{140} = \$1.50$ per pet chia

(We do not need to solve for q.)

45. Demand: $D = 85 - 5P$

Supply: $S = 25 + 5P$

For the equilibrium, we can equate the supply and demand:

Demand = Supply

$85 - 5P = 25 + 5P$

$-10P = -60$

$P = \$6$ per widget

Substituting $P = \$6$ in the demand curve gives

$D = 85 - 5(6) = 85 - 30 = 55$ widgets

47. Demand: The given points are

$(p, q) = (8, 15)$ and $(11, 3)$

$m = \dfrac{q_2 - q_1}{p_2 - p_1} = \dfrac{3 - 15}{11 - 8} = \dfrac{-12}{3} = -4$

$b = q_1 - mp_1 = 15 - (-4)(8) = 15 + 32 = 47$

Thus, the demand equation is

$q = mp + b$

$q = -4p + 47$

Supply: The given points are

$(p, q) = (8, 3)$ and $(11, 15)$

$m = \dfrac{q_2 - q_1}{p_2 - p_1} = \dfrac{15 - 3}{11 - 8} = \dfrac{12}{3} = 4$

$b = q_1 - mp_1 = 3 - (4)(8) = 3 - 32 = -29$

Thus, the supply equation is

$q = mp + b$

$q = 4p - 29$

For the equilibrium price, we can equate the supply and demand:

Supply = Demand

$-4p + 47 = 4p - 29$

$-8p = -76$

$p = \dfrac{76}{8} = \$9.50$

49. Unknowns:

x = number of pairs of dirty socks

y = number of T shirts

We are given two pieces of information:

(1) A total of 44 items were washed:

$x + y = 44$

(2) There were three times as many pairs of dirty socks as T shirts, Rephrasing this gives:

The number of pairs of dirty socks was three times the number of T shirts.

$x = 3y$

Thus we have two equations:

$x + y = 44$

$x - 3y = 0$

Subtracting (or multiplying the second by –1 and adding) gives

$4y = 44$

$y = 11$ T shirts

Substituting $y = 11$ in the first equation gives

$x + 11 = 44$

$x = 44 - 11 = 33$ pairs of dirty socks.

Solution: Joe's roommate threw out 33 pairs of dirty socks and 11 T shirts.

51. Unknowns:

x = size of raise for each full-time employee

y = size of raise for each part-time employee

We are given two pieces of information:

(1) Total budget = $6000. There are 4 full-time employees each getting a raise of x and 2 part-time employees each getting a raise of y. Thus,

$4x + 2y = 6000$

(2) The raise received by each full-time employee is twice the raise that each of the part-time employees receives.

$x = 2y$

Thus we have two equations:

$4x + 2y = 6000$

$x - 2y = 0$

Adding gives

$5x = 6000$

$x = \$1200$ per full time employee

(We are not asked for the value of y.)

53. The three lines in a plane must intersect in a single point for there to be a unique solution. This can happen in two ways: (1) the three lines intersect in a single point, or (2) two of the lines are the same, and the third line intersects it in a single point.

55. The equilibrium price occurs at the point where the demand and supply lines cross. Even if two lines have negative slope, they will still intersect if the slopes differ. Therefore, there can be an equilibrium price.

57. You cannot round both of them up, since there will not be sufficient eggs and cream. Rounding both answers down will ensure that you will not run out of ingredients. It may be possible to round one answer down and the other up, and still have sufficient eggs and cream. and this should be tried.

59. Since multiplying both sides of an equation by a non-zero number has no effect on its solutions, the graph (which represents the set of all solutions) is unchanged: (B).

61. The associated system has no solutions, and so the lines do not intersect. Thus, they must be parallel: (B).

65. Choosing two lines at random gives two random slopes (which are numbers). Since two randomly selected numbers are unlikely to be the same, it follows that two randomly chosen straight lines are very unlikely to be parallel (or the same line). Thus, the two lines are very likely to intersect in a point, giving a unique solution.

2.2

1. $x + y = 4$

$x - y = 2$

$$\begin{bmatrix} 1 & 1 & 4 \\ 1 & -1 & 2 \end{bmatrix} R_2 - R_1$$

$$\begin{bmatrix} 1 & 1 & 4 \\ 0 & -2 & -2 \end{bmatrix} (1/2)R_2$$

$$\begin{bmatrix} 1 & 1 & 4 \\ 0 & -1 & -1 \end{bmatrix} R_1 + R_2$$

$$\begin{bmatrix} 1 & 0 & 3 \\ 0 & -1 & -1 \end{bmatrix} -R_2 \rightarrow \begin{bmatrix} 1 & 0 & 3 \\ 0 & 1 & 1 \end{bmatrix}$$

Solution: $x = 3$, $y = 1$, or $(3, 1)$

3. $3x - 2y = 6$

$2x - 3y = -6$

$$\begin{bmatrix} 3 & -2 & 6 \\ 2 & -3 & -6 \end{bmatrix} 3R_2 - 2R_1$$

$$\begin{bmatrix} 3 & -2 & 6 \\ 0 & -5 & -30 \end{bmatrix} (1/5)R_2$$

$$\begin{bmatrix} 3 & -2 & 6 \\ 0 & -1 & -6 \end{bmatrix} R_1 - 2R_2$$

$$\begin{bmatrix} 3 & 0 & 18 \\ 0 & -1 & -6 \end{bmatrix} (1/3)R_1$$

$$\begin{bmatrix} 1 & 0 & 6 \\ 0 & -1 & -6 \end{bmatrix} -R_2 \rightarrow \begin{bmatrix} 1 & 0 & 6 \\ 0 & 1 & 6 \end{bmatrix}$$

Solution: $x = 6$, $y = 6$, or $(6, 6)$

5. $2x + 3y = 1$

$-x - \dfrac{3y}{2} = -\dfrac{1}{2}$

$$\begin{bmatrix} 2 & 3 & 1 \\ -1 & -3/2 & -1/2 \end{bmatrix} 2R_2$$

$$\begin{bmatrix} 2 & 3 & 1 \\ -2 & -3 & -1 \end{bmatrix} R_2 + R_1$$

$$\begin{bmatrix} 2 & 3 & 1 \\ 0 & 0 & 0 \end{bmatrix} \begin{matrix} (1/2)R_1 \\ (1/2)R_2 \end{matrix} \rightarrow \begin{bmatrix} 1 & 3/2 & 1/2 \\ 0 & 0 & 0 \end{bmatrix}$$

Converting back to equations:

$$x + \frac{3}{2}y = \frac{1}{2}$$

Solve for x:

$$x = \frac{1}{2} - \frac{3}{2}y$$

General Solution:

$$x = \frac{1}{2} - \frac{3}{2}y = \frac{1}{2}(1-3y);$$

y is arbitrary

or $(\frac{1}{2}[1-3y], y)$; y arbitrary

7. $2x + 3y = 2$

$-x - \dfrac{3y}{2} = -\dfrac{1}{2}$

$$\begin{bmatrix} 2 & 3 & 2 \\ -1 & -3/2 & -1/2 \end{bmatrix} 2R_2$$

$$\begin{bmatrix} 2 & 3 & 2 \\ -2 & -3 & -1 \end{bmatrix} R_2 + R_1 \rightarrow \begin{bmatrix} 2 & 3 & 2 \\ 0 & 0 & 1 \end{bmatrix}$$

Since the bottom row translates to the false statement $0 = 1$, there is no solution.

9 $x + y = 1$

$3x - y = 0$

$x - 3y = -2$

$$\begin{bmatrix} 1 & 1 & 1 \\ 3 & -1 & 0 \\ 1 & -3 & -2 \end{bmatrix} \begin{matrix} \\ R_2 - 3R_1 \\ R_3 - R_1 \end{matrix}$$

$$\begin{bmatrix} 1 & 1 & 1 \\ 0 & -4 & -3 \\ 0 & -4 & -3 \end{bmatrix} \begin{matrix} 4R_1 + R_2 \\ \\ R_3 - R_2 \end{matrix}$$

$$\begin{bmatrix} 4 & 0 & 1 \\ 0 & -4 & -3 \\ 0 & 0 & 0 \end{bmatrix} \begin{matrix} (1/4)R_1 \\ -(1/4)R_2 \\ -(1/4)R_3 \end{matrix}$$

$$\begin{bmatrix} 1 & 0 & 1/4 \\ 0 & 1 & 3/4 \\ 0 & 0 & 0 \end{bmatrix}$$

Solution: $x = 1/4$, $y = 3/4$, or $(1/4. 3/4)$

11. $x + y = 0$

$\quad 3x - y = 1$

$\quad x - y = -1$

$$\begin{bmatrix} 1 & 1 & 0 \\ 3 & -1 & 1 \\ 1 & -1 & -1 \end{bmatrix} \begin{matrix} \\ R_2 - 3R_1 \\ R_3 - R_1 \end{matrix}$$

$$\begin{bmatrix} 1 & 1 & 0 \\ 0 & -4 & 1 \\ 0 & -2 & -1 \end{bmatrix} \begin{matrix} 4R_1 + R_2 \\ \\ 2R_3 - R_2 \end{matrix}$$

$$\begin{bmatrix} 4 & 0 & 1 \\ 0 & -4 & 1 \\ 0 & 0 & -3 \end{bmatrix}$$

Since the bottom row translates to the false statement $0 = -3$, there is no solution.

13. $0.5x + 0.1y = 1.7$

$\quad 0.1x - 0.1y = 0.3$

$\quad x + y = \dfrac{11}{3}$

$$\begin{bmatrix} 0.5 & 0.1 & 1.7 \\ 0.1 & -0.1 & 0.3 \\ 1 & 1 & 11/3 \end{bmatrix} \begin{matrix} 10R_1 \\ 10R_2 \\ 3R_3 \end{matrix}$$

$$\begin{bmatrix} 5 & 1 & 17 \\ 1 & -1 & 3 \\ 3 & 3 & 11 \end{bmatrix} \begin{matrix} \\ 5R_2 - R_1 \\ 5R_3 - 3R_1 \end{matrix}$$

$$\begin{bmatrix} 5 & 1 & 17 \\ 0 & -6 & -2 \\ 0 & 12 & 4 \end{bmatrix} \begin{matrix} \\ (1/2)R_2 \\ (1/4)R_3 \end{matrix}$$

$$\begin{bmatrix} 5 & 1 & 17 \\ 0 & -3 & -1 \\ 0 & 3 & 1 \end{bmatrix} \begin{matrix} 3R_1 + R_2 \\ \\ R_3 + R_2 \end{matrix}$$

$$\begin{bmatrix} 15 & 0 & 50 \\ 0 & -3 & -1 \\ 0 & 0 & 0 \end{bmatrix} \begin{matrix} (1/15)R_1 \\ -(1/3)R_2 \\ \end{matrix}$$

$$\begin{bmatrix} 1 & 0 & 10/3 \\ 0 & 1 & 1/3 \\ 0 & 0 & 0 \end{bmatrix}$$

Solution: $x = 10/3$, $y = 1/3$

15. $-x + 2y - z = 0$

$\quad -x - y + 2z = 0$

$\quad 2x \quad - z = 4$

$$\begin{bmatrix} -1 & 2 & -1 & 0 \\ -1 & -1 & 2 & 0 \\ 2 & 0 & -1 & 4 \end{bmatrix} \begin{matrix} \\ R_2 - R_1 \\ R_3 + 2R_1 \end{matrix}$$

$$\begin{bmatrix} -1 & 2 & -1 & 0 \\ 0 & -3 & 3 & 0 \\ 0 & 4 & -3 & 4 \end{bmatrix} \begin{matrix} \\ (1/3)R_2 \\ \end{matrix}$$

$$\begin{bmatrix} -1 & 2 & -1 & 0 \\ 0 & -1 & 1 & 0 \\ 0 & 4 & -3 & 4 \end{bmatrix} \begin{matrix} R_1 + 2R_2 \\ \\ R_3 + 4R_2 \end{matrix}$$

$$\begin{bmatrix} -1 & 0 & 1 & 0 \\ 0 & -1 & 1 & 0 \\ 0 & 0 & 1 & 4 \end{bmatrix} \begin{matrix} R_1 - R_3 \\ R_2 - R_3 \\ \end{matrix}$$

$$\begin{bmatrix} -1 & 0 & 0 & -4 \\ 0 & -1 & 0 & -4 \\ 0 & 0 & 1 & 4 \end{bmatrix} \begin{matrix} -R_1 \\ -R_2 \\ \end{matrix}$$

$$\begin{bmatrix} 1 & 0 & 0 & 4 \\ 0 & 1 & 0 & 4 \\ 0 & 0 & 1 & 4 \end{bmatrix}$$

Solution: $x = 4$, $y = 4$, $z = 4$

17. $x + y + 6z = -1$

$\quad \frac{1}{3}x - \frac{1}{3}y + \frac{2}{3}z = 1$

$\quad \frac{1}{2}x \quad + z = 0$

$$\begin{bmatrix} 1 & 1 & 6 & -1 \\ 1/3 & -1/3 & 2/3 & 1 \\ 1/2 & 0 & 1 & 0 \end{bmatrix} \begin{matrix} \\ 3R_2 \\ 2R_3 \end{matrix}$$

59

$$\begin{bmatrix} 1 & 1 & 6 & -1 \\ 0 & -2 & -4 & 4 \\ 0 & -1 & -4 & 1 \end{bmatrix} (1/2)R_2$$

$$\begin{bmatrix} 1 & 1 & 6 & -1 \\ 0 & -1 & -2 & 2 \\ 0 & -1 & -4 & 1 \end{bmatrix} \begin{matrix} R_1 + R_2 \\ \\ R_3 - R_2 \end{matrix}$$

$$\begin{bmatrix} 1 & 0 & 4 & 1 \\ 0 & -1 & -2 & 2 \\ 0 & 0 & -2 & -1 \end{bmatrix} \begin{matrix} R_1 + 2R_3 \\ R_2 - R_3 \end{matrix}$$

$$\begin{bmatrix} 1 & 0 & 0 & -1 \\ 0 & -1 & 0 & 3 \\ 0 & 0 & -2 & -1 \end{bmatrix} \begin{matrix} \\ -R_2 \\ -(1/2)R_3 \end{matrix}$$

$$\begin{bmatrix} 1 & 0 & 0 & -1 \\ 0 & 1 & 0 & -3 \\ 0 & 0 & 1 & 1/2 \end{bmatrix}$$

Solution: $x = -1$, $y = -3$, $z = 1/2$

19. $-\frac{1}{2}x + y - \frac{1}{2}z = 0$

$-\frac{1}{2}x - \frac{1}{2}y + z = 0$

$x - \frac{1}{2}y - \frac{1}{2}z = 0$

$$\begin{bmatrix} -1/2 & 1 & -1/2 & 0 \\ -1/2 & -1/2 & 1 & 0 \\ 1 & -1/2 & -1/2 & 0 \end{bmatrix} \begin{matrix} 2R_1 \\ 2R_2 \\ 2R_3 \end{matrix}$$

$$\begin{bmatrix} -1 & 2 & -1 & 0 \\ -1 & -1 & 2 & 0 \\ 2 & -1 & -1 & 0 \end{bmatrix} \begin{matrix} \\ R_2 - R_1 \\ R_3 + 2R_1 \end{matrix}$$

$$\begin{bmatrix} -1 & 2 & -1 & 0 \\ 0 & -3 & 3 & 0 \\ 0 & 3 & -3 & 0 \end{bmatrix} \begin{matrix} \\ (1/3)R_2 \\ (1/3)R_3 \end{matrix}$$

$$\begin{bmatrix} -1 & 2 & -1 & 0 \\ 0 & -1 & 1 & 0 \\ 0 & 1 & -1 & 0 \end{bmatrix} \begin{matrix} R_1 + 2R_2 \\ \\ R_3 + R_2 \end{matrix}$$

$$\begin{bmatrix} -1 & 0 & 1 & 0 \\ 0 & -1 & 1 & 0 \\ 0 & 0 & 0 & 0 \end{bmatrix} \begin{matrix} -R_1 \\ -R_2 \end{matrix}$$

$$\begin{bmatrix} 1 & 0 & -1 & 0 \\ 0 & 1 & -1 & 0 \\ 0 & 0 & 0 & 0 \end{bmatrix}$$

Translating back to equations gives

$x - z = 0$

$y - z = 0$

Thus, the general solution is

$x = z, y = z, z$ arbitrary,

or(z, z, z), z arbitrary

21. $x + y + 2z = -1$

$2x + 2y + 2z = 2$

$\frac{3}{5}x + \frac{3}{5}y + \frac{3}{5}z = \frac{2}{5}$

$$\begin{bmatrix} 1 & 1 & 2 & -1 \\ 2 & 2 & 2 & 2 \\ 3/5 & 3/5 & 3/5 & 2/5 \end{bmatrix} \begin{matrix} \\ (1/2)R_2 \\ 5R_3 \end{matrix}$$

$$\begin{bmatrix} 1 & 1 & 2 & -1 \\ 1 & 1 & 1 & 1 \\ 3 & 3 & 3 & 2 \end{bmatrix} \begin{matrix} \\ R_2 - R_1 \\ R_3 - 3R_1 \end{matrix}$$

$$\begin{bmatrix} 1 & 1 & 2 & -1 \\ 0 & 0 & -1 & 2 \\ 0 & 0 & -3 & 5 \end{bmatrix} \begin{matrix} R_1 + 2R_2 \\ \\ R_3 - 3R_2 \end{matrix}$$

$$\begin{bmatrix} 1 & 1 & 0 & 3 \\ 0 & 0 & -1 & 2 \\ 0 & 0 & 0 & -1 \end{bmatrix}$$

Since the bottom row translates to the false statement $0 = -1$, there is no solution.

23. $-0.5x + 0.5y + 0.5z = 1.5$

$4.2x + 2.1y + 2.1z = 0$

$0.2x \qquad + 0.2z = 0$

$$\begin{bmatrix} -.5 & .5 & .5 & 1.5 \\ 4.2 & 2.1 & 2.1 & 0 \\ .2 & 0 & .2 & 0 \end{bmatrix} \begin{matrix} 2R_1 \\ 10R_2 \\ 5R_3 \end{matrix}$$

$$\begin{bmatrix} -1 & 1 & 1 & 3 \\ 42 & 21 & 21 & 0 \\ 1 & 0 & 1 & 0 \end{bmatrix} (1/21)R_2$$

$$\begin{bmatrix} -1 & 1 & 1 & 3 \\ 2 & 1 & 1 & 0 \\ 1 & 0 & 1 & 0 \end{bmatrix} \begin{matrix} \\ R_2 + 2R_1 \\ R_3 + R_1 \end{matrix}$$

$$\begin{bmatrix} -1 & 1 & 1 & 3 \\ 0 & 3 & 3 & 6 \\ 0 & 1 & 2 & 3 \end{bmatrix} (1/3)R_2$$

$$\begin{bmatrix} -1 & 1 & 1 & 3 \\ 0 & 1 & 1 & 2 \\ 0 & 1 & 2 & 3 \end{bmatrix} \begin{matrix} R_1 - R_2 \\ \\ R_3 - R_2 \end{matrix}$$

$$\begin{bmatrix} -1 & 0 & 0 & 1 \\ 0 & 1 & 1 & 2 \\ 0 & 0 & 1 & 1 \end{bmatrix} \begin{matrix} \\ R_2 - R_3 \\ \end{matrix}$$

$$\begin{bmatrix} -1 & 0 & 0 & 1 \\ 0 & 1 & 0 & 1 \\ 0 & 0 & 1 & 1 \end{bmatrix} -R_1$$

$$\begin{bmatrix} 1 & 0 & 0 & -1 \\ 0 & 1 & 0 & 1 \\ 0 & 0 & 1 & 1 \end{bmatrix}$$

Solution: $x = -1$, $y = 1$, $z = 1$

25. $2x - y + z = 4$
$\qquad 3x - y + z = 5$

$$\begin{bmatrix} 2 & -1 & 1 & 4 \\ 3 & -1 & 1 & 5 \end{bmatrix} 2R_2 - 3R_1$$

$$\begin{bmatrix} 2 & -1 & 1 & 4 \\ 0 & 1 & -1 & -2 \end{bmatrix} R_1 + R_2$$

$$\begin{bmatrix} 2 & 0 & 0 & 2 \\ 0 & 1 & -1 & -2 \end{bmatrix} (1/2)R_1$$

$$\begin{bmatrix} 1 & 0 & 0 & 1 \\ 0 & 1 & -1 & -2 \end{bmatrix}$$

Translating back to equations gives
$\qquad x = 1$
$\qquad y - z = -2$
Thus, the general solution is
$\qquad x = 1$
$\qquad y = z - 2$
$\qquad z$ arbitrary,
or
$\qquad (1, z-2, z).$ z arbitrary

27. $0.75x - 0.75y - z = 4$
$\qquad x - y + 4z = 0$

$$\begin{bmatrix} 0.75 & -.075 & -1 & 4 \\ 1 & -1 & 4 & 0 \end{bmatrix} 4R_1$$

$$\begin{bmatrix} 3 & -3 & -4 & 16 \\ 1 & -1 & 4 & 0 \end{bmatrix} 3R_2 - R_1$$

$$\begin{bmatrix} 3 & -3 & -4 & 16 \\ 0 & 0 & 16 & -16 \end{bmatrix} (1/16)R_2$$

$$\begin{bmatrix} 3 & -3 & -4 & 16 \\ 0 & 0 & 1 & -1 \end{bmatrix} R_1 + 4R_2$$

$$\begin{bmatrix} 3 & -3 & 0 & 12 \\ 0 & 0 & 1 & -1 \end{bmatrix} (1/3)R_1$$

$$\begin{bmatrix} 1 & -1 & 0 & 4 \\ 0 & 0 & 1 & -1 \end{bmatrix}$$

Translating back to equations gives
$\qquad x - y = 4$
$\qquad z = -1$
Thus, the general solution is
$\qquad x = y + 4$
$\qquad y$ arbitrary
$\qquad z = -1,$
or $(y + 4, y, -1)$; y arbitrary

29. $3x + y - z = 12$

$\begin{bmatrix} 3 & 1 & -1 & 12 \end{bmatrix}$ $(1/3)R_1$

$\begin{bmatrix} 1 & 1/3 & -1/3 & 4 \end{bmatrix}$

Translating back to equations gives

$x + y/3 - z/3 = 4$

General Solution:

$x = 4 - y/3 + z/3$

y arbitrary

z arbitrary,

or $(4 - y/3 + z/3, y, z)$; y, z arbitrary

31. $\quad x + y + 2z = -1$

$\qquad 2x + 2y + 2z = 2$

$\qquad 0.75x + 0.75y + z = 0.25$

$\qquad -x - 2z = 21$

$\begin{bmatrix} 1 & 1 & 2 & -1 \\ 2 & 2 & 2 & 2 \\ .75 & .75 & 1 & .25 \\ -1 & 0 & -2 & 21 \end{bmatrix} \begin{array}{l} (1/2)R_2 \\ 4R_3 \end{array}$

$\begin{bmatrix} 1 & 1 & 2 & -1 \\ 1 & 1 & 1 & 1 \\ 3 & 3 & 4 & 1 \\ -1 & 0 & -2 & 21 \end{bmatrix} \begin{array}{l} R_2 - R_1 \\ R_3 - 3R_1 \\ R_4 + R_1 \end{array}$

$\begin{bmatrix} 1 & 1 & 2 & -1 \\ 0 & 0 & -1 & 2 \\ 0 & 0 & -2 & 4 \\ 0 & 1 & 0 & 20 \end{bmatrix} (1/2)R_3$

$\begin{bmatrix} 1 & 1 & 2 & -1 \\ 0 & 0 & -1 & 2 \\ 0 & 0 & -1 & 2 \\ 0 & 1 & 0 & 20 \end{bmatrix} \begin{array}{l} R_1 + 2R_2 \\ \\ R_3 - R_2 \end{array}$

$\begin{bmatrix} 1 & 1 & 0 & 3 \\ 0 & 0 & -1 & 2 \\ 0 & 0 & 0 & 0 \\ 0 & 1 & 0 & 20 \end{bmatrix} R_1 - R_4$

$\begin{bmatrix} 1 & 0 & 0 & -17 \\ 0 & 0 & -1 & 2 \\ 0 & 0 & 0 & 0 \\ 0 & 1 & 0 & 20 \end{bmatrix} -R_2$

$\begin{bmatrix} 1 & 0 & 0 & -17 \\ 0 & 0 & 1 & -2 \\ 0 & 0 & 0 & 0 \\ 0 & 1 & 0 & 20 \end{bmatrix}$ *Rearrange Rows*

$\begin{bmatrix} 1 & 0 & 0 & -17 \\ 0 & 1 & 0 & 20 \\ 0 & 0 & 1 & -2 \\ 0 & 0 & 0 & 0 \end{bmatrix}$

Solution: $x = -17$, $y = 20$, $z = -2$

33. $\quad x + y + 5z = 1$

$\qquad y + 2z + w = 1$

$\qquad x + 3y + 7z + 2w = 2$

$\qquad x + y + 5z + w = 1$

$\begin{bmatrix} 1 & 1 & 5 & 0 & 1 \\ 0 & 1 & 2 & 1 & 1 \\ 1 & 3 & 7 & 2 & 2 \\ 1 & 1 & 5 & 1 & 1 \end{bmatrix} \begin{array}{l} R_3 - R_1 \\ R_4 - R_1 \end{array}$

$\begin{bmatrix} 1 & 1 & 5 & 0 & 1 \\ 0 & 1 & 2 & 1 & 1 \\ 0 & 2 & 2 & 2 & 1 \\ 0 & 0 & 0 & 1 & 0 \end{bmatrix} \begin{array}{l} R_1 - R_2 \\ \\ R_3 - 2R_2 \end{array}$

$$\begin{bmatrix} 1 & 0 & 3 & -1 & 0 \\ 0 & 1 & 2 & 1 & 1 \\ 0 & 0 & -2 & 0 & -1 \\ 0 & 0 & 0 & 1 & 0 \end{bmatrix} \begin{matrix} 2R_1 + 3R_3 \\ R_2 + R_3 \\ \\ \end{matrix}$$

$$\begin{bmatrix} 2 & 0 & 0 & -2 & -3 \\ 0 & 1 & 0 & 1 & 0 \\ 0 & 0 & -2 & 0 & -1 \\ 0 & 0 & 0 & 1 & 0 \end{bmatrix} \begin{matrix} R_1 + 2R_4 \\ R_2 - R_4 \\ \\ \end{matrix}$$

$$\begin{bmatrix} 2 & 0 & 0 & 0 & -3 \\ 0 & 1 & 0 & 0 & 0 \\ 0 & 0 & -2 & 0 & -1 \\ 0 & 0 & 0 & 1 & 0 \end{bmatrix} \begin{matrix} (1/2)R_1 \\ \\ -(1/2)R_3 \\ \end{matrix}$$

$$\begin{bmatrix} 1 & 0 & 0 & 0 & -3/2 \\ 0 & 1 & 0 & 0 & 0 \\ 0 & 0 & 1 & 0 & 1/2 \\ 0 & 0 & 0 & 1 & 0 \end{bmatrix}$$

Solution: $x = -3/2$, $y = 0$, $z = 1/2$, $w = 0$

35. $x + y + 5z = 1$
$y + 2z + w = 1$
$x + y + 5z + w = 1$
$x + 2y + 7z + 2w = 2$

$$\begin{bmatrix} 1 & 1 & 5 & 0 & 1 \\ 0 & 1 & 2 & 1 & 1 \\ 1 & 1 & 5 & 1 & 1 \\ 1 & 2 & 7 & 2 & 2 \end{bmatrix} \begin{matrix} \\ \\ R_3 - R_1 \\ R_4 - R_1 \end{matrix}$$

$$\begin{bmatrix} 1 & 1 & 5 & 0 & 1 \\ 0 & 1 & 2 & 1 & 1 \\ 0 & 0 & 0 & 1 & 0 \\ 0 & 1 & 2 & 2 & 1 \end{bmatrix} \begin{matrix} R_1 - R_2 \\ \\ \\ R_4 - R_2 \end{matrix}$$

$$\begin{bmatrix} 1 & 0 & 3 & -1 & 0 \\ 0 & 1 & 2 & 1 & 1 \\ 0 & 0 & 0 & 1 & 0 \\ 0 & 0 & 0 & 1 & 0 \end{bmatrix} \begin{matrix} R_1 + R_3 \\ R_2 - R_3 \\ \\ R_4 - R_3 \end{matrix}$$

$$\begin{bmatrix} 1 & 0 & 3 & 0 & 0 \\ 0 & 1 & 2 & 0 & 1 \\ 0 & 0 & 0 & 1 & 0 \\ 0 & 0 & 0 & 0 & 0 \end{bmatrix}$$

Translating back to equations gives
$x + 3z = 0$
$y + 2z = 1$
$w = 0$
General solution:
$x = -3z$
$y = 1 - 2z$
z arbitrary
$w = 0$,
or $(-3z, 1 - 2z. z. 0)$; z arbitrary

37. $x - 2y + z - 4w = 1$
$x + 3y + 7z + 2w = 2$
$2x + y + 8z - 2w = 3$

$$\begin{bmatrix} 1 & -2 & 1 & -4 & 1 \\ 1 & 3 & 7 & 2 & 2 \\ 2 & 1 & 8 & -2 & 3 \end{bmatrix} \begin{matrix} \\ R_2 - R_1 \\ R_3 - 2R_1 \end{matrix}$$

$$\begin{bmatrix} 1 & -2 & 1 & -4 & 1 \\ 0 & 5 & 6 & 6 & 1 \\ 0 & 5 & 6 & 6 & 1 \end{bmatrix} \begin{matrix} 5R_1 + 2R_2 \\ \\ R_3 - R_2 \end{matrix}$$

$$\begin{bmatrix} 5 & 0 & 17 & -8 & 7 \\ 0 & 5 & 6 & 6 & 1 \\ 0 & 0 & 0 & 0 & 0 \end{bmatrix} \begin{matrix} (1/5)R_1 \\ (1/5)R_2 \\ \end{matrix}$$

$$\begin{bmatrix} 1 & 0 & 17/5 & -8/5 & 7/5 \\ 0 & 1 & 6/5 & 6/5 & 1/5 \\ 0 & 0 & 0 & 0 & 0 \end{bmatrix}$$

$x + 17z/5 - 8w/5 = 7/5$
$y + 6z/5 + 6w/5 = 1/5$
General solution:
$x = 7/5 - 17z/5 + 8w/5$
$y = 1/5 - 6z/5 - 6w/5$
z, w arbitrary,

or $(7/5-17z/5+8w/5, 1/5-6z/5-6w/5, z, w)$

z, w arbitrary

or $(\frac{1}{5}(7 - 17z + 8w), \frac{1}{5}(1 - 6z - 6w), z, w)$, z, w

arbitrary

39. $x + y + z + u + v = 15$

$\qquad y - z + u - v = -2$

$\qquad\quad z + u + v = 12$

$\qquad\qquad\quad u - v = -1$

$\qquad\qquad\qquad\quad v = 5$

$$\begin{bmatrix} 1 & 1 & 1 & 1 & 1 & 15 \\ 0 & 1 & -1 & 1 & -1 & -2 \\ 0 & 0 & 1 & 1 & 1 & 12 \\ 0 & 0 & 0 & 1 & -1 & -1 \\ 0 & 0 & 0 & 0 & 1 & 5 \end{bmatrix} \begin{matrix} R_1 - R_2 \\ \\ \\ \\ \end{matrix}$$

$$\begin{bmatrix} 1 & 0 & 2 & 0 & 2 & 17 \\ 0 & 1 & -1 & 1 & -1 & -2 \\ 0 & 0 & 1 & 1 & 1 & 12 \\ 0 & 0 & 0 & 1 & -1 & -1 \\ 0 & 0 & 0 & 0 & 1 & 5 \end{bmatrix} \begin{matrix} R_1 - 2R_3 \\ R_2 + R_3 \\ \\ \\ \end{matrix}$$

$$\begin{bmatrix} 1 & 0 & 0 & -2 & 0 & -7 \\ 0 & 1 & 0 & 2 & 0 & 10 \\ 0 & 0 & 1 & 1 & 1 & 12 \\ 0 & 0 & 0 & 1 & -1 & -1 \\ 0 & 0 & 0 & 0 & 1 & 5 \end{bmatrix} \begin{matrix} R_1 + 2R_4 \\ R_2 - 2R_4 \\ R_3 - R_4 \\ \\ \end{matrix}$$

$$\begin{bmatrix} 1 & 0 & 0 & 0 & -2 & -9 \\ 0 & 1 & 0 & 0 & 2 & 12 \\ 0 & 0 & 1 & 0 & 2 & 13 \\ 0 & 0 & 0 & 1 & -1 & -1 \\ 0 & 0 & 0 & 0 & 1 & 5 \end{bmatrix} \begin{matrix} R_1 + 2R_5 \\ R_2 - 2R_5 \\ R_3 - 2R_5 \\ R_4 + R_5 \\ \end{matrix}$$

$$\begin{bmatrix} 1 & 0 & 0 & 0 & 0 & 1 \\ 0 & 1 & 0 & 0 & 0 & 2 \\ 0 & 0 & 1 & 0 & 0 & 3 \\ 0 & 0 & 0 & 1 & 0 & 4 \\ 0 & 0 & 0 & 0 & 1 & 5 \end{bmatrix}$$

Solution: $x = 1, y = 2, z = 3, u = 4, v = 5$

41. $x - y + z - u + v = 0$

$\qquad y - z + u - v = -2$

$x \qquad\qquad\qquad - 2v = -2$

$2x - y + z - u - 3v = -2$

$4x - y + z - u - 7v = -6$

$$\begin{bmatrix} 1 & -1 & 1 & -1 & 1 & 0 \\ 0 & 1 & -1 & 1 & -1 & -2 \\ 1 & 0 & 0 & 0 & -2 & -2 \\ 2 & -1 & 1 & -1 & -3 & -2 \\ 4 & -1 & 1 & -1 & -7 & -6 \end{bmatrix} \begin{matrix} \\ \\ R_3 - R_1 \\ R_4 - 2R_1 \\ R_5 - 4R_1 \end{matrix}$$

$$\begin{bmatrix} 1 & -1 & 1 & -1 & 1 & 0 \\ 0 & 1 & -1 & 1 & -1 & -2 \\ 0 & 1 & -1 & 1 & -3 & -2 \\ 0 & 1 & -1 & 1 & -5 & -2 \\ 0 & 3 & -3 & 3 & -11 & -6 \end{bmatrix} \begin{matrix} R_1 + R_2 \\ \\ R_3 - R_2 \\ R_4 - R_2 \\ R_5 - 3R_2 \end{matrix}$$

$$\begin{bmatrix} 1 & 0 & 0 & 0 & 0 & -2 \\ 0 & 1 & -1 & 1 & -1 & -2 \\ 0 & 0 & 0 & 0 & -2 & 0 \\ 0 & 0 & 0 & 0 & -4 & 0 \\ 0 & 0 & 0 & 0 & -8 & 0 \end{bmatrix} \begin{matrix} \\ \\ (1/2)R_3 \\ (1/4)R_4 \\ (1/8)R_5 \end{matrix}$$

$$\begin{bmatrix} 1 & 0 & 0 & 0 & 0 & -2 \\ 0 & 1 & -1 & 1 & -1 & -2 \\ 0 & 0 & 0 & 0 & -1 & 0 \\ 0 & 0 & 0 & 0 & -1 & 0 \\ 0 & 0 & 0 & 0 & -1 & 0 \end{bmatrix} \begin{matrix} \\ R_2 - R_3 \\ \\ R_4 - R_3 \\ R_5 - R_3 \end{matrix}$$

$$\begin{bmatrix} 1 & 0 & 0 & 0 & 0 & -2 \\ 0 & 1 & -1 & 1 & 0 & -2 \\ 0 & 0 & 0 & 0 & -1 & 0 \\ 0 & 0 & 0 & 0 & 0 & 0 \\ 0 & 0 & 0 & 0 & 0 & 0 \end{bmatrix} -R_3$$

$$\begin{bmatrix} 1 & 0 & 0 & 0 & 0 & -2 \\ 0 & 1 & -1 & 1 & 0 & -2 \\ 0 & 0 & 0 & 0 & 1 & 0 \\ 0 & 0 & 0 & 0 & 0 & 0 \\ 0 & 0 & 0 & 0 & 0 & 0 \end{bmatrix}$$

$x = -2$

$y - z + u = -2$

$v = 0$

General solution:

$x = -2$

$y = -2 + z - u$

z is arbitrary

u is arbitrary

$v = 0$

or

$(-2, -2+z-u. z. u. 0)$; z, u arbitrary

43. $x + 2y - z + w = 30$

$2x \quad - z + 2w = 30$

$x + 3y + 3z - 4w = 2$

$2x - 9y \quad + w = 4$

Using technology:

Matrix #1

x	y	z	w	
1	2	-1	1	30
2	0	-1	2	30
1	3	3	-4	2
2	-9	0	1	4

Matrix #2

x	y	z	w	
1	2	-1	1	30
0	-4	1	0	-30
0	1	4	-5	-28
0	-13	2	-1	-56

Matrix #3

x	y	z	w	
2	0	-1	2	30
0	-4	1	0	-30
0	0	17	-20	-142
0	0	-5	-4	166

Matrix #4

x	y	z	w	
34	0	0	14	368
0	-68	0	20	-368
0	0	17	-20	-142
0	0	0	-168	2112

Matrix #5

x	y	z	w	
17	0	0	7	184
0	-17	0	5	-92
0	0	17	-20	-142
0	0	0	-7	88

Matrix #6

x	y	z	w	
17	0	0	0	272
0	-119	0	0	-204
0	0	119	0	-2754
0	0	0	-7	88

Matrix #7

x	y	z	w	
1	0	0	0	16
0	1	0	0	12/7
0	0	1	0	-162/7
0	0	0	1	-88/7

Solution: $(16, 12/7, -162/7, -88/7)$

45. $x + 2y + 3z + 4w + 5t = 6$

$2x + 3y + 4z + 5w + t = 5$

$3x + 4y + 5z + w + 2t = 4$

$4x + 5y + z + 2w + 3t = 3$

$5x + y + 2z + 3w + 4t = 2$

Using technology:

Matrix #1

x	y	z	w	t	
1	2	3	4	5	6
2	3	4	5	1	5
3	4	5	1	2	4
4	5	1	2	3	3
5	1	2	3	4	2

Section 2.2

Matrix #2

x	y	z	w	t	
1	2	3	4	5	6
0	-1	-2	-3	-9	-7
0	-2	-4	-11	-13	-14
0	-3	-11	-14	-17	-21
0	-9	-13	-17	-21	-28

Matrix #3

x	y	z	w	t	
1	0	-1	-2	-13	-8
0	-1	-2	-3	-9	-7
0	0	0	-5	5	0
0	0	-5	-5	10	0
0	0	5	10	60	35

Matrix #4

x	y	z	w	t	
1	0	-1	-2	-13	-8
0	-1	-2	-3	-9	-7
0	0	0	-1	1	0
0	0	-1	-1	2	0
0	0	1	2	12	7

Matrix #5

x	y	z	w	t	
1	0	-1	0	-15	-8
0	-1	-2	0	-12	-7
0	0	0	-1	1	0
0	0	-1	0	1	0
0	0	1	0	14	7

Matrix #6

x	y	z	w	t	
1	0	0	0	-16	-8
0	-1	0	0	-14	-7
0	0	0	-1	1	0
0	0	-1	0	1	0
0	0	0	0	15	7

Matrix #7

x	y	z	w	t	
15	0	0	0	0	-8
0	-15	0	0	0	-7
0	0	0	-15	0	-7
0	0	-15	0	0	-7
0	0	0	0	15	7

Matrix #8

x	y	z	w	t	
1	0	0	0	0	-8/15
0	1	0	0	0	7/15
0	0	0	1	0	7/15
0	0	1	0	0	7/15
0	0	0	0	1	7/15

Solution: $(-8/15, 7/15, 7/15, 7/15, 7/15)$

47. $1.6x + 2.4y - 3.2z = 4.4$
$5.1x - 6.3y + 0.6z = -3.2$
$4.2x + 3.5y + 4.9z = 10.1$

We use the Excel Matrix Pivot Tool (available online):

x	y	z	
1.6	2.4	-3.2	4.4
5.1	-6.3	0.6	-3.2
4.2	3.5	4.9	10.1

x	y	z	
1	1.5	-2	2.75
0	-13.95	10.8	-17.225
0	-2.8	13.3	-1.45

x	y	z	
1	0	-0.8387097	0.89784946
0	1	-0.7741935	1.23476703
0	0	11.1322581	2.00734767

x	y	z	
1	0	0	1.049084
0	1	0	1.37436814
0	0	1	0.1803181

Solution (rounded to 1 decimal place):
$(1.0, 1.4, 0.2)$

49. $-0.2x + 0.3y + 0.4z - t = 4.5$
$2.2x + 1.1y - 4.7z + 2t = 8.3$
$9.2y - 1.3t = 0$
$3.4x + 0.5z - 3.4t = 0.1$

We use the Excel Matrix Pivot Tool (available online):

x	y	z	t	
-0.2	0.3	0.4	-1	4.5
2.2	1.1	-4.7	2	8.3
0	9.2	0	-1.3	0
3.4	0	0.5	-3.4	0.1

x	y	z	t	
1	-1.5	-2	5	-22.5
0	4.4	-0.3	-9	57.8
0	9.2	0	-1.3	0
0	5.1	7.3	-20.4	76.6

x	y	z	t	
1	0	-2.1022727	1.93181818	-2.7954545
0	1	-0.0681818	-2.0454545	13.1363636
0	0	0.62727273	17.5181818	-120.85455
0	0	7.64772727	-9.9681818	9.60454545

x	y	z	t	
1	0	0	60.6431159	-407.83333
0	1	0	-0.1413043	1.7764E-15
0	0	1	27.9275362	-192.66667
0	0	0	-223.55036	1483.06667

x	y	z	t	
1	0	0	0	-5.5177974
0	1	0	0	-0.9374343
0	0	1	0	-7.3911984
0	0	0	1	-6.6341501

Solution (rounded to 1 decimal place):

$(-5.5, -0.9, -7.4, -6.6)$

51. A pivot is an entry in a matrix that is selected to "clear a column;" that is, use the row operations of a certain type to obtain zeros everywhere above and below it. "Pivoting" is the procedure of clearing a column using a designated pivot.

53. $2R_1 + 5R_4$, or $6R_1 + 15R_4$ (which is less desirable, since it will produce a row in which every entry is divisible by 3)

55. It will include a row of zeros. (Subtracting the two rows produces a row of zeros.)

57. The claim is wrong. If there are more equations than unknowns, there can be a unique solution as well as row(s) of zeros in the reduced matrix, as in Example 6.

59. Since there are 5 columns, there are 4 unknowns. (The last column is for the answers.) Since there are 5 rows of which 3 are zero, that leaves 2 rows with pivots. Thus, there are 2 unknowns that are *not* parameters. The remaining 2 unknowns are arbitrary (parameters).

61. The number of pivots must equal the number of variables, since no variable will be used as a parameter.

63. A simple example is:

$x = 1; y - z = 1; x + y - z = 2$

2.3

1. Unknowns:

x = the number of batches of vanilla

y = the number of batches of mocha

z = the number of batches of strawberry

Arrange the given information in a table with unknowns across the top:

	Vanilla (x)	Mocha (y)	Strawberry (z)	Avail.
Eggs	2	1	1	350
Milk	1	1	2	350
Cream	2	2	1	400

We can now set up an equation for each of the items listed on the left:

Eggs: $2x + y + z = 350$

Milk: $x + y + 2z = 350$

Cream: $2x + 2y + z = 400$

$$\begin{bmatrix} 2 & 1 & 1 & 350 \\ 1 & 1 & 2 & 350 \\ 2 & 2 & 1 & 400 \end{bmatrix} \begin{matrix} \\ 2R_2 - R_1 \\ R_3 - R_1 \end{matrix}$$

$$\begin{bmatrix} 2 & 1 & 1 & 350 \\ 0 & 1 & 3 & 350 \\ 0 & 1 & 0 & 50 \end{bmatrix} \begin{matrix} R_1 - R_2 \\ \\ R_3 - R_2 \end{matrix}$$

$$\begin{bmatrix} 2 & 0 & -2 & 0 \\ 0 & 1 & 3 & 350 \\ 0 & 0 & -3 & -300 \end{bmatrix} \begin{matrix} (1/2)R_1 \\ \\ (1/3)R_3 \end{matrix}$$

$$\begin{bmatrix} 1 & 0 & -1 & 0 \\ 0 & 1 & 3 & 350 \\ 0 & 0 & -1 & -100 \end{bmatrix} \begin{matrix} R_1 - R_3 \\ R_2 + 3R_3 \\ \\ \end{matrix}$$

$$\begin{bmatrix} 1 & 0 & 0 & 100 \\ 0 & 1 & 0 & 50 \\ 0 & 0 & -1 & -100 \end{bmatrix} \begin{matrix} \\ \\ -R_3 \end{matrix}$$

$$\begin{bmatrix} 1 & 0 & 0 & 100 \\ 0 & 1 & 0 & 50 \\ 0 & 0 & 1 & 100 \end{bmatrix}$$

$x = 100$, $y = 50$, $z = 50$

Solution: Make 100 batches of vanilla, 50 batches of mocha, and 100 batches of strawberry

3. Unknowns:

x = the number of sections of Finite Math

y = the number of sections of Applied Calculus

z = the number of sections of Computer Methods.

We are given three pieces of information:

(1) There are a total of 6 sections

$x + y + z = 6$

(2) The total number of students is 210:

$40x + 40y + 10z = 210$

(3) The total revenue is \$260,000:

$40,000x + 60,000y + 20,000z = 260,000$

or, working in thousands of dollars,

$40x + 60y + 20z = 260$

Thus, we have a system of 3 equations in 3 unknowns:

$x + y + z = 6$

$40x + 40y + 10z = 210$

$40x + 60y + 20z = 260$

$$\begin{bmatrix} 1 & 1 & 1 & 6 \\ 40 & 40 & 10 & 210 \\ 40 & 60 & 20 & 260 \end{bmatrix} \begin{matrix} \\ (1/10)R_2 \\ (1/20)R_3 \end{matrix}$$

$$\begin{bmatrix} 1 & 1 & 1 & 6 \\ 4 & 4 & 1 & 21 \\ 2 & 3 & 1 & 13 \end{bmatrix} \begin{matrix} \\ R_2 - 4R_1 \\ R_3 - 2R_1 \end{matrix}$$

$$\begin{bmatrix} 1 & 1 & 1 & 6 \\ 0 & 0 & -3 & -3 \\ 0 & 1 & -1 & 1 \end{bmatrix} \begin{matrix} \\ (1/3)R_2 \\ \\ \end{matrix}$$

$$\begin{bmatrix} 1 & 1 & 1 & 6 \\ 0 & 0 & -1 & -1 \\ 0 & 1 & -1 & 1 \end{bmatrix} \begin{matrix} R_1 + R_2 \\ \\ R_3 - R_2 \end{matrix}$$

$$\begin{bmatrix} 1 & 1 & 0 & 5 \\ 0 & 0 & -1 & -1 \\ 0 & 1 & 0 & 2 \end{bmatrix} \begin{matrix} R_1 - R_3 \\ \\ \\ \end{matrix}$$

$$\begin{bmatrix} 1 & 0 & 0 & 3 \\ 0 & 0 & -1 & -1 \\ 0 & 1 & 0 & 2 \end{bmatrix} -R_2$$

$$\begin{bmatrix} 1 & 0 & 0 & 3 \\ 0 & 0 & 1 & 1 \\ 0 & 1 & 0 & 2 \end{bmatrix} R_2 \leftrightarrow R_3$$

$$\begin{bmatrix} 1 & 0 & 0 & 3 \\ 0 & 1 & 0 & 2 \\ 0 & 0 & 1 & 1 \end{bmatrix}$$

$x = 3, y = 2, z = 1$

Solution: Offer 3 sections of Finite Math, 2 sections of Applied Calculus and 1 section of Computer Methods

5. Unknowns:

x = the number of Boeing 747s

y = the number of Boeing 777s

z = the number of Airbus A330s

Passengers: $400x + 300y + 300z = 5000$

Cost: $200x + 160y + 120z = 2400$

The number of US aircraft is twice the number of foreign aircraft: $x + y = 2z$, or

$\quad x + y - 2z = 0$

Solving:

$$\begin{bmatrix} 400 & 300 & 300 & 5000 \\ 200 & 160 & 120 & 2400 \\ 1 & 1 & -2 & 0 \end{bmatrix} \begin{matrix} (1/100)R_1 \\ (1/40)R_2 \\ \end{matrix}$$

$$\begin{bmatrix} 4 & 3 & 3 & 50 \\ 5 & 4 & 3 & 60 \\ 1 & 1 & -2 & 0 \end{bmatrix} \begin{matrix} 4R_2 - 5R_1 \\ 4R_3 - R_1 \end{matrix}$$

$$\begin{bmatrix} 4 & 3 & 3 & 50 \\ 0 & 1 & -3 & -10 \\ 0 & 1 & -11 & -50 \end{bmatrix} \begin{matrix} R_1 - 3R_2 \\ \\ R_3 - R_2 \end{matrix}$$

$$\begin{bmatrix} 4 & 0 & 12 & 80 \\ 0 & 1 & -3 & -10 \\ 0 & 0 & -8 & -40 \end{bmatrix} \begin{matrix} (1/4)R_1 \\ \\ (1/8)R_3 \end{matrix}$$

$$\begin{bmatrix} 1 & 0 & 3 & 20 \\ 0 & 1 & -3 & -10 \\ 0 & 0 & -1 & -5 \end{bmatrix} \begin{matrix} R_1 + 3R_3 \\ R_2 - 3R_3 \end{matrix}$$

$$\begin{bmatrix} 1 & 0 & 0 & 5 \\ 0 & 1 & 0 & 5 \\ 0 & 0 & -1 & -5 \end{bmatrix} -R_3$$

$$\begin{bmatrix} 1 & 0 & 0 & 5 \\ 0 & 1 & 0 & 5 \\ 0 & 0 & 1 & 5 \end{bmatrix}$$

$x = y = z = 5$

Solution: Order 5 of each aircraft.

7. Unknowns:

x = the number of tons from CCC

y = the number of tons from SSS

z = the number of tons from BBF

Total order of cheese is 100 tons:

$\quad x + y + z = 100$

Total cost = \$5990:

$\quad 80x + 50y + 65z = 5990$

Same amount from CCC and BBF:

$\quad x = z$, or

$\quad x - z = 0$

Solving:

$$\begin{bmatrix} 1 & 1 & 1 & 100 \\ 80 & 50 & 65 & 5990 \\ 1 & 0 & -1 & 0 \end{bmatrix} (1/5)R_2$$

$$\begin{bmatrix} 1 & 1 & 1 & 100 \\ 16 & 10 & 13 & 1198 \\ 1 & 0 & -1 & 0 \end{bmatrix} \begin{matrix} R_2 - 16R_1 \\ R_3 - R_1 \end{matrix}$$

$$\begin{bmatrix} 1 & 1 & 1 & 100 \\ 0 & -6 & -3 & -402 \\ 0 & -1 & -2 & -100 \end{bmatrix} (1/3)R_2$$

$$\begin{bmatrix} 1 & 1 & 1 & 100 \\ 0 & -2 & -1 & -134 \\ 0 & -1 & -2 & -100 \end{bmatrix} \begin{matrix} 2R_1 + R_2 \\ \\ 2R_3 - R_2 \end{matrix}$$

$$\begin{bmatrix} 2 & 0 & 1 & 66 \\ 0 & -2 & -1 & -134 \\ 0 & 0 & -3 & -66 \end{bmatrix} (1/3)R_3$$

$$\begin{bmatrix} 2 & 0 & 1 & 66 \\ 0 & -2 & -1 & -134 \\ 0 & 0 & -1 & -22 \end{bmatrix} \begin{matrix} R_1 + R_3 \\ R_2 - R_3 \end{matrix}$$

$$\begin{bmatrix} 2 & 0 & 0 & 44 \\ 0 & -2 & 0 & -112 \\ 0 & 0 & -1 & -22 \end{bmatrix} \begin{matrix} (1/2)R_1 \\ -(1/2)R_2 \\ -R_3 \end{matrix}$$

$$\begin{bmatrix} 1 & 0 & 0 & 22 \\ 0 & 1 & 0 & 56 \\ 0 & 0 & 1 & 22 \end{bmatrix}$$

$x = 22, y = 56, z = 22$

Solution: The store ordered 22 tons from Cheesy Cream, 56 tons from Super Smooth & Sons, and 22 tons from Bagel's Best Friend.

9. Unknowns:

x = the number of evil sorcerers slain

y = the number of trolls slain

z = the number of orcs slain

Total number slain was 560:

$x + y + z = 560$

Total number of sword thrusts was 620:

$2x + 2y + z = 620$

The number of trolls slain is five times the number of evil sorcerers slain:

$y = 5x$, or

$-5x + y = 0$

Solving:

$$\begin{bmatrix} 1 & 1 & 1 & 560 \\ 2 & 2 & 1 & 620 \\ -5 & 1 & 0 & 0 \end{bmatrix} \begin{matrix} \\ R_2 - 2R_1 \\ R_3 + 5R_1 \end{matrix}$$

$$\begin{bmatrix} 1 & 1 & 1 & 560 \\ 0 & 0 & -1 & -500 \\ 0 & 6 & 5 & 2800 \end{bmatrix} \begin{matrix} R_1 + R_2 \\ \\ R_3 + 5R_2 \end{matrix}$$

$$\begin{bmatrix} 1 & 1 & 0 & 60 \\ 0 & 0 & -1 & -500 \\ 0 & 6 & 0 & 300 \end{bmatrix} (1/6)R_3$$

$$\begin{bmatrix} 1 & 1 & 0 & 60 \\ 0 & 0 & -1 & -500 \\ 0 & 1 & 0 & 50 \end{bmatrix} R_1 - R_3$$

$$\begin{bmatrix} 1 & 0 & 0 & 10 \\ 0 & 0 & -1 & -500 \\ 0 & 1 & 0 & 50 \end{bmatrix} -R_2$$

$$\begin{bmatrix} 1 & 0 & 0 & 10 \\ 0 & 0 & 1 & 500 \\ 0 & 1 & 0 & 50 \end{bmatrix} R_2 \leftrightarrow R_3$$

$$\begin{bmatrix} 1 & 0 & 0 & 10 \\ 0 & 1 & 0 & 50 \\ 0 & 0 & 1 & 500 \end{bmatrix}$$

$x = 10, y = 50, z = 500$

Solution: Conan has slain 10 evil sorcerers, 50 trolls and 500 orcs.

11. Unknowns:

x = revenue (in billions) earned from rock music

y = revenue (in billions) earned from rap music

z = revenue (in billions) earned from classical music

Total revenues were 5.8 billion:

$x + y + z = 5.8$

Rock music brought in twice as much revenue as rap music. Reword this follows:

The revenue earned from rock music was twice the revenue earned from rap music:

$x = 2y$, or

$x - 2y = 0$

Rock music brought in 900% the revenue of classical music. Reword this follows:

The revenue earned from rock music was 9 times the revenue earned from classical music:

$x = 9z$, or

$x - 9z = 0$

We thus solve the system:

$x + y + z = 5.8$

$x - 2y = 0$

$x - 9z = 0$

$$\begin{bmatrix} 1 & 1 & 1 & 5.8 \\ 1 & -2 & 0 & 0 \\ 1 & 0 & -9 & 0 \end{bmatrix} \begin{matrix} 5R_1 \\ \\ \end{matrix}$$

$$\begin{bmatrix} 5 & 5 & 5 & 29 \\ 1 & -2 & 0 & 0 \\ 1 & 0 & -9 & 0 \end{bmatrix} \begin{matrix} \\ 5R_2 - R_1 \\ 5R_3 - R_1 \end{matrix}$$

$$\begin{bmatrix} 5 & 5 & 5 & 29 \\ 0 & -15 & -5 & -29 \\ 0 & -5 & -50 & -29 \end{bmatrix} \begin{matrix} 3R_1 + R_2 \\ \\ 3R_3 - R_2 \end{matrix}$$

$$\begin{bmatrix} 15 & 0 & 10 & 58 \\ 0 & -15 & -5 & -29 \\ 0 & 0 & -145 & -58 \end{bmatrix} \begin{matrix} \\ \\ (1/29)R_3 \end{matrix}$$

$$\begin{bmatrix} 15 & 0 & 10 & 58 \\ 0 & -15 & -5 & -29 \\ 0 & 0 & -5 & -2 \end{bmatrix} \begin{matrix} R_1 + 2R_3 \\ R_2 - R_3 \\ \end{matrix}$$

$$\begin{bmatrix} 15 & 0 & 0 & 54 \\ 0 & -15 & 0 & -27 \\ 0 & 0 & -5 & -2 \end{bmatrix} \begin{matrix} (1/15)R_1 \\ -(1/15)R_2 \\ -(1/5)R_3 \end{matrix}$$

$$\begin{bmatrix} 1 & 0 & 0 & 18/5 \\ 0 & 1 & 0 & 9/5 \\ 0 & 0 & 1 & 2/5 \end{bmatrix}$$

$x = 18/5 = 3.6, y = 9/5 = 1.8, z = 2/5 = 0.4$

Solution: Revenues were \$3.6 billion for rock music, \$1.8 billion for rap music, and \$0.4 billion for classical music.

13. Unknowns:

$x =$ amount of money donated to the MPBF

$y =$ amount of money donated to the SCN

$z =$ amount of money donated to the NY Jets

Given information:

(1) The society donated twice as much to the NY Jets as to the MPBF. Rephrase this as follows: The amount of money donated to the NY Jets was equal to twice the amount of money donated to the MPBF:

$z = 2x$, or $2x - z = 0$

(2) The society donated equal amounts to the first two funds:

$x = y$, or $x - y$ 0

(3) Money donated back to the society:

$x + 2y + 2z = 4200$

Solving:

$$\begin{bmatrix} 2 & 0 & -1 & 0 \\ 1 & -1 & 0 & 0 \\ 1 & 2 & 2 & 4200 \end{bmatrix} \begin{matrix} \\ 2R_2 - R_1 \\ 2R_3 - R_1 \end{matrix}$$

$$\begin{bmatrix} 2 & 0 & -1 & 0 \\ 0 & -2 & 1 & 0 \\ 0 & 4 & 5 & 8400 \end{bmatrix} \begin{matrix} \\ \\ R_3 + 2R_2 \end{matrix}$$

$$\begin{bmatrix} 2 & 0 & -1 & 0 \\ 0 & -2 & 1 & 0 \\ 0 & 0 & 7 & 8400 \end{bmatrix} \begin{matrix} \\ \\ (1/7)R_3 \end{matrix}$$

$$\begin{bmatrix} 2 & 0 & -1 & 0 \\ 0 & -2 & 1 & 0 \\ 0 & 0 & 1 & 1200 \end{bmatrix} \begin{matrix} R_1 + R_3 \\ R_2 - R_3 \\ \end{matrix}$$

$$\begin{bmatrix} 2 & 0 & 0 & 1200 \\ 0 & -2 & 0 & -1200 \\ 0 & 0 & 1 & 1200 \end{bmatrix} \begin{matrix} (1/2)R_1 \\ -(1/2)R_2 \\ \end{matrix}$$

$$\begin{bmatrix} 1 & 0 & 0 & 600 \\ 0 & 1 & 0 & 600 \\ 0 & 0 & 1 & 1200 \end{bmatrix}$$

Solution: It donated $600 to each of the MPBF and the SCN, and $1200 to the Jets

15. Unknowns:

x = the number of empty seats on United

y = the number of empty seats on American

z = the number of empty seats on SouthWest

We are given the following information:

(1) United, American and SouthWest flew a total of 210 empty seats:

$x + y + z = 210$

(2) The total cost of these seats was $65,580. (the figures on the table are for one mile, but the trip was 3000 miles):

$(3000)0.113x + (3000)0.11y + (3000)0.078z$
$= 65,580$

That is,

$339x + 330y + 234z = 65,580$

(3) United had three times as many empty seats as American

$x = 3y$, or $x - 3y = 0$

We use the Excel Matrix Pivot Tool (available online):

x	y	z	
1	1	1	210
339	330	234	65580
1	-3	0	0

x	y	z	
1	1	1	210
0	-9	-105	-5610
0	-4	-1	-210

x	y	z	
1	0	-10.666667	-413.33333
0	1	11.6666667	623.333333
0	0	45.6666667	2283.33333

x	y	z	
1	0	0	120
0	1	0	40
0	0	1	50

Solution: United: 120; American: 40; SouthWest: 50

17. x = amount invested in PNF

y = amount invested in FDMMX

z = amount invested in FFLIX

The total investment was $9000:

$x + y + z = 9000$

You invested an equal amount in FDMMX and FFLIX:

$y = z$, or

$y - z = 0$

Interest for the year from the first two stocks was $400:

$0.06x + 0.05y = 400$

Solving:

$$\begin{bmatrix} 1 & 1 & 1 & 9000 \\ 0 & 1 & -1 & 0 \\ 0.06 & 0.05 & 0 & 400 \end{bmatrix} 100R_3$$

$$\begin{bmatrix} 1 & 1 & 1 & 9000 \\ 0 & 1 & -1 & 0 \\ 6 & 5 & 0 & 40000 \end{bmatrix} R_3 - 6R_1$$

$$\begin{bmatrix} 1 & 1 & 1 & 9000 \\ 0 & 1 & -1 & 0 \\ 0 & -1 & -6 & -14000 \end{bmatrix} \begin{matrix} R_1 - R_2 \\ \\ R_3 + R_2 \end{matrix}$$

$$\begin{bmatrix} 1 & 0 & 2 & 9000 \\ 0 & 1 & -1 & 0 \\ 0 & 0 & -7 & -14000 \end{bmatrix} (1/7)R_3$$

$$\begin{bmatrix} 1 & 0 & 2 & 9000 \\ 0 & 1 & -1 & 0 \\ 0 & 0 & -1 & -2000 \end{bmatrix} \begin{matrix} R_1 + 2R_3 \\ R_2 - R_3 \end{matrix}$$

$$\begin{bmatrix} 1 & 0 & 0 & 5000 \\ 0 & 1 & 0 & 2000 \\ 0 & 0 & -1 & -2000 \end{bmatrix} -R_3$$

$$\begin{bmatrix} 1 & 0 & 0 & 5000 \\ 0 & 1 & 0 & 2000 \\ 0 & 0 & 1 & 2000 \end{bmatrix}$$

$x = 5000, y = 2000, z = 2000$

Solution: You invested \$5000 in PNF, \$2000 in FDMMX, \$2000 in FFLIX.

19. x = the number of shares of APPL
y = the number of shares of HPQ
z = the number of shares of DELL
The total investment was \$5,800
Investment in APPL = x shares @ \$25 = $25x$
Investment in HPQ = y shares @ \$25 = $25y$
Investment in DELL = z shares @ \$35 = $35z$

Thus,
$$25x + 25y + 35z = 5800$$
You earned \$21 in dividends:
APPL dividend = 0.6% of $25x$ invested
$\quad = 0.006(25x) = 0.15x$
HPQ dividend = 1.2% of $25y$ invested
$\quad = 0.012(25y) = 0.3y$
DELL dividend = 0
Thus,
$$0.15x + 0.3y = 21$$
You purchased a total of 200 shares:
$$x + y + z = 200$$
We therefore have the following system:
$$x + y + z = 200$$
$$25x + 25y + 35z = 5800$$
$$0.15x + 0.3y = 21$$
Solving:

$$\begin{bmatrix} 1 & 1 & 1 & 200 \\ 25 & 25 & 35 & 5800 \\ 0.15 & 0.3 & 0 & 21 \end{bmatrix} 20R_3$$

$$\begin{bmatrix} 1 & 1 & 1 & 200 \\ 25 & 25 & 35 & 5800 \\ 3 & 6 & 0 & 420 \end{bmatrix} \begin{array}{l} (1/5)R_2 \\ (1/3)R_3 \end{array}$$

$$\begin{bmatrix} 1 & 1 & 1 & 200 \\ 5 & 5 & 7 & 1160 \\ 1 & 2 & 0 & 140 \end{bmatrix} \begin{array}{l} R_2 - 5R_1 \\ R_3 - R_1 \end{array}$$

$$\begin{bmatrix} 1 & 1 & 1 & 200 \\ 0 & 0 & 2 & 160 \\ 0 & 1 & -1 & -60 \end{bmatrix} (1/2)R_2$$

$$\begin{bmatrix} 1 & 1 & 1 & 200 \\ 0 & 0 & 1 & 80 \\ 0 & 1 & -1 & -60 \end{bmatrix} \begin{array}{l} R_1 - R_2 \\ R_3 + R_2 \end{array}$$

$$\begin{bmatrix} 1 & 1 & 0 & 120 \\ 0 & 0 & 1 & 80 \\ 0 & 1 & 0 & 20 \end{bmatrix} R_1 - R_3$$

$$\begin{bmatrix} 1 & 0 & 0 & 100 \\ 0 & 0 & 1 & 80 \\ 0 & 1 & 0 & 20 \end{bmatrix} R_2 \leftrightarrow R_3$$

$$\begin{bmatrix} 1 & 0 & 0 & 100 \\ 0 & 1 & 0 & 20 \\ 0 & 0 & 1 & 80 \end{bmatrix}$$

$x = 100, y = 20, z = 80$

Solution: You purchased 100 shares of APPL, 20 shares of HPQ, and 80 shares of DELL.

21. Solution: With x, y, z, and u as indicated, the first piece of information we have is that
$$x + y + z + u = 284$$
The remaining equations must be written in standard form:
$$-3x + 3y + z - u = 6$$
$$x + y - z - u = 50$$
$$x - y + z - u = 42$$

Section 2.3

Using pivoting technology, we obtain:

Matrix #1

x	y	z	u	
1	1	1	1	284
-3	3	1	-1	6
1	1	-1	-1	50
1	-1	1	-1	42

Matrix #2

x	y	z	u	
1	1	1	1	284
0	6	4	2	858
0	0	-2	-2	-234
0	-2	0	-2	-242

Matrix #3

x	y	z	u	
1	1	1	1	284
0	3	2	1	429
0	0	-1	-1	-117
0	-1	0	-1	-121

Matrix #4

x	y	z	u	
3	0	1	2	423
0	3	2	1	429
0	0	-1	-1	-117
0	0	2	-2	66

Matrix #5

x	y	z	u	
3	0	1	2	423
0	3	2	1	429
0	0	-1	-1	-117
0	0	1	-1	33

Matrix #6

x	y	z	u	
3	0	0	1	306
0	3	0	-1	195
0	0	-1	-1	-117
0	0	0	-2	-84

Matrix #7

x	y	z	u	
3	0	0	1	306
0	3	0	-1	195
0	0	-1	-1	-117
0	0	0	-1	-42

74

Matrix #8

x	y	z	u	
3	0	0	0	264
0	3	0	0	237
0	0	-1	0	-75
0	0	0	-1	-42

Matrix #9

x	y	z	u	
1	0	0	0	88
0	1	0	0	79
0	0	1	0	75
0	0	0	1	42

Solution: $x = 88$, $y = 79$, $z = 75$, $u = 42$

Microsoft: 88 million, Time Warner: 79 million, Yahoo: 75 million, Google: 42 million

23. Since the market shares add up to 1, the third equation is

$$x + y + z + w = 1$$

If we rewrite the given equations in standard form, we get the second and third equations:

$$x + 0.85y + 0.71z = 0.48$$
$$z + 1.9w = 0.95$$

Let us solve this system using the Excel Pivot Tool (available online). The outlined cells indicate where we pivot at each step to reduce the matrix:

x	y	z	w	Ans
1	1	1	1	1
1	0.85	0.71	0	0.48
0	0	1	1.9	0.95

x	y	z	w	Ans
1	1	1	1	1
0	-0.15	-0.29	-1	-0.52
0	0	1	1.9	0.95

x	y	z	w	Ans
1	0	-0.9333333	-5.6666667	-2.4666667
0	1	1.93333333	6.66666667	3.46666667
0	0	1	1.9	0.95

x	y	z	w	Ans
1	0	0	-3.8933333	-1.58
0	1	0	2.99333333	1.63
0	0	1	1.9	0.95

Translating back to equations (and rounding to two decimal places) gives

$$x - 3.89w = -1.58$$
$$y + 2.99w = 1.63$$
$$z + 1.9w = 0.95$$

Solving for x, y, and z in terms of w gives the general solution:

$$x = -1.58 + 3.89w$$
$$y = 1.63 - 2.99w$$
$$z = 0.95 - 1.9w$$
$$w \text{ arbitrary}$$

We now answer the question: Which of the three companies' market share is most impacted by the share held by Other? Since Other is represented by w, we determine which of the four unknowns has the coefficient of w with the greatest absolute value—namely, x, representing State Farm. Thus, State Farm is most impacted by Other.

25. Unknowns:

x = the number of books sent from Brooklyn to Long Island

y = the number of books sent from Queens to Long Island

z = the number of books sent from Brooklyn to Manhattan

w = the number of books sent from Queens to Manhattan

We represent the given information in a diagram:

Note that, since a total of 3000 books are ordered and there are a total of 3000 in stock, both warehouses need to clear all their stocks.

Books to Long Island: $x + y = 1500$

Books to Manhattan Order: $z + w = 1500$

Books from Brooklyn: $x + z = 1000$

Books from Queens: $y + w = 2000$

(a) Transportation budget:

$$5x + 4y + z + 2w = 9000$$

We have five equations in 4 unknowns. Solving:

$$\begin{bmatrix} 1 & 1 & 0 & 0 & 1500 \\ 0 & 0 & 1 & 1 & 1500 \\ 1 & 0 & 1 & 0 & 1000 \\ 0 & 1 & 0 & 1 & 2000 \\ 5 & 4 & 1 & 2 & 9000 \end{bmatrix} \begin{matrix} \\ \\ R_3 - R_1 \\ \\ R_5 - 5R_1 \end{matrix}$$

$$\begin{bmatrix} 1 & 1 & 0 & 0 & 1500 \\ 0 & 0 & 1 & 1 & 1500 \\ 0 & -1 & 1 & 0 & -500 \\ 0 & 1 & 0 & 1 & 2000 \\ 0 & -1 & 1 & 2 & 1500 \end{bmatrix} \begin{matrix} \\ \\ R_3 - R_2 \\ \\ R_5 - R_2 \end{matrix}$$

$$\begin{bmatrix} 1 & 1 & 0 & 0 & 1500 \\ 0 & 0 & 1 & 1 & 1500 \\ 0 & -1 & 0 & -1 & -2000 \\ 0 & 1 & 0 & 1 & 2000 \\ 0 & -1 & 0 & 1 & 0 \end{bmatrix} \begin{matrix} R_1 + R_3 \\ \\ \\ R_4 + R_3 \\ R_5 - R_3 \end{matrix}$$

$$\begin{bmatrix} 1 & 0 & 0 & -1 & -500 \\ 0 & 0 & 1 & 1 & 1500 \\ 0 & -1 & 0 & -1 & -2000 \\ 0 & 0 & 0 & 0 & 0 \\ 0 & 0 & 0 & 2 & 2000 \end{bmatrix} \begin{matrix} \\ \\ \\ \\ (1/2)R_5 \end{matrix}$$

$$\begin{bmatrix} 1 & 0 & 0 & -1 & -500 \\ 0 & 0 & 1 & 1 & 1500 \\ 0 & -1 & 0 & -1 & -2000 \\ 0 & 0 & 0 & 0 & 0 \\ 0 & 0 & 0 & 1 & 1000 \end{bmatrix} \begin{matrix} R_1 + R_5 \\ R_2 - R_5 \\ R_3 + R_5 \\ \\ \end{matrix}$$

$$\begin{bmatrix} 1 & 0 & 0 & 0 & 500 \\ 0 & 0 & 1 & 0 & 500 \\ 0 & -1 & 0 & 0 & -1000 \\ 0 & 0 & 0 & 0 & 0 \\ 0 & 0 & 0 & 1 & 1000 \end{bmatrix} -R_3$$

$$\begin{bmatrix} 1 & 0 & 0 & 0 & 500 \\ 0 & 0 & 1 & 0 & 500 \\ 0 & 1 & 0 & 0 & 1000 \\ 0 & 0 & 0 & 0 & 0 \\ 0 & 0 & 0 & 1 & 1000 \end{bmatrix} \text{Rearrange rows}$$

$$\begin{bmatrix} 1 & 0 & 0 & 0 & 500 \\ 0 & 1 & 0 & 0 & 1000 \\ 0 & 0 & 1 & 0 & 500 \\ 0 & 0 & 0 & 1 & 1000 \\ 0 & 0 & 0 & 0 & 0 \end{bmatrix}$$

Solution: Brooklyn to Long Island: 500 books; Queens to Long Island: 1000 books; Brooklyn to Manhattan: 500 books; Queens to Manhattan: 1000 books

(b) We try setting $x = 0$ to avoid the expensive $5 per book cost. This reduces the system to:

$0 + y = 1500$, so $y = 1500$

$z + w = 1500$

$0 + z = 1000$, so $z = 1000$

$y + w = 2000$

Substituting the values of y and z in the second equation gives $w = 1500 - 1000 = 500$. The total transportation cost is

$5x + 4y + z + 2w$

$= 5(0) + 4(1500) + 1000 + 2(500)$

$= \$8000$,

which is less than the $9000 budget. Thus, a solution is:

Brooklyn to Long Island: no books; Queens to Long Island: 1500 books; Brooklyn to Manhattan:

1000 books; Queens to Manhattan: 500 books, for a total cost of $8000.

27. Unknowns:

x = the number of tourists from North America to Australia

y = the number of tourists from North America to South Africa

z = the number of tourists from Europe to Australia

w = the number of tourists from Europe to South Africa

We represent the given information in a diagram:

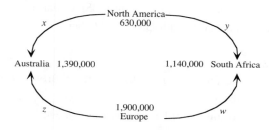

(a)

North America: $x + y = 630{,}000$

Europe: $z + w = 1{,}900{,}000$

Australia: $x + z = 1{,}390{,}000$

South Africa: $y + w = 1{,}140{,}000$

Reducing the matrix form:

$$\begin{bmatrix} 1 & 1 & 0 & 0 & 630000 \\ 0 & 0 & 1 & 1 & 1900000 \\ 1 & 0 & 1 & 0 & 1390000 \\ 0 & 1 & 0 & 1 & 1140000 \end{bmatrix} R_3 - R_1$$

$$\begin{bmatrix} 1 & 1 & 0 & 0 & 630000 \\ 0 & 0 & 1 & 1 & 1900000 \\ 0 & -1 & 1 & 0 & 760000 \\ 0 & 1 & 0 & 1 & 1140000 \end{bmatrix} R_3 - R_2$$

$$\left[\begin{array}{cccc|c} 1 & 1 & 0 & 0 & 630000 \\ 0 & 0 & 1 & 1 & 1900000 \\ 0 & -1 & 0 & -1 & -1140000 \\ 0 & 1 & 0 & 1 & 1140000 \end{array}\right] \begin{array}{l} R_1 + R_3 \\ \\ \\ R_4 + R_3 \end{array}$$

$$\left[\begin{array}{cccc|c} 1 & 0 & 0 & -1 & -510000 \\ 0 & 0 & 1 & 1 & 1900000 \\ 0 & -1 & 0 & -1 & -1140000 \\ 0 & 0 & 0 & 0 & 0 \end{array}\right] \begin{array}{l} \\ \\ -R_3 \\ \\ \end{array}$$

$$\left[\begin{array}{cccc|c} 1 & 0 & 0 & -1 & -510000 \\ 0 & 0 & 1 & 1 & 1900000 \\ 0 & 1 & 0 & 1 & 1140000 \\ 0 & 0 & 0 & 0 & 0 \end{array}\right]$$

This system has infinitely many solutions, which is why the given information is not sufficient to determine the number of tourists from each region to each destination.

(b) We are told that that

$$x + y + z + y = 2{,}530{,}000$$

However, this equation can be obtained by adding the first two equations in part (a). Thus, the new equation gives us no additional information, and so the associated system of linear equations will have the same infinite solution set as part (a).

(c) The additional information is, rephrased:

The number of people from Europe to Australia = the number of people from Europe to South Africa: $z = w$, or

$$z - w = 0.$$

Adding this to our list of equations and solving gives:

$$\left[\begin{array}{cccc|c} 1 & 1 & 0 & 0 & 630000 \\ 0 & 0 & 1 & 1 & 1900000 \\ 1 & 0 & 1 & 0 & 1390000 \\ 0 & 1 & 0 & 1 & 1140000 \\ 0 & 0 & 1 & -1 & 0 \end{array}\right] \begin{array}{l} \\ \\ R_3 - R_1 \\ \\ \\ \end{array}$$

$$\left[\begin{array}{cccc|c} 1 & 1 & 0 & 0 & 630000 \\ 0 & 0 & 1 & 1 & 1900000 \\ 0 & -1 & 1 & 0 & 760000 \\ 0 & 1 & 0 & 1 & 1140000 \\ 0 & 0 & 1 & -1 & 0 \end{array}\right] \begin{array}{l} \\ \\ R_3 - R_2 \\ \\ R_5 - R_2 \end{array}$$

$$\left[\begin{array}{cccc|c} 1 & 1 & 0 & 0 & 630000 \\ 0 & 0 & 1 & 1 & 1900000 \\ 0 & -1 & 0 & -1 & -1140000 \\ 0 & 1 & 0 & 1 & 1140000 \\ 0 & 0 & 0 & -2 & -1900000 \end{array}\right] \begin{array}{l} \\ \\ \\ \\ (1/2)R_5 \end{array}$$

$$\left[\begin{array}{cccc|c} 1 & 1 & 0 & 0 & 630000 \\ 0 & 0 & 1 & 1 & 1900000 \\ 0 & -1 & 0 & -1 & -1140000 \\ 0 & 1 & 0 & 1 & 1140000 \\ 0 & 0 & 0 & -1 & -950000 \end{array}\right] \begin{array}{l} R_1 + R_3 \\ \\ \\ R_4 + R_3 \\ \\ \end{array}$$

$$\left[\begin{array}{cccc|c} 1 & 0 & 0 & -1 & -510000 \\ 0 & 0 & 1 & 1 & 1900000 \\ 0 & -1 & 0 & -1 & -1140000 \\ 0 & 0 & 0 & 0 & 0 \\ 0 & 0 & 0 & -1 & -950000 \end{array}\right] \begin{array}{l} R_1 - R_5 \\ R_2 + R_5 \\ R_3 - R_5 \\ \\ \\ \end{array}$$

$$\left[\begin{array}{cccc|c} 1 & 0 & 0 & 0 & 440000 \\ 0 & 0 & 1 & 0 & 950000 \\ 0 & -1 & 0 & 0 & -190000 \\ 0 & 0 & 0 & 0 & 0 \\ 0 & 0 & 0 & -1 & -950000 \end{array}\right] \begin{array}{l} \\ \\ -R_3 \\ \\ -R_5 \end{array}$$

$$\left[\begin{array}{cccc|c} 1 & 0 & 0 & 0 & 440000 \\ 0 & 0 & 1 & 0 & 950000 \\ 0 & 1 & 0 & 0 & 190000 \\ 0 & 0 & 0 & 0 & 0 \\ 0 & 0 & 0 & 1 & 950000 \end{array}\right] \text{Rearrange rows}$$

$$\begin{bmatrix} 1 & 0 & 0 & 0 & 440000 \\ 0 & 1 & 0 & 0 & 190000 \\ 0 & 0 & 1 & 0 & 950000 \\ 0 & 0 & 0 & 1 & 950000 \end{bmatrix}$$

$x = 440,000, y = 190,000, z = 950,000, w = 950,000$

Since this is a unique solution, we can determine the numbers from each country to each destination: North America to Australia: 440,000, North America to South Africa: 190,000, Europe to Australia: 950,000, Europe to South Africa: 950,000

29.

	Used alcohol	Alcohol-free	Totals
US	x	y	14,000
Europe	z	w	95,000
Totals	63,550	45,450	

(a) Using the row and column totals, we get four equations:

U.S.: $x + y = 14,000$

Europe: $z + w = 95,000$

Used Alcohol: $x + z = 63,550$

Alcohol-free: $y + w = 45,450$

Row-reducing the augmented matrix gives:

$$\begin{bmatrix} 1 & 1 & 0 & 0 & 14000 \\ 0 & 0 & 1 & 1 & 95000 \\ 1 & 0 & 1 & 0 & 63550 \\ 0 & 1 & 0 & 1 & 45450 \end{bmatrix} \begin{matrix} \\ \\ R_3 - R_1 \\ \end{matrix}$$

$$\begin{bmatrix} 1 & 1 & 0 & 0 & 14000 \\ 0 & 0 & 1 & 1 & 95000 \\ 0 & -1 & 1 & 0 & 49550 \\ 0 & 1 & 0 & 1 & 45450 \end{bmatrix} \begin{matrix} \\ \\ R_3 - R_2 \\ \end{matrix}$$

$$\begin{bmatrix} 1 & 1 & 0 & 0 & 14000 \\ 0 & 0 & 1 & 1 & 95000 \\ 0 & -1 & 0 & -1 & -45450 \\ 0 & 1 & 0 & 1 & 45450 \end{bmatrix} \begin{matrix} R_1 + R_3 \\ \\ \\ R_4 + R_3 \end{matrix}$$

$$\begin{bmatrix} 1 & 0 & 0 & -1 & -31450 \\ 0 & 0 & 1 & 1 & 95000 \\ 0 & -1 & 0 & -1 & -45450 \\ 0 & 0 & 0 & 0 & 0 \end{bmatrix} \begin{matrix} \\ \\ -R_3 \\ \end{matrix}$$

$$\begin{bmatrix} 1 & 0 & 0 & -1 & -31450 \\ 0 & 0 & 1 & 1 & 95000 \\ 0 & 1 & 0 & 1 & 45450 \\ 0 & 0 & 0 & 0 & 0 \end{bmatrix}$$

The row-reduced matrix shows that there are infinitely many solutions, and hence no unique solution. Thus, the given data are insufficient to obtain the missing data (x, y, z and w).

(b) The number of US 10th graders who were alcohol-free was 50% more than the number who had used alcohol:

$y = x + 0.50x = 1.50x$, or

$-1.50x + y = 0$

If we include this additional equation, we get the system

$x + y = 14.000$

$z + w = 95.000$

$x + z = 63.550$

$y + w = 45.450$

$-1.50x + y = 0$

Solving:

$$\begin{bmatrix} 1 & 1 & 0 & 0 & 14000 \\ 0 & 0 & 1 & 1 & 95000 \\ 1 & 0 & 1 & 0 & 63550 \\ 0 & 1 & 0 & 1 & 45450 \\ -1.5 & 1 & 0 & 0 & 0 \end{bmatrix} \begin{matrix} \\ \\ \\ \\ 2R_5 \end{matrix}$$

$$\begin{bmatrix} 1 & 1 & 0 & 0 & 14000 \\ 0 & 0 & 1 & 1 & 95000 \\ 1 & 0 & 1 & 0 & 63550 \\ 0 & 1 & 0 & 1 & 45450 \\ -3 & 2 & 0 & 0 & 0 \end{bmatrix} \begin{array}{l} \\ \\ R_3 - R_1 \\ \\ R_5 + 3R_1 \end{array}$$

$$\begin{bmatrix} 1 & 1 & 0 & 0 & 14000 \\ 0 & 0 & 1 & 1 & 95000 \\ 0 & -1 & 1 & 0 & 49550 \\ 0 & 1 & 0 & 1 & 45450 \\ 0 & 5 & 0 & 0 & 42000 \end{bmatrix} \begin{array}{l} \\ \\ \\ \\ (1/5)R_5 \end{array}$$

$$\begin{bmatrix} 1 & 1 & 0 & 0 & 14000 \\ 0 & 0 & 1 & 1 & 95000 \\ 0 & -1 & 1 & 0 & 49550 \\ 0 & 1 & 0 & 1 & 45450 \\ 0 & 1 & 0 & 0 & 8400 \end{bmatrix} \begin{array}{l} \\ \\ R_3 - R_2 \\ \\ \end{array}$$

$$\begin{bmatrix} 1 & 1 & 0 & 0 & 14000 \\ 0 & 0 & 1 & 1 & 95000 \\ 0 & -1 & 0 & -1 & -45450 \\ 0 & 1 & 0 & 1 & 45450 \\ 0 & 1 & 0 & 0 & 8400 \end{bmatrix} \begin{array}{l} R_1 + R_3 \\ \\ \\ R_4 + R_3 \\ R_5 + R_3 \end{array}$$

$$\begin{bmatrix} 1 & 0 & 0 & -1 & -31450 \\ 0 & 0 & 1 & 1 & 95000 \\ 0 & -1 & 0 & -1 & -45450 \\ 0 & 0 & 0 & 0 & 0 \\ 0 & 0 & 0 & -1 & -37050 \end{bmatrix} \begin{array}{l} R_1 - R_5 \\ R_2 + R_5 \\ R_3 - R_5 \\ \\ \end{array}$$

$$\begin{bmatrix} 1 & 0 & 0 & 0 & 5600 \\ 0 & 0 & 1 & 0 & 57950 \\ 0 & -1 & 0 & 0 & -8400 \\ 0 & 0 & 0 & 0 & 0 \\ 0 & 0 & 0 & -1 & -37050 \end{bmatrix} \begin{array}{l} \\ \\ -R_3 \\ \\ -R_5 \end{array}$$

$$\begin{bmatrix} 1 & 0 & 0 & 0 & 5600 \\ 0 & 0 & 1 & 0 & 57950 \\ 0 & 1 & 0 & 0 & 8400 \\ 0 & 0 & 0 & 0 & 0 \\ 0 & 0 & 0 & 1 & 37050 \end{bmatrix} \text{Rearrange Rows}$$

$$\begin{bmatrix} 1 & 0 & 0 & 0 & 5600 \\ 0 & 1 & 0 & 0 & 8400 \\ 0 & 0 & 1 & 0 & 57950 \\ 0 & 0 & 0 & 1 & 37050 \\ 0 & 0 & 0 & 0 & 0 \end{bmatrix}$$

Thus, the missing data is:
$x = 5600$, $y = 8400$, $z = 57{,}950$, $w = 37{,}050$

31. x = daily traffic flow along Eastward Blvd.
y = daily traffic flow along Northwest La.
z = daily traffic flow along Southwest La.

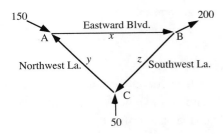

(a) Intersection A: Traffic in = Traffic out:
$150 + y = x$, or $x - y = 150$
Intersection B: Traffic in = Traffic out:
$x = 200 + z$, or $x - z = 200$
Intersection C: Traffic in = Traffic out:
$50 + z = y$, or $y - z = 50$
This gives us a system of 3 linear equations:
$x - y = 150$
$x - z = 200$
$y - z = 50$
Solving:

$$\begin{bmatrix} 1 & -1 & 0 & 150 \\ 1 & 0 & -1 & 200 \\ 0 & 1 & -1 & 50 \end{bmatrix} R_2 - R_1$$

$$\begin{bmatrix} 1 & -1 & 0 & 150 \\ 0 & 1 & -1 & 50 \\ 0 & 1 & -1 & 50 \end{bmatrix} \begin{matrix} R_1 + R_2 \\ \\ R_3 - R_2 \end{matrix}$$

$$\begin{bmatrix} 1 & 0 & -1 & 200 \\ 0 & 1 & -1 & 50 \\ 0 & 0 & 0 & 0 \end{bmatrix}$$

Since there are infinitely many solutions, it is not possible to determine the daily flow of traffic along each of the three streets from the information given.

Translating the row-reduced matrix back into equations gives

$x - z = 200$, or $x = 200 + z$

$y - z = 50$, or $y = 50 + z$

thus, the general solution is:

$x = 200 + z$

$y = 50 + z$

$z \geq 0$ arbitrary,

where z is traffic along Southwest Lane. Thus, it would suffice to know the traffic along Southwest to obtain the other traffic flows.

(b) If we set $z = 60$, we obtain, from the general solution in part (a),

$x = 200 + 60 = 260$ vehicles along Eastward Blvd., $y = 50 + 60 = 110$ vehicles along Northwest La., $z = 60$ vehicles along Southwest La.

(c) From the general solution in part (a), the traffic flow along Northwest La. is given by

$y = 50 + z$

Since $z \geq 0$, the value of y must be at least 50 vehicles per day.

33. Let us take the unknowns to be the net traffic flow going east on the three stretches of Broadway as shown in the figure. (If any of the unknowns is

negative, it indicates a net positive flow in the opposite direction.)

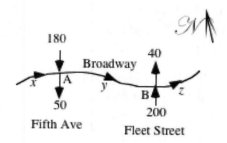

(a) Intersection A: Traffic in = Traffic out:

$180 + x = 50 + y$, or $x - y = -130$

Intersection B: Traffic in = Traffic out:

$200 + y = 40 + z$, or $y - z = -160$

Solving:

$$\begin{bmatrix} 1 & -1 & 0 & -130 \\ 0 & 1 & -1 & -160 \end{bmatrix} R_1 + R_2$$

$$\begin{bmatrix} 1 & 0 & -1 & -290 \\ 0 & 1 & -1 & -160 \end{bmatrix}$$

The general solution is:

$x - z = -290$, or $x = z - 290$

$y - z = -160$, or $y = z - 160$

z is arbitrary

Since there are infinitely many solutions, we cannot determine the traffic flow along each stretch of Broadway. Knowing any one of z, y, or z would enable us to solve for the other two unknowns uniquely.

(b) The general solution from part (a) is:

$x = z - 290$

$y = z - 160$

z is arbitrary

East of Fleet Street, the traffic flow is z. If z is smaller than 160, the above solution shows that values of z and y are negative, indicating a net flow to the west on those stretches.

35. We rewrite each given equation in standard form, using the given information that $M = 120$ billion dollars:

$$120 = C + D \quad \rightarrow \quad C + D = 120$$
$$C = 0.2D \quad \rightarrow \quad C - 0.2D = 0$$
$$R = 0.1D \quad \rightarrow \quad R - 0.1D = 0$$
$$H = R + C \quad \rightarrow \quad H - R - C = 0$$

The matrices below are set up with the unknowns in the order: C, R, D, H

$$\begin{bmatrix} 1 & 0 & 1 & 0 & 120 \\ 1 & 0 & -.2 & 0 & 0 \\ 0 & 1 & -.1 & 0 & 0 \\ -1 & -1 & 0 & 1 & 0 \end{bmatrix} \begin{matrix} \\ 5R_2 \\ 10R_3 \\ \\ \end{matrix}$$

$$\begin{bmatrix} 1 & 0 & 1 & 0 & 120 \\ 5 & 0 & -1 & 0 & 0 \\ 0 & 10 & -1 & 0 & 0 \\ -1 & -1 & 0 & 1 & 0 \end{bmatrix} \begin{matrix} \\ R_2 - 5R_1 \\ \\ R_4 + R_1 \end{matrix}$$

$$\begin{bmatrix} 1 & 0 & 1 & 0 & 120 \\ 0 & 0 & -6 & 0 & -600 \\ 0 & 10 & -1 & 0 & 0 \\ 0 & -1 & 1 & 1 & 120 \end{bmatrix} \begin{matrix} \\ (1/6)R_2 \\ \\ \\ \end{matrix}$$

$$\begin{bmatrix} 1 & 0 & 1 & 0 & 120 \\ 0 & 0 & -1 & 0 & -100 \\ 0 & 10 & -1 & 0 & 0 \\ 0 & -1 & 1 & 1 & 120 \end{bmatrix} \begin{matrix} R_1 + R_2 \\ \\ R_3 - R_2 \\ R_4 + R_2 \end{matrix}$$

$$\begin{bmatrix} 1 & 0 & 0 & 0 & 20 \\ 0 & 0 & -1 & 0 & -100 \\ 0 & 10 & 0 & 0 & 100 \\ 0 & -1 & 0 & 1 & 20 \end{bmatrix} \begin{matrix} \\ \\ (1/10)R_3 \\ \\ \end{matrix}$$

$$\begin{bmatrix} 1 & 0 & 0 & 0 & 20 \\ 0 & 0 & -1 & 0 & -100 \\ 0 & 1 & 0 & 0 & 10 \\ 0 & -1 & 0 & 1 & 20 \end{bmatrix} \begin{matrix} \\ \\ \\ R_4 + R_3 \end{matrix}$$

$$\begin{bmatrix} 1 & 0 & 0 & 0 & 20 \\ 0 & 0 & -1 & 0 & -100 \\ 0 & 1 & 0 & 0 & 10 \\ 0 & 0 & 0 & 1 & 30 \end{bmatrix} -R_2$$

$$\begin{bmatrix} 1 & 0 & 0 & 0 & 20 \\ 0 & 0 & 1 & 0 & 100 \\ 0 & 1 & 0 & 0 & 10 \\ 0 & 0 & 0 & 1 & 30 \end{bmatrix} R_2 \leftrightarrow R_3$$

$$\begin{bmatrix} 1 & 0 & 0 & 0 & 20 \\ 0 & 1 & 0 & 0 & 10 \\ 0 & 0 & 1 & 0 & 100 \\ 0 & 0 & 0 & 1 & 30 \end{bmatrix}$$

$C = 20$, $R = 10$, $D = 100$, $H = 30$

We are asked for bank reserves R, which are therefore $10 billion.

37.

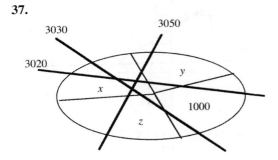

Adding the values along each beam gives

$$x + y + z = 3050$$
$$x + z + 1000 = 3030, \text{ or } x + z = 2030$$
$$x + y + 1000 = 3020, \text{ or } x + y = 2020$$

$$\begin{bmatrix} 1 & 1 & 1 & 3050 \\ 1 & 0 & 1 & 2030 \\ 1 & 1 & 0 & 2020 \end{bmatrix} \begin{matrix} \\ R_2 - R_1 \\ R_3 - R_1 \end{matrix}$$

$$\begin{bmatrix} 1 & 1 & 1 & 3050 \\ 0 & -1 & 0 & -1020 \\ 0 & 0 & -1 & -1030 \end{bmatrix} R_1 + R_2$$

$$\begin{bmatrix} 1 & 0 & 1 & 2030 \\ 0 & -1 & 0 & -1020 \\ 0 & 0 & -1 & -1030 \end{bmatrix} \begin{matrix} R_1 + R_3 \\ \\ \end{matrix}$$

$$\begin{bmatrix} 1 & 0 & 0 & 1000 \\ 0 & -1 & 0 & -1020 \\ 0 & 0 & -1 & -1030 \end{bmatrix} \begin{matrix} \\ -R_2 \\ -R_3 \end{matrix}$$

$$\begin{bmatrix} 1 & 0 & 0 & 1000 \\ 0 & 1 & 0 & 1020 \\ 0 & 0 & 1 & 1030 \end{bmatrix}$$

Solution: $x = 1000$, $y = 1020$, $z = 1030$

From the table, we find the corresponding components:

x = water, y = gray matter, z = tumor

39.

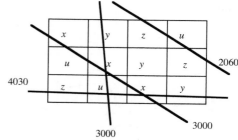

Adding the values along each beam gives

$x + y + z + u = 4030$

$x + y + u = 3000$

$3x + 2u = 3000$

$2z + u = 2060$

Solving:

$$\begin{bmatrix} 1 & 1 & 1 & 1 & 4030 \\ 1 & 1 & 0 & 1 & 3000 \\ 3 & 0 & 0 & 2 & 3000 \\ 0 & 0 & 2 & 1 & 2060 \end{bmatrix} \begin{matrix} \\ R_2 - R_1 \\ R_3 - 3R_1 \\ \end{matrix}$$

$$\begin{bmatrix} 1 & 1 & 1 & 1 & 4030 \\ 0 & 0 & -1 & 0 & -1030 \\ 0 & -3 & -3 & -1 & -9090 \\ 0 & 0 & 2 & 1 & 2060 \end{bmatrix} \begin{matrix} R_1 + R_2 \\ \\ R_3 - 3R_2 \\ R_4 + 2R_2 \end{matrix}$$

$$\begin{bmatrix} 1 & 1 & 0 & 1 & 3000 \\ 0 & 0 & -1 & 0 & -1030 \\ 0 & -3 & 0 & -1 & -6000 \\ 0 & 0 & 0 & 1 & 0 \end{bmatrix} \begin{matrix} 3R_1 + R_3 \\ \\ \\ \end{matrix}$$

$$\begin{bmatrix} 3 & 0 & 0 & 2 & 3000 \\ 0 & 0 & -1 & 0 & -1030 \\ 0 & -3 & 0 & -1 & -6000 \\ 0 & 0 & 0 & 1 & 0 \end{bmatrix} \begin{matrix} R_1 - 2R_4 \\ \\ R_3 + R_4 \\ \end{matrix}$$

$$\begin{bmatrix} 3 & 0 & 0 & 0 & 3000 \\ 0 & 0 & -1 & 0 & -1030 \\ 0 & -3 & 0 & 0 & -6000 \\ 0 & 0 & 0 & 1 & 0 \end{bmatrix} \begin{matrix} (1/3)R_1 \\ -R_2 \\ -(1/3)R_3 \\ \end{matrix}$$

$$\begin{bmatrix} 1 & 0 & 0 & 0 & 1000 \\ 0 & 0 & 1 & 0 & 1030 \\ 0 & 1 & 0 & 0 & 2000 \\ 0 & 0 & 0 & 1 & 0 \end{bmatrix} \begin{matrix} \\ \\ R_2 \leftrightarrow R_3 \\ \end{matrix}$$

$$\begin{bmatrix} 1 & 0 & 0 & 0 & 1000 \\ 0 & 1 & 0 & 0 & 2000 \\ 0 & 0 & 1 & 0 & 1030 \\ 0 & 0 & 0 & 1 & 0 \end{bmatrix}$$

$x = 1000$, $y = 2000$, $z = 1030$, $u = 0$

From the table, we find the corresponding components:

x = water, y = bone, z = tumor, u = air

41. Since the three beams extending from left to right pass though the same four squares on the left, let us label those squares y, The vertical beam passes through an additional two squares, which we will label as z and u as shown:

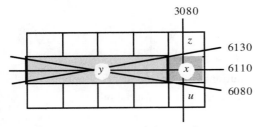

Adding the values along each beam gives

$x + z + u = 3080$

$y + z = 6130$

$x + y = 6110$

$y + u = 6080$

Solving:

$$\begin{bmatrix} 1 & 0 & 1 & 1 & 3080 \\ 0 & 1 & 1 & 0 & 6130 \\ 1 & 1 & 0 & 0 & 6110 \\ 0 & 1 & 0 & 1 & 6080 \end{bmatrix} \begin{matrix} \\ \\ R_3 - R_1 \\ \\ \end{matrix}$$

$$\begin{bmatrix} 1 & 0 & 1 & 1 & 3080 \\ 0 & 1 & 1 & 0 & 6130 \\ 0 & 1 & -1 & -1 & 3030 \\ 0 & 1 & 0 & 1 & 6080 \end{bmatrix} \begin{matrix} \\ \\ R_3 - R_2 \\ R_4 - R_2 \end{matrix}$$

$$\begin{bmatrix} 1 & 0 & 1 & 1 & 3080 \\ 0 & 1 & 1 & 0 & 6130 \\ 0 & 0 & -2 & -1 & -3100 \\ 0 & 0 & -1 & 1 & -50 \end{bmatrix} \begin{matrix} 2R_1 + R_3 \\ 2R_2 + R_3 \\ \\ 2R_4 - R_3 \end{matrix}$$

$$\begin{bmatrix} 2 & 0 & 0 & 1 & 3060 \\ 0 & 2 & 0 & -1 & 9160 \\ 0 & 0 & -2 & -1 & -3100 \\ 0 & 0 & 0 & 3 & 3000 \end{bmatrix} \begin{matrix} \\ \\ \\ (1/3)R_4 \end{matrix}$$

$$\begin{bmatrix} 2 & 0 & 0 & 1 & 3060 \\ 0 & 2 & 0 & -1 & 9160 \\ 0 & 0 & -2 & -1 & -3100 \\ 0 & 0 & 0 & 1 & 1000 \end{bmatrix} \begin{matrix} R_1 - R_4 \\ R_2 + R_4 \\ R_3 + R_4 \\ \end{matrix}$$

$$\begin{bmatrix} 2 & 0 & 0 & 0 & 2060 \\ 0 & 2 & 0 & 0 & 10160 \\ 0 & 0 & -2 & 0 & -2100 \\ 0 & 0 & 0 & 1 & 1000 \end{bmatrix} \begin{matrix} (1/2)R_1 \\ (1/2)R_2 \\ -(1/2)R_3 \\ \end{matrix}$$

$$\begin{bmatrix} 1 & 0 & 0 & 0 & 1030 \\ 0 & 1 & 0 & 0 & 5080 \\ 0 & 0 & 1 & 0 & 1050 \\ 0 & 0 & 0 & 1 & 1000 \end{bmatrix}$$

All we need is the value of x: 1030, which corresponds to tumor

43. Unknowns:

x = the number of Democrats who voted in favor

y = the number of Republicans who voted in favor

z = the number of Others who voted in favor

Note:

   333–x Democrats voted against,

   89–y Republicans voted against, and

   13–z Others voted against.

Given information:

(1) There were 31 more votes in favor than against:

The total number of votes in favor = $x+y+z$

Total number of votes against

   $= (333–x) + (89–y) + (13–z)$

   $= 435 – x – y – z$

Thus,

   $x + y + z –(435 – x – y – z) = 31$, or

   $2x + 2y + 2z = 466$, or

   $x + y + z = 233$

(2) 10 times as many Democrats voted for the bill as Republicans. Rephrasing this:

The number of Democrats voting for the bill was 10 times the number of Republicans voting for the bill:

$$x = 10y, \text{ or}$$
$$x - 10y = 0$$

(3) 36 more non-Democrats voted against the bill than for it. Rephrasing this: The number of non-Democrats voting against the bill exceeded the number of non-Democrats voting for by 36

$$(89-y) + (13-z) - (y+z) = 36, \text{ or}$$
$$-2y - 2z = -66, \text{ or}$$
$$y + z = 33$$

Thus, we have 3 equations with 3 unknowns:

$$x + y + z = 233$$
$$x - 10y = 0$$
$$y + z = 33$$

Solving:

$$\begin{bmatrix} 1 & 1 & 1 & 233 \\ 1 & -10 & 0 & 0 \\ 0 & 1 & 1 & 33 \end{bmatrix} R_2 - R_1$$

$$\begin{bmatrix} 1 & 1 & 1 & 233 \\ 0 & -11 & -1 & -233 \\ 0 & 1 & 1 & 33 \end{bmatrix} \begin{matrix} 11R_1 + R_2 \\ \\ 11R_3 + R_2 \end{matrix}$$

$$\begin{bmatrix} 11 & 0 & 10 & 2330 \\ 0 & -11 & -1 & -233 \\ 0 & 0 & 10 & 130 \end{bmatrix} (1/10)R_3$$

$$\begin{bmatrix} 11 & 0 & 10 & 2330 \\ 0 & -11 & -1 & -233 \\ 0 & 0 & 1 & 13 \end{bmatrix} \begin{matrix} R_1 - 10R_3 \\ R_2 + R_3 \end{matrix}$$

$$\begin{bmatrix} 11 & 0 & 0 & 2200 \\ 0 & -11 & 0 & -220 \\ 0 & 0 & 1 & 13 \end{bmatrix} \begin{matrix} (1/11)R_1 \\ -(1/11)R_2 \end{matrix}$$

$$\begin{bmatrix} 1 & 0 & 0 & 200 \\ 0 & 1 & 0 & 20 \\ 0 & 0 & 1 & 13 \end{bmatrix}$$

Solution: 200 Democrats, 20 Republicans, 13 of other parties voted for the bill.

45. Unknowns:

$x =$ amount invested in company X
$y =$ amount invested in company Y
$z =$ amount invested in company Z
$w =$ amount invested in company W

Investments totaled \$65 million:

$$x + y + z + w = 65$$

Total return on investments was \$8:

$$0.15x - 0.20y + 0.20w = 8$$

Colossal invested twice as much in company X as in company Z. Rewording this: The amount invested in company X was twice the amount invested in company Z:

$$x = 2z, \text{ or}$$
$$x - 2z = 0$$

The amount invested in company W was 3 times the amount invested in company Z:

$$w = 3z, \text{ or}$$
$$-3z + w = 0$$

$$\begin{bmatrix} 1 & 1 & 1 & 1 & 65 \\ 0.15 & -.2 & 0 & 0.2 & 8 \\ 1 & 0 & -2 & 0 & 0 \\ 0 & 0 & -3 & 1 & 0 \end{bmatrix} 20R_2$$

$$\begin{bmatrix} 1 & 1 & 1 & 1 & 65 \\ 3 & -4 & 0 & 4 & 160 \\ 1 & 0 & -2 & 0 & 0 \\ 0 & 0 & -3 & 1 & 0 \end{bmatrix} \begin{matrix} R_2 - 3R_1 \\ R_3 - R_1 \end{matrix}$$

$$\begin{bmatrix} 1 & 1 & 1 & 1 & 65 \\ 0 & -7 & -3 & 1 & -35 \\ 0 & -1 & -3 & -1 & -65 \\ 0 & 0 & -3 & 1 & 0 \end{bmatrix} \begin{matrix} 7R_1 + R_2 \\ \\ 7R_3 - R_2 \end{matrix}$$

$$\begin{bmatrix} 7 & 0 & 4 & 8 & 420 \\ 0 & -7 & -3 & 1 & -35 \\ 0 & 0 & -18 & -8 & -420 \\ 0 & 0 & -3 & 1 & 0 \end{bmatrix} (1/2)R_3$$

$$\begin{bmatrix} 7 & 0 & 4 & 8 & 420 \\ 0 & -7 & -3 & 1 & -35 \\ 0 & 0 & -9 & -4 & -210 \\ 0 & 0 & -3 & 1 & 0 \end{bmatrix} \begin{matrix} 9R_1 + 4R_3 \\ 3R_2 - R_3 \\ \\ 3R_4 - R_3 \end{matrix}$$

$$\begin{bmatrix} 63 & 0 & 0 & 56 & 2940 \\ 0 & -21 & 0 & 7 & 105 \\ 0 & 0 & -9 & -4 & -210 \\ 0 & 0 & 0 & 7 & 210 \end{bmatrix} \begin{matrix} (1/7)R_1 \\ (1/7)R_2 \\ \\ (1/7)R_4 \end{matrix}$$

$$\begin{bmatrix} 9 & 0 & 0 & 8 & 420 \\ 0 & -3 & 0 & 1 & 15 \\ 0 & 0 & -9 & -4 & -210 \\ 0 & 0 & 0 & 1 & 30 \end{bmatrix} \begin{matrix} R_1 - 8R_4 \\ R_2 - R_4 \\ R_3 + 4R_4 \end{matrix}$$

$$\begin{bmatrix} 9 & 0 & 0 & 0 & 180 \\ 0 & -3 & 0 & 0 & -15 \\ 0 & 0 & -9 & 0 & -90 \\ 0 & 0 & 0 & 1 & 30 \end{bmatrix} \begin{matrix} (1/9)R_1 \\ -(1/3)R_2 \\ -(1/9)R_3 \end{matrix}$$

$$\begin{bmatrix} 1 & 0 & 0 & 0 & 20 \\ 0 & 1 & 0 & 0 & 5 \\ 0 & 0 & 1 & 0 & 10 \\ 0 & 0 & 0 & 1 & 30 \end{bmatrix}$$

$x = 20, y = 5, z = 10, w = 30$

Since this is a unique solution, Smiley has sufficient information to piece together Colossal's investment portfolio: Colossal invested \$20m in company X; \$5m in company Y, \$10m in company Z, and \$30m in company W.

47. It is not realistic to expect to use exactly all of the ingredients. Solutions of the associated system may involve negative numbers or not exist. Only solutions with nonnegative values for all the unknowns correspond to being able to use up all of the ingredients.

49. The blend consists of 100 pounds of ingredient X. This says that $x = 100$, which is a linear equation.

51. The blend contains 30% ingredient Y by weight. Rephrasing: The weight of ingredient Y is 30% of the combined weights of X, Y, and Z:

$y = 0.30(x + y + z)$
$y = 0.30x + 0.30y + 0.30z$
$0.3x - 0.7y + 0.3z = 0,$

which is a linear equation.

53. There is at least 30% ingredient Y by weight. Rephrasing: the weight of ingredient Y is at least 30% of the combined weights of X, Y, and Z:

$y \geq 0.30(x + y + z)$

This gives a linear inequality—not an equation.

Chapter 2 Review Exercises

1. To graph with technology, solve the equations for y:

First line: $y = -x/2 + 2$

Second line: $y = 2x - 1$

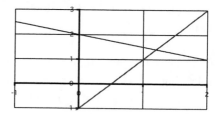

One solution

2. To graph with technology, solve the equations for y:

First line: $y = 2x - 3$

Second line: $y = -x + 2$

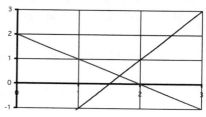

One solution

3. To graph with technology, solve the equations for y:

First line: $y = 2x/3$

Second line: $y = 2x/3$

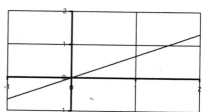

Infinitely many solutions

4. To graph with technology, solve the equations for y:

First line: $y = -2x/3 + 2/3$

Second line: $y = -2x/3 - 1/3$

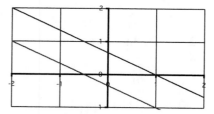

No solution

5. To graph with technology, solve the equations for y:

First line: $y = -x + 1$

Second line: $y = -2x + 0.3$

Third line: $y = -3x/2 + 13/20$

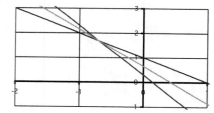

One solution

6. To graph with technology, solve the equations for y:

First line: $y = -6x + 0.2$

Second line: $y = -6x + 0.2$

Third line: $y = 6x - 0.2$

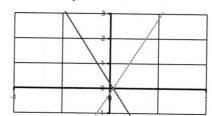

One solution

7. $x + 2y = 4$

$\qquad 2x - y = 1$

Multiply the second equation by 2:

$\qquad x + 2y = 4$

$\qquad 4x - 2y = 2$

Adding gives

$\qquad 5x = 6$, so $x = \dfrac{6}{5}$

Substituting $x = 6/5$ in the first equation gives

$\qquad \dfrac{6}{5} + 2y = 4$

$\qquad 2y = 4 - \dfrac{6}{5} = \dfrac{14}{5}$

$\qquad y = \dfrac{7}{5}$

Solution: $x = \dfrac{6}{5}$, $y = \dfrac{7}{5}$

8. $\quad 0.2x - 0.1y = 0.3$

$\qquad 0.2x + 0.2y = 0.4$

Multiply the first equation by 10 and the second by -10:

$\qquad 2x - y = 3$

$\qquad -2x - 2y = -4$

Adding gives

$\qquad -3y = -1$

$\qquad y = \dfrac{1}{3}$

Substituting $y = \dfrac{1}{3}$ in $2x - y = 3$ gives

$\qquad 2x - \dfrac{1}{3} = 3$

$\qquad 2x = 3 + \dfrac{1}{3} = \dfrac{10}{3}$

$\qquad x = \dfrac{5}{3}$

Solution: $x = \dfrac{5}{3}$, $y = \dfrac{1}{3}$

9. $\quad \frac{1}{2}x - \frac{3}{4}y = 0$

$\qquad 6x - 9y = 0$

Multiply the first equation by 4 and divide the second by -3:

$\qquad 2x - 3y = 0$

$\qquad -2x + 3y = 0$

Adding gives $0 = 0$, so the system is redundant (the equations give the same lines).

To get the general solution, we solve for x:

$\qquad 2x = 3y$

$\qquad x = \dfrac{3y}{2}$

General solution:

$\qquad x = \dfrac{3y}{2}$, y arbitrary, or

$\qquad \left(\dfrac{3y}{2}, y\right)$; y arbitrary

10. $2x + 3y = 2$

$\qquad -x - 3y/2 = 1/2$

Multiply the second equation by 2:

$\qquad 2x + 3y = 2$

$\qquad -2x - 3y = 1$

Adding gives

$\qquad 0 = 3$,

showing that the given system is inconsistent, and therefore has no solution.

11. $\quad x + y = 1$

$\qquad 2x + y = 0.3$

$\qquad 3x + 2y = \dfrac{13}{10}$

$$\left[\begin{array}{ccc|c} 1 & 1 & 1 & \\ 2 & 1 & 3/10 & 10R_2 \\ 3 & 2 & 13/10 & 10R_3 \end{array}\right]$$

$$\left[\begin{array}{ccc|c} 1 & 1 & 1 & \\ 20 & 10 & 3 & R_2 - 20R_1 \\ 30 & 20 & 13 & R_3 - 30R_1 \end{array}\right]$$

$$\begin{bmatrix} 1 & 1 & 1 \\ 0 & -10 & -17 \\ 0 & -10 & -17 \end{bmatrix} \begin{matrix} 10R_1 + R_2 \\ \\ R_3 - R_2 \end{matrix}$$

$$\begin{bmatrix} 10 & 0 & -7 \\ 0 & -10 & -17 \\ 0 & 0 & 0 \end{bmatrix} \begin{matrix} (1/10)R_1 \\ -(1/10)R_2 \\ \end{matrix}$$

$$\begin{bmatrix} 1 & 0 & -7/10 \\ 0 & 1 & 17/10 \\ 0 & 0 & 0 \end{bmatrix}$$

Solution: $x = -0.7$, $y = 1.7$

12. $3x + 0.5y = 0.1$
$6x + y = 0.2$
$\dfrac{3x}{10} - 0.05y = 0.01$

$$\begin{bmatrix} 3 & 0.5 & 0.1 \\ 6 & 1 & 0.2 \\ 0.3 & -0.05 & 0.01 \end{bmatrix} \begin{matrix} 10R_1 \\ 5R_2 \\ 100R_3 \end{matrix}$$

$$\begin{bmatrix} 30 & 5 & 1 \\ 30 & 5 & 1 \\ 30 & -5 & 1 \end{bmatrix} \begin{matrix} \\ R_2 - R_1 \\ R_3 - R_1 \end{matrix}$$

$$\begin{bmatrix} 30 & 5 & 1 \\ 0 & 0 & 0 \\ 0 & -10 & 0 \end{bmatrix} \begin{matrix} \\ \\ (1/10)R_3 \end{matrix}$$

$$\begin{bmatrix} 30 & 5 & 1 \\ 0 & 0 & 0 \\ 0 & -1 & 0 \end{bmatrix} \begin{matrix} R_1 + 5R_3 \\ \\ \end{matrix}$$

$$\begin{bmatrix} 30 & 0 & 1 \\ 0 & 0 & 0 \\ 0 & -1 & 0 \end{bmatrix} \begin{matrix} (1/30)R_1 \\ \\ -R_3 \end{matrix}$$

$$\begin{bmatrix} 1 & 0 & 1/30 \\ 0 & 0 & 0 \\ 0 & 1 & 0 \end{bmatrix}$$

Solution: $x = 1/30$, $y = 0$

13. $x + 2y \quad = -3$
$\quad x \quad - z = 0$
$\quad x + 3y - 2z = -2$

$$\begin{bmatrix} 1 & 2 & 0 & -3 \\ 1 & 0 & -1 & 0 \\ 1 & 3 & -2 & -2 \end{bmatrix} \begin{matrix} \\ R_2 - R_1 \\ R_3 - R_1 \end{matrix}$$

$$\begin{bmatrix} 1 & 2 & 0 & -3 \\ 0 & -2 & -1 & 3 \\ 0 & 1 & -2 & 1 \end{bmatrix} \begin{matrix} R_1 + R_2 \\ \\ 2R_3 + R_2 \end{matrix}$$

$$\begin{bmatrix} 1 & 0 & -1 & 0 \\ 0 & -2 & -1 & 3 \\ 0 & 0 & -5 & 5 \end{bmatrix} \begin{matrix} \\ \\ (1/5)R_3 \end{matrix}$$

$$\begin{bmatrix} 1 & 0 & -1 & 0 \\ 0 & -2 & -1 & 3 \\ 0 & 0 & -1 & 1 \end{bmatrix} \begin{matrix} R_1 - R_3 \\ R_2 - R_3 \\ \end{matrix}$$

$$\begin{bmatrix} 1 & 0 & 0 & -1 \\ 0 & -2 & 0 & 2 \\ 0 & 0 & -1 & 1 \end{bmatrix} \begin{matrix} \\ -(1/2)R_2 \\ -R_3 \end{matrix}$$

$$\begin{bmatrix} 1 & 0 & 0 & -1 \\ 0 & 1 & 0 & -1 \\ 0 & 0 & 1 & -1 \end{bmatrix}$$

Solution: $x = -1$, $y = -1$, $z = -1$

14. $x - y + z = 2$
$7x + y - z = 6$
$x - \frac{1}{2}y + \frac{1}{3}z = 1$
$x + y + z = 6$

$$\begin{bmatrix} 1 & -1 & 1 & 2 \\ 7 & 1 & -1 & 6 \\ 1 & -1/2 & 1/3 & 1 \\ 1 & 1 & 1 & 6 \end{bmatrix} \begin{matrix} \\ \\ 6R_3 \\ \end{matrix}$$

$$\begin{bmatrix} 1 & -1 & 1 & 2 \\ 7 & 1 & -1 & 6 \\ 6 & -3 & 2 & 6 \\ 1 & 1 & 1 & 6 \end{bmatrix} \begin{matrix} \\ R_2 - 7R_1 \\ R_3 - 6R_1 \\ R_4 - R_1 \end{matrix}$$

$$\begin{bmatrix} 1 & -1 & 1 & 2 \\ 0 & 8 & -8 & -8 \\ 0 & 3 & -4 & -6 \\ 0 & 2 & 0 & 4 \end{bmatrix} \begin{matrix} \\ (1/8)R_2 \\ \\ (1/2)R_4 \end{matrix}$$

$$\begin{bmatrix} 1 & -1 & 1 & 2 \\ 0 & 1 & -1 & -1 \\ 0 & 3 & -4 & -6 \\ 0 & 1 & 0 & 2 \end{bmatrix} \begin{matrix} R_1 + R_2 \\ \\ R_3 - 3R_2 \\ R_4 - R_2 \end{matrix}$$

$$\begin{bmatrix} 1 & 0 & 0 & 1 \\ 0 & 1 & -1 & -1 \\ 0 & 0 & -1 & -3 \\ 0 & 0 & 1 & 3 \end{bmatrix} \begin{matrix} \\ R_2 - R_3 \\ \\ R_4 + R_3 \end{matrix}$$

$$\begin{bmatrix} 1 & 0 & 0 & 1 \\ 0 & 1 & 0 & 2 \\ 0 & 0 & -1 & -3 \\ 0 & 0 & 0 & 0 \end{bmatrix} \begin{matrix} \\ \\ -R_3 \\ \end{matrix}$$

$$\begin{bmatrix} 1 & 0 & 0 & 1 \\ 0 & 1 & 0 & 2 \\ 0 & 0 & 1 & 3 \\ 0 & 0 & 0 & 0 \end{bmatrix}$$

Solution: $x = 1, y = 2, z = 3$

15. $x - \frac{1}{2}y + z = 0$

$\quad \frac{1}{2}x \quad - \frac{1}{2}z = -1$

$\quad \frac{3}{2}x - \frac{1}{2}y + \frac{1}{2}z = -1$

$$\begin{bmatrix} 1 & -1/2 & 1 & 0 \\ 1/2 & 0 & -1/2 & -1 \\ 3/2 & -1/2 & 1/2 & -1 \end{bmatrix} \begin{matrix} 2R_1 \\ 2R_2 \\ 2R_3 \end{matrix}$$

$$\begin{bmatrix} 2 & -1 & 2 & 0 \\ 1 & 0 & -1 & -2 \\ 3 & -1 & 1 & -2 \end{bmatrix} \begin{matrix} \\ 2R_2 - R_1 \\ 2R_3 - 3R_1 \end{matrix}$$

$$\begin{bmatrix} 2 & -1 & 2 & 0 \\ 0 & 1 & -4 & -4 \\ 0 & 1 & -4 & -4 \end{bmatrix} \begin{matrix} R_1 + R_2 \\ \\ R_3 - R_2 \end{matrix}$$

$$\begin{bmatrix} 2 & 0 & -2 & -4 \\ 0 & 1 & -4 & -4 \\ 0 & 0 & 0 & 0 \end{bmatrix} (1/2)R_1$$

$$\begin{bmatrix} 1 & 0 & -1 & -2 \\ 0 & 1 & -4 & -4 \\ 0 & 0 & 0 & 0 \end{bmatrix}$$

Translating back to equations:

$\quad x - z = -2$

$\quad y - 4z = -4$

General solution:

$\quad x = z - 2$

$\quad y = 4z - 4 = 4(z-1)$

$\quad z$ arbitrary.

or

$\quad (z-2, 4(z-1), z); z$ arbitrary

16. $x + y - 2z = -1$

$\quad -2x - 2y + 4z = 2$

$\quad 0.75x + 0.75y - 1.5z = -0.75$

$$\begin{bmatrix} 1 & 1 & -2 & -1 \\ -2 & -2 & 4 & 2 \\ 0.75 & 0.75 & -1.5 & -.75 \end{bmatrix} \begin{matrix} \\ \\ 4R_3 \end{matrix}$$

$$\begin{bmatrix} 1 & 1 & -2 & -1 \\ -2 & -2 & 4 & 2 \\ 3 & 3 & -6 & -3 \end{bmatrix} \begin{matrix} \\ (1/2)R_2 \\ (1/3)R_3 \end{matrix}$$

$$\begin{bmatrix} 1 & 1 & -2 & -1 \\ -1 & -1 & 2 & 1 \\ 1 & 1 & -2 & -1 \end{bmatrix} \begin{matrix} \\ R_2 + R_1 \\ R_3 - R_1 \end{matrix}$$

$$\begin{bmatrix} 1 & 1 & -2 & -1 \\ 0 & 0 & 0 & 0 \\ 0 & 0 & 0 & 0 \end{bmatrix}$$

Translating back to equations:

$x + y - 2z = -1$

General Solution:

$x = -2 - y + 2z$

y arbitrary

z arbitrary,

or

$(-1 - y + 2z, y, z)$; y, z arbitrary

17. Rewrite the given equations in standard form:

$x - \frac{1}{2}y = 0$

$\frac{1}{2}x + \frac{1}{2}z = 2$

$3x - y + z = 0$

$$\left[\begin{array}{ccc|c} 1 & -1/2 & 0 & 0 \\ 1/2 & 0 & 1/2 & 2 \\ 3 & -1 & 1 & 0 \end{array}\right] \begin{array}{l} 2R_1 \\ 2R_2 \\ \end{array}$$

$$\left[\begin{array}{ccc|c} 2 & -1 & 0 & 0 \\ 1 & 0 & 1 & 4 \\ 3 & -1 & 1 & 0 \end{array}\right] \begin{array}{l} \\ 2R_2 - R_1 \\ 2R_3 - 3R_1 \end{array}$$

$$\left[\begin{array}{ccc|c} 2 & -1 & 0 & 0 \\ 0 & 1 & 2 & 8 \\ 0 & 1 & 2 & 0 \end{array}\right] \begin{array}{l} R_1 + R_2 \\ \\ R_3 - R_2 \end{array}$$

$$\left[\begin{array}{ccc|c} 2 & 0 & 2 & 8 \\ 0 & 1 & 2 & 8 \\ 0 & 0 & 0 & -8 \end{array}\right]$$

The last row translates to the false statement $0 = -8$, showing that there is no solution for the system.

18. $x - y + z = 1$

$ y - z + w = 1$

$x + z - w = 1$

$2x + z = 3$

$$\left[\begin{array}{cccc|c} 1 & -1 & 1 & 0 & 1 \\ 0 & 1 & -1 & 1 & 1 \\ 1 & 0 & 1 & -1 & 1 \\ 2 & 0 & 1 & 0 & 3 \end{array}\right] \begin{array}{l} \\ \\ R_3 - R_1 \\ R_4 - 2R_1 \end{array}$$

$$\left[\begin{array}{cccc|c} 1 & -1 & 1 & 0 & 1 \\ 0 & 1 & -1 & 1 & 1 \\ 0 & 1 & 0 & -1 & 0 \\ 0 & 2 & -1 & 0 & 1 \end{array}\right] \begin{array}{l} R_1 + R_2 \\ \\ R_3 - R_2 \\ R_4 - 2R_2 \end{array}$$

$$\left[\begin{array}{cccc|c} 1 & 0 & 0 & 1 & 2 \\ 0 & 1 & -1 & 1 & 1 \\ 0 & 0 & 1 & -2 & -1 \\ 0 & 0 & 1 & -2 & -1 \end{array}\right] \begin{array}{l} \\ R_2 + R_3 \\ \\ R_4 - R_3 \end{array}$$

$$\left[\begin{array}{cccc|c} 1 & 0 & 0 & 1 & 2 \\ 0 & 1 & 0 & -1 & 0 \\ 0 & 0 & 1 & -2 & -1 \\ 0 & 0 & 0 & 0 & 0 \end{array}\right]$$

Translating back to equations:

$x + w = 2$

$y - w = 0$

$z - 2w = -1$

General solution:

$x = 2 - w$

$y = w$

$z = -1 + 2w$

w arbitrary,

or

$(2 - w, w, -1 + 2w, w)$; w arbitrary

19. $5F - 9C = 160$,

(a) We are given the additional information

$F = C$, or $F - C = 0$

giving us a second linear equation.

We can solve the resulting system of 2 equations in 2 unknowns by multiplying the second equation by -5:

$5F - 9C = 160$,

$-5F + 5C = 0$

adding gives

$-4C = 160$

$C = -40°$

Since $C = F$, $F = -40°$ also.

(b) We are told that the Celsius temperature is half the Fahrenheit temperature: $C = \frac{1}{2}F$, giving us the system

$$5F - 9C = 160$$
$$\tfrac{1}{2}F - C = 0$$

Multiply the first equation by 2 and the second by −20:

$$10F - 18C = 320$$
$$-10F + 20C = 0$$

Adding:

$$2C = 320$$
$$C = 160$$

Substituting $C = 160$ in the equation $C = \frac{1}{2}F$ gives

$$F = 2C = 320.$$

Thus, the temperature is 160°C, or 320°F.

(c) We are told that the Fahrenheit temperature is 1.8 times the Celsius temperature: $F = 1.8C$, or $F - 1.8C = 0$, giving us the system

$$5F - 9C = 160$$
$$F - 1.8C = 0$$

Multiplying the second equation by −5 gives

$$5F - 9C = 160$$
$$-5F + 9C = 0$$

Adding gives the false statement $0 = 160$, showing that the given system is inconsistent (has no solution). Thus, it is not possible for the Fahrenheit temperature of an object to be 1.8 times its Celsius temperature.

20. Unknowns:

x = the population (in millions) of city A
y = the population (in millions) of city B
z = the population (in millions) of city C
w = the population (in millions) of city D

(a) The total population of the four cities is 10 million people

$$x + y + z + w = 10$$

This is a linear equation.

(b) City A has three times as many people as cities B and C combined.

Rephrasing: the population of City A equals 3 times the sum of the populations of cities B and C:

$$x = 3(y + z), \text{ or}$$
$$x - 3y - 3z = 0$$

This is a linear equation.

(c) there are no people living in city D:

$$w = 0$$

This is a linear equation.

(d) The population of city A is the sum of the squares of the populations of the other three cities:

$$x = y^2 + z^2 + w^2, \text{ or}$$
$$x - y^2 - z^2 - w^2 = 0$$

This is not a linear equation because of the squared terms.

(e) City C has 30% more people than City B.

Rephrasing: the population of City C is 130% of the population of City B:

$$z = 1.30y, \text{ or}$$
$$-1.30y + z = 0$$

This is a linear equation.

(f) City C has 30% fewer people than City B.

Rephrasing: the population of City C is 70% of the population of City B:

$$z = 0.70y, \text{ or}$$
$$-0.70y + z = 0$$

This is a linear equation.

21. Unknowns:

x = the number of packages from Duffin House
y = the number of packages from Higgins Press

Arrange the given information in a table:

	Duffin (x)	Higgins (y)	Desired totals
Horror	5	5	4500
Romance	5	11	6600

Horror: $5x + 5y = 4500$

Romance: $5x + 11y = 6600$

To solve, multiply the first equation by –1 and add, to obtain

$6y = 2100$

$y = 350$

The first equation can be divided by 5 and rewritten as $x + y = 900$ Substituting $y = 350$ gives us

$x + 350 = 900$

$x = 550$

Solution: Purchase 550 packages from Duffin House, 350 from Higgins Press

22. Unknowns:

x = the number of packages from Duffin House

y = the number of packages from Higgins Press

Cost: $50x + 150y = 50,000$

Also, you have promised to buy twice as many packages from Duffin as from Higgins. Rephrasing: The number of packages from Duffin equals twice the number from Higgins:

$x = 2y$, or $x - 2y = 0$

Dividing the first equation by 50 gives:

$x + 3y = 1000$

$x - 2y = 0$

Subtracting the second from the first gives

$5y = 1000,$

$y = 200$

Substituting $y = 200$ in the second equation gives

$x - 2(200) = 0$

$x = 400$

Solution: Purchase 400 packages from Duffin House, 200 from Higgins Press.

23. Unknowns:

x = the number of packages from Duffin House

y = the number of packages from Higgins Press

Cost: $50x + 150y = 60,000$

Also, you have promised to spend the same amount on both publishers:

Amount spent on Duffin = amount spent on Higgins

$50x = 150y$, or

$50x - 150y = 0$

Dividing each of these equations by 50 gives:

$x + 3y = 1200$

$x - 3y = 0$

Adding gives

$2x = 1200$

$x = 600$

Substituting $x = 600$ in the second equation gives

$600 - 3y = 0$

$3y = 600$

$y = 200$

Solution: Purchase 600 packages from Duffin House, 200 from Higgins Press.

24. Unknowns:

x = the number of packages from Duffin House

y = the number of packages from Higgins Press

Cost: $50x + 150y = 90,000$

Also, you have promised to spend twice as much money for books from Duffin as from Higgins:

Amount spent on Duffin = 2× amount spent on Higgins

$50x = 2(150)y$, or

$50x - 300y = 0$

Dividing each of these equations by 50 gives:

$x + 3y = 1800$

$x - 6y = 0$

Multiplying the first equation by 2 and adding gives:

$3x = 3600$

$x = 1200$

Substituting $x = 1200$ in the second equation gives

$1200 - 6y = 0$

$6y = 1200$

$y = 200$

Solution: Purchase 1200 packages from Duffin House, 200 from Higgins Press.

25. Demand: $q = -1000p + 140,000$

   Supply: $q = 2000p + 20,000$

For the equilibrium price, we can equate the supply and demand:

   Demand = Supply

   $-1000p + 140,000 = 2000p + 20,000$

   $-3000p = -120,000$

   $p = \dfrac{120,000}{3000} = \$40$ per book

26. Demand: $q = -2p + 18$

   Supply: $q = 3p + 3$

For the equilibrium price,

   Demand = Supply

   $-2p + 18 = 3p + 3$

   $5p = 15$

   $p = \$3$ million

27. Unknowns:

$x =$ the number of baby sharks

$y =$ the number of piranhas

$z =$ the number of squids

Arrange the given data in a table with the unknowns across the top:

	Sharks (x)	Piranha (y)	Squid (z)	Total Consumed
Goldfish	1	1	1	21
Angelfish	2	0	1	21
Butterfly fish	2	3	0	35

Goldfish: $x + y + z = 21$

Angelfish: $2x + z = 21$

Butterfly fish: $2x + 3y = 35$

$$\begin{bmatrix} 1 & 1 & 1 & 21 \\ 2 & 0 & 1 & 21 \\ 2 & 3 & 0 & 35 \end{bmatrix} \begin{matrix} \\ R_2 - 2R_1 \\ R_3 - 2R_1 \end{matrix}$$

$$\begin{bmatrix} 1 & 1 & 1 & 21 \\ 0 & -2 & -1 & -21 \\ 0 & 1 & -2 & -7 \end{bmatrix} \begin{matrix} 2R_1 + R_2 \\ \\ 2R_3 + R_2 \end{matrix}$$

$$\begin{bmatrix} 2 & 0 & 1 & 21 \\ 0 & -2 & -1 & -21 \\ 0 & 0 & -5 & -35 \end{bmatrix} \begin{matrix} \\ \\ (1/5)R_3 \end{matrix}$$

$$\begin{bmatrix} 2 & 0 & 1 & 21 \\ 0 & -2 & -1 & -21 \\ 0 & 0 & -1 & -7 \end{bmatrix} \begin{matrix} R_1 + R_3 \\ R_2 - R_3 \\ \end{matrix}$$

$$\begin{bmatrix} 2 & 0 & 0 & 14 \\ 0 & -2 & 0 & -14 \\ 0 & 0 & -1 & -7 \end{bmatrix} \begin{matrix} (1/2)R_1 \\ -(1/2)R_2 \\ -R_3 \end{matrix}$$

$$\begin{bmatrix} 1 & 0 & 0 & 7 \\ 0 & 1 & 0 & 7 \\ 0 & 0 & 1 & 7 \end{bmatrix}$$

$x = 7, y = 7, z = 7$

Solution: You have 7 of each type of carnivorous creature.

28. Unknowns:

$x =$ the number of servings of granola

$y =$ the number of servings of nutty granola

$z =$ the number of servings of nuttiest granola

From the data in the given table:

   $x + y + 5z = 1500$

   $4x + 8y + 8z = 10,000$

   $2x + 4y + 8z = 4000$

$$\begin{bmatrix} 1 & 1 & 5 & 1500 \\ 4 & 8 & 8 & 10000 \\ 2 & 4 & 8 & 4000 \end{bmatrix} \begin{matrix} \\ (1/4)R_2 \\ (1/2)R_3 \end{matrix}$$

$$\begin{bmatrix} 1 & 1 & 5 & 1500 \\ 1 & 2 & 2 & 2500 \\ 1 & 2 & 4 & 2000 \end{bmatrix} \begin{matrix} \\ R_2 - R_1 \\ R_3 - R_1 \end{matrix}$$

$$\begin{bmatrix} 1 & 1 & 5 & 1500 \\ 0 & 1 & -3 & 1000 \\ 0 & 1 & -1 & 500 \end{bmatrix} \begin{matrix} R_1 - R_2 \\ \\ R_3 - R_2 \end{matrix}$$

$$\begin{bmatrix} 1 & 0 & 8 & 500 \\ 0 & 1 & -3 & 1000 \\ 0 & 0 & 2 & -500 \end{bmatrix} \begin{matrix} \\ \\ (1/2)R_3 \end{matrix}$$

$$\begin{bmatrix} 1 & 0 & 8 & 500 \\ 0 & 1 & -3 & 1000 \\ 0 & 0 & 1 & -250 \end{bmatrix} \begin{matrix} R_1 - 8R_3 \\ R_2 + 3R_3 \\ \\ \end{matrix}$$

$$\begin{bmatrix} 1 & 0 & 0 & 2500 \\ 0 & 1 & 0 & 250 \\ 0 & 0 & 1 & -250 \end{bmatrix}$$

The solution of the system is: 2500 granola treats, 250 nutty granola treats, and –250 nuttiest granola treats. Since the last number is negative, it is impossible for them to use up all the ingredients.

29. Unknowns:

x = The number of hits at OHaganBooks.com

y = The number of hits at JungleBooks.com

z = The number of hits at FarmerBooks.com

We are given the following information:

(1) Combined web site traffic at the three sites is estimated at 10,000 hits per day.

$x + y + z = 10,000$

(2) The total number of orders is 1500 per day:

$0.10x + 0.20y + 0.20z = 1500$

(3) FarmerBooks.com gets as many book orders as the other two combined.

$0.20z = 0.10x + 0.20y$, or

$0.10x + 0.20y - 0.20z = 0$

Solving this system of 3 linear equations in 2 unknowns:

$$\begin{bmatrix} 1 & 1 & 1 & 10000 \\ 0.1 & 0.2 & 0.2 & 1500 \\ 0.1 & 0.2 & -.2 & 0 \end{bmatrix} \begin{matrix} \\ 10R_2 \\ 10R_3 \end{matrix}$$

$$\begin{bmatrix} 1 & 1 & 1 & 10000 \\ 1 & 2 & 2 & 15000 \\ 1 & 2 & -2 & 0 \end{bmatrix} \begin{matrix} \\ R_2 - R_1 \\ R_3 - R_1 \end{matrix}$$

$$\begin{bmatrix} 1 & 1 & 1 & 10000 \\ 0 & 1 & 1 & 5000 \\ 0 & 1 & -3 & -10000 \end{bmatrix} \begin{matrix} R_1 - R_2 \\ \\ R_3 - R_2 \end{matrix}$$

$$\begin{bmatrix} 1 & 0 & 0 & 5000 \\ 0 & 1 & 1 & 5000 \\ 0 & 0 & -4 & -15000 \end{bmatrix} \begin{matrix} \\ \\ (1/4)R_3 \end{matrix}$$

$$\begin{bmatrix} 1 & 0 & 0 & 5000 \\ 0 & 1 & 1 & 5000 \\ 0 & 0 & -1 & -3750 \end{bmatrix} \begin{matrix} \\ R_2 + R_3 \\ \\ \end{matrix}$$

$$\begin{bmatrix} 1 & 0 & 0 & 5000 \\ 0 & 1 & 0 & 1250 \\ 0 & 0 & -1 & -3750 \end{bmatrix} \begin{matrix} \\ \\ -R_3 \end{matrix}$$

$$\begin{bmatrix} 1 & 0 & 0 & 5000 \\ 0 & 1 & 0 & 1250 \\ 0 & 0 & 1 & 3750 \end{bmatrix}$$

Solution: 5000 hits per day at OHaganBooks.com, 1250 at JungleBooks.com, 3750 at FarmerBooks.com

30. Unknowns:

x = the number of shares of HAL

y = the number of shares of POM

z = the number of shares of WELL

The total investment was $12,400:

Investment in HAL = x shares @ $100 = $100x$

Investment in POM = y shares @ $20 = $20y$

Investment in WELL = z shares @ $25 = $25z$

Thus,

$100x + 20y + 25z = 12,400$

You earned $56 in dividends:

HAL dividend = 0.5% of $100x$ invested

$\quad = 0.005(100x) = 0.5x$

POM dividend = 1.5% of $20y$ invested

$\quad = 0.015(20y) = 0.3y$

WELL dividend = 0

Thus,

$0.5x + 0.3y = 56$

You purchased a total of 200 shares:

$\quad x + y + z = 200$

We therefore have the following system:

$\quad 100x + 20y + 25z = 12{,}400$

$\quad 0.5x + 0.3y = 56$

$\quad x + y + z = 200$

Solving:

$$\left[\begin{array}{ccc|c} 100 & 20 & 25 & 12400 \\ 0.5 & 0.3 & 0 & 56 \\ 1 & 1 & 1 & 200 \end{array}\right] 10R_2$$

$$\left[\begin{array}{ccc|c} 100 & 20 & 25 & 12400 \\ 5 & 3 & 0 & 560 \\ 1 & 1 & 1 & 200 \end{array}\right] (1/5)R_1$$

$$\left[\begin{array}{ccc|c} 20 & 4 & 5 & 2480 \\ 5 & 3 & 0 & 560 \\ 1 & 1 & 1 & 200 \end{array}\right] \begin{array}{l} \\ 4R_2 - R_1 \\ 20R_3 - R_1 \end{array}$$

$$\left[\begin{array}{ccc|c} 20 & 4 & 5 & 2480 \\ 0 & 8 & -5 & -240 \\ 0 & 16 & 15 & 1520 \end{array}\right] \begin{array}{l} 2R_1 - R_2 \\ \\ R_3 - 2R_2 \end{array}$$

$$\left[\begin{array}{ccc|c} 40 & 0 & 15 & 5200 \\ 0 & 8 & -5 & -240 \\ 0 & 0 & 25 & 2000 \end{array}\right] \begin{array}{l} (1/5)R_1 \\ \\ (1/25)R_3 \end{array}$$

$$\left[\begin{array}{ccc|c} 8 & 0 & 3 & 1040 \\ 0 & 8 & -5 & -240 \\ 0 & 0 & 1 & 80 \end{array}\right] \begin{array}{l} R_1 - 3R_3 \\ R_2 + 5R_3 \end{array}$$

$$\left[\begin{array}{ccc|c} 8 & 0 & 0 & 800 \\ 0 & 8 & 0 & 160 \\ 0 & 0 & 1 & 80 \end{array}\right] \begin{array}{l} (1/8)R_1 \\ (1/8)R_2 \end{array}$$

$$\left[\begin{array}{ccc|c} 1 & 0 & 0 & 100 \\ 0 & 1 & 0 & 20 \\ 0 & 0 & 1 & 80 \end{array}\right]$$

$x = 100,\ y = 20,\ z = 80$

Solution: He purchased 100 shares of HAL, 20 shares of POM, and 80 shares of WELL.

31. Unknowns:

x = the number of DHS shares purchased

y = the number of HPR shares purchased

z = the number of SPUB shares purchased

He purchased a total of 2000 shares:

$\quad x + y + z = 2000$

The total investment was $20,000

$\quad 8x + 10y + 15z = 20{,}000$

The total profit was $3400:

$\quad 0.20(8x) + 0.15(10y) + 0.15(15z) = 3400$

$\quad 1.6x + 1.5y + 2.25z = 3400$

Solving:

$$\left[\begin{array}{ccc|c} 1 & 1 & 1 & 2000 \\ 8 & 10 & 15 & 20000 \\ 1.6 & 1.5 & 2.25 & 3400 \end{array}\right] 40R_3$$

$$\left[\begin{array}{ccc|c} 1 & 1 & 1 & 2000 \\ 8 & 10 & 15 & 20000 \\ 64 & 60 & 90 & 136000 \end{array}\right] (1/2)R_3$$

$$\left[\begin{array}{ccc|c} 1 & 1 & 1 & 2000 \\ 8 & 10 & 15 & 20000 \\ 32 & 30 & 45 & 68000 \end{array}\right] \begin{array}{l} \\ R_2 - 8R_1 \\ R_3 - 32R_1 \end{array}$$

$$\left[\begin{array}{ccc|c} 1 & 1 & 1 & 2000 \\ 0 & 2 & 7 & 4000 \\ 0 & -2 & 13 & 4000 \end{array}\right] \begin{array}{l} 2R_1 - R_2 \\ \\ R_3 + R_2 \end{array}$$

$$\begin{bmatrix} 2 & 0 & -5 & 0 \\ 0 & 2 & 7 & 4000 \\ 0 & 0 & 20 & 8000 \end{bmatrix} (1/20)R_3$$

$$\begin{bmatrix} 2 & 0 & -5 & 0 \\ 0 & 2 & 7 & 4000 \\ 0 & 0 & 1 & 400 \end{bmatrix} \begin{matrix} R_1 + 5R_3 \\ R_2 - 7R_3 \end{matrix}$$

$$\begin{bmatrix} 2 & 0 & 0 & 2000 \\ 0 & 2 & 0 & 1200 \\ 0 & 0 & 1 & 400 \end{bmatrix} \begin{matrix} (1/2)R_1 \\ (1/2)R_2 \end{matrix}$$

$$\begin{bmatrix} 1 & 0 & 0 & 1000 \\ 0 & 1 & 0 & 600 \\ 0 & 0 & 1 & 400 \end{bmatrix}$$

Solution: He purchased 1000 DHS shares, 600 HPR shares, and 400 SPUB shares.

32. Unknowns:

$x =$ the number of credits of Liberal Arts

$y =$ the number of credits of Sciences

$z =$ the number of credits of Fine Arts

$w =$ the number of credits of Mathematics

Given information:

(1) The total number of credits is 124:

$x + y + z + w = 124$

(2) An equal number of Science and Fine Arts credit:

$y = z$, or $y - z = 0$

(3) Twice as many Mathematics credits as Science credits and Fine Arts credits combined. Rephrasing: The number of Mathematics credits is twice the sum of the numbers of Science and Fine Arts credits:

$w = 2(y + z)$, or

$2y + 2z - w = 0$

(4) Liberal Arts credits exceed Mathematics credits by one third of the number of Fine Arts credits. Rephrasing: the number of Liberal Arts credits minus the number of Mathematics credits is one third of the number of Fine Arts credits:

$x - w = \frac{1}{3}z$, or

$x - \frac{1}{3}z - w = 0$

Solving:

$$\begin{bmatrix} 1 & 1 & 1 & 1 & 124 \\ 0 & 1 & -1 & 0 & 0 \\ 0 & 2 & 2 & -1 & 0 \\ 1 & 0 & -1/3 & -1 & 0 \end{bmatrix} 3R_4$$

$$\begin{bmatrix} 1 & 1 & 1 & 1 & 124 \\ 0 & 1 & -1 & 0 & 0 \\ 0 & 2 & 2 & -1 & 0 \\ 3 & 0 & -1 & -3 & 0 \end{bmatrix} R_4 - 3R_1$$

$$\begin{bmatrix} 1 & 1 & 1 & 1 & 124 \\ 0 & 1 & -1 & 0 & 0 \\ 0 & 2 & 2 & -1 & 0 \\ 0 & -3 & -4 & -6 & -372 \end{bmatrix} \begin{matrix} R_1 - R_2 \\ \\ R_3 - 2R_2 \\ R_4 + 3R_2 \end{matrix}$$

$$\begin{bmatrix} 1 & 0 & 2 & 1 & 124 \\ 0 & 1 & -1 & 0 & 0 \\ 0 & 0 & 4 & -1 & 0 \\ 0 & 0 & -7 & -6 & -372 \end{bmatrix} \begin{matrix} 2R_1 - R_3 \\ 4R_2 + R_3 \\ \\ 4R_4 + 7R_3 \end{matrix}$$

$$\begin{bmatrix} 2 & 0 & 0 & 3 & 248 \\ 0 & 4 & 0 & -1 & 0 \\ 0 & 0 & 4 & -1 & 0 \\ 0 & 0 & 0 & -31 & -1488 \end{bmatrix} (1/31)R_4$$

$$\begin{bmatrix} 2 & 0 & 0 & 3 & 248 \\ 0 & 4 & 0 & -1 & 0 \\ 0 & 0 & 4 & -1 & 0 \\ 0 & 0 & 0 & -1 & -48 \end{bmatrix} \begin{matrix} R_1 + 3R_4 \\ R_2 - R_4 \\ R_3 - R_4 \end{matrix}$$

$$\begin{bmatrix} 2 & 0 & 0 & 0 & 104 \\ 0 & 4 & 0 & 0 & 48 \\ 0 & 0 & 4 & 0 & 48 \\ 0 & 0 & 0 & -1 & -48 \end{bmatrix} \begin{matrix} (1/2)R_1 \\ (1/4)R_2 \\ (1/4)R_3 \\ -R_4 \end{matrix}$$

$$\begin{bmatrix} 1 & 0 & 0 & 0 & 52 \\ 0 & 1 & 0 & 0 & 12 \\ 0 & 0 & 1 & 0 & 12 \\ 0 & 0 & 0 & 1 & 48 \end{bmatrix}$$

Solution: Billy Sean is forced to take exactly the following combination: Liberal Arts: 52 credits, Sciences: 12 credits, Fine Arts: 12 credits, Mathematics: 48 credits.

33.

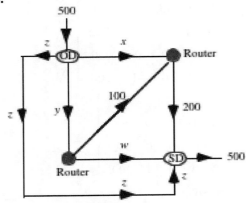

OD = Order Department
SD = Shipping Department

(a) Order Department: Traffic in = Traffic out
$500 = x + y + z$
Top right router: Traffic in = Traffic out
$100 + x = 200$, or $x = 100$
Bottom left router: Traffic in = Traffic out
$y = 100 + w$, or $y - w = 100$
Shipping Department: Traffic in = Traffic out
$z + w + 200 = 500$, or $z + w = 300$
Solving:

$$\begin{bmatrix} 1 & 1 & 1 & 0 & 500 \\ 1 & 0 & 0 & 0 & 100 \\ 0 & 1 & 0 & -1 & 100 \\ 0 & 0 & 1 & 1 & 300 \end{bmatrix} \quad R_2 - R_1$$

$$\begin{bmatrix} 1 & 1 & 1 & 0 & 500 \\ 0 & -1 & -1 & 0 & -400 \\ 0 & 1 & 0 & -1 & 100 \\ 0 & 0 & 1 & 1 & 300 \end{bmatrix} \quad \begin{matrix} R_1 + R_2 \\ \\ R_3 + R_2 \\ \end{matrix}$$

$$\begin{bmatrix} 1 & 0 & 0 & 0 & 100 \\ 0 & -1 & -1 & 0 & -400 \\ 0 & 0 & -1 & -1 & -300 \\ 0 & 0 & 1 & 1 & 300 \end{bmatrix} \quad \begin{matrix} \\ R_2 - R_3 \\ \\ R_4 + R_3 \end{matrix}$$

$$\begin{bmatrix} 1 & 0 & 0 & 0 & 100 \\ 0 & -1 & 0 & 1 & -100 \\ 0 & 0 & -1 & -1 & -300 \\ 0 & 0 & 0 & 0 & 0 \end{bmatrix} \quad \begin{matrix} \\ -R_2 \\ -R_3 \\ \end{matrix}$$

$$\begin{bmatrix} 1 & 0 & 0 & 0 & 100 \\ 0 & 1 & 0 & -1 & 100 \\ 0 & 0 & 1 & 1 & 300 \\ 0 & 0 & 0 & 0 & 0 \end{bmatrix}$$

Translating back to equations:
$x = 100$
$y - w = 100$
$z + w = 300$
General solution:
$x = 100$
$y = 100 + w$
$z = 300 - w$
w arbitrary

(b) Since $y = 100 + w$, and w can be any number ≥ 0 (w cannot be negative because it represents the number of book orders along any route). Therefore, the smallest possible value of y is 100 books per day.

(c) The equation $z = 300 - w$ tells us that w cannot exceed 300 books per day., or else x would become negative.

(d) If there is no traffic along z, then $z = 0$, giving:

$x = 100$

$y = 100 + w$

$0 = 300 - w$

Thus, from the third equation, $w = 300$, giving us the particular solution

$x = 100$, $y = 100+300 = 400$, $z = 0$, $w = 300$

(e) If there is the same volume of traffic along y and z, then $y = z$, and so

$100 + w = 300 - w$

Thus,

$2w = 200,$

so $w = 100$ books per day.

34. Solution:

(a) At each intersection, equate traffic entering with traffic leaving

Lagoon Drive/North Beach Way Intersection:

$200 + u = x$

Intersection of Smith St., Beach View St., and Lagoon Drive

$50 = z + u$

Beach View St./North Beach Way Intersection:

$x + z = y$

South Beach Way, Dock Rd. and Smith St Intersection:

$y = 50 + v$

Rewriting these equations in standard form gives:

$x - u = 200$

$z + u = 50$

$x - y + z = 0$

$y - v = 50$

$$\begin{bmatrix} 1 & 0 & 0 & -1 & 0 & 200 \\ 0 & 0 & 1 & 1 & 0 & 50 \\ 1 & -1 & 1 & 0 & 0 & 0 \\ 0 & 1 & 0 & 0 & -1 & 50 \end{bmatrix} \begin{matrix} \\ \\ R_3 - R_1 \\ \\ \end{matrix}$$

$$\begin{bmatrix} 1 & 0 & 0 & -1 & 0 & 200 \\ 0 & 0 & 1 & 1 & 0 & 50 \\ 0 & -1 & 1 & 1 & 0 & -200 \\ 0 & 1 & 0 & 0 & -1 & 50 \end{bmatrix} \begin{matrix} \\ \\ R_3 - R_2 \\ \\ \end{matrix}$$

$$\begin{bmatrix} 1 & 0 & 0 & -1 & 0 & 200 \\ 0 & 0 & 1 & 1 & 0 & 50 \\ 0 & -1 & 0 & 0 & 0 & -250 \\ 0 & 1 & 0 & 0 & -1 & 50 \end{bmatrix} \begin{matrix} \\ \\ \\ R_4 + R_3 \end{matrix}$$

$$\begin{bmatrix} 1 & 0 & 0 & -1 & 0 & 200 \\ 0 & 0 & 1 & 1 & 0 & 50 \\ 0 & -1 & 0 & 0 & 0 & -250 \\ 0 & 0 & 0 & 0 & -1 & -200 \end{bmatrix} \begin{matrix} \\ \\ -R_3 \\ -R_4 \end{matrix}$$

$$\begin{bmatrix} 1 & 0 & 0 & -1 & 0 & 200 \\ 0 & 0 & 1 & 1 & 0 & 50 \\ 0 & 1 & 0 & 0 & 0 & 250 \\ 0 & 0 & 0 & 0 & 1 & 200 \end{bmatrix}$$

General Solution:

$x = 200 + u$

$y = 250$

$z = 50 - u$

u is arbitrary

$v = 250$

(b) The Lagoon Drive traffic is represented by u, which is arbitrary in the general solution. However, the equation for z, namely $z = 50 - u$, tells us that u cannot exceed 50 (or else z becomes negative). Thus, the maximum traffic along Lagoon Drive is 50 cars every 5 minutes.

(c) This time, z is allowed to be negative (corresponding to net traffic flow in the opposite direction) and so there is no upper limit to the value of u, since none of the other variables become negative for large u. Hence, there is no limit to the possible traffic along Lagoon Drive.

35. Unknowns:

x = the number of packages from New York to Texas

y = the number of packages from New York to California

z = the number of packages from Illinois to Texas

w = the number of packages from Illinois to California

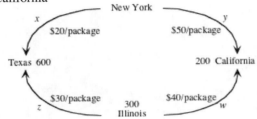

Books to Texas: $x + z = 600$

Books to California: $y + w = 200$

Books from Illinois: $z + w = 300$

Budget: $20x + 50y + 30z + 40w = 22{,}000$

Solving:

$$\begin{bmatrix} 1 & 0 & 1 & 0 & 600 \\ 0 & 1 & 0 & 1 & 200 \\ 0 & 0 & 1 & 1 & 300 \\ 20 & 50 & 30 & 40 & 22000 \end{bmatrix} (1/10)R_4$$

$$\begin{bmatrix} 1 & 0 & 1 & 0 & 600 \\ 0 & 1 & 0 & 1 & 200 \\ 0 & 0 & 1 & 1 & 300 \\ 2 & 5 & 3 & 4 & 2200 \end{bmatrix} R_4 - 2R_1$$

$$\begin{bmatrix} 1 & 0 & 1 & 0 & 600 \\ 0 & 1 & 0 & 1 & 200 \\ 0 & 0 & 1 & 1 & 300 \\ 0 & 5 & 1 & 4 & 1000 \end{bmatrix} R_4 - 5R_2$$

$$\begin{bmatrix} 1 & 0 & 1 & 0 & 600 \\ 0 & 1 & 0 & 1 & 200 \\ 0 & 0 & 1 & 1 & 300 \\ 0 & 0 & 1 & -1 & 0 \end{bmatrix} \begin{matrix} R_1 - R_3 \\ \\ \\ R_4 - R_3 \end{matrix}$$

$$\begin{bmatrix} 1 & 0 & 0 & -1 & 300 \\ 0 & 1 & 0 & 1 & 200 \\ 0 & 0 & 1 & 1 & 300 \\ 0 & 0 & 0 & -2 & -300 \end{bmatrix} (1/2)R_4$$

$$\begin{bmatrix} 1 & 0 & 0 & -1 & 300 \\ 0 & 1 & 0 & 1 & 200 \\ 0 & 0 & 1 & 1 & 300 \\ 0 & 0 & 0 & -1 & -150 \end{bmatrix} \begin{matrix} R_1 - R_4 \\ R_2 + R_4 \\ R_3 + R_4 \end{matrix}$$

$$\begin{bmatrix} 1 & 0 & 0 & 0 & 450 \\ 0 & 1 & 0 & 0 & 50 \\ 0 & 0 & 1 & 0 & 150 \\ 0 & 0 & 0 & -1 & -150 \end{bmatrix} -R_4$$

$$\begin{bmatrix} 1 & 0 & 0 & 0 & 450 \\ 0 & 1 & 0 & 0 & 50 \\ 0 & 0 & 1 & 0 & 150 \\ 0 & 0 & 0 & 1 & 150 \end{bmatrix}$$

Solution: New York to Texas: 450 packages, New York to California: 50 packages, Illinois to Texas: 150 packages, Illinois to California: 150 packages.

Chapter 3
3.1

1. $A = [1 \quad 5 \quad 0 \quad \frac{1}{4}]$

A has 1 row and 4 columns. Therefore, it is a 1×4 matrix. a_{13} is the entry in row 1 and column 3, so $a_{13} = 0$.

3. $C = \begin{bmatrix} \frac{5}{2} \\ 1 \\ -2 \\ 8 \end{bmatrix}$

C has 4 rows and 1 column. Therefore, it is a 4×1 matrix. C_{11} is the entry in row 1 and column 1, so $C_{11} = \frac{5}{2}$.

5. $E = \begin{bmatrix} e_{11} & e_{12} & e_{13} & \cdots & e_{1q} \\ e_{21} & e_{22} & e_{23} & \cdots & e_{2q} \\ \vdots & \vdots & \vdots & \ddots & \vdots \\ e_{p1} & e_{p2} & e_{p3} & \cdots & e_{pq} \end{bmatrix}$

E has p rows and q columns. Therefore, it is a $p \times q$ matrix. E_{22} is the entry in row 2 and column 2, so $E_{22} = c_{22}$.

7. $B = \begin{bmatrix} 1 & 3 \\ 5 & -6 \end{bmatrix}$

B has 2 rows and 2 columns. Therefore, it is a 2×2 matrix. b_{12} is the entry in row 1 and column 2, so $b_{12} = 3$.

9. $D = [d_1 \quad d_2 \quad \cdots \quad d_n]$

D has 1 row and n columns. Therefore, it is a $1 \times n$ matrix. D_{1r} is the entry in row 1 and column r, so $D_{1r} = d_r$.

11. $\begin{bmatrix} x+y & x+z \\ y+z & w \end{bmatrix} = \begin{bmatrix} 3 & 4 \\ 5 & 4 \end{bmatrix}$

Equating corresponding entries gives

$x + y = 3$

$x + z = 4$

$y + z = 5$

$w = 4$

The first three equations can be solved in various ways. Let us use row reduction:

$\begin{bmatrix} 1 & 1 & 0 & 3 \\ 1 & 0 & 1 & 4 \\ 0 & 1 & 1 & 5 \end{bmatrix} \begin{matrix} \\ R_2 - R_1 \\ \end{matrix}$

$\begin{bmatrix} 1 & 1 & 0 & 3 \\ 0 & -1 & 1 & 1 \\ 0 & 1 & 1 & 5 \end{bmatrix} \begin{matrix} R_1 + R_2 \\ \\ R_3 + R_2 \end{matrix}$

$\begin{bmatrix} 1 & 0 & 1 & 4 \\ 0 & -1 & 1 & 1 \\ 0 & 0 & 2 & 6 \end{bmatrix} \begin{matrix} \\ \\ (1/2)R_3 \end{matrix}$

$\begin{bmatrix} 1 & 0 & 1 & 4 \\ 0 & -1 & 1 & 1 \\ 0 & 0 & 1 & 3 \end{bmatrix} \begin{matrix} R_1 - R_3 \\ R_2 - R_3 \\ \end{matrix}$

$\begin{bmatrix} 1 & 0 & 0 & 1 \\ 0 & -1 & 0 & -2 \\ 0 & 0 & 1 & 3 \end{bmatrix} \begin{matrix} \\ -R_2 \\ \end{matrix}$

$\begin{bmatrix} 1 & 0 & 0 & 1 \\ 0 & 1 & 0 & 2 \\ 0 & 0 & 1 & 3 \end{bmatrix}$

Thus, $x = 1$, $y = 2$, $z = 3$, $w = 4$.

13. We obtain $A+B$ by adding corresponding entries:

$A+B = \begin{bmatrix} 0 & -1 \\ 1 & 0 \\ -1 & 2 \end{bmatrix} + \begin{bmatrix} 0.25 & -1 \\ 0 & 0.5 \\ -1 & 3 \end{bmatrix}$

$= \begin{bmatrix} 0.25 & -2 \\ 1 & 0.5 \\ -2 & 5 \end{bmatrix}$

15. To obtain $A+B-C$, add corresponding entries of A and B and subtract those of C:

$$A+B-C = \begin{bmatrix} 0 & -1 \\ 1 & 0 \\ -1 & 2 \end{bmatrix} + \begin{bmatrix} 0.25 & -1 \\ 0 & 0.5 \\ -1 & 3 \end{bmatrix}$$

$$- \begin{bmatrix} 1 & -1 \\ 1 & 1 \\ -1 & -1 \end{bmatrix}$$

$$= \begin{bmatrix} -0.75 & -1 \\ 0 & -0.5 \\ -1 & 6 \end{bmatrix}$$

17. $2A-C = 2\begin{bmatrix} 0 & -1 \\ 1 & 0 \\ -1 & 2 \end{bmatrix} - \begin{bmatrix} 1 & -1 \\ 1 & 1 \\ -1 & -1 \end{bmatrix}$

$$= \begin{bmatrix} -1 & -1 \\ 1 & -1 \\ -1 & 5 \end{bmatrix}$$

19. The transpose, A^T, of A is obtained by writing the rows of A as columns:

$$A = \begin{bmatrix} 0 & -1 \\ 1 & 0 \\ -1 & 2 \end{bmatrix}; A^T = \begin{bmatrix} 0 & 1 & -1 \\ -1 & 0 & 2 \end{bmatrix}.$$

Thus, $2A^T = 2\begin{bmatrix} 0 & 1 & -1 \\ -1 & 0 & 2 \end{bmatrix} = \begin{bmatrix} 0 & 2 & -2 \\ -2 & 0 & 4 \end{bmatrix}$

21. $A+B = \begin{bmatrix} 1 & -1 & 0 \\ 0 & 2 & -1 \end{bmatrix} + \begin{bmatrix} 3 & 0 & -1 \\ 5 & -1 & 1 \end{bmatrix}$

$$= \begin{bmatrix} 4 & -1 & -1 \\ 5 & 1 & 0 \end{bmatrix}$$

23. $A-B+C = \begin{bmatrix} 1 & -1 & 0 \\ 0 & 2 & -1 \end{bmatrix} - \begin{bmatrix} 3 & 0 & -1 \\ 5 & -1 & 1 \end{bmatrix}$

$$+ \begin{bmatrix} x & 1 & w \\ z & r & 4 \end{bmatrix}$$

$$= \begin{bmatrix} -2+x & 0 & 1+w \\ -5+z & 3+r & 2 \end{bmatrix}$$

25. $2A-B = 2\begin{bmatrix} 1 & -1 & 0 \\ 0 & 2 & -1 \end{bmatrix} - \begin{bmatrix} 3 & 0 & -1 \\ 5 & -1 & 1 \end{bmatrix}$

$$= \begin{bmatrix} -1 & -2 & 1 \\ -5 & 5 & -3 \end{bmatrix}$$

27. $B = \begin{bmatrix} 3 & 0 & -1 \\ 5 & -1 & 1 \end{bmatrix}; B^T = \begin{bmatrix} 3 & 5 \\ 0 & -1 \\ -1 & 1 \end{bmatrix}$

Therefore, $3B^T = 3\begin{bmatrix} 3 & 5 \\ 0 & -1 \\ -1 & 1 \end{bmatrix} = \begin{bmatrix} 9 & 15 \\ 0 & -3 \\ -3 & 3 \end{bmatrix}$

29. $A-C = \begin{bmatrix} 1.5 & -2.35 & 5.6 \\ 44.2 & 0 & 12.2 \end{bmatrix}$

$$- \begin{bmatrix} 10 & 20 & 30 \\ -10 & -20 & -30 \end{bmatrix}$$

$$= \begin{bmatrix} -8.5 & -22.35 & -24.4 \\ 54.2 & 20 & 42.2 \end{bmatrix}$$

TI-83/84 format: `[A]-[C]`

Online Matrix Algebra Tool format: `A-C`

31. $1.1B = 1.1\begin{bmatrix} 1.4 & 7.8 \\ 5.4 & 0 \\ 5.6 & 6.6 \end{bmatrix} = \begin{bmatrix} 1.54 & 8.58 \\ 5.94 & 0 \\ 6.16 & 7.26 \end{bmatrix}$

TI-83/84 format: `1.1[B]`

Online Matrix Algebra Tool format: `1.1*B`

33. $A^T+4.2B = \begin{bmatrix} 1.5 & 44.2 \\ -2.35 & 0 \\ 5.6 & 12.2 \end{bmatrix} + 4.2\begin{bmatrix} 1.4 & 7.8 \\ 5.4 & 0 \\ 5.6 & 6.6 \end{bmatrix}$

$$= \begin{bmatrix} 7.38 & 76.96 \\ 20.33 & 0 \\ 29.12 & 39.92 \end{bmatrix}$$

TI-83/84 format: `[A]T+4.2[B]`

Online Matrix Algebra Tool format:
`A^T+4.2*B`

35. $(2.1A-2.3C)^T = \begin{bmatrix} -19.85 & 115.82 \\ -50.935 & 46 \\ -57.24 & 94.62 \end{bmatrix}$

TI-83/84 format: `(2.1[A]-2.3[C])T`

Online Matrix Algebra Tool format:
`(2.1*A-2.3*C)^T`

37.

	New York	London	Hong Kong
2001	600	620	400
Change in 2002	120	60	−50
Change in 2003	40	120	−50

Regard each row of the table as a matrix:

Cost in 2001 = [600 620 400]

Increase in 2002 = [120 60 −50]

Increase in 2003 = [40 120 −50]

(a) Cost in 2002 = Cost in 2001 + Increase in 2002

= [600 620 400] + [120 60 −50]

= [720 680 350]

(b) Cost in 2003 = Cost in 2002 + Increase in 2003

= [720 680 350] + [40 120 −50]

= [760 800 300]

39. Write the given inventory table as a 2×3 matrix:

$$\text{Inventory} = \begin{bmatrix} 1000 & 2000 & 5000 \\ 1000 & 5000 & 2000 \end{bmatrix}$$

Write the given sales figures as a similar matrix:

$$\text{Sales} = \begin{bmatrix} 700 & 1300 & 2000 \\ 400 & 300 & 500 \end{bmatrix}$$

We then compute the remaining inventory by subtracting the sales:

Remaining Inventory = Inventory − Sales

$$= \begin{bmatrix} 1000 & 2000 & 5000 \\ 1000 & 5000 & 2000 \end{bmatrix}$$

$$- \begin{bmatrix} 700 & 1300 & 2000 \\ 400 & 300 & 500 \end{bmatrix}$$

$$= \begin{bmatrix} 300 & 700 & 3000 \\ 600 & 4700 & 1500 \end{bmatrix}$$

41.

	A	B	C	D
1	Revenue			
2		2004	2004	2006
3	Full Boots	$10,000	$9,000	$11,000
4	Half Boots	$8,000	$7,200	$8,800
5	Sandals	$4,000	$5,000	$6,000
6				
7	Production Costs			
8		2004	2004	2006
9	Full Boots	$2,000	$1,800	$2,200
10	Half Boots	$2,400	$1,440	$1,760
11	Sandals	$1,200	$1,500	$2,000

Arrange the revenue and costs in two 3×3 matrices

$$\text{Revenue} = \begin{bmatrix} 10{,}000 & 9000 & 11{,}000 \\ 8000 & 7200 & 8800 \\ 4000 & 5000 & 6000 \end{bmatrix}$$

$$\text{Cost} = \begin{bmatrix} 2000 & 1800 & 2200 \\ 2400 & 1440 & 1760 \\ 1200 & 1500 & 2000 \end{bmatrix}$$

Profit = Revenue − Cost

$$= \begin{bmatrix} 10{,}000 & 9000 & 11{,}000 \\ 8000 & 7200 & 8800 \\ 4000 & 5000 & 6000 \end{bmatrix} - \begin{bmatrix} 2000 & 1800 & 2200 \\ 2400 & 1440 & 1760 \\ 1200 & 1500 & 2000 \end{bmatrix}$$

$$= \begin{bmatrix} 8000 & 7200 & 8800 \\ 5600 & 5760 & 7040 \\ 2800 & 3500 & 4000 \end{bmatrix}$$

43. Row vectors for the populations, in millions:

1980 distribution = A = [49.1 58.9 75.4 43.2]

1990 distribution = B = [50.8 59.7 85.4 52.8]

Net change 1980 to 1990 = $B - A$ = [1.7 0.8 10 9.6]

Since all the net changes are positive, they represent net increases in the population.

45.

	Jan 01	Apr	Jul	Oct	Jan 02
Manhattan	150	250	150	100	150
Brooklyn	300	400	300	200	250
Newark	250	400	250	200	200

Regard each row in the table as a matrix:

Filings in Manhattan = [150 250 150 100 150]

Filings in Brooklyn = [300 400 300 200 250]

Filings in Newark = [250 400 250 200 200]

Total filings = Filings in Manhattan

+ Filings in Brooklyn + Filings in Newark

= [150 250 150 100 150]

+ [300 400 300 200 250]

+ [250 400 250 200 200]

= [700 1050 700 500 600]

47. (Refer to the solution to Exercise 45.) The difference between the number of bankruptcy filings in Brooklyn and in Newark is given by

Difference = Filings in Brooklyn − Filings in Newark

= [300 400 300 200 250] − [250 400 250 200 200]

= [50 0 50 0 50]

The difference was greatest in January 01, July 01, and January 02.

49. First, organize the starting inventory as a 2×3 matrix:

$$\begin{array}{c} \text{Proc} \ \text{Mem} \ \text{Tubes} \\ \begin{array}{c} \text{Pom II} \\ \text{Pom Classic} \end{array} \left[\begin{array}{ccc} 500 & 5000 & 10{,}000 \\ 200 & 2000 & 20{,}000 \end{array} \right] \end{array}$$

Thus,

$$\text{Inventory} = \left[\begin{array}{ccc} 500 & 5000 & 10{,}000 \\ 200 & 2000 & 20{,}000 \end{array} \right]$$

The parts used in making one of each computer can also be arranged in a matrix:

$$\text{Use} = \left[\begin{array}{ccc} 2 & 16 & 20 \\ 1 & 4 & 40 \end{array} \right]$$

(a) After two months, the company has made 100 of each computer, so the inventory remaining is

Inventory Remaining = Inventory − 100·Use

$$= \left[\begin{array}{ccc} 500 & 5000 & 10{,}000 \\ 200 & 2000 & 20{,}000 \end{array} \right] - 100 \left[\begin{array}{ccc} 2 & 16 & 20 \\ 1 & 4 & 40 \end{array} \right]$$

$$= \left[\begin{array}{ccc} 300 & 3400 & 8000 \\ 100 & 1600 & 16{,}000 \end{array} \right]$$

(b) After n months, the company has made $50n$ of each computer. Therefore, the inventory remaining is

$$= \left[\begin{array}{ccc} 500 & 5000 & 10{,}000 \\ 200 & 2000 & 20{,}000 \end{array} \right] - 50n \left[\begin{array}{ccc} 2 & 16 & 20 \\ 1 & 4 & 40 \end{array} \right]$$

$$= \left[\begin{array}{ccc} 500{-}100n & 5000{-}800n & 10{,}000{-}1000n \\ 200{-}50n & 2000{-}200n & 20{,}000{-}2000n \end{array} \right]$$

The 1,1 entry is zero after $n = 5$ months

The 1,2 entry is zero after $n = 6.25$ months

The 1,3 entry is zero after $n = 10$ months

The 2,1 entry is zero after $n = 4$ months

The 2,2 entry is zero after $n = 10$ months

The 2,3 entry is zero after $n = 10$ months

Thus, after 4 months, the first entry becomes zero, meaning that the company will run out of Pom Classic processor chips after 4 months.

51.

TO	Australia	South Africa
FROM		
North America	440	190
Europe	950	950
Asia	1790	200

(a) Arrange the above 1998 tourism figures in a matrix

$$A = \left[\begin{array}{cc} 440 & 190 \\ 950 & 950 \\ 1790 & 200 \end{array} \right]$$

Then, arrange the predicted changes from 1998 to 2008 in a new matrix:

$$D = \left[\begin{array}{cc} -20 & 40 \\ 50 & 50 \\ 0 & 100 \end{array} \right]$$

Tourism in 2008 = Tourism in 1998 + Change

$$= A + D = \left[\begin{array}{cc} 420 & 230 \\ 1000 & 1000 \\ 1790 & 300 \end{array} \right]$$

(b) The average pf the tourism figures is obtained by taking the average of the entries in A and B. To obtain the average, add and divide by 2:

Average $= \frac{1}{2}(A+B)$

$$A = 1998 \text{ figures} = \begin{bmatrix} 440 & 190 \\ 950 & 950 \\ 1790 & 200 \end{bmatrix}$$

$$B = 2008 \text{ figures} = \begin{bmatrix} 420 & 230 \\ 1000 & 1000 \\ 1790 & 300 \end{bmatrix}$$

(From part (a))

$$\text{Average} = \frac{1}{2}(A+B) = \frac{1}{2}\begin{bmatrix} 860 & 420 \\ 1950 & 1950 \\ 3580 & 500 \end{bmatrix}$$

$$= \begin{bmatrix} 430 & 210 \\ 975 & 975 \\ 1790 & 250 \end{bmatrix}$$

53. $(A+B)_{ij} = A_{ij} + B_{ij}$

The left-hand side is the ijth entry of $A+B$. The right-hand side is the sum of the ijth entries of A and B. Thus, the equation tells us that the ij th entry of the sum $A+B$ is obtained by adding the ij th entries of A and B.

55. If A is any matrix, then A_{11}, A_{22}, ..., A_{ii}, ... denote the entries going down the main diagonal (top left to bottom right). Thus, if $A_{ii} = 0$ for every i, then A would have zeros down the diagonal:

$$A = \begin{bmatrix} 0 & \# & \# & \# & \# \\ \# & 0 & \# & \# & \# \\ \# & \# & 0 & \# & \# \\ \# & \# & \# & 0 & \# \\ \# & \# & \# & \# & 0 \end{bmatrix}$$

(The symbols # indicate arbitrary numbers.)

57. The transpose of an $m \times n$ matrix is the $n \times m$ matrix obtained by writing its rows as columns. Thus, the entry originally in Row j and Column i winds up in Row i and Column j when a matrix is transposed. In other words,

ijth entry of the transpose $= ji$th entry of the original matrix

$(A^T)_{ij} = A_{ji}$

59. In a skew-symmetric matrix, the entry in the ij position is the negative of that in the ji position. In other words, each entry is the negative of its mirror image in the diagonal.

What about the diagonal entries? Each diagonal entry stays the same when the matrix is transposed. So, in a skew-symmetric matrix, each diagonal entry must equal its own negative. The only way this can happen is if the diagonal entries are zero.

Examples will vary.

(a) $\begin{bmatrix} 0 & -4 \\ 4 & 0 \end{bmatrix}$ **(b)** $\begin{bmatrix} 0 & -4 & 5 \\ 4 & 0 & 1 \\ -5 & -1 & 0 \end{bmatrix}$

61. The associativity of matrix addition is a consequence of the associativity of addition of numbers, since we add matrices by adding the corresponding entries (which are real numbers).

3.2

1. $[1 \quad 3 \quad -1]\begin{bmatrix} 9 \\ 1 \\ -1 \end{bmatrix} = [9+3+1] = [13]$

3. $[-1 \quad \frac{1}{2}]\begin{bmatrix} -\frac{1}{3} \\ 1 \end{bmatrix} = [\frac{1}{3}+\frac{1}{2}] = [\frac{5}{6}]$

5. $[0 \quad -2 \quad 1]\begin{bmatrix} x \\ y \\ z \end{bmatrix} = [-2y+z]$

7. $[1 \quad 3 \quad 2]\begin{bmatrix} 1 \\ -1 \end{bmatrix}$ is undefined.

(The dimensions are 1×3 and 2×1; the 3 and the 2 do not match.)

9. $[-1 \quad 1]\begin{bmatrix} -3 & 1 & 4 & 3 \\ 0 & 1 & -2 & 1 \end{bmatrix}$

$= [(3+0) \quad (-1+1) \quad (-4-2) \quad (-3+1)]$

$= [3 \quad 0 \quad -6 \quad -2]$

11. $[1 \quad -1 \quad 2 \quad 3]\begin{bmatrix} -1 & 2 & 0 \\ 2 & -1 & 0 \\ 0 & 5 & 2 \\ -1 & 8 & 1 \end{bmatrix}$

$= [(-1-2+0-3) \quad (2+1+10+24) \quad (0+0+4+3)]$

$= [-6 \quad 37 \quad 7]$

13. $\begin{bmatrix} 1 & 0 & -1 \\ 1 & 1 & 2 \end{bmatrix}\begin{bmatrix} 0 & 1 & -1 \\ 1 & 0 & 1 \\ 4 & 8 & 0 \end{bmatrix}$

$= \begin{bmatrix} (0+0-4) & (1+0-8) & (-1+0+0) \\ (0+1+8) & (1+0+16) & (-1+1+0) \end{bmatrix}$

$= \begin{bmatrix} -4 & -7 & -1 \\ 9 & 17 & 0 \end{bmatrix}$

15. $\begin{bmatrix} 1 & 0 \\ 1 & -1 \end{bmatrix}\begin{bmatrix} 0 & 1 \\ 0 & 1 \end{bmatrix}$

$= \begin{bmatrix} (0+0) & (1+0) \\ (0+0) & (1-1) \end{bmatrix} = \begin{bmatrix} 0 & 1 \\ 0 & 0 \end{bmatrix}$

17. $\begin{bmatrix} 0 & 1 \\ 0 & 1 \end{bmatrix}\begin{bmatrix} 1 & 0 \\ 1 & -1 \end{bmatrix}$

$= \begin{bmatrix} (0+1) & (0-1) \\ (0+1) & (0-1) \end{bmatrix} = \begin{bmatrix} 1 & -1 \\ 1 & -1 \end{bmatrix}$

19. $\begin{bmatrix} 1 & -1 \\ 1 & -1 \end{bmatrix}\begin{bmatrix} 2 & 3 \\ 2 & 3 \end{bmatrix}$

$= \begin{bmatrix} (2-2) & (3-3) \\ (2-2) & (3-3) \end{bmatrix} = \begin{bmatrix} 0 & 0 \\ 0 & 0 \end{bmatrix}$

21. $\begin{bmatrix} 1 & -1 \\ -1 & 1 \end{bmatrix}\begin{bmatrix} 2 & 3 \\ 2 & 3 \\ 1 & 1 \end{bmatrix}$ is undefined.

(The dimensions are 2×2 and 3×2; the 2 and the 3 do not match.)

23. $\begin{bmatrix} 1 & 0 & -1 \\ 2 & -2 & 1 \\ 0 & 0 & 1 \end{bmatrix}\begin{bmatrix} 1 & -1 & 4 \\ 1 & 1 & 0 \\ 0 & 4 & 1 \end{bmatrix}$

$= \begin{bmatrix} (1+0+0) & (-1+0-4) & (4+0-1) \\ (2-2+0) & (-2-2+4) & (8+0+1) \\ (0+0+0) & (0+0+4) & (0+0+1) \end{bmatrix}$

$= \begin{bmatrix} 1 & -5 & 3 \\ 0 & 0 & 9 \\ 0 & 4 & 1 \end{bmatrix}$

25. $\begin{bmatrix} 1 & 0 & 1 & 0 \\ -1 & 1 & 0 & 1 \\ -2 & 0 & 1 & 4 \\ 0 & -1 & 0 & 1 \end{bmatrix}\begin{bmatrix} 1 \\ -3 \\ 2 \\ 0 \end{bmatrix} = \begin{bmatrix} (1+0+2+0) \\ (-1-3+0+0) \\ (-2+0+2+0) \\ (0+3+0+0) \end{bmatrix}$

$= \begin{bmatrix} 3 \\ -4 \\ 0 \\ 3 \end{bmatrix}$

27. A matrix algebra utility is available online.

$\begin{bmatrix} 1.1 & 2.3 & 3.4 & -1.2 \\ 3.4 & 4.4 & 2.3 & 1.1 \\ 2.3 & 0 & -2.2 & 1.1 \\ 1.2 & 1.3 & 1.1 & 1.1 \end{bmatrix}\begin{bmatrix} -2.1 & 0 & -3.3 \\ -3.4 & -4.8 & -4.2 \\ 3.4 & 5.6 & 1 \\ 1 & 2.2 & 9.8 \end{bmatrix}$

$= \begin{bmatrix} 0.23 & 5.36 & -21.65 \\ -13.18 & -5.82 & -16.62 \\ -11.21 & -9.9 & 0.99 \\ -2.1 & 2.34 & 2.46 \end{bmatrix}$

TI-83/84 format: [A]*[B]

Online Matrix Algebra Tool format: A*B

29. $A = \begin{bmatrix} 0 & 1 & 1 & 1 \\ 0 & 0 & 1 & 1 \\ 0 & 0 & 0 & 1 \\ 0 & 0 & 0 & 0 \end{bmatrix}$

$A^2 = A \cdot A = \begin{bmatrix} 0 & 1 & 1 & 1 \\ 0 & 0 & 1 & 1 \\ 0 & 0 & 0 & 1 \\ 0 & 0 & 0 & 0 \end{bmatrix}\begin{bmatrix} 0 & 1 & 1 & 1 \\ 0 & 0 & 1 & 1 \\ 0 & 0 & 0 & 1 \\ 0 & 0 & 0 & 0 \end{bmatrix}$

$= \begin{bmatrix} (0+0+0+0) & (0+0+0+0) & (0+1+0+0) & (0+1+1+0) \\ (0+0+0+0) & (0+0+0+0) & (0+0+0+0) & (0+0+1+0) \\ (0+0+0+0) & (0+0+0+0) & (0+0+0+0) & (0+0+0+0) \\ (0+0+0+0) & (0+0+0+0) & (0+0+0+0) & (0+0+0+0) \\ (0+0+0+0) & (0+0+0+0) & (0+0+0+0) & (0+0+0+0) \end{bmatrix}$

$= \begin{bmatrix} 0 & 0 & 1 & 2 \\ 0 & 0 & 0 & 1 \\ 0 & 0 & 0 & 0 \\ 0 & 0 & 0 & 0 \end{bmatrix}$

$A^3 = A \cdot A^2 = \begin{bmatrix} 0 & 1 & 1 & 1 \\ 0 & 0 & 1 & 1 \\ 0 & 0 & 0 & 1 \\ 0 & 0 & 0 & 0 \end{bmatrix}\begin{bmatrix} 0 & 0 & 1 & 2 \\ 0 & 0 & 0 & 1 \\ 0 & 0 & 0 & 0 \\ 0 & 0 & 0 & 0 \end{bmatrix}$

$= \begin{bmatrix} (0+0+0+0) & (0+0+0+0) & (0+0+0+0) & (0+1+0+0) \\ (0+0+0+0) & (0+0+0+0) & (0+0+0+0) & (0+0+0+0) \\ (0+0+0+0) & (0+0+0+0) & (0+0+0+0) & (0+0+0+0) \\ (0+0+0+0) & (0+0+0+0) & (0+0+0+0) & (0+0+0+0) \\ (0+0+0+0) & (0+0+0+0) & (0+0+0+0) & (0+0+0+0) \end{bmatrix}$

$= \begin{bmatrix} 0 & 0 & 0 & 1 \\ 0 & 0 & 0 & 0 \\ 0 & 0 & 0 & 0 \\ 0 & 0 & 0 & 0 \end{bmatrix}$

$A^4 = A \cdot A^3 = \begin{bmatrix} 0 & 1 & 1 & 1 \\ 0 & 0 & 1 & 1 \\ 0 & 0 & 0 & 1 \\ 0 & 0 & 0 & 0 \end{bmatrix}\begin{bmatrix} 0 & 0 & 0 & 1 \\ 0 & 0 & 0 & 0 \\ 0 & 0 & 0 & 0 \\ 0 & 0 & 0 & 0 \end{bmatrix}$

$= \begin{bmatrix} 0 & 0 & 0 & 0 \\ 0 & 0 & 0 & 0 \\ 0 & 0 & 0 & 0 \\ 0 & 0 & 0 & 0 \end{bmatrix}.$

Continuing to multiply by A continues to yield zero, so $A^{100} = 0$.

31. $AB = \begin{bmatrix} 0 & -1 & 0 & 1 \\ 10 & 0 & 1 & 0 \end{bmatrix}\begin{bmatrix} 0 & -1 \\ 1 & 1 \\ -1 & 3 \\ 5 & 0 \end{bmatrix}$

$= \begin{bmatrix} (0-1+0+5) & (0-1+0+0) \\ (0+0-1+0) & (-10+0+3+0) \end{bmatrix} = \begin{bmatrix} 4 & -1 \\ -1 & -7 \end{bmatrix}$

TI-83/84 format: [A]*[B]

Online Matrix Algebra Tool format: A*B

33. $A(B-C)$

$= \begin{bmatrix} 0 & -1 & 0 & 1 \\ 10 & 0 & 1 & 0 \end{bmatrix}\left(\begin{bmatrix} 0 & -1 \\ 1 & 1 \\ -1 & 3 \\ 5 & 0 \end{bmatrix} - \begin{bmatrix} 1 & -1 \\ 1 & 1 \\ 1 & 1 \\ 1 & 1 \end{bmatrix}\right)$

$= \begin{bmatrix} 0 & -1 & 0 & 1 \\ 10 & 0 & 1 & 0 \end{bmatrix}\begin{bmatrix} -1 & 0 \\ 0 & 0 \\ -2 & 2 \\ 4 & -1 \end{bmatrix} = \begin{bmatrix} 4 & -1 \\ -12 & 2 \end{bmatrix}$

TI-83/84 format: [A]*([B]-[C])

Online Matrix Algebra Tool format:

A*(B-C)

35. $AB = \begin{bmatrix} 1 & -1 \\ 0 & 2 \\ 0 & -2 \end{bmatrix}\begin{bmatrix} 3 & 0 & -1 \\ 5 & -1 & 1 \end{bmatrix}$

$= \begin{bmatrix} -2 & 1 & -2 \\ 10 & -2 & 2 \\ -10 & 2 & -2 \end{bmatrix}$

TI-83/84 format: [A]*[B]

Online Matrix Algebra Tool format: A*B

37. $A(B+C)$

$= \begin{bmatrix} 1 & -1 \\ 0 & 2 \\ 0 & -2 \end{bmatrix}\left(\begin{bmatrix} 3 & 0 & -1 \\ 5 & -1 & 1 \end{bmatrix} + \begin{bmatrix} x & 1 & w \\ z & r & 4 \end{bmatrix}\right)$

$= \begin{bmatrix} 1 & -1 \\ 0 & 2 \\ 0 & -2 \end{bmatrix}\begin{bmatrix} 3+x & 1 & -1+w \\ 5+z & -1+r & 5 \end{bmatrix}$

$= \begin{bmatrix} -2+x-z & 2-r & -6+w \\ 10+2z & -2+2r & 10 \\ -10-2z & 2-2r & -10 \end{bmatrix}$

39.

(a) $P^2 = P \cdot P = \begin{bmatrix} 0.2 & 0.8 \\ 0.2 & 0.8 \end{bmatrix}\begin{bmatrix} 0.2 & 0.8 \\ 0.2 & 0.8 \end{bmatrix}$

$= \begin{bmatrix} (0.04+0.16) & (0.16+0.64) \\ (0.04+0.16) & (0.16+0.64) \end{bmatrix}$

$= \begin{bmatrix} 0.2 & 0.8 \\ 0.2 & 0.8 \end{bmatrix}$

(b) $P^4 = P^2 \cdot P^2 = \begin{bmatrix} 0.2 & 0.8 \\ 0.2 & 0.8 \end{bmatrix}\begin{bmatrix} 0.2 & 0.8 \\ 0.2 & 0.8 \end{bmatrix}$

$= \begin{bmatrix} 0.2 & 0.8 \\ 0.2 & 0.8 \end{bmatrix}$ again.

(c) $P^8 = P^4 \cdot P^4 = \begin{bmatrix} 0.2 & 0.8 \\ 0.2 & 0.8 \end{bmatrix}\begin{bmatrix} 0.2 & 0.8 \\ 0.2 & 0.8 \end{bmatrix}$

$= \begin{bmatrix} 0.2 & 0.8 \\ 0.2 & 0.8 \end{bmatrix}$ again.

(d) $P^{1000} = P^8 \cdot P^8 \cdot P^8 \cdot \ldots P^8$ (125 times)

$= \begin{bmatrix} 0.2 & 0.8 \\ 0.2 & 0.8 \end{bmatrix}.$

41.

(a) $P^2 = P \cdot P = \begin{bmatrix} 0.1 & 0.9 \\ 0 & 1 \end{bmatrix}\begin{bmatrix} 0.1 & 0.9 \\ 0 & 1 \end{bmatrix}$

$= \begin{bmatrix} 0.01 & 0.99 \\ 0 & 1 \end{bmatrix}$

(b) $P^4 = P^2 \cdot P^2 = \begin{bmatrix} 0.01 & 0.99 \\ 0 & 1 \end{bmatrix}\begin{bmatrix} 0.01 & 0.99 \\ 0 & 1 \end{bmatrix}$

$= \begin{bmatrix} 0.0001 & 0.9999 \\ 0 & 1 \end{bmatrix}$

(c) $P^8 = P^4 \cdot P^4$

$= \begin{bmatrix} 0.0001 & 0.9999 \\ 0 & 1 \end{bmatrix}\begin{bmatrix} 0.0001 & 0.9999 \\ 0 & 1 \end{bmatrix}$

$\approx \begin{bmatrix} 0 & 1 \\ 0 & 1 \end{bmatrix}$ when we round to 4 decimal places.

(d) $P^{1000} = P^8 \cdot P^8 \cdot P^8 \cdot \ldots P^8$ (125 times)

$\approx \begin{bmatrix} 0 & 1 \\ 0 & 1 \end{bmatrix}\begin{bmatrix} 0 & 1 \\ 0 & 1 \end{bmatrix}\begin{bmatrix} 0 & 1 \\ 0 & 1 \end{bmatrix}\ldots\begin{bmatrix} 0 & 1 \\ 0 & 1 \end{bmatrix}$

$= \begin{bmatrix} 0 & 1 \\ 0 & 1 \end{bmatrix}$

43.

(a) $P^2 = \begin{bmatrix} 0.25 & 0.25 & 0.50 \\ 0.25 & 0.25 & 0.50 \\ 0.25 & 0.25 & 0.50 \end{bmatrix}\begin{bmatrix} 0.25 & 0.25 & 0.50 \\ 0.25 & 0.25 & 0.50 \\ 0.25 & 0.25 & 0.50 \end{bmatrix}$

$= \begin{bmatrix} .0625+.0625+.125 & .0625+.0625+.125 & .125+.125+.25 \\ .0625+.0625+.125 & .0625+.0625+.125 & .125+.125+.25 \\ .0625+.0625+.125 & .0625+.0625+.125 & .125+.125+.25 \end{bmatrix}$

$= \begin{bmatrix} 0.25 & 0.25 & 0.50 \\ 0.25 & 0.25 & 0.50 \\ 0.25 & 0.25 & 0.50 \end{bmatrix}$

(b) $P^4 = P^2 \cdot P^2$

$= \begin{bmatrix} 0.25 & 0.25 & 0.50 \\ 0.25 & 0.25 & 0.50 \\ 0.25 & 0.25 & 0.50 \end{bmatrix}\begin{bmatrix} 0.25 & 0.25 & 0.50 \\ 0.25 & 0.25 & 0.50 \\ 0.25 & 0.25 & 0.50 \end{bmatrix}$

$= \begin{bmatrix} 0.25 & 0.25 & 0.50 \\ 0.25 & 0.25 & 0.50 \\ 0.25 & 0.25 & 0.50 \end{bmatrix}$ again

(c) $P^8 = P^4 \cdot P^4$

$= \begin{bmatrix} 0.25 & 0.25 & 0.50 \\ 0.25 & 0.25 & 0.50 \\ 0.25 & 0.25 & 0.50 \end{bmatrix}\begin{bmatrix} 0.25 & 0.25 & 0.50 \\ 0.25 & 0.25 & 0.50 \\ 0.25 & 0.25 & 0.50 \end{bmatrix}$

$= \begin{bmatrix} 0.25 & 0.25 & 0.50 \\ 0.25 & 0.25 & 0.50 \\ 0.25 & 0.25 & 0.50 \end{bmatrix}$ again.

(d) $P^{1000} = P^8 \cdot P^8 \cdot P^8 \cdot \ldots P^8$ (125 times)

$= \begin{bmatrix} 0.25 & 0.25 & 0.50 \\ 0.25 & 0.25 & 0.50 \\ 0.25 & 0.25 & 0.50 \end{bmatrix}.$

45. $\begin{bmatrix} 2 & -1 & 4 \\ -4 & \frac{3}{4} & \frac{1}{3} \\ -3 & 0 & 0 \end{bmatrix}\begin{bmatrix} x \\ y \\ z \end{bmatrix} = \begin{bmatrix} 3 \\ -1 \\ 0 \end{bmatrix}$

$\begin{bmatrix} 2x-y+4z \\ -4x+\frac{3}{4}y+\frac{1}{3}z \\ -3x \end{bmatrix} = \begin{bmatrix} 3 \\ -1 \\ 0 \end{bmatrix}$

Equating entries gives the following system of linear equations:

$2x - y + 4z = 3$

$-4x + \frac{3}{4}y + \frac{1}{3}z = -1$

$-3x = 0$

47. $\begin{bmatrix} 1 & -1 & 0 & 1 \\ 1 & 1 & 2 & 4 \end{bmatrix}\begin{bmatrix} x \\ y \\ z \\ w \end{bmatrix} = \begin{bmatrix} -1 \\ 2 \end{bmatrix}$

$$\left[\begin{array}{c} x-y+w \\ x+y+2z+4w \end{array}\right]=\left[\begin{array}{c} -1 \\ 2 \end{array}\right]$$

Equating entries gives the following system of linear equations:

$$x - y + w = -1$$
$$x + y + 2z + 4w = 2$$

49. $x - y = 4$
$2x - y = 0$

The matrix form is $AX = B$, where A is the matrix of coefficients, X is the column matrix of unknowns, and B is the column matrix of right-hand sides.

$$A = \left[\begin{array}{cc} 1 & -1 \\ 2 & -1 \end{array}\right], X = \left[\begin{array}{c} x \\ y \end{array}\right], B = \left[\begin{array}{c} 4 \\ 0 \end{array}\right]$$

Thus, the matrix system is

$$\left[\begin{array}{cc} 1 & -1 \\ 2 & -1 \end{array}\right]\left[\begin{array}{c} x \\ y \end{array}\right] = \left[\begin{array}{c} 4 \\ 0 \end{array}\right]$$

51. $x + y - z = 8$
$2x + y + z = 4$
$\dfrac{3x}{4} + \dfrac{z}{2} = 1$

The matrix form is $AX = B$, where A is the matrix of coefficients, X is the column matrix of unknowns, and B is the column matrix of right-hand sides.

$$A = \left[\begin{array}{ccc} 1 & 1 & -1 \\ 2 & 1 & 1 \\ \frac{3}{4} & 0 & \frac{1}{2} \end{array}\right] X = \left[\begin{array}{c} x \\ y \\ z \end{array}\right], B = \left[\begin{array}{c} 8 \\ 4 \\ 1 \end{array}\right]$$

Thus, the matrix system is

$$\left[\begin{array}{ccc} 1 & 1 & -1 \\ 2 & 1 & 1 \\ \frac{3}{4} & 0 & \frac{1}{2} \end{array}\right]\left[\begin{array}{c} x \\ y \\ z \end{array}\right] = \left[\begin{array}{c} 8 \\ 4 \\ 1 \end{array}\right]$$

53. We have three prices and three quantities. Arrange the prices as a row matrix and quantities as a column matrix (so that we can multiply them):

$$\text{Price} = [15\ 10\ 12] \quad \text{Quantity} = \left[\begin{array}{c} 50 \\ 40 \\ 30 \end{array}\right]$$

$$\text{Revenue} = \text{Price} \times \text{Quantity} = [15\ 10\ 12]\left[\begin{array}{c} 50 \\ 40 \\ 30 \end{array}\right]$$

$$= [750+400+360] = [1510]$$

55. We are asked to write the prices as a column matrix:

$$\text{Price:} \begin{array}{c} \text{Hard} \\ \text{Soft} \\ \text{Plastic} \end{array}\left[\begin{array}{c} 30 \\ 10 \\ 15 \end{array}\right]$$

We are asked to find the revenue, Since the prices are given as a column matrix, we expect it to go on the right when we multiply rows by columns. So we write the formula for revenue with prices on the right:

Revenue = Quantity × Price Note the reversal of the usual order

This means that the quantities (hard, soft, plastic) should appear as rows in the quantity matrix:

$$\text{Quantity:} \left[\begin{array}{ccc} 700 & 1300 & 2000 \\ 400 & 300 & 500 \end{array}\right], \text{Price} = \left[\begin{array}{c} 30 \\ 10 \\ 15 \end{array}\right]$$

Revenue = Quantity × Price

$$= \left[\begin{array}{ccc} 700 & 1300 & 2000 \\ 400 & 300 & 500 \end{array}\right]\left[\begin{array}{c} 30 \\ 10 \\ 15 \end{array}\right] = \left[\begin{array}{c} \$64,000 \\ \$22,500 \end{array}\right]$$

57. We are given the number of books per editor and the number of editors each year. The total number of books is

Number of books = Number of books per editor × Number of editors

Thus, we should write the number of books per editor as a row matrix and the number of editors as a column matrix:

Number of books per editor = [3 3.5 5 5.2]

$$\text{Number of editors} = \begin{bmatrix} 16,000 \\ 15,000 \\ 12,500 \\ 13,000 \end{bmatrix}$$

Number of books = Number of books per editor × Number of editors

$$= [3 \quad 3.5 \quad 5 \quad 5.2] \begin{bmatrix} 16,000 \\ 15,000 \\ 12,500 \\ 13,000 \end{bmatrix}$$

$$= [230,600],$$

or 230,600 new books.

59. We are given the mean income per person for each age group as well as the number people.

Total income = Income per person × Number of females in 2004

Thus, we could write the income per person as a row matrix and the number of females in 2004 as a column matrix:

Income per person = [12.5 31.7 40.4 43.0 38.0 25.2] thousand dollars

$$\text{Number of females in 2004} = \begin{bmatrix} 20 \\ 19 \\ 22 \\ 21 \\ 16 \\ 42 \end{bmatrix} \text{ million}$$

Total income = Income per person × Number of females in 2004

$$= [12.5 \quad 31.7 \quad 40.4 \quad 43.0 \quad 38.0 \quad 25.2] \begin{bmatrix} 20 \\ 19 \\ 22 \\ 21 \\ 16 \\ 42 \end{bmatrix}$$

$$= \$4310.5 \text{ thousand million,}$$

Since a thousand million is a billion, the total income, rounded to two significant digits, is $4300 billion (or $4.3 trillion).

61. Referring to Exercise 59, Take N to be the income per person

$$N = [12.5 \quad 31.7 \quad 40.4 \quad 43.0 \quad 38.0 \quad 25.2],$$

take F and M to be, respectively, the female and male populations in 2005:

$$F = \begin{bmatrix} 20 \\ 19 \\ 22 \\ 21 \\ 15 \\ 42 \end{bmatrix} \quad M = \begin{bmatrix} 21 \\ 18 \\ 21 \\ 20 \\ 14 \\ 42 \end{bmatrix}$$

Then the female income is NF and the male income is NM. The difference is therefore

$$D = NF - NM$$

$$= N(F - M),$$

which is the single formula required. Computing,

$$N(F - M)$$

$$= [12.5 \quad 31.7 \quad 40.4 \quad 43.0 \quad 38.0 \quad 25.2]$$

$$\times \left(\begin{bmatrix} 20 \\ 19 \\ 22 \\ 21 \\ 15 \\ 42 \end{bmatrix} - \begin{bmatrix} 21 \\ 18 \\ 21 \\ 20 \\ 14 \\ 42 \end{bmatrix} \right)$$

$$= [12.5 \quad 31.7 \quad 40.4 \quad 43.0 \quad 38.0 \quad 25.2] \begin{bmatrix} -1 \\ 1 \\ 1 \\ 1 \\ 1 \\ 0 \end{bmatrix}$$

$$= \$140.6 \text{ thousand million (or billion)}$$

$$\approx \$140 \text{ billion}$$

63. $P = \begin{bmatrix} 2.7 & 3.0 \\ 3.9 & 4.0 \end{bmatrix}$

$$[-1 \quad 1] P = [-1 \quad 1] \begin{bmatrix} 2.7 & 3.0 \\ 3.9 & 4.0 \end{bmatrix}$$

$$= [(-2.7+3.9) \quad (-3.0+4.0)] = [1.2 \quad 1.0]$$

To interpret this, look at what we are doing when we multiply the matrices: Multiplying [−1, 1] by the first column of P computes the north central states total for 1999 – the western states total for

1999, giving the amount, in billions of pounds, by which cheese production in north central states exceeded that in western states in 1999. Multiplying by the second columns gives a similar calculation for 2000. Thus, $[-1 \quad 1]\,P$ represents the amount, in billions of pounds, by which cheese production in north central states exceeded that in western states.

65. Take A to be the matrix given by the table:

$$A = \text{Total number of filings} = \begin{bmatrix} 150 & 150 & 150 \\ 300 & 300 & 250 \\ 250 & 250 & 200 \end{bmatrix}$$

The percentages for the three New York/New Jersey regions handled by the firm can be represented by a row matrix:

Percentage handled by firm = $[0.10\ \ 0.05\ \ 0.20]$

(We used a row so that the three percentages match the three regions down each column of A.)

Number of bankruptcy filings handled by firm

= Percentage handled by firm × Total number

$$= [0.10\ \ 0.05\ \ 0.20]\begin{bmatrix} 150 & 150 & 150 \\ 300 & 300 & 250 \\ 250 & 250 & 200 \end{bmatrix}$$

$$= [80\ \ 80\ \ 67.5]$$

67. $B = [1\ \ 1\ \ 0]\,,\ A = \begin{bmatrix} 150 & 150 & 150 \\ 300 & 300 & 250 \\ 250 & 250 & 200 \end{bmatrix}$

When we multiply B by a column of A, we are adding the filings in Manhattan and Brooklyn for the corresponding month. Therefore, BA gives number of filings in Manhattan and Brooklyn combined in each of the months shown.

69. To compute the amount by which the combined filings in Manhattan and Newark exceeded the filings in Brooklyn each year, we add the Manhattan and Newark entries and subtract the Brooklyn entry in each column. This

may be accomplished by multiplying each column by the row matrix $[1\ \ -1\ \ 1]$:

$$[1\ \ -1\ \ 1]\begin{bmatrix} 150 & 150 & 150 \\ 300 & 300 & 250 \\ 250 & 250 & 200 \end{bmatrix}$$

$$= [100\ \ 100\ \ 100]$$

This product gives the result for each of the three months. To add them up, we can multiply on the right by a column matrix whose entries are all 1:

$$[100\ \ 100\ \ 100]\begin{bmatrix} 1 \\ 1 \\ 1 \end{bmatrix} = [300]$$

Therefore, the matrix product we used was

$$[1\ \ -1\ \ 1]\begin{bmatrix} 150 & 150 & 150 \\ 300 & 300 & 250 \\ 250 & 250 & 200 \end{bmatrix}\begin{bmatrix} 1 \\ 1 \\ 1 \end{bmatrix} = [300]$$

71. We are asked to organize the parts required data in a matrix, and the prices per part from each supplier in another,

Quantity of parts: $\begin{array}{c} \\ \text{Pom II} \\ \text{Pom Classic} \end{array} \begin{array}{ccc} \text{Proc} & \text{Mem} & \text{Tubes} \\ \begin{bmatrix} 2 & 16 & 20 \\ 1 & 4 & 40 \end{bmatrix} \end{array}$

$= Q$

We multiply this by the prices per part matrix to obtain the total costs. Thus, we should write the price per parts matrix with the prices for each component as columns (to match the rows above):

Prices: $\begin{array}{c} \\ \text{Proc} \\ \text{Mem} \\ \text{Tubes} \end{array}\begin{array}{cc} \text{Motorel} & \text{Intola} \\ \begin{bmatrix} 100 & 150 \\ 50 & 40 \\ 10 & 15 \end{bmatrix} \end{array} = P$

Total cost = Quantity × Price = QP

$$= \begin{bmatrix} 2 & 16 & 20 \\ 1 & 4 & 40 \end{bmatrix}\begin{bmatrix} 100 & 150 \\ 50 & 40 \\ 10 & 15 \end{bmatrix}$$

$$= \begin{bmatrix} 1200 & 1240 \\ 700 & 910 \end{bmatrix}$$

How are these costs organized? The clue is that the rows of the product QP correspond to the rows of the left-hand matrix, Q. Here Q has rows corresponding to Pom II and Classic. On the other hand, the *columns* of the product QP correspond to the columns of P, which are Motorel and Intola. Thus, the prices are organized as follows:

$$\begin{array}{c} \\ \text{Pom II} \\ \text{Pom Classic} \end{array} \begin{array}{cc} \text{Motorel} & \text{Intola} \\ \left[\begin{array}{cc} \$1200 & \$1240 \\ \$700 & \$910 \end{array} \right] \end{array}$$

Solution:

Motorel parts on the Pom II: $1200
Intola parts on the Pom II: $1240
Motorel parts on the Pom Classic: $700
Intola parts on the Pom Classic: $910

73. $A = \begin{bmatrix} 440 & 190 \\ 950 & 950 \\ 1790 & 200 \end{bmatrix}$

$AB = \begin{bmatrix} 440 & 190 \\ 950 & 950 \\ 1790 & 200 \end{bmatrix} \begin{bmatrix} 0.05 \\ 0.04 \end{bmatrix} = \begin{bmatrix} 29.6 \\ 85.5 \\ 97.5 \end{bmatrix}$

In computing this product, we are adding 5% of the number of people visiting Australia and 4% of the number visiting South Africa for each of the three region. These are exactly the percentages that decide to settle there. Thus, the entries of AB give the number of people from each of the three regions who settle in Australia or South Africa.

$AC = \begin{bmatrix} 440 & 190 \\ 950 & 950 \\ 1790 & 200 \end{bmatrix} \begin{bmatrix} 0.05 & 0 \\ 0 & 0.04 \end{bmatrix}$

$= \begin{bmatrix} 22 & 7.6 \\ 47.5 & 38 \\ 89.5 & 8 \end{bmatrix}$

Each row of the product is computed by taking 5% of all the tourists to Australia and 4% of those to South Africa separately, Thus, the entries of AC give the number of people from each of the three regions who settle in Australia, and the number that settle in South Africa.

75. $P = \begin{bmatrix} 0.9879 & 0.0020 & 0.0081 & 0.0019 \\ 0.0014 & 0.9895 & 0.0063 & 0.0028 \\ 0.0027 & 0.0025 & 0.9927 & 0.0022 \\ 0.0010 & 0.0030 & 0.0050 & 0.9909 \end{bmatrix}$

2003 Distribution = A = [53.3 64.0 101.6 65.4]
To compute the population distribution in 2004, compute $AP \approx$ [53.1 63.9 102.0 65.3]

77. Answers will vary.
One example: $A = [1 \ 2] \ B = \begin{bmatrix} 1 & 2 & 3 \\ 4 & 5 & 6 \end{bmatrix}$
Another example: $A = [1] \quad B = [1 \ 2]$

79. We find that the addition and multiplication of 1×1 matrices is identical to the addition and multiplication of numbers.

81. The claim is correct. Every matrix equation represents the equality of two matrices. When two matrices are equal, each of their corresponding entries must be equal. Equating the corresponding entries gives a system of equations.

83. Here is a possible scenario:
Costs of items A, B and C in 1995 = [10 20 30]
Percentage increases in these costs in 1996
   = [0.5 0.1 0.20]
Actual increases in costs
   = [10×0.5 20×0.1 30×0.20]

85. It produces a matrix whose ij entry is the product of the ij entries of the two matrices.

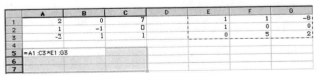

3.3

1. $A = \begin{bmatrix} 0 & 1 \\ 1 & 0 \end{bmatrix}, B = \begin{bmatrix} 0 & 1 \\ 1 & 0 \end{bmatrix}$

$AB = \begin{bmatrix} 0 & 1 \\ 1 & 0 \end{bmatrix}\begin{bmatrix} 0 & 1 \\ 1 & 0 \end{bmatrix} = \begin{bmatrix} 1 & 0 \\ 0 & 1 \end{bmatrix} = I$

Since A and B are square matrices with $AB = I$, A and B are inverses. (We do not have to check that $BA = I$ as well—see the note in the text after the definition of inverse.)

3. $A = \begin{bmatrix} 2 & 1 & 1 \\ 0 & 1 & 1 \\ 0 & 0 & 1 \end{bmatrix}, B = \begin{bmatrix} \frac{1}{2} & -\frac{1}{2} & 0 \\ 0 & 1 & -1 \\ 0 & 0 & 1 \end{bmatrix}$

$AB = \begin{bmatrix} 2 & 1 & 1 \\ 0 & 1 & 1 \\ 0 & 0 & 1 \end{bmatrix}\begin{bmatrix} \frac{1}{2} & -\frac{1}{2} & 0 \\ 0 & 1 & -1 \\ 0 & 0 & 1 \end{bmatrix}$

$= \begin{bmatrix} 1 & 0 & 0 \\ 0 & 1 & 0 \\ 0 & 0 & 1 \end{bmatrix} = I$

Since A and B are square matrices with $AB = I$, A and B are inverses. (We do not have to check that $BA = I$ as well—see the note in the text after the definition of inverse.)

5. $A = \begin{bmatrix} a & 0 & 0 \\ 0 & b & 0 \\ 0 & 0 & 0 \end{bmatrix}, B = \begin{bmatrix} a^{-1} & 0 & 0 \\ 0 & b^{-1} & 0 \\ 0 & 0 & 0 \end{bmatrix} (a, b \neq 0)$

$AB = \begin{bmatrix} a & 0 & 0 \\ 0 & b & 0 \\ 0 & 0 & 0 \end{bmatrix}\begin{bmatrix} a^{-1} & 0 & 0 \\ 0 & b^{-1} & 0 \\ 0 & 0 & 0 \end{bmatrix}$

$= \begin{bmatrix} 1 & 0 & 0 \\ 0 & 1 & 0 \\ 0 & 0 & 0 \end{bmatrix} \neq I$

Since $AB \neq I$, A and B are not inverses.

7. To find the inverse of $\begin{bmatrix} 1 & 1 \\ 2 & 1 \end{bmatrix}$, we augment with the 2×2 identity matrix and row-reduce:

$\begin{bmatrix} 1 & 1 & 1 & 0 \\ 2 & 1 & 0 & 1 \end{bmatrix}_{R_2 - 2R_1}$

$\begin{bmatrix} 1 & 1 & 1 & 0 \\ 0 & -1 & -2 & 1 \end{bmatrix}_{R_1 + R_2}$

$\begin{bmatrix} 1 & 0 & -1 & 1 \\ 0 & -1 & -2 & 1 \end{bmatrix}_{-R_2}$

$\begin{bmatrix} 1 & 0 & -1 & 1 \\ 0 & 1 & 2 & -1 \end{bmatrix}$

The right-hand 2×2 block is the desired inverse:

$\begin{bmatrix} 1 & 1 \\ 2 & 1 \end{bmatrix}^{-1} = \begin{bmatrix} -1 & 1 \\ 2 & -1 \end{bmatrix}$

9. To find the inverse of $\begin{bmatrix} 0 & 1 \\ 1 & 0 \end{bmatrix}$, we augment with the 2×2 identity matrix and row-reduce:

$\begin{bmatrix} 0 & 1 & 1 & 0 \\ 1 & 0 & 0 & 1 \end{bmatrix}_{R_1 \leftrightarrow R_2} \rightarrow \begin{bmatrix} 1 & 0 & 0 & 1 \\ 0 & 1 & 1 & 0 \end{bmatrix}$

The right-hand 2×2 block is the desired inverse:

$\begin{bmatrix} 0 & 1 \\ 1 & 0 \end{bmatrix}^{-1} = \begin{bmatrix} 0 & 1 \\ 1 & 0 \end{bmatrix}$

11. $\begin{bmatrix} 2 & 1 & 1 & 0 \\ 1 & 1 & 0 & 1 \end{bmatrix}_{2R_2 - R_1}$

$\begin{bmatrix} 2 & 1 & 1 & 0 \\ 0 & 1 & -1 & 2 \end{bmatrix}_{R_1 - R_2}$

$\begin{bmatrix} 2 & 0 & 2 & -2 \\ 0 & 1 & -1 & 2 \end{bmatrix}_{(1/2)R_1}$

$\begin{bmatrix} 1 & 0 & 1 & -1 \\ 0 & 1 & -1 & 2 \end{bmatrix}$

The right-hand 2×2 block is the desired inverse:

$\begin{bmatrix} 2 & 1 \\ 1 & 1 \end{bmatrix}^{-1} = \begin{bmatrix} 1 & -1 \\ -1 & 2 \end{bmatrix}$

13.

$\begin{bmatrix} 2 & 1 & 1 & 0 \\ 4 & 2 & 0 & 1 \end{bmatrix}_{R_2 - 2R_1} \rightarrow \begin{bmatrix} 2 & 1 & 1 & 0 \\ 0 & 0 & -2 & 1 \end{bmatrix}$

Since the left-hand 2×2 block has a row of zeros, we cannot reduce the matrix to obtain the 2×2 identity on the left. Therefore, the matrix is singular.

15. To find the inverse of a 3×3 matrix, we augment it with the 3×3 identity matrix and row-reduce:

$$\begin{bmatrix} 1 & 1 & 1 & 1 & 0 & 0 \\ 0 & 1 & 1 & 0 & 1 & 0 \\ 0 & 0 & 1 & 0 & 0 & 1 \end{bmatrix} R_1 - R_2$$

$$\begin{bmatrix} 1 & 0 & 0 & 1 & -1 & 0 \\ 0 & 1 & 1 & 0 & 1 & 0 \\ 0 & 0 & 1 & 0 & 0 & 1 \end{bmatrix} R_2 - R_3$$

$$\begin{bmatrix} 1 & 0 & 0 & 1 & -1 & 0 \\ 0 & 1 & 0 & 0 & 1 & -1 \\ 0 & 0 & 1 & 0 & 0 & 1 \end{bmatrix}$$

The right-hand 3×3 block is the desired inverse:

$$\begin{bmatrix} 1 & 1 & 1 \\ 0 & 1 & 1 \\ 0 & 0 & 1 \end{bmatrix}^{-1} = \begin{bmatrix} 1 & -1 & 0 \\ 0 & 1 & -1 \\ 0 & 0 & 1 \end{bmatrix}$$

17.

$$\begin{bmatrix} 1 & 1 & 1 & 1 & 0 & 0 \\ 1 & 0 & 2 & 0 & 1 & 0 \\ 1 & -1 & 1 & 0 & 0 & 1 \end{bmatrix} \begin{matrix} \\ R_2 - R_1 \\ R_3 - R_1 \end{matrix}$$

$$\begin{bmatrix} 1 & 1 & 1 & 1 & 0 & 0 \\ 0 & -1 & 1 & -1 & 1 & 0 \\ 0 & -2 & 0 & -1 & 0 & 1 \end{bmatrix} \begin{matrix} R_1 + R_2 \\ \\ R_3 - 2R_2 \end{matrix}$$

$$\begin{bmatrix} 1 & 0 & 2 & 0 & 1 & 0 \\ 0 & -1 & 1 & -1 & 1 & 0 \\ 0 & 0 & -2 & 1 & -2 & 1 \end{bmatrix} \begin{matrix} R_1 + R_3 \\ 2R_2 + R_3 \\ \end{matrix}$$

$$\begin{bmatrix} 1 & 0 & 0 & 1 & -1 & 1 \\ 0 & -2 & 0 & -1 & 0 & 1 \\ 0 & 0 & -2 & 1 & -2 & 1 \end{bmatrix} \begin{matrix} \\ -(1/2)R_2 \\ -(1/2)R_3 \end{matrix}$$

$$\begin{bmatrix} 1 & 0 & 0 & 1 & -1 & 1 \\ 0 & 1 & 0 & 1/2 & 0 & -1/2 \\ 0 & 0 & 1 & -1/2 & 1 & -1/2 \end{bmatrix}$$

The right-hand 3×3 block is the desired inverse:

$$\begin{bmatrix} 1 & 1 & 1 \\ 1 & 0 & 2 \\ 1 & -1 & 1 \end{bmatrix}^{-1} = \begin{bmatrix} 1 & -1 & 1 \\ 1/2 & 0 & -1/2 \\ -1/2 & 1 & -1/2 \end{bmatrix}$$

19.

$$\begin{bmatrix} 1 & 1 & 1 & 1 & 0 & 0 \\ 1 & -1 & 0 & 0 & 1 & 0 \\ 1 & 2 & 3 & 0 & 0 & 1 \end{bmatrix} \begin{matrix} \\ R_2 - R_1 \\ R_3 - R_1 \end{matrix}$$

$$\begin{bmatrix} 1 & 1 & 1 & 1 & 0 & 0 \\ 0 & -2 & -1 & -1 & 1 & 0 \\ 0 & 1 & 2 & -1 & 0 & 1 \end{bmatrix} \begin{matrix} 2R_1 + R_2 \\ \\ 2R_3 + R_2 \end{matrix}$$

$$\begin{bmatrix} 2 & 0 & 1 & 1 & 1 & 0 \\ 0 & -2 & -1 & -1 & 1 & 0 \\ 0 & 0 & 3 & -3 & 1 & 2 \end{bmatrix} \begin{matrix} 3R_1 - R_3 \\ 3R_2 + R_3 \\ \end{matrix}$$

$$\begin{bmatrix} 6 & 0 & 0 & 6 & 2 & -2 \\ 0 & -6 & 0 & -6 & 4 & 2 \\ 0 & 0 & 3 & -3 & 1 & 2 \end{bmatrix} \begin{matrix} (1/6)R_1 \\ -(1/6)R_2 \\ (1/3)R_3 \end{matrix}$$

$$\begin{bmatrix} 1 & 0 & 0 & 1 & 1/3 & -1/3 \\ 0 & 1 & 0 & 1 & -2/3 & -1/3 \\ 0 & 0 & 1 & -1 & 1/3 & 2/3 \end{bmatrix}$$

The right-hand 3×3 block is the desired inverse:

$$\begin{bmatrix} 1 & 1 & 1 \\ 1 & -1 & 0 \\ 1 & 2 & 3 \end{bmatrix}^{-1} = \begin{bmatrix} 1 & 1/3 & -1/3 \\ 1 & -2/3 & -1/3 \\ -1 & 1/3 & 2/3 \end{bmatrix}$$

21.

$$\begin{bmatrix} 1 & 1 & 1 & 1 & 0 & 0 \\ 1 & 0 & 1 & 0 & 1 & 0 \\ 1 & -1 & 1 & 0 & 0 & 1 \end{bmatrix} \begin{matrix} \\ R_2 - R_1 \\ R_3 - R_1 \end{matrix}$$

$$\begin{bmatrix} 1 & 1 & 1 & 1 & 0 & 0 \\ 0 & -1 & 0 & -1 & 1 & 0 \\ 0 & -2 & 0 & -1 & 0 & 1 \end{bmatrix} \begin{matrix} R_1 + R_2 \\ \\ R_3 - 2R_2 \end{matrix}$$

$$\begin{bmatrix} 1 & 0 & 1 & 0 & 1 & 0 \\ 0 & -1 & 0 & -1 & 1 & 0 \\ 0 & 0 & 0 & 1 & -2 & 1 \end{bmatrix}$$

Since the left-hand 3×3 block has a row of zeros, we cannot reduce the matrix to obtain the 3×3 identity on the left. Therefore, the matrix is singular.

113

23. To find the inverse of a 4×4 matrix, we augment it with the 4×4 identity matrix and row-reduce:

$$\begin{bmatrix} 1 & 0 & 1 & 0 & 1 & 0 & 0 & 0 \\ -1 & 1 & 0 & 1 & 0 & 1 & 0 & 0 \\ -1 & 0 & 0 & 1 & 0 & 0 & 1 & 0 \\ 0 & -1 & 0 & 1 & 0 & 0 & 0 & 1 \end{bmatrix} \begin{matrix} \\ R_2 + R_1 \\ R_3 + R_1 \\ \\ \end{matrix}$$

$$\begin{bmatrix} 1 & 0 & 1 & 0 & 1 & 0 & 0 & 0 \\ 0 & 1 & 1 & 1 & 1 & 1 & 0 & 0 \\ 0 & 0 & 1 & 1 & 1 & 0 & 1 & 0 \\ 0 & -1 & 0 & 1 & 0 & 0 & 0 & 1 \end{bmatrix} \begin{matrix} \\ \\ \\ R_4 + R_2 \end{matrix}$$

$$\begin{bmatrix} 1 & 0 & 1 & 0 & 1 & 0 & 0 & 0 \\ 0 & 1 & 1 & 1 & 1 & 1 & 0 & 0 \\ 0 & 0 & 1 & 1 & 1 & 0 & 1 & 0 \\ 0 & 0 & 1 & 2 & 1 & 1 & 0 & 1 \end{bmatrix} \begin{matrix} R_1 - R_3 \\ R_2 - R_3 \\ \\ R_4 - R_3 \end{matrix}$$

$$\begin{bmatrix} 1 & 0 & 0 & -1 & 0 & 0 & -1 & 0 \\ 0 & 1 & 0 & 0 & 0 & 1 & -1 & 0 \\ 0 & 0 & 1 & 1 & 1 & 0 & 1 & 0 \\ 0 & 0 & 0 & 1 & 0 & 1 & -1 & 1 \end{bmatrix} \begin{matrix} R_1 + R_4 \\ \\ R_3 - R_4 \\ \end{matrix}$$

$$\begin{bmatrix} 1 & 0 & 0 & 0 & 0 & 1 & -2 & 1 \\ 0 & 1 & 0 & 0 & 0 & 1 & -1 & 0 \\ 0 & 0 & 1 & 0 & 1 & -1 & 2 & -1 \\ 0 & 0 & 0 & 1 & 0 & 1 & -1 & 1 \end{bmatrix}$$

The right-hand 4×4 block is the desired inverse:

$$\begin{bmatrix} 1 & 0 & 1 & 0 \\ -1 & 1 & 0 & 1 \\ -1 & 0 & 0 & 1 \\ 0 & -1 & 0 & 1 \end{bmatrix}^{-1} = \begin{bmatrix} 0 & 1 & -2 & 1 \\ 0 & 1 & -1 & 0 \\ 1 & -1 & 2 & -1 \\ 0 & 1 & -1 & 1 \end{bmatrix}$$

25.

$$\begin{bmatrix} 1 & 2 & 3 & 4 & 1 & 0 & 0 & 0 \\ 0 & 1 & 2 & 3 & 0 & 1 & 0 & 0 \\ 0 & 0 & 1 & 2 & 0 & 0 & 1 & 0 \\ 0 & 0 & 0 & 1 & 0 & 0 & 0 & 1 \end{bmatrix} \begin{matrix} R_1 - 2R_2 \\ \\ \\ \end{matrix}$$

$$\begin{bmatrix} 1 & 0 & -1 & -2 & 1 & -2 & 0 & 0 \\ 0 & 1 & 2 & 3 & 0 & 1 & 0 & 0 \\ 0 & 0 & 1 & 2 & 0 & 0 & 1 & 0 \\ 0 & 0 & 0 & 1 & 0 & 0 & 0 & 1 \end{bmatrix} \begin{matrix} R_1 + R_3 \\ R_2 - 2R_3 \\ \\ \end{matrix}$$

$$\begin{bmatrix} 1 & 0 & 0 & 0 & 1 & -2 & 1 & 0 \\ 0 & 1 & 0 & -1 & 0 & 1 & -2 & 0 \\ 0 & 0 & 1 & 2 & 0 & 0 & 1 & 0 \\ 0 & 0 & 0 & 1 & 0 & 0 & 0 & 1 \end{bmatrix} \begin{matrix} \\ R_2 + R_4 \\ R_3 - 2R_4 \\ \end{matrix}$$

$$\begin{bmatrix} 1 & 0 & 0 & 0 & 1 & -2 & 1 & 0 \\ 0 & 1 & 0 & 0 & 0 & 1 & -2 & 1 \\ 0 & 0 & 1 & 0 & 0 & 0 & 1 & -2 \\ 0 & 0 & 0 & 1 & 0 & 0 & 0 & 1 \end{bmatrix}$$

The right-hand 4×4 block is the desired inverse:

$$\begin{bmatrix} 1 & 2 & 3 & 4 \\ 0 & 1 & 2 & 3 \\ 0 & 0 & 1 & 2 \\ 0 & 0 & 0 & 1 \end{bmatrix}^{-1} = \begin{bmatrix} 1 & -2 & 1 & 0 \\ 0 & 1 & -2 & 1 \\ 0 & 0 & 1 & -2 \\ 0 & 0 & 0 & 1 \end{bmatrix}$$

27. $\begin{bmatrix} a & b \\ c & d \end{bmatrix} = \begin{bmatrix} 1 & 1 \\ 1 & -1 \end{bmatrix}$

$\det \begin{bmatrix} 1 & 1 \\ 1 & -1 \end{bmatrix} = ad - bc$

$\quad = (1)(-1) - (1)(1) = -2$

$\begin{bmatrix} 1 & 1 \\ 1 & -1 \end{bmatrix}^{-1} = \dfrac{1}{ad - bc} \begin{bmatrix} d & -b \\ -c & a \end{bmatrix}$

$\quad = \dfrac{1}{-2} \begin{bmatrix} -1 & -1 \\ -1 & 1 \end{bmatrix} = \begin{bmatrix} 1/2 & 1/2 \\ 1/2 & -1/2 \end{bmatrix}$

29. $\begin{bmatrix} a & b \\ c & d \end{bmatrix} = \begin{bmatrix} 1 & 2 \\ 3 & 4 \end{bmatrix}$

$\det \begin{bmatrix} 1 & 2 \\ 3 & 4 \end{bmatrix} = ad - bc$

$\quad = (1)(4) - (2)(3) = -2$

$\begin{bmatrix} 1 & 2 \\ 3 & 4 \end{bmatrix}^{-1} = \dfrac{1}{ad - bc} \begin{bmatrix} d & -b \\ -c & a \end{bmatrix}$

$\quad = \dfrac{1}{-2} \begin{bmatrix} 4 & -2 \\ -3 & 1 \end{bmatrix} = \begin{bmatrix} -2 & 1 \\ 3/2 & -1/2 \end{bmatrix}$

31. $\begin{bmatrix} a & b \\ c & d \end{bmatrix} = \begin{bmatrix} 1/6 & -1/6 \\ 0 & 1/6 \end{bmatrix}$

$\det \begin{bmatrix} 1/6 & -1/6 \\ 0 & 1/6 \end{bmatrix} = ad - bc$

$= (1/6)(1/6) - (-1/6)(0) = 1/36$

$\begin{bmatrix} 1/6 & -1/6 \\ 0 & 1/6 \end{bmatrix}^{-1} = \dfrac{1}{ad-bc} \begin{bmatrix} d & -b \\ -c & a \end{bmatrix}$

$= 36 \begin{bmatrix} 1/6 & 1/6 \\ 0 & 1/6 \end{bmatrix} = \begin{bmatrix} 6 & 6 \\ 0 & 6 \end{bmatrix}$

33. $\begin{bmatrix} a & b \\ c & d \end{bmatrix} = \begin{bmatrix} 1 & 0 \\ 3/4 & 0 \end{bmatrix}$

$\det \begin{bmatrix} 1 & 0 \\ 3/4 & 0 \end{bmatrix} = ad - bc$

$= (1)(0) - (0)(3/4) = 0$

Singular matrix

35. $\begin{bmatrix} 1.1 & 1.2 \\ 1.3 & -1 \end{bmatrix}^{-1} = \begin{bmatrix} 0.38 & 0.45 \\ 0.49 & -0.41 \end{bmatrix}$

TI-83/84 format: $[A]^{-1}$
Online Matrix Algebra Tool format: `A^(-1)`
Excel: `=MINVERSE(A1:B2)`
(Assuming the matrix is in cells A1→B2)

37. $\begin{bmatrix} 3.56 & 1.23 \\ -1.01 & 0 \end{bmatrix}^{-1} = \begin{bmatrix} 0.00 & -0.99 \\ 0.81 & 2.87 \end{bmatrix}$

TI-83/84 format: $[A]^{-1}$
Online Matrix Algebra Tool format: `A^(-1)`
Excel: `=MINVERSE(A1:B2)`
(Assuming the matrix is in cells A1→B2)

39. $\begin{bmatrix} 1.1 & 3.1 & 2.4 \\ 1.7 & 2.4 & 2.3 \\ 0.6 & -0.7 & -0.1 \end{bmatrix}$ is singular

TI-83/84 format: $[A]^{-1}$
Online Matrix Algebra Tool format: `A^(-1)`
Excel: `=MINVERSE(A1:C3)`
(Assuming the matrix is in cells A1→C3)

41. $\begin{bmatrix} 0.01 & 0.32 & 0 & 0.04 \\ -0.01 & 0 & 0 & 0.34 \\ 0 & 0.32 & -0.23 & 0.23 \\ 0 & 0.41 & 0 & 0.01 \end{bmatrix}^{-1}$

$= \begin{bmatrix} 91.35 & -8.65 & 0 & -71.30 \\ -0.07 & -0.07 & 0 & 2.49 \\ 2.60 & 2.60 & -4.35 & 1.37 \\ 2.69 & 2.69 & 0 & -2.10 \end{bmatrix}$

TI-83/84 format: $[A]^{-1}$
Online Matrix Algebra Tool format: `A^(-1)`
Excel: `=MINVERSE(A1:D4)`
(Assuming the matrix is in cells A1→D4)

43. $x + y = 4$
$x - y = 1$

Matrix form:

$\begin{bmatrix} 1 & 1 \\ 1 & -1 \end{bmatrix} \begin{bmatrix} x \\ y \end{bmatrix} = \begin{bmatrix} 4 \\ 1 \end{bmatrix}$

$AX = B$
$X = A^{-1}B$

$A^{-1} = \begin{bmatrix} 1 & 1 \\ 1 & -1 \end{bmatrix}^{-1} = \begin{bmatrix} 1/2 & 1/2 \\ 1/2 & -1/2 \end{bmatrix}$

See Exercise 27
Thus,

$X = A^{-1}B = \begin{bmatrix} 1/2 & 1/2 \\ 1/2 & -1/2 \end{bmatrix} \begin{bmatrix} 4 \\ 1 \end{bmatrix} = \begin{bmatrix} 5/2 \\ 3/2 \end{bmatrix}$

So, $(x, y) = (5/2, 3/2)$

45. $\dfrac{x}{3} + \dfrac{y}{2} = 0$

$\dfrac{x}{2} + y = -1$

Matrix form:

$\begin{bmatrix} 1/3 & 1/2 \\ 1/2 & 1 \end{bmatrix} \begin{bmatrix} x \\ y \end{bmatrix} = \begin{bmatrix} 0 \\ -1 \end{bmatrix}$

$AX = B$
$X = A^{-1}B$

$A^{-1} = \begin{bmatrix} 1/3 & 1/2 \\ 1/2 & 1 \end{bmatrix}^{-1} = \begin{bmatrix} 12 & -6 \\ -6 & 4 \end{bmatrix}$

Thus,

$$X = A^{-1}B = \begin{bmatrix} 12 & -6 \\ -6 & 4 \end{bmatrix} \begin{bmatrix} 0 \\ -1 \end{bmatrix} = \begin{bmatrix} 6 \\ -4 \end{bmatrix}$$

So, $(x, y) = (6, -4)$

47. $-x + 2y - z = 0$

$-x - y + 2z = 0$

$2x \qquad - z = 6$

Matrix form:

$$\begin{bmatrix} -1 & 2 & -1 \\ -1 & -1 & 2 \\ 2 & 0 & -1 \end{bmatrix} \begin{bmatrix} x \\ y \\ z \end{bmatrix} = \begin{bmatrix} 0 \\ 0 \\ 6 \end{bmatrix}$$

$AX = B$

$X = A^{-1}B$

$$A^{-1} = \begin{bmatrix} -1 & 2 & -1 \\ -1 & -1 & 2 \\ 2 & 0 & -1 \end{bmatrix}^{-1} = \begin{bmatrix} 1/3 & 2/3 & 1 \\ 1 & 1 & 1 \\ 2/3 & 4/3 & 1 \end{bmatrix}$$

Thus,

$$X = A^{-1}B = \begin{bmatrix} 1/3 & 2/3 & 1 \\ 1 & 1 & 1 \\ 2/3 & 4/3 & 1 \end{bmatrix} \begin{bmatrix} 0 \\ 0 \\ 6 \end{bmatrix} = \begin{bmatrix} 6 \\ 6 \\ 6 \end{bmatrix}$$

So, $(x, y) = (6, 6, 6)$

49. The three systems of equations have the matrix form $AX = B$ where

$$A = \begin{bmatrix} -1 & -4 & 2 \\ 1 & 2 & -1 \\ 1 & 1 & -1 \end{bmatrix}; A^{-1} = \begin{bmatrix} 1 & 2 & 0 \\ 0 & 1 & -1 \\ 1 & 3 & -2 \end{bmatrix}$$

(a) $B = \begin{bmatrix} 4 \\ 3 \\ 8 \end{bmatrix}$

$$X = A^{-1}B = \begin{bmatrix} 1 & 2 & 0 \\ 0 & 1 & -1 \\ 1 & 3 & -2 \end{bmatrix} \begin{bmatrix} 4 \\ 3 \\ 8 \end{bmatrix} = \begin{bmatrix} 10 \\ -5 \\ -3 \end{bmatrix}$$

$(x, y, z) = (10, -5, -3)$

(b) $B = \begin{bmatrix} 0 \\ 3 \\ 2 \end{bmatrix}$

$$X = A^{-1}B = \begin{bmatrix} 1 & 2 & 0 \\ 0 & 1 & -1 \\ 1 & 3 & -2 \end{bmatrix} \begin{bmatrix} 0 \\ 3 \\ 2 \end{bmatrix} = \begin{bmatrix} 6 \\ 1 \\ 5 \end{bmatrix}$$

$(x, y, z) = (6, 1, 5)$

(c) $B = \begin{bmatrix} 0 \\ 0 \\ 0 \end{bmatrix}$

$$X = A^{-1}B = \begin{bmatrix} 1 & 2 & 0 \\ 0 & 1 & -1 \\ 1 & 3 & -2 \end{bmatrix} \begin{bmatrix} 0 \\ 0 \\ 0 \end{bmatrix} = \begin{bmatrix} 0 \\ 0 \\ 0 \end{bmatrix}$$

$(x, y, z) = (0, 0, 0)$

51. Unknowns:

x = the number of servings of Pork & Beans

y = the number of slices of bread

(a) Arrange the given information in a table with unknowns across the top:

	Pork & Beans (x)	Bread (y)	Desired
Protein	5	4	20
Carbs.	21	12	80

We can now set up an equation for each of the items listed on the left:

Protein $\qquad 5x + 4y = 20$

Carbs: $\qquad 21x + 12y = 80$

We put this system in matrix form:

$$\begin{bmatrix} 5 & 4 \\ 21 & 12 \end{bmatrix} \begin{bmatrix} x \\ y \end{bmatrix} = \begin{bmatrix} 20 \\ 80 \end{bmatrix}$$

$AX = B$

$X = A^{-1}B$

$$A^{-1} = \begin{bmatrix} 5 & 4 \\ 21 & 12 \end{bmatrix}^{-1} = \begin{bmatrix} -1/2 & 1/6 \\ 7/8 & -5/24 \end{bmatrix}$$

Thus,

$$X = A^{-1}B = \begin{bmatrix} -1/2 & 1/6 \\ 7/8 & -5/24 \end{bmatrix} \begin{bmatrix} 20 \\ 80 \end{bmatrix} = \begin{bmatrix} 10/3 \\ 5/6 \end{bmatrix}$$

So, $(x, y) = (10/3, 5/6)$

Solution: Prepare 10/3 servings of beans, and 5/6 slices of bread.

(b) We must solve the following system:

$5x + 4y = A$

$21x + 12y = B$

We put this system in matrix form:

$$\begin{bmatrix} 5 & 4 \\ 21 & 12 \end{bmatrix} \begin{bmatrix} x \\ y \end{bmatrix} = \begin{bmatrix} A \\ B \end{bmatrix}$$

Solving gives

$$\begin{bmatrix} x \\ y \end{bmatrix} = \begin{bmatrix} 5 & 4 \\ 21 & 12 \end{bmatrix}^{-1} \begin{bmatrix} A \\ B \end{bmatrix}$$

$$= \begin{bmatrix} -1/2 & 1/6 \\ 7/8 & -5/24 \end{bmatrix} \begin{bmatrix} A \\ B \end{bmatrix}$$

$$= \begin{bmatrix} -A/2 + B/6 \\ 7A/8 - 5B/24 \end{bmatrix}$$

Solution: Prepare $-A/2 + B/6$ servings of beans and $7A/8 - 5B/24$ slices of bread.

53. Unknowns:

x = the number of batches of vanilla

y = the number of batches of mocha

z = the number of batches of strawberry

Arrange the given information in a table with unknowns across the top:

	Vanilla (x)	Mocha (y)	Strawberry (z)
Eggs	2	1	1
Milk	1	1	2
Cream	2	2	1

(a)

Eggs: $2x + y + z = 350$

Milk: $x + y + 2z = 350$

Cream: $2x + 2y + z = 400$

Matrix form:

$$\begin{bmatrix} 2 & 1 & 1 \\ 1 & 1 & 2 \\ 2 & 2 & 1 \end{bmatrix} \begin{bmatrix} x \\ y \\ z \end{bmatrix} = \begin{bmatrix} 350 \\ 350 \\ 400 \end{bmatrix}$$

$$AX = B \Rightarrow X = A^{-1}B$$

$$X = \begin{bmatrix} 2 & 1 & 1 \\ 1 & 1 & 2 \\ 2 & 2 & 1 \end{bmatrix}^{-1} \begin{bmatrix} 350 \\ 350 \\ 400 \end{bmatrix}$$

$$= \begin{bmatrix} 1 & -1/3 & -1/3 \\ -1 & 0 & 1 \\ 0 & 2/3 & -1/3 \end{bmatrix} \begin{bmatrix} 350 \\ 350 \\ 400 \end{bmatrix} = \begin{bmatrix} 100 \\ 50 \\ 100 \end{bmatrix}$$

Solution: Use 100 batches of vanilla, 50 batches of mocha, and 100 batches of strawberry.

(b) The requirements lead to the following matrix equation:

$$\begin{bmatrix} 2 & 1 & 1 \\ 1 & 1 & 2 \\ 2 & 2 & 1 \end{bmatrix} \begin{bmatrix} x \\ y \\ z \end{bmatrix} = \begin{bmatrix} 400 \\ 500 \\ 400 \end{bmatrix}$$

$$AX = B \Rightarrow X = A^{-1}B$$

$$X = \begin{bmatrix} 2 & 1 & 1 \\ 1 & 1 & 2 \\ 2 & 2 & 1 \end{bmatrix}^{-1} \begin{bmatrix} 400 \\ 500 \\ 400 \end{bmatrix}$$

$$= \begin{bmatrix} 1 & -1/3 & -1/3 \\ -1 & 0 & 1 \\ 0 & 2/3 & -1/3 \end{bmatrix} \begin{bmatrix} 400 \\ 500 \\ 400 \end{bmatrix} = \begin{bmatrix} 100 \\ 0 \\ 200 \end{bmatrix}$$

Solution: Use 100 batches of vanilla, no mocha, and 200 batches of strawberry.

(c) The requirements lead to the following matrix equation:

$$\begin{bmatrix} 2 & 1 & 1 \\ 1 & 1 & 2 \\ 2 & 2 & 1 \end{bmatrix} \begin{bmatrix} x \\ y \\ z \end{bmatrix} = \begin{bmatrix} A \\ B \\ C \end{bmatrix}$$

$$AX = B \Rightarrow X = A^{-1}B$$

$$X = \begin{bmatrix} 2 & 1 & 1 \\ 1 & 1 & 2 \\ 2 & 2 & 1 \end{bmatrix}^{-1} \begin{bmatrix} A \\ B \\ C \end{bmatrix} =$$

$$= \begin{bmatrix} 1 & -1/3 & -1/3 \\ -1 & 0 & 1 \\ 0 & 2/3 & -1/3 \end{bmatrix} \begin{bmatrix} A \\ B \\ C \end{bmatrix} = \begin{bmatrix} A - B/3 - C/3 \\ -A + C \\ 2B/3 - C/3 \end{bmatrix}$$

Solution: Use $A - B/3 - C/3$ batches of Vanilla, $-A + C$ batches of mocha, and $2B/3 - C/3$ batches of strawberry.

55. Unknowns:

x = amount invested in PNF

y = amount invested in FDMMX

z = amount invested in FFLIX

The total investment was \$9000:

$x + y + z = 9000$

You invested an equal amount in FDMMX and FFLIX:

$y = z$, or

$y - z = 0$

Interest for the year from the first two stocks was $400:

$$0.06x + 0.05y = 400$$

Solving:

Matrix form:

$$\begin{bmatrix} 1 & 1 & 1 \\ 0 & 1 & -1 \\ 0.06 & 0.05 & 0 \end{bmatrix} \begin{bmatrix} x \\ y \\ z \end{bmatrix} = \begin{bmatrix} 9000 \\ 0 \\ 400 \end{bmatrix}$$

$$AX = B \Rightarrow X = A^{-1}B$$

$$X = \begin{bmatrix} 1 & 1 & 1 \\ 0 & 1 & -1 \\ 0.06 & 0.05 & 0 \end{bmatrix}^{-1} \begin{bmatrix} 9000 \\ 0 \\ 400 \end{bmatrix}$$

$$= \begin{bmatrix} -5/7 & -5/7 & 200/7 \\ 6/7 & 6/7 & -100/7 \\ 6/7 & -1/7 & -100/7 \end{bmatrix} \begin{bmatrix} 9000 \\ 0 \\ 400 \end{bmatrix} = \begin{bmatrix} 5000 \\ 2000 \\ 2000 \end{bmatrix}$$

Solution: You invested $5000 in PNF, $2000 in FDMMX, $2000 in FFLIX

57. Unknowns:

x = the number of shares of APPL

y = the number of shares of HPQ

z = the number of shares of DELL

The total investment was $12,400:

Investment in APPL = x shares @ $25 = $25x$

Investment in HPQ = y shares @ $25 = $25y$

Investment in DELL = z shares @ $35 = $35z$

Thus,

$$25x + 25y + 35z = 5800$$

You earned $21 in dividends:

APPL dividend = 0.6% of $25x$ invested

$\quad = 0.006(25x) = 0.15x$

HPQ dividend = 1.2% of $25y$ invested

$\quad = 0.012(25y) = 0.3y$

DELL dividend = 0

Thus,

$$0.15x + 0.3y = 21$$

You purchased a total of 200 shares:

$$x + y + z = 200$$

We therefore have the following system:

$$x + y + z = 200$$
$$25x + 25y + 35z = 5800$$
$$0.15x + 0.3y = 21$$

Matrix form:

$$\begin{bmatrix} 1 & 1 & 1 \\ 25 & 25 & 35 \\ 0.15 & 0.3 & 0 \end{bmatrix} \begin{bmatrix} x \\ y \\ z \end{bmatrix} = \begin{bmatrix} 200 \\ 5800 \\ 21 \end{bmatrix}$$

$$AX = B \Rightarrow X = A^{-1}B$$

$$X = \begin{bmatrix} 1 & 1 & 1 \\ 25 & 25 & 35 \\ 0.15 & 0.3 & 0 \end{bmatrix}^{-1} \begin{bmatrix} 200 \\ 5800 \\ 21 \end{bmatrix}$$

$$= \begin{bmatrix} 7 & -1/5 & -20/3 \\ -7/2 & 1/10 & 20/3 \\ -5/2 & 1/10 & 0 \end{bmatrix} \begin{bmatrix} 200 \\ 5800 \\ 21 \end{bmatrix} = \begin{bmatrix} 100 \\ 20 \\ 80 \end{bmatrix}$$

Solution: You purchased 100 shares of APPL, 20 shares of HPQ, and 80 shares of DELL.

59.

2003 Distribution = A = [53.3 64.0 101.6 65.4]

$$P = \begin{bmatrix} 0.9879 & 0.002 & 0.0081 & 0.0019 \\ 0.0014 & 0.9895 & 0.0063 & 0.0028 \\ 0.0027 & 0.0025 & 0.9927 & 0.0022 \\ 0.001 & 0.003 & 0.005 & 0.9909 \end{bmatrix}$$

$$P^{-1} \approx \begin{bmatrix} 1.0123 & -0.002 & -0.0082 & -0.0019 \\ -0.0014 & 1.0106 & -0.0064 & -0.0028 \\ -0.0027 & -0.0025 & 1.0074 & -0.0022 \\ -0.001 & -0.003 & -0.0051 & 1.0092 \end{bmatrix}$$

Distribution in 2002 = $A \cdot P^{-1} \approx$ [53.5 64.1 101.2 65.5]

61. $R = \begin{bmatrix} \sqrt{1/2} & -\sqrt{1/2} \\ \sqrt{1/2} & \sqrt{1/2} \end{bmatrix} \approx \begin{bmatrix} 0.7071 & -0.7071 \\ 0.7071 & 0.7071 \end{bmatrix}$

(a) The coordinates of a rotated point are given by

$$\begin{bmatrix} x' \\ y' \end{bmatrix} = R \begin{bmatrix} x \\ y \end{bmatrix} \approx \begin{bmatrix} 0.7071 & -0.7071 \\ 0.7071 & 0.7071 \end{bmatrix} \begin{bmatrix} 2 \\ 3 \end{bmatrix}$$

$$= \begin{bmatrix} -0.7071 \\ 3.5355 \end{bmatrix}$$

Thus, $(x', y') \approx (-0.7071, 3.5355)$

(b) Multiplication by R rotates points through $45°$. To rotate through $90°$, multiply by R again, obtaining

$$R\left(R\left[\begin{array}{c} x \\ y \end{array}\right]\right) = R^2\left[\begin{array}{c} x \\ y \end{array}\right]$$

In other words, multiplying by R^2 will result in a counterclockwise rotation of $90°$.

To rotate by $135°$, multiply yet again by R. This amounts to multiplying the original column vector by R^3.

(c) Let S be the matrix representing a clockwise rotation through $45°$. Since a rotation of $45°$ clockwise followed by a rotation of $45°$ counterclockwise results in no change,

$$R\left(S\left[\begin{array}{c} x \\ y \end{array}\right]\right) = \left[\begin{array}{c} x \\ y \end{array}\right] = I\left[\begin{array}{c} x \\ y \end{array}\right]$$

In other words, multiplication by RS should have the same effect as multiplication by the 2×2 identity matrix. This will happen if $S = R^{-1}$.

63. First, write the uncoded message as a string of numbers using A = 1, B = 2, C = 3, and so on:
"GO TO PLAN B" = [7 15 0 20 15 0 16 12 1 14 0 2] .

First, arrange the numbers in a matrix with 2 rows:

Uncoded message $= \left[\begin{array}{cccccc} 7 & 0 & 15 & 16 & 1 & 0 \\ 15 & 20 & 0 & 12 & 14 & 2 \end{array}\right]$

Then encode by multiplying by $A = \left[\begin{array}{cc} 1 & 2 \\ 3 & 4 \end{array}\right]$:

Coded message

$= \left[\begin{array}{cc} 1 & 2 \\ 3 & 4 \end{array}\right]\left[\begin{array}{cccccc} 7 & 0 & 15 & 16 & 1 & 0 \\ 15 & 20 & 0 & 12 & 14 & 2 \end{array}\right]$

$= \left[\begin{array}{cccccc} 37 & 40 & 15 & 40 & 29 & 4 \\ 81 & 80 & 45 & 96 & 59 & 8 \end{array}\right]$

Arrange as a single row:
[37 81 40 80 15 45 40 96 29 59 4 8]

65. Coded message = [33 69 54 126 11 27 20 60 29 59 65 149 41 87]

First, arrange the numbers in a matrix with 2 rows:

Coded message $= \left[\begin{array}{ccccccc} 33 & 54 & 11 & 20 & 29 & 65 & 41 \\ 69 & 126 & 27 & 60 & 59 & 149 & 87 \end{array}\right]$

To decode the message, multiply by A^{-1}

$A^{-1} = \left[\begin{array}{cc} 1 & 2 \\ 3 & 4 \end{array}\right]^{-1} = \left[\begin{array}{cc} -2 & 1 \\ 1.5 & -0.5 \end{array}\right]$

Decoded message

$\left[\begin{array}{cc} -2 & 1 \\ 1.5 & -0.5 \end{array}\right]\left[\begin{array}{ccccccc} 33 & 54 & 11 & 20 & 29 & 65 & 41 \\ 69 & 126 & 27 & 60 & 59 & 149 & 87 \end{array}\right]$

$= \left[\begin{array}{ccccccc} 3 & 18 & 5 & 20 & 1 & 19 & 5 \\ 15 & 18 & 3 & 0 & 14 & 23 & 18 \end{array}\right]$

Arrange as a single row:
[3 15 18 18 5 3 20 0 1 14 19 23 5 18]
Translate to letters using A = 1, B = 2, C = 3, and so on:

Decoded message = "CORRECT ANSWER"

67. If $AB = I$ and $BA = I$, then A and B are inverse matrices. (See the definition of the inverse matrix in the text.) Thus, choice (A) is the correct choice.

69. The given matrix has two identical rows. If two rows of a matrix are the same, then row reducing it will lead to a row of zeros, and so it cannot be reduced to the identity. Thus, the inverse of the given matrix does not exist—the matrix is singular.

71. We check that the given matrix is the inverse of $\left[\begin{array}{cc} a & b \\ c & d \end{array}\right]$ by multiplying the two matrices:

$\left(\dfrac{1}{ad - bc}\left[\begin{array}{cc} d & -b \\ -c & a \end{array}\right]\right)\left[\begin{array}{cc} a & b \\ c & d \end{array}\right]$

$= \dfrac{1}{ad - bc}\left[\begin{array}{cc} ad{-}bc & db{-}bd \\ -ac{+}ca & -bc{+}ad \end{array}\right]$

$= \dfrac{1}{ad - bc}\left[\begin{array}{cc} ad{-}bc & 0 \\ 0 & ad{-}bc \end{array}\right]$

$= \left[\begin{array}{cc} 1 & 0 \\ 0 & 1 \end{array}\right].$

Since the product of the two matrices is the identity, they must be inverse matrices as claimed. In other words,

$$\begin{bmatrix} a & b \\ c & d \end{bmatrix}^{-1} = \frac{1}{ad - bc} \begin{bmatrix} d & -b \\ -c & a \end{bmatrix}$$

(provided $ad - bc \neq 0$).

73. D is singular when one or more of the d_i are zero. If that is the case, then the matrix $[D \mid I]$ easily reduces to a matrix that has a row of zeros on the left-hand portion, so that D is singular, Conversely, if none of the d_i are zero, then $[D \mid I]$ easily reduces to a matrix of the form $[I \mid E]$, showing that D is invertible.

75. To check that $B^{-1}A^{-1}$ is the inverse of AB, we multiply these two matrices and check that we obtain the identity matrix:

$$(AB)(B^{-1}A^{-1}) = A(BB^{-1})A^{-1} \text{ Associative law}$$
$$= AIA^{-1} \quad \text{Since } BB^{-1} = I$$
$$= AA^{-1} = I$$

Since the product of the given matrices is the identity, they must be inverses as claimed.

77. If a square matrix A reduces to one with a row of zeros, then it cannot have an inverse. The reason is that, if A has an inverse, then every system of equations $AX = B$ has a unique solution, namely $X = A^{-1}B$. But if A reduces to a matrix with a row of zeros, then such a system has either infinitely many solutions or no solution at all.

3.4

1. Checking the rows: Neither of rows *a* and *b* dominate the other: 1 is worse than 2, but 10 is better than –4.

Checking the columns: Column *p* dominates column *q* because each of the payoffs in column *p* is lower than or equal to the corresponding payoff in column *q*. We therefore eliminate column *q* to obtain

$$\mathbf{A} \begin{array}{c} a \\ b \end{array} \begin{array}{cc} & \mathbf{B} \\ & \begin{array}{cc} p & r \end{array} \\ \left[\begin{array}{cc} 1 & 10 \\ 2 & -4 \end{array} \right] \end{array}$$

This matrix cannot be reduced further.

3. Checking the rows: Row 3 dominates both rows 1 and 2 because each of its payoffs is greater than or equal to the corresponding payoff in row 1 and 2 so we eliminate rows 1 and 2 to obtain

$$\mathbf{A} \quad 3 \begin{array}{c} \mathbf{B} \\ \begin{array}{ccc} a & b & c \end{array} \\ \left[\begin{array}{ccc} 5 & 0 & -1 \end{array} \right] \end{array}$$

Checking the columns: Column *c* dominates columns *a* and b because –1 is less than 0 and 5*p*. We therefore eliminate columns *a* and *b* to obtain

$$\mathbf{A} \quad 3 \begin{array}{c} \mathbf{B} \\ c \\ \left[\begin{array}{c} -1 \end{array} \right] \end{array}$$

5. Checking the rows: Row *q* dominates rows *p* and *s* because each of its payoffs is greater than or equal to the corresponding payoff in row *p* and *s* so we eliminate rows *p* and *s* to obtain

$$\mathbf{A} \begin{array}{c} q \\ r \end{array} \begin{array}{c} \mathbf{B} \\ \begin{array}{ccc} a & b & c \end{array} \\ \left[\begin{array}{ccc} 4 & 0 & 2 \\ 3 & -3 & 10 \end{array} \right] \end{array}$$

Checking the columns: Column *b* dominates the other two columns (each of its payoffs is less than or equal to the corresponding payoff in the other two columns). We therefore eliminate columns *a* and *c* to obtain

$$\mathbf{A} \quad q \begin{array}{c} \mathbf{B} \\ b \\ \left[\begin{array}{c} 0 \end{array} \right] \end{array}$$

7. Circle the row minima:

$$\mathbf{A} \begin{array}{c} a \\ b \end{array} \begin{array}{c} \mathbf{B} \\ \begin{array}{cc} p & q \end{array} \\ \left[\begin{array}{cc} ① & ① \\ 2 & ④ \end{array} \right] \end{array}$$

The largest row minimum is the 1 occurring in row *a*.

Box the column maxima:

$$\mathbf{A} \begin{array}{c} a \\ b \end{array} \begin{array}{c} \mathbf{B} \\ \begin{array}{cc} p & q \end{array} \\ \left[\begin{array}{cc} 1 & \boxed{1} \\ \boxed{2} & -4 \end{array} \right] \end{array}$$

The smallest column maximum is the 1 occurring in column *q*.

Since the *a,q*-entry 1 was both circled and boxed, it is a saddle point, and the game is strictly determined with A's optimal strategy being *a*, B's optimal strategy being *q*, and the value = saddle point = 1.

9. Circle the row minima:

$$\mathbf{A} \begin{array}{c} a \\ b \end{array} \begin{array}{c} \mathbf{B} \\ \begin{array}{ccc} p & q & r \end{array} \\ \left[\begin{array}{ccc} 2 & 0 & ⊖2 \\ ⊖1 & 3 & 0 \end{array} \right] \end{array}$$

The largest row minimum is the –1 occurring in row *b*. Box the column maxima:

$$\mathbf{A} \begin{array}{c} a \\ b \end{array} \begin{array}{c} \mathbf{B} \\ \begin{array}{ccc} p & q & r \end{array} \\ \left[\begin{array}{ccc} \boxed{2} & 0 & -2 \\ -1 & \boxed{3} & \boxed{0} \end{array} \right] \end{array}$$

The smallest column maximum is the 0 occurring in column r.

Since no entry is both circled and boxed, there are no saddle points, and so the game is not strictly determined.

11. Circle the row minima:

$$
A \quad
\begin{array}{c}
 \\
P \\
Q \\
R \\
S
\end{array}
\overset{\begin{array}{ccc} & \mathbf{B} & \\ a & b & c \end{array}}{
\left[
\begin{array}{ccc}
1 & -1 & \boxed{-5} \\
4 & \boxed{-4} & 2 \\
3 & -3 & \boxed{-10} \\
5 & \boxed{-5} & -4
\end{array}
\right]}
$$

The largest row minimum is the -4 occurring in row Q.

Box the column maxima:

$$
A \quad
\begin{array}{c}
 \\
P \\
Q \\
R \\
S
\end{array}
\overset{\begin{array}{ccc} & \mathbf{B} & \\ a & b & c \end{array}}{
\left[
\begin{array}{ccc}
1 & \boxed{-1} & -5 \\
4 & -4 & \boxed{2} \\
3 & -3 & -10 \\
\boxed{5} & -5 & -4
\end{array}
\right]}
$$

The smallest column maximum is the -1 occurring in column b.

Since no entry is both circled and boxed, there are no saddle points, and so the game is not strictly determined.

13. $e = RPC$

$$
= [0 \ 1 \ 0 \ 0]
\begin{bmatrix}
2 & 0 & -1 & 2 \\
-1 & 0 & 0 & -2 \\
-2 & 0 & 0 & 1 \\
3 & 1 & -1 & 1
\end{bmatrix}
\begin{bmatrix}
1 \\ 0 \\ 0 \\ 0
\end{bmatrix}
$$

$$
= -1
$$

15. $e = RPC$

$$
= [0.5 \ 0.5 \ 0 \ 0]
\begin{bmatrix}
2 & 0 & -1 & 2 \\
-1 & 0 & 0 & -2 \\
-2 & 0 & 0 & 1 \\
3 & 1 & -1 & 1
\end{bmatrix}
\begin{bmatrix}
0 \\ 0 \\ 0.5 \\ 0.5
\end{bmatrix}
$$

$$
= -0.25
$$

17. Since we are given a column matrix, we take our row strategy to be

$R = [x \ y \ z \ t]$.

$e = RPC$

$$
= [x \ y \ z \ t]
\begin{bmatrix}
0 & -1 & 5 \\
2 & -2 & 4 \\
0 & 3 & 0 \\
1 & 0 & -5
\end{bmatrix}
\begin{bmatrix}
0.25 \\ 0.75 \\ 0
\end{bmatrix}
$$

$$
= [x \ y \ z \ t]
\begin{bmatrix}
-0.75 \\ -1 \\ 2.25 \\ 0.25
\end{bmatrix}
$$

$$
= -0.74x - y + 2.25z + 0.25t.
$$

The greatest coefficient is the coefficient of z, so we take $z = 1$ and $x = y = t = 0$. This gives the strategy

$R = [0 \ 0 \ 1 \ 0]$

and resulting expected value of the game

$e = -0.74(0) - 0 + 2.25(1) + 0.25(0)$

$= 2.25.$

19. Since we are given a row matrix, we take our column strategy to be

$C = [x \ y \ z]^T$.

$e = RPC$

$$
= [1/2 \ 0 \ 1/4 \ 1/4]
\begin{bmatrix}
0 & -1 & 5 \\
2 & -2 & 4 \\
0 & 3 & 0 \\
1 & 0 & -5
\end{bmatrix}
\begin{bmatrix}
x \\ y \\ z
\end{bmatrix}
$$

$$
= [1/4 \ 1/4 \ 5/4]
\begin{bmatrix}
x \\ y \\ z
\end{bmatrix}
$$

$$
= (1/4)x + (1/4)y + (5/4)z
$$

The lowest coefficients are the coefficients of x and y, so we can either take $x = 1$ and $y = z = 0$,

giving the strategy $C = [1 \ 0 \ 0]^T$, or we could take $y = 1$ and $x = z = 0$, giving the strategy
$$C = [0 \ 1 \ 0]^T$$
and resulting expected value of the game
$$e = (1/4)(1) + (1/4)(0) + (5/4)(0)$$
$$= 1/4$$

21. (a) Row strategy:
Take $R = [x, \ 1-x]$, $C = [1 \ 0]^T$
$$e = RPC = [x \ 1-x] \begin{bmatrix} -1 & 2 \\ 0 & -1 \end{bmatrix} \begin{bmatrix} 1 \\ 0 \end{bmatrix}$$
$$= -x$$
Take $R = [x \ 1-x]$, $C = [0 \ 1]^T$
$$f = RPC = [x \ 1-x] \begin{bmatrix} -1 & 2 \\ 0 & -1 \end{bmatrix} \begin{bmatrix} 0 \\ 1 \end{bmatrix}$$
$$= 2x - (1-x) = 3x - 1$$
Graphs of e and f:

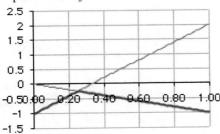

Lower (heavy) portion has its highest point at the intersection of the two lines. The x-coordinate of the intersection is given when $e = f$:
$$-x = 3x - 1$$
$$4x = 1$$
$$x = 1/4$$
So the optimal row strategy is
$$R = [x \ 1-x]$$
$$= [1/4 \ 3/4]$$
(b) Column Strategy:
Take $R = [1 \ 0]$, $C = [x \ 1-x]^T$
$$e = RPC = [1 \ 0] \begin{bmatrix} -1 & 2 \\ 0 & -1 \end{bmatrix} \begin{bmatrix} x \\ 1-x \end{bmatrix}$$
$$= -x + 2(1-x)$$
$$= -3x + 2$$

Take $R = [0 \ 1]$, $C = [x \ 1-x]^T$
$$f = RPC = [0 \ 1] \begin{bmatrix} -1 & 2 \\ 0 & -1 \end{bmatrix} \begin{bmatrix} x \\ 1-x \end{bmatrix}$$
$$= x - 1$$
Graphs of e and f:

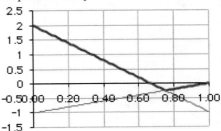

The upper (heavy) portion has its lowest point at the intersection of the two lines. The x-coordinate of the intersection is given when $e = f$:
$$-3x + 2 = x - 1$$
$$4x = 3$$
$$x = 3/4$$
So the optimal column strategy is
$$C = [x \ 1-x]^T$$
$$= [3/4 \ 1/4]^T$$
(c) To compute the expected value of the game, compute the y-coordinate of the point on either graph with the given x-coordinate. For instance, using the second graph (part (b)),
$$e = x - 1 = 3/4 - 1 = -1/4.$$
Solution: $R = [1/4 \ 3/4]$, $C = [3/4 \ 1/4]^T$,
$$e = -1/4$$

23. (a) Row strategy:
Take $R = [x, \ 1-x]$, $C = [1 \ 0]^T$
$$e = RPC = [x \ 1-x] \begin{bmatrix} -1 & -2 \\ -2 & 1 \end{bmatrix} \begin{bmatrix} 1 \\ 0 \end{bmatrix}$$
$$= -x - 2(1-x) = x - 2$$
Take $R = [x \ 1-x]$, $C = [0 \ 1]^T$
$$f = RPC = [x \ 1-x] \begin{bmatrix} -1 & -2 \\ -2 & 1 \end{bmatrix} \begin{bmatrix} 0 \\ 1 \end{bmatrix}$$
$$= -2x + (1-x) = -3x + 1$$

Graphs of e and f:

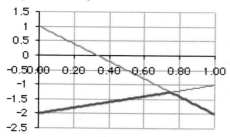

Lower (heavy) portion has its highest point at the intersection of the two lines. The x-coordinate of the intersection is given when $e = f$:

$$x - 2 = -3x + 1$$
$$4x = 3$$
$$x = 3/4$$

So the optimal row strategy is

$$R = [x \quad 1-x]$$
$$= [3/4 \quad 1/4]$$

(b) Column Strategy:

Take $R = [1 \quad 0]$, $C = [x \quad 1-x]^T$

$$e = RPC = [1 \quad 0] \begin{bmatrix} -1 & -2 \\ -2 & 1 \end{bmatrix} \begin{bmatrix} x \\ 1-x \end{bmatrix}$$

$$= -x - 2(1-x) = x - 2$$

Take $R = [0 \quad 1]$, $C = [x \quad 1-x]^T$

$$f = RPC = [0 \quad 1] \begin{bmatrix} -1 & -2 \\ -2 & 1 \end{bmatrix} \begin{bmatrix} x \\ 1-x \end{bmatrix}$$

$$= -2x + (1-x) = -3x + 1$$

Note that e and f are the same as for part (a). Therefore, their graphs are the same, except that this time we are interested in the lowest point of the upper region:

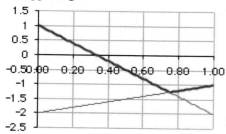

This is the same intersection point as in part (a), so $x = 3/4$ and the optimal column strategy is

124

$$C = [x \quad 1-x]^T$$
$$= [3/4 \quad 1/4]^T$$

(c) To compute the expected value of the game, compute the y-coordinate of the point on either graph with the given x-coordinate. For instance, using the second graph (part (b)),

$$e = x - 2 = 3/4 - 2 = -5/4.$$

Solution: $R = [3/4 \quad 1/4]$, $C = [3/4 \quad 1/4]^T$, $e = -5/4$

25. The possible outcomes are:

HH: You lose 1 point. Payoff: -1

HT: You win 1 point. Payoff: 1

TH: You win 1 point. Payoff: 1

TT: You lose 1 point. Payoff: -1

Payoff matrix:

$$\begin{array}{cc} & \text{Friend} \\ & H \quad T \\ \text{You} \begin{array}{c} H \\ T \end{array} & \begin{bmatrix} -1 & 1 \\ 1 & -1 \end{bmatrix} \end{array}$$

27. Your strategies are to attack by sea or air. Your opponent's strategies are to defend by sea, air, or both. The possible outcomes are:

(Sea, Sea) or (Air, Air): Your attack is met by an all-out defense: you lose 200 points.

Payoff: -200

(Sea, Air) or (Air, Sea): Your attack is met by no defense: you win 100 points.

Payoff: 100

(Sea, Both) or (Air, Both): Your attack is met by a shared defense: you win 50 points.

Payoff: 50

Payoff matrix:

		Your Opponent Defends		
		Sea	Air	Both
You Attack	Sea	-200	100	50
	Air	100	-200	50

29. Take B = Brakpan; N = Nigel; S = Springs. Your strategies are to locate in one of B, N, S, and your opponent's strategies are the same. The possible outcomes are:

BB, SS, or SS: You share the total business. No net gain or loss of sales. Payoff: 0

BN or NB: One of you locates at B, the other at N. These two cities provide the same potential market, so again there is no net gain or loss. Payoff: 0

SB or SN: You locate at S (total market: 1000 burgers/day), your opponent locates at N or B (total market: 2000 burgers/day). Your net loss is then 1000 burgers/day. Payoff: −1000.

BS or NS: You locate at N or B (total market: 2000 burgers/day), your opponent locates at. S (total market: 1000 burgers/day) Your net gain is then 1000 burgers/day. Payoff: 1000.

Payoff matrix:

$$\begin{array}{c} \textbf{Your Opponent} \\ \begin{array}{cc} & \begin{array}{ccc} \text{B} & \text{N} & \text{S} \end{array} \\ \textbf{You} \begin{array}{c} \text{B} \\ \text{N} \\ \text{S} \end{array} & \left[\begin{array}{ccc} 0 & 0 & 1000 \\ 0 & 0 & 1000 \\ -1000 & -1000 & 0 \end{array}\right] \end{array} \end{array}$$

31. Let P = PleasantTap; T = Thunder Rumble; S = Strike the Gold, N = None. Your strategies: P, T, S. Your "opponent" is Nature, which decides the winner. Opponent's strategies: P, T, S, N. The possible outcomes are:

PP: You bet on P and P wins. Odds: 5-2. $2 wins you $5. Therefore $10 wins you 5×$5 = $25. Payoff: 25

TT: You bet on T and T wins. Odds: 7-2. $2 wins you $7. Therefore $10 wins you 5×$7 = $35. Payoff: 35

SS: You bet on S and S wins. Odds: 4-1. $1 wins you $4. Therefore $10 wins you 10×$4 = $40. Payoff: 40

All other outcomes: You lose your $10 because your horse fails to win. Payoff: −10

Payoff matrix:

$$\begin{array}{c} \textbf{Winner} \\ \begin{array}{cc} & \begin{array}{cccc} \text{P} & \text{T} & \text{S} & \text{N} \end{array} \\ \textbf{You Bet} \begin{array}{c} \text{P} \\ \text{T} \\ \text{S} \end{array} & \left[\begin{array}{cccc} 25 & -10 & -10 & -10 \\ -10 & 35 & -10 & -10 \\ -10 & -10 & 40 & -10 \end{array}\right] \end{array} \end{array}$$

33. (a) The payoff matrix is

$$P = \left[\begin{array}{ccc} 15 & 60 & 80 \\ 15 & 60 & 60 \\ 10 & 20 & 40 \end{array}\right]$$

Checking the rows: Row 1 dominates each of the other rows, so we eliminate rows 2 and 3, leaving

[15 60 80]

Checking the columns: Column 1 dominates each of the other columns, so we eliminate columns 2 and 3, leaving the 1×1 game

[15],

which corresponds to the first CE strategy (charge $1000) and the first GCS strategy (charge $900). So, CE should charge $1000 and GCS should charge $900. Since the payoff is 15, a 15% gain in market share for CE results.

(b) Look at the original (unreduced) payoff matrix. CE is aware that GCS is planning to charge $900. For the best market share, CE should charge either $1000 or $1200 because either will result in the best market share (15% gain) under the circumstances. Thus, in terms of market share, the added information has no effect. However, in terms of revenue, the better of the two options would be to charge the larger price: $1200. (the more CE can charge for the same market, the better!).

35. Take CCC as the row player and MMA as the column player. Thus, CCC's strategies are Pablo, Sal and Edison, while MMA's strategies are Carlos, Marcus and Noto. To set up the payoff matrix, enter 1 for every combination in which the CCC wrestler beats the MMA wrestler, −1 for every combination in which the MMA wrestler beats the CCC wrestler, and 0 in all the evenly matched combinations. The resulting payoff matrix is

$$\begin{array}{c} \\ \\ \mathbf{CCC} \end{array} \begin{array}{c} \\ \begin{array}{ccc} \text{Carlos} & \text{Marcus} & \text{Noto} \end{array} \\ \begin{array}{c} \text{Pablo} \\ \text{Sal} \\ \text{Edison} \end{array} \left[\begin{array}{ccc} 1 & 1 & 0 \\ 0 & -1 & 0 \\ 0 & -1 & -1 \end{array} \right] \end{array}$$

Comparing rows, we see that Row 1 dominates the other two rows, so we eliminate Rows 2 and 3 leaving

$$\begin{array}{c} \\ \text{Pablo} \end{array} \begin{array}{c} \begin{array}{ccc} \text{Carlos} & \text{Marcus} & \text{Noto} \end{array} \\ \left[\begin{array}{ccc} 1 & 1 & 0 \end{array} \right] \end{array}$$

Comparing columns, we see that Column 3 dominates the other two columns, so we eliminate Columns 1 and 2, leaving us with

$$\begin{array}{c} \\ \text{Pablo} \end{array} \begin{array}{c} \text{Noto} \\ \left[\begin{array}{c} 0 \end{array} \right] \end{array}$$

Thus the game is reduced to Pablo vs. Noto. Since the payoff is 0, the game is evenly matched.

37. We first reduce by dominance (following the "FAQ" in the textbook):

Comparing the rows, we find that neither row dominates the other.

Comparing the columns, we see than Column 1 dominates Column 2. So we eliminate Column 2, leaving

$$\begin{array}{c} \\ \begin{array}{c} \text{Northern Route} \\ \text{Southern Route} \end{array} \end{array} \begin{array}{c} \text{Northern Route} \\ \left[\begin{array}{c} 2 \\ 1 \end{array} \right] \end{array}$$

Comparing the rows, we now see that Row 1 dominates Row 2, so we eliminate Row 2, leaving us with

$$\begin{array}{c} \\ \text{Northern Route} \end{array} \begin{array}{c} \text{Northern Route} \\ \left[\begin{array}{c} 1 \end{array} \right] \end{array}$$

Thus, both commanders should use the northern route, resulting in an estimate of one day's bombing time.

39. Take C = Confess and N = Do not confess. If we take the (negative) payoffs as the amount of time Slim faces behind bars, we get:

$$\begin{array}{c} \\ \\ \mathbf{Slim} \end{array} \begin{array}{c} \mathbf{Joe} \\ \begin{array}{cc} \text{C} & \text{N} \end{array} \\ \begin{array}{c} \text{C} \\ \text{N} \end{array} \left[\begin{array}{cc} -2 & 0 \\ -10 & -5 \end{array} \right] \end{array}$$

Since Row 1 dominates Row 2, Slim's optimal strategy is to confess.

41. (a) Let F represent a visit to Florida and O a visit to Ohio. The outcomes are:

FF and OO: Both candidates visit the same state, so Bush still has a 24% chance of winning, so the payoff is 24 in both cases.

FO: Bush visits Florida, giving him a 60+10 = 70% chance in that state, and Kerry visits Ohio, reducing Bush's chances there to 40−10 = 30%. Thus, the probability of Bush winning both states is 0.70×0.30 = 0.21

OF: Bush visits Ohio, giving him a 40+10 = 50% chance in that state, and Kerry visits Florida, reducing Bush's chances there to 60−10 = 50%.

Thus, the probability of Bush winning both states is
$0.50 \times 0.50 = 0.25$

This gives the following payoff matrix:

$$
\begin{array}{c}
\textbf{Kerry} \\
\begin{array}{cc} F & O \end{array} \\
\textbf{Bush} \begin{array}{c} F \\ O \end{array}
\left[\begin{array}{cc} 24 & 21 \\ 25 & 24 \end{array}\right]
\end{array}
$$

(**b**) In the payoff matrix in part (a), Row 2 dominates Row 1 (meaning that Bush should visit Ohio. in that case):

$$
\begin{array}{c}
\textbf{Kerry} \\
\begin{array}{cc} F & O \end{array} \\
\textbf{Bush} \quad O \left[\begin{array}{cc} 25 & 24 \end{array}\right]
\end{array}
$$

Column 2 now dominates Column 1, leaving us with the following solution

$$
\begin{array}{c}
\textbf{Kerry} \\
O \\
\textbf{Bush} \quad O \left[\begin{array}{c} 24 \end{array}\right]
\end{array}
$$

meaning that both candidates should visit Ohio, leaving Bush with a 21% chance of winning the election.

43. $P = \left[\begin{array}{cc} -60 & -40 \\ 30 & -50 \end{array}\right], R = [.50 \quad .50], C = \left[\begin{array}{c} 0.20 \\ 0.80 \end{array}\right]$

$e = RPC$

$\quad = [.50 \quad .50] \left[\begin{array}{cc} -60 & -40 \\ 30 & -50 \end{array}\right]\left[\begin{array}{c} 0.20 \\ 0.80 \end{array}\right]$

$\quad = [-15 \quad -45]\left[\begin{array}{c} 0.20 \\ 0.80 \end{array}\right] = -39$

You can expect to lose 39 customers.

45. $P = \left[\begin{array}{ccc} 200 & 150 & 140 \\ 130 & 220 & 130 \\ 110 & 110 & 220 \end{array}\right], R = [x \quad y \quad z],$

$C = \left[\begin{array}{c} 0.2 \\ 0.5 \\ 0.3 \end{array}\right]$

$e = RPC$

$\quad = [x \quad y \quad z]\left[\begin{array}{ccc} 200 & 150 & 140 \\ 130 & 220 & 130 \\ 110 & 110 & 220 \end{array}\right]\left[\begin{array}{c} 0.2 \\ 0.5 \\ 0.3 \end{array}\right]$

$\quad = [x \quad y \quad z]\left[\begin{array}{c} 157 \\ 175 \\ 143 \end{array}\right] = 157x + 175y + 143z$

The largest coefficient is the coefficient of y, so we take

$\quad R = [0 \quad 1 \quad 0]$

meaning that Option 2: Move to the suburbs (corresponding to y) is the best option.

47. (a) $P = \left[\begin{array}{ccc} 90 & 70 & 70 \\ 40 & 90 & 40 \\ 60 & 40 & 90 \end{array}\right],$

$\quad R = [0.25 \quad 0.50 \quad 0.25], C = \left[\begin{array}{c} 0.25 \\ 0.50 \\ 0.25 \end{array}\right]$

$e = RPC$

$\quad = [0.25 \quad 0.50 \quad 0.25]\left[\begin{array}{ccc} 90 & 70 & 70 \\ 40 & 90 & 40 \\ 60 & 40 & 90 \end{array}\right]\left[\begin{array}{c} 0.25 \\ 0.50 \\ 0.25 \end{array}\right]$

$\quad = [0.25 \quad 0.50 \quad 0.25]\left[\begin{array}{c} 75 \\ 65 \\ 57.5 \end{array}\right]$

$\quad = [65.625]$

You can expect to get about 66% on the test.

(b) $P = \left[\begin{array}{ccc} 90 & 70 & 70 \\ 40 & 90 & 40 \\ 60 & 40 & 90 \end{array}\right], R = [x \quad y \quad z],$

$\quad C = \left[\begin{array}{c} 0.25 \\ 0.50 \\ 0.25 \end{array}\right]$

$e = RPC$

$\quad = [x \quad y \quad z]\left[\begin{array}{ccc} 90 & 70 & 70 \\ 40 & 90 & 40 \\ 60 & 40 & 90 \end{array}\right]\left[\begin{array}{c} 0.25 \\ 0.50 \\ 0.25 \end{array}\right]$

$$= [x \ y \ z] \begin{bmatrix} 75 \\ 65 \\ 57.5 \end{bmatrix} = 75x + 65y + 57.5z$$

The largest coefficient is the coefficient of x, so we take

$$R = [1 \ \ 0 \ \ 0]$$

meaning that Game Theory (corresponding to x) is the best option. Your expected score is then

$$75(1) + 65(0) + 57.5(0) = 75\%$$

(c) $P = \begin{bmatrix} 90 & 70 & 70 \\ 40 & 90 & 40 \\ 60 & 40 & 90 \end{bmatrix}$, $R = [0.25 \ \ 0.50 \ \ 0.25]$,

$$C = \begin{bmatrix} x \\ y \\ z \end{bmatrix}$$

$e = RPC$

$$= [0.25 \ \ 0.50 \ \ 0.25] \begin{bmatrix} 90 & 70 & 70 \\ 40 & 90 & 40 \\ 60 & 40 & 90 \end{bmatrix} \begin{bmatrix} x \\ y \\ z \end{bmatrix}$$

$$= [57.5 \ \ 72.5 \ \ 60] \begin{bmatrix} x \\ y \\ z \end{bmatrix} = 57.5x + 72.5y + 60z$$

The lowest coefficient is the coefficient of x, so we take

$$C = [1 \ \ 0 \ \ 0]^T$$

meaning that game theory would be worst for you, and you could expect to earn

$$57.5(1) + 72.5(0) + 60(0) = 57.5\%$$

for the test.

49. (a)

$$P = \begin{bmatrix} -500,000 & -200,000 & 10,000 & 200,000 \\ -200,000 & 0 & 0 & 0 \\ -100,000 & 10,000 & -200,000 & -300,000 \end{bmatrix}$$

$$R = [x \ \ y \ \ z], C = \begin{bmatrix} 0.2 \\ 0.2 \\ 0.3 \\ 0.3 \end{bmatrix},$$

(based on the past 10 years)

$e = RPC$

$$= [x \ y \ z] \times$$

$$\begin{bmatrix} -500,000 & -200,000 & 10,000 & 200,000 \\ -200,000 & 0 & 0 & 0 \\ -100,000 & 10,000 & -200,000 & -300,000 \end{bmatrix}$$

$$\times \begin{bmatrix} 0.2 \\ 0.2 \\ 0.3 \\ 0.3 \end{bmatrix}$$

$$= [x \ y \ z] \begin{bmatrix} -77,000 \\ -40,000 \\ -168,000 \end{bmatrix}$$

$$= -77,000x - 40,000y - 168,000z$$

The largest coefficient is the coefficient of y, so we take

$$R = [0 \ 1 \ 0]$$

meaning that laying off 10 workers (corresponding to y) is the best option. The expected payoff is

$$-77,000(0) - 40,000(1) - 168,000(0) =$$
$$-40,000,$$

corresponding to a cost of \$40,000.

(b)

$$P = \begin{bmatrix} -500,000 & -200,000 & 10,000 & 200,000 \\ -200,000 & 0 & 0 & 0 \\ -100,000 & 10,000 & -200,000 & -300,000 \end{bmatrix}$$

$$R = [0.50 \ \ 0.50 \ \ 0], C = \begin{bmatrix} x \\ y \\ z \\ t \end{bmatrix}$$

$e = RPC$

$$= [0.50 \ \ 0.50 \ \ 0] \times$$

$$\begin{bmatrix} -500,000 & -200,000 & 10,000 & 200,000 \\ -200,000 & 0 & 0 & 0 \\ -100,000 & 10,000 & -200,000 & -300,000 \end{bmatrix} \times$$

$$\begin{bmatrix} x \\ y \\ z \\ t \end{bmatrix}$$

$$= [-350{,}000 \quad -100{,}000 \quad 5000 \quad 100{,}000] \begin{bmatrix} x \\ y \\ z \\ t \end{bmatrix}$$

$$= -350{,}000x - 100{,}000y + 5000z + 100{,}000t.$$

The lowest coefficient is the coefficient of x, so we take

$$C = [1 \quad 0 \quad 0 \quad 0]^T$$

meaning that 60 inches of snow would be worst for him, costing him $350,000.

(c) The strategy from part (a) is

$$R = [0 \quad 1 \quad 0];$$

laying off 10 workers. If he does so, the worst that could happen is 0 inches of snow, (which would cost him $20,000). Since he feels that the Gods of Chaos are planning on 0 inches of snow, his best option would be to lay off 15 workers (according to the payoff matrix) and cut his losses to $100,000.

51. The payoff matrix (payoffs in thousands of dollars) is

$$\begin{array}{c} & \textbf{Splish} \\ & \text{WISH} \quad \text{WASH} \\ \textbf{Softex} \begin{array}{c} \text{WISH} \\ \text{WASH} \end{array} & \begin{bmatrix} -20 & 100 \\ 0 & -20 \end{bmatrix} \end{array}$$

We are asked to provide the optimal row strategy.
Take $R = [x, \ 1{-}x]$, $C = [1 \quad 0]^T$

$$e = RPC = [x \ \ 1{-}x] \begin{bmatrix} -20 & 100 \\ 0 & -20 \end{bmatrix} \begin{bmatrix} 1 \\ 0 \end{bmatrix}$$

$$= -20x + 0(1{-}x) = -20x$$

Take $R = [x \ \ 1{-}x]$, $C = [0 \quad 1]^T$

$$f = RPC = [x \ \ 1{-}x] \begin{bmatrix} -20 & 100 \\ 0 & -20 \end{bmatrix} \begin{bmatrix} 0 \\ 1 \end{bmatrix}$$

$$= 100x - 20(1{-}x) = 120x - 20$$

Graphs of e and f:

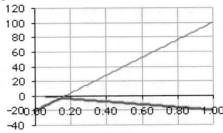

Lower (heavy) portion has its highest point at the intersection of the two lines. The x-coordinate of the intersection is given when $e = f$:

$$-20x = 120x - 20$$
$$140x = 20$$
$$x = 1/7$$

So the optimal row strategy is

$$R = [x \ \ 1{-}x]$$
$$= [1/7 \ \ 6/7]$$

This solution corresponds to allocating 1/7 of its advertising budget to WISH and the rest (6/7) to WASH.

The value of the game is then

$$e = -20x = -20/7 \approx -2.86$$

So Softex can expect to lose approximately $2,860.

53. Like a saddle point in a payoff matrix, the center of a saddle is a low point (minimum height) in one direction and a high point (maximum) in a perpendicular direction.

55. Although there is a saddle point in the 2,4 position, you would be wrong to use saddle points (based on the minimax criterion) to reach the conclusion that row strategy 2 is best. One reason is that the entries in the matrix do not represent payoffs, since high numbers of employees in an area do not necessarily represent benefit to the row player. Another reason for this is that there is no opponent deciding what your job will be in such a way as to force you into the least populated job.

57. If you strictly alternate the two strategies the column player will know which pure strategy you will play on each move, and can choose a pure strategy accordingly. For example, consider the game

$$\begin{array}{cc} & \begin{array}{cc} a & b \end{array} \\ \begin{array}{c} A \\ B \end{array} & \left[\begin{array}{cc} 1 & 0 \\ 0 & 1 \end{array} \right]. \end{array}$$

By the analysis of Example 3 (or the symmetry of the game), the best strategy for the row player is $[0.5 \quad 0.5]$ and the best strategy for the column player is $[0.5 \quad 0.5]^T$. This gives an expected value of 0.5 for the game. However, suppose that the row player alternates A and B strictly and that the column player catches on to this. Then, whenever the row player plays A the column player will play b and whenever the row player plays B the column player will play a. This gives a payoff of 0 each time, worse for the row player than the expected value of 0.5.

3.5

1. $A = \begin{bmatrix} 0.2 & 0.05 \\ 0.8 & 0.01 \end{bmatrix}$

Sector 1 = Paper; Sector 2 = Wood

(a) Wood → Paper = Sector 2→Sector 1
The corresponding entry of the technology matrix is
$$a_{21} = 0.8$$

(b) Paper → Paper = Sector 1→Sector 1
The corresponding entry of the technology matrix is
$$a_{11} = 0.2$$

(c) Paper → Wood = Sector 1→Sector 2
The corresponding entry of the technology matrix is
$$a_{12} = 0.05$$

3. Sector 1 = Television news, Sector 2 = Radio news
a_{11} = units of Television news needed to produce one unit of Television news = 0.2
a_{12} = units of Television news needed to produce one unit of Radio news = 0.1
a_{21} = units of Radio news needed to produce one unit of Television news = 0.5
a_{22} = units of Radio news needed to produce one unit of Radio news = 0
Thus, $A = \begin{bmatrix} 0.2 & 0.1 \\ 0.5 & 0 \end{bmatrix}$

5. $X = (I\text{--}A)^{-1}D$
$$= \left(\begin{bmatrix} 1 & 0 \\ 0 & 1 \end{bmatrix} - \begin{bmatrix} 0.5 & 0.4 \\ 0 & 0.5 \end{bmatrix} \right)^{-1} \begin{bmatrix} 10,000 \\ 20,000 \end{bmatrix}$$
$$= \begin{bmatrix} 0.5 & -0.4 \\ 0 & 0.5 \end{bmatrix}^{-1} \begin{bmatrix} 10,000 \\ 20,000 \end{bmatrix}$$
$$= \begin{bmatrix} 2 & 1.6 \\ 0 & 2 \end{bmatrix} \begin{bmatrix} 10,000 \\ 20,000 \end{bmatrix} = \begin{bmatrix} 52,000 \\ 40,000 \end{bmatrix}$$

7. $X = (I\text{--}A)^{-1}D$
$$= \left(\begin{bmatrix} 1 & 0 \\ 0 & 1 \end{bmatrix} - \begin{bmatrix} 0.1 & 0.4 \\ 0.2 & 0.5 \end{bmatrix} \right)^{-1} \begin{bmatrix} 25,000 \\ 15,000 \end{bmatrix}$$
$$= \begin{bmatrix} 0.9 & -0.4 \\ -0.2 & 0.5 \end{bmatrix}^{-1} \begin{bmatrix} 25,000 \\ 15,000 \end{bmatrix}$$
$$= \begin{bmatrix} 50/37 & 40/37 \\ 20/37 & 90/37 \end{bmatrix} \begin{bmatrix} 25,000 \\ 15,000 \end{bmatrix} = \begin{bmatrix} 50,000 \\ 50,000 \end{bmatrix}$$

9. $X = (I\text{--}A)^{-1}D$
$$= \left(\begin{bmatrix} 1 & 0 & 0 \\ 0 & 1 & 0 \\ 0 & 0 & 1 \end{bmatrix} - \begin{bmatrix} 0.5 & 0.1 & 0 \\ 0 & 0.5 & 0.1 \\ 0 & 0 & 0.5 \end{bmatrix} \right) \begin{bmatrix} 1000 \\ 1000 \\ 2000 \end{bmatrix}$$
$$= \begin{bmatrix} 0.5 & -.1 & 0 \\ 0 & 0.5 & -.1 \\ 0 & 0 & 0.5 \end{bmatrix}^{-1} \begin{bmatrix} 1000 \\ 1000 \\ 2000 \end{bmatrix}$$
$$= \begin{bmatrix} 2 & 0.4 & 0.08 \\ 0 & 2 & 0.4 \\ 0 & 0 & 2 \end{bmatrix} \begin{bmatrix} 1000 \\ 1000 \\ 2000 \end{bmatrix} = \begin{bmatrix} 2560 \\ 2800 \\ 4000 \end{bmatrix}$$

11. $X = (I\text{--}A)^{-1}D$
$$= \left(\begin{bmatrix} 1 & 0 & 0 \\ 0 & 1 & 0 \\ 0 & 0 & 1 \end{bmatrix} - \begin{bmatrix} 0.2 & 0.2 & 0 \\ 0.2 & 0.4 & 0.2 \\ 0 & 0.2 & 0.2 \end{bmatrix} \right)^{-1} \begin{bmatrix} 16,000 \\ 8000 \\ 8000 \end{bmatrix}$$
$$= \begin{bmatrix} 0.8 & -.2 & 0 \\ -.2 & 0.6 & -.2 \\ 0 & -.2 & 0.8 \end{bmatrix}^{-1} \begin{bmatrix} 16,000 \\ 8000 \\ 8000 \end{bmatrix}$$
$$= \begin{bmatrix} 11/8 & 1/2 & 1/8 \\ 1/2 & 2 & 1/2 \\ 1/8 & 1/2 & 11/8 \end{bmatrix} \begin{bmatrix} 16,000 \\ 8000 \\ 8000 \end{bmatrix} = \begin{bmatrix} 27,000 \\ 28,000 \\ 17,000 \end{bmatrix}$$

13. Change in Production
$= (I\text{--}A)^{-1} \times$ Change in Demand
$$= \left(\begin{bmatrix} 1 & 0 \\ 0 & 1 \end{bmatrix} - \begin{bmatrix} 0.1 & 0.4 \\ 0.2 & 0.5 \end{bmatrix} \right)^{-1} \begin{bmatrix} 50 \\ 30 \end{bmatrix}$$
$$= \begin{bmatrix} 0.9 & -0.4 \\ -0.2 & 0.5 \end{bmatrix}^{-1} \begin{bmatrix} 50 \\ 30 \end{bmatrix}$$
$$= \begin{bmatrix} 50/37 & 40/37 \\ 20/37 & 90/37 \end{bmatrix} \begin{bmatrix} 50 \\ 30 \end{bmatrix} = \begin{bmatrix} 100 \\ 100 \end{bmatrix}$$

15. Change in Production

$= (I{-}A)^{-1} \times$ Change in Demand

$$= \begin{bmatrix} 1.5 & 0.1 & 0 \\ 0.2 & 1.2 & 0.1 \\ 0.1 & 0.7 & 1.6 \end{bmatrix} \begin{bmatrix} 1 \\ 0 \\ 0 \end{bmatrix} = \begin{bmatrix} 1.5 \\ 0.2 \\ 0.1 \end{bmatrix}$$

Note that this is the first column of $(I{-}A)^{-1}$.

In general, the ith column of $(I{-}A)^{-1}$ gives the change in production necessary to meet an increase in external demand of one unit for the product of Sector i.

17.

To	A	B	C
From A	1000	2000	3000
B	0	4000	0
C	0	1000	3000
Total Output	5000	5000	6000

We obtain the technology matrix from the input-output table by dividing each column by its total:

$$A = \begin{bmatrix} 1000/5000 & 2000/5000 & 3000/6000 \\ 0 & 4000/5000 & 0 \\ 0 & 1000/5000 & 3000/6000 \end{bmatrix}$$

$$= \begin{bmatrix} 0.2 & 0.4 & 0.5 \\ 0 & 0.8 & 0 \\ 0 & 0.2 & 0.5 \end{bmatrix}$$

19. First obtain the technology matrix from the input-output table by dividing each column by its total:

$$A = \begin{bmatrix} 10,000/50,000 & 20,000/40,000 \\ 5000/50,000 & 0 \end{bmatrix}$$

$$= \begin{bmatrix} 0.2 & 0.5 \\ 0.1 & 0 \end{bmatrix}$$

Production $= (I{-}A)^{-1} \times$ Demand

$$= \left(\begin{bmatrix} 1 & 0 \\ 0 & 1 \end{bmatrix} - \begin{bmatrix} 0.2 & 0.5 \\ 0.1 & 0 \end{bmatrix} \right)^{-1} \begin{bmatrix} 45,000 \\ 30,000 \end{bmatrix}$$

$$= \begin{bmatrix} 0.8 & -.5 \\ -.1 & 1 \end{bmatrix}^{-1} \begin{bmatrix} 45,000 \\ 30,000 \end{bmatrix}$$

$$= \begin{bmatrix} 4/3 & 2/3 \\ 2/15 & 16/15 \end{bmatrix} \begin{bmatrix} 45,000 \\ 30,000 \end{bmatrix} = \begin{bmatrix} 80,000 \\ 38,000 \end{bmatrix}$$

Solution: The Main DR had to produce $80,000 worth of food, while Bits & Bytes had to produce $38,000 worth of food.

21. First obtain the technology matrix from the input-output table by dividing each column by its total:

$$A = \begin{bmatrix} 6000/90,000 & 500/140,000 \\ 24,000/90,000 & 30,000/140,000 \end{bmatrix}$$

$$= \begin{bmatrix} 1/15 & 1/280 \\ 4/15 & 3/14 \end{bmatrix}$$

Production $= (I{-}A)^{-1} \times$ Demand

$$= \left(\begin{bmatrix} 1 & 0 \\ 0 & 1 \end{bmatrix} - \begin{bmatrix} 1/15 & 1/280 \\ 4/15 & 3/14 \end{bmatrix} \right)^{-1} \begin{bmatrix} 80,000 \\ 90,000 \end{bmatrix}$$

$$= \begin{bmatrix} 14/15 & -1/280 \\ -4/15 & 11/14 \end{bmatrix}^{-1} \begin{bmatrix} 80,000 \\ 90,000 \end{bmatrix}$$

$$\approx \begin{bmatrix} 1.07282 & 0.00488 \\ 0.36411 & 1.27438 \end{bmatrix} \begin{bmatrix} 80,000 \\ 90,000 \end{bmatrix}$$

$$\approx \begin{bmatrix} 86265 \\ 143823 \end{bmatrix}$$

Solution: Equipment Sector production approximately $86,000 million, Components Sector production approximately $140,000 million.

23. $(I{-}A)^{-1} = \begin{bmatrix} 1.228 & 0.182 \\ 0.006 & 1.1676 \end{bmatrix}$

Sector 1 = Textiles, Sector 2 = Clothing & footwear

(a) The missing term is the number of units of Sector 2 needed to produce one additional unit of Sector 1. This quantity is given by the 2,1-entry of $(I{-}A)^{-1}$, or 0.006.

(b) The missing terms refer to the 1,2-entry of $(I{-}A)^{-1}$, which gives the number of units of Sector 1 needed to produce one additional unit of Sector 2—in other words, the additional dollars worth of <u>textiles</u> that must be produced one-dollar increase in the demand for <u>clothing and footwear</u>.

In Exercises 25–28, the technology we show in the solutions is the online Excel tutorial found at Chapter 3 → Excel Tutorials for Chapter 3 → Section 3.4.

25.

To	1	2	3	4
From 1	11,937	9	109	855
2	26,649	4,285	0	4,744
3	0	0	439	61
4	5,423	10,952	3,002	216
Total	97,795	120,594	14,642	47,473

To determine how each sector would need to react to an increase in demand, we compute the columns of $(I–A)^{-1}$. First, we obtain the technology matrix A by dividing each column by its total. (This process is automated in the Excel tutorial):

0.12206	0.00007	0.00744	0.01801
0.27250	0.03553	0.00000	0.09993
0.00000	0.00000	0.02998	0.00128
0.05545	0.09082	0.20503	0.00455

Then compute the matrix $(I–A)^{-1}$

TI-83/84 Format: `(Identity(4)-[A])`$^{-1}$
Online Matrix Algebra Tool: `(I-A)^-1`

1.14099	0.00205	0.01317	0.02087
0.33210	1.04734	0.02605	0.11118
0.00012	0.00013	1.03119	0.00135
0.09388	0.09569	0.21550	1.01615

We are given external demand increases of $1000 million in each sector, so we multiply $(I–A)^{-1}$ by the column matrix

$$D = \begin{bmatrix} 1000 \\ 1000 \\ 1000 \\ 1000 \end{bmatrix}$$

and obtain

1140.99	2.05	13.17	20.87
332.10	1047.34	26.05	111.18
0.12	0.13	1031.19	1.35
93.88	95.69	215.50	1016.15

The columns of this matrix show the amounts, in millions of dollars, by which each sector needs to increase production to meet the additional demand.

27.

To	1	2	3	4
From 1	678.4	3.7	3341.5	1023.5
2	15.5	6.9	17.1	124.5
3	47.3	4.3	893.1	145.8
4	312.5	22.1	83.2	693.5
Total	9401.3	685.8	6997.3	4818.3

To determine how each sector would need to react to an increase in demand, we compute the columns of $(I–A)^{-1}$. First, we obtain the technology matrix A by dividing each column by its total. (This process is automated in the Excel tutorial):

0.07216	0.00540	0.47754	0.21242
0.00165	0.01006	0.00244	0.02584
0.00503	0.00627	0.12763	0.03026
0.03324	0.03223	0.01189	0.14393

Then compute the matrix $(I–A)^{-1}$

TI-83/84 Format: `(Identity(4)-[A])`$^{-1}$
Online Matrix Algebra Tool: `(I-A)^-1`

1.09155	0.01929	0.60157	0.29270
0.00295	1.01123	0.00488	0.03143
0.00779	0.00873	1.15118	0.04289
0.04260	0.03894	0.03953	1.18127

(a) We are given an external demand increase of $100 in Sector 1, so we multiply $(I{-}A)^{-1}$ by the column matrix

$$D = \begin{bmatrix} 100 \\ 0 \\ 0 \\ 0 \end{bmatrix}$$

and obtain

$$X = \begin{bmatrix} 109.155 \\ 0.295 \\ 0.779 \\ 4.260 \end{bmatrix}$$

The additional production required from the meat and milk sector (Sector 3) is the 1,3 entry: $0.78 (rounded to the nearest 1¢).

(b) The diagonal entries in $(I{-}A)^{-1}$ show the additional production required from that sector to meet a $1 increase for the product of that sector. Since the largest diagonal entry is the 4,4-entry: 1.18127, we conclude that Sector 4 (other food products) requires the most of its own product in order to meet a $1 increase in external demand for that product.

29. It would mean that all of the sectors require neither their own product or the product of any other sector.

31. The sum of the entries in a row of an input-output table gives the total internal demand for that sector's products. If that total was equal to the total output for that sector, it would mean that all of the output of that sector was used internally in the economy. Thus, none of the output was available for export and no importing was necessary.

33. If an entry in the matrix $(I{-}A)^{-1}$ is zero, then an increase in demand for one sector (the column sector) has no effect on the production of another sector (the row sector).

35. The off-diagonal entries in $(I{-}A)^{-1}$ show the additional production required by each sector to meet a one-unit increase for the product of some other sector. Usually, to produce one unit of one sector requires less than one unit of input from another. We would expect then that an increase in demand of one unit for one sector would require a smaller increase in production in another sector.

Chapter 3 Review Exercises

1. The sum of two matrices is defined only when they have the same dimensions. Since A is 2×3 and B is 2×2, their dimensions differ, so the sum $A+B$ is undefined.

2. $A-D = \begin{bmatrix} 1 & 2 & 3 \\ 4 & 5 & 6 \end{bmatrix} - \begin{bmatrix} -3 & -2 & -1 \\ 1 & 2 & 3 \end{bmatrix}$

$\quad = \begin{bmatrix} 1+3 & 2+2 & 3+1 \\ 4-1 & 5-2 & 6-3 \end{bmatrix} = \begin{bmatrix} 4 & 4 & 4 \\ 3 & 3 & 3 \end{bmatrix}$

3. $A^{\mathrm{T}} = \begin{bmatrix} 1 & 4 \\ 2 & 5 \\ 3 & 6 \end{bmatrix}$, so

$\quad 2A^{T}+C = 2\begin{bmatrix} 1 & 4 \\ 2 & 5 \\ 3 & 6 \end{bmatrix} + \begin{bmatrix} -1 & 0 \\ 1 & 1 \\ 0 & 1 \end{bmatrix} = \begin{bmatrix} 1 & 8 \\ 5 & 11 \\ 6 & 13 \end{bmatrix}$

4. Computing AB would require taking the product $(2\times3)(2\times2)$. Since the number of columns of A do not match the number of rows of B, the product is undefined.

5. $A^{T}B = \begin{bmatrix} 1 & 4 \\ 2 & 5 \\ 3 & 6 \end{bmatrix}\begin{bmatrix} 1 & -1 \\ 0 & 1 \end{bmatrix}$

$\quad = \begin{bmatrix} (1+0) & (-1+4) \\ (2+0) & (-2+5) \\ (3+0) & (-3+6) \end{bmatrix} = \begin{bmatrix} 1 & 3 \\ 2 & 3 \\ 3 & 3 \end{bmatrix}$

6. Computing A^2 would require taking the product $(2\times3)(2\times3)$. Since the number of columns do not match the number of rows, the product is undefined.

7. $B^{2} = \begin{bmatrix} 1 & -1 \\ 0 & 1 \end{bmatrix}\begin{bmatrix} 1 & -1 \\ 0 & 1 \end{bmatrix}$

$\quad = \begin{bmatrix} (1+0) & (-1-1) \\ (0+0) & (0+1) \end{bmatrix} = \begin{bmatrix} 1 & -2 \\ 0 & 1 \end{bmatrix}$

8. $B^{3} = B^{2}B = \begin{bmatrix} 1 & -2 \\ 0 & 1 \end{bmatrix}\begin{bmatrix} 1 & -1 \\ 0 & 1 \end{bmatrix}$

$\quad = \begin{bmatrix} (1+0) & (-1-2) \\ (0+0) & (0+1) \end{bmatrix} = \begin{bmatrix} 1 & -3 \\ 0 & 1 \end{bmatrix}$

9. $AC+B = \begin{bmatrix} 1 & 2 & 3 \\ 4 & 5 & 6 \end{bmatrix}\begin{bmatrix} -1 & 0 \\ 1 & 1 \\ 0 & 1 \end{bmatrix} + \begin{bmatrix} 1 & -1 \\ 0 & 1 \end{bmatrix}$

$\quad = \begin{bmatrix} 1 & 5 \\ 1 & 11 \end{bmatrix} + \begin{bmatrix} 1 & -1 \\ 0 & 1 \end{bmatrix} = \begin{bmatrix} 2 & 4 \\ 1 & 12 \end{bmatrix}$

10. $CD = \begin{bmatrix} -1 & 0 \\ 1 & 1 \\ 0 & 1 \end{bmatrix}\begin{bmatrix} -3 & -2 & -1 \\ 1 & 2 & 3 \end{bmatrix}$

$\quad = \begin{bmatrix} -4 & -4 & -4 \\ 1 & 2 & 3 \end{bmatrix}$

However, B is 2×2, so the sum $CD + B$ is undefined.

11. To find the inverse of $\begin{bmatrix} 1 & -1 \\ 0 & 1 \end{bmatrix}$, we augment with the 2×2 identity matrix and row-reduce:

$\begin{bmatrix} 1 & -1 & 1 & 0 \\ 0 & 1 & 0 & 1 \end{bmatrix}\begin{array}{c} R_1 + R_2 \\ {} \end{array} \rightarrow \begin{bmatrix} 1 & 0 & 1 & 1 \\ 0 & 1 & 0 & 1 \end{bmatrix}$

The right-hand 2×2 block is the desired inverse:

$\begin{bmatrix} 1 & -1 \\ 0 & 1 \end{bmatrix}^{-1} = \begin{bmatrix} 1 & 1 \\ 0 & 1 \end{bmatrix}$

12. $\begin{bmatrix} 1 & 2 \\ 0 & 0 \end{bmatrix}$ is singular, since it has a row of zeros and therefore cannot be row-reduced to the identity.

13. To find the inverse of a 3×3 matrix, we augment it with the 3×3 identity matrix and row-reduce:

$\begin{bmatrix} 1 & 2 & 3 & 1 & 0 & 0 \\ 0 & 4 & 1 & 0 & 1 & 0 \\ 0 & 0 & 1 & 0 & 0 & 1 \end{bmatrix}\begin{array}{c} 2R_1 - R_2 \\ {} \\ {} \end{array}$

135

$$\begin{bmatrix} 2 & 0 & 5 & 2 & -1 & 0 \\ 0 & 4 & 1 & 0 & 1 & 0 \\ 0 & 0 & 1 & 0 & 0 & 1 \end{bmatrix} \begin{matrix} R_1 - 5R_3 \\ R_2 - R_3 \\ \ \end{matrix}$$

$$\begin{bmatrix} 2 & 0 & 0 & 2 & -1 & -5 \\ 0 & 4 & 0 & 0 & 1 & -1 \\ 0 & 0 & 1 & 0 & 0 & 1 \end{bmatrix} \begin{matrix} (1/2)R_1 \\ (1/4)R_2 \\ \ \end{matrix}$$

$$\begin{bmatrix} 1 & 0 & 0 & 1 & -1/2 & -5/2 \\ 0 & 1 & 0 & 0 & 1/4 & -1/4 \\ 0 & 0 & 1 & 0 & 0 & 1 \end{bmatrix}$$

The right-hand 3×3 block is the desired inverse:

$$\begin{bmatrix} 1 & 2 & 3 \\ 0 & 4 & 1 \\ 0 & 0 & 1 \end{bmatrix}^{-1} = \begin{bmatrix} 1 & -1/2 & -5/2 \\ 0 & 1/4 & -1/4 \\ 0 & 0 & 1 \end{bmatrix}$$

14. To find the inverse of a 4×4 matrix, we augment it with the 4×4 identity matrix and row-reduce:

$$\begin{bmatrix} 1 & 2 & 3 & 4 & 1 & 0 & 0 & 0 \\ 1 & 3 & 4 & 2 & 0 & 1 & 0 & 0 \\ 0 & 1 & 2 & 3 & 0 & 0 & 1 & 0 \\ 0 & 0 & 1 & 2 & 0 & 0 & 0 & 1 \end{bmatrix} \begin{matrix} \ \\ R_2 - R_1 \\ \ \\ \ \end{matrix}$$

$$\begin{bmatrix} 1 & 2 & 3 & 4 & 1 & 0 & 0 & 0 \\ 0 & 1 & 1 & -2 & -1 & 1 & 0 & 0 \\ 0 & 1 & 2 & 3 & 0 & 0 & 1 & 0 \\ 0 & 0 & 1 & 2 & 0 & 0 & 0 & 1 \end{bmatrix} \begin{matrix} R_1 - 2R_2 \\ \ \\ R_3 - R_2 \\ \ \end{matrix}$$

$$\begin{bmatrix} 1 & 0 & 1 & 8 & 3 & -2 & 0 & 0 \\ 0 & 1 & 1 & -2 & -1 & 1 & 0 & 0 \\ 0 & 0 & 1 & 5 & 1 & -1 & 1 & 0 \\ 0 & 0 & 1 & 2 & 0 & 0 & 0 & 1 \end{bmatrix} \begin{matrix} R_1 - R_3 \\ R_2 - R_3 \\ \ \\ R_4 - R_3 \end{matrix}$$

$$\begin{bmatrix} 1 & 0 & 0 & 3 & 2 & -1 & -1 & 0 \\ 0 & 1 & 0 & -7 & -2 & 2 & -1 & 0 \\ 0 & 0 & 1 & 5 & 1 & -1 & 1 & 0 \\ 0 & 0 & 0 & -3 & -1 & 1 & -1 & 1 \end{bmatrix} \begin{matrix} R_1 + R_4 \\ 3R_2 - 7R_4 \\ 3R_3 + 5R_4 \\ \ \end{matrix}$$

$$\begin{bmatrix} 1 & 0 & 0 & 0 & 1 & 0 & -2 & 1 \\ 0 & 3 & 0 & 0 & 1 & -1 & 4 & -7 \\ 0 & 0 & 3 & 0 & -2 & 2 & -2 & 5 \\ 0 & 0 & 0 & -3 & -1 & 1 & -1 & 1 \end{bmatrix} \begin{matrix} \ \\ (1/3)R_2 \\ (1/3)R_3 \\ -(1/3)R_4 \end{matrix}$$

$$\begin{bmatrix} 1 & 0 & 0 & 0 & 1 & 0 & -2 & 1 \\ 0 & 1 & 0 & 0 & 1/3 & -1/3 & 4/3 & -7/3 \\ 0 & 0 & 1 & 0 & -2/3 & 2/3 & -2/3 & 5/3 \\ 0 & 0 & 0 & 1 & 1/3 & -1/3 & 1/3 & -1/3 \end{bmatrix}$$

The right-hand 4×4 block is the desired inverse:

$$\begin{bmatrix} 1 & 2 & 3 & 4 \\ 1 & 3 & 4 & 2 \\ 0 & 1 & 2 & 3 \\ 0 & 0 & 1 & 2 \end{bmatrix}^{-1} = \begin{bmatrix} 1 & 0 & -2 & 1 \\ 1/3 & -1/3 & 4/3 & -7/3 \\ -2/3 & 2/3 & -2/3 & 5/3 \\ 1/3 & -1/3 & 1/3 & -1/3 \end{bmatrix}$$

15.
$$\begin{bmatrix} 1 & 2 & 3 & 4 & 1 & 0 & 0 & 0 \\ 2 & 3 & 3 & 3 & 0 & 1 & 0 & 0 \\ 0 & 1 & 2 & 3 & 0 & 0 & 1 & 0 \\ 0 & 0 & 1 & 2 & 0 & 0 & 0 & 1 \end{bmatrix} \begin{matrix} \ \\ R_2 - 2R_1 \\ \ \\ \ \end{matrix}$$

$$\begin{bmatrix} 1 & 2 & 3 & 4 & 1 & 0 & 0 & 0 \\ 0 & -1 & -3 & -5 & -2 & 1 & 0 & 0 \\ 0 & 1 & 2 & 3 & 0 & 0 & 1 & 0 \\ 0 & 0 & 1 & 2 & 0 & 0 & 0 & 1 \end{bmatrix} \begin{matrix} R_1 + 2R_2 \\ \ \\ R_3 + R_2 \\ \ \end{matrix}$$

$$\begin{bmatrix} 1 & 0 & -3 & -6 & -3 & 2 & 0 & 0 \\ 0 & -1 & -3 & -5 & -2 & 1 & 0 & 0 \\ 0 & 0 & -1 & -2 & -2 & 1 & 1 & 0 \\ 0 & 0 & 1 & 2 & 0 & 0 & 0 & 1 \end{bmatrix} \begin{matrix} R_1 - 3R_3 \\ R_2 - 3R_3 \\ \ \\ R_4 + R_3 \end{matrix}$$

$$\begin{bmatrix} 1 & 0 & 0 & 0 & 3 & -1 & -3 & 0 \\ 0 & -1 & 0 & 1 & 4 & -2 & -3 & 0 \\ 0 & 0 & -1 & -2 & -2 & 1 & 1 & 0 \\ 0 & 0 & 0 & 0 & -2 & 1 & 1 & 1 \end{bmatrix}$$

Since the left-hand 4×4 block has a row of zeros, we cannot reduce the matrix to obtain the 4×4 identity on the left. Therefore, the matrix is singular.

16.
$$\begin{bmatrix} 0 & 1 & 0 & 0 & 1 & 0 & 0 & 0 \\ 1 & 0 & 0 & 0 & 0 & 1 & 0 & 0 \\ 0 & 0 & 0 & 1 & 0 & 0 & 1 & 0 \\ 0 & 0 & 1 & 0 & 0 & 0 & 0 & 1 \end{bmatrix}$$

Rearrange rows

$$\begin{bmatrix} 1 & 0 & 0 & 0 & 0 & 1 & 0 & 0 \\ 0 & 1 & 0 & 0 & 1 & 0 & 0 & 0 \\ 0 & 0 & 1 & 0 & 0 & 0 & 0 & 1 \\ 0 & 0 & 0 & 1 & 0 & 0 & 1 & 0 \end{bmatrix}$$

Thus, $\begin{bmatrix} 0 & 1 & 0 & 0 \\ 1 & 0 & 0 & 0 \\ 0 & 0 & 0 & 1 \\ 0 & 0 & 1 & 0 \end{bmatrix}$ is its own inverse.

17. $x + 2y = 0$
$\quad 3x + 4y = 2$

Matrix form:
$$\begin{bmatrix} 1 & 2 \\ 3 & 4 \end{bmatrix}\begin{bmatrix} x \\ y \end{bmatrix} = \begin{bmatrix} 0 \\ 2 \end{bmatrix}$$

Solving gives
$$\begin{bmatrix} x \\ y \end{bmatrix} = \begin{bmatrix} 1 & 2 \\ 3 & 4 \end{bmatrix}^{-1}\begin{bmatrix} 0 \\ 2 \end{bmatrix}$$
$$= \begin{bmatrix} -2 & 1 \\ 3/2 & -1/2 \end{bmatrix}\begin{bmatrix} 0 \\ 2 \end{bmatrix}$$
$$= \begin{bmatrix} 2 \\ -1 \end{bmatrix}$$

Solution: $(x, y) = (2, -1)$

18. $x + y + z = 3$
$\quad y + 2z = 4$
$\quad y - z = 1$

Matrix form:
$$\begin{bmatrix} 1 & 1 & 1 \\ 0 & 1 & 2 \\ 0 & 1 & -1 \end{bmatrix}\begin{bmatrix} x \\ y \\ z \end{bmatrix} = \begin{bmatrix} 3 \\ 4 \\ 1 \end{bmatrix}$$

Solving gives
$$\begin{bmatrix} x \\ y \\ z \end{bmatrix} = \begin{bmatrix} 1 & 1 & 1 \\ 0 & 1 & 2 \\ 0 & 1 & -1 \end{bmatrix}^{-1}\begin{bmatrix} 3 \\ 4 \\ 1 \end{bmatrix}$$
$$= \begin{bmatrix} 1 & -2/3 & -1/3 \\ 0 & 1/3 & 2/3 \\ 0 & 1/3 & -1/3 \end{bmatrix}\begin{bmatrix} 3 \\ 4 \\ 1 \end{bmatrix}$$

$$= \begin{bmatrix} 0 \\ 2 \\ 1 \end{bmatrix}$$
Solution: $(x, y, z) = (0, 2, 1)$

19. $x + y + z = 2$
$\quad x + 2y + z = 3$
$\quad x + y + 2z = 1$

Matrix form:
$$\begin{bmatrix} 1 & 1 & 1 \\ 1 & 2 & 1 \\ 1 & 1 & 2 \end{bmatrix}\begin{bmatrix} x \\ y \\ z \end{bmatrix} = \begin{bmatrix} 2 \\ 3 \\ 1 \end{bmatrix}$$

Solving gives
$$\begin{bmatrix} x \\ y \\ z \end{bmatrix} = \begin{bmatrix} 1 & 1 & 1 \\ 1 & 2 & 1 \\ 1 & 1 & 2 \end{bmatrix}^{-1}\begin{bmatrix} 2 \\ 3 \\ 1 \end{bmatrix}$$
$$= \begin{bmatrix} 3 & -1 & -1 \\ -1 & 1 & 0 \\ -1 & 0 & 1 \end{bmatrix}\begin{bmatrix} 2 \\ 3 \\ 1 \end{bmatrix}$$
$$= \begin{bmatrix} 2 \\ 1 \\ -1 \end{bmatrix}$$
Solution: $(x, y, z) = (2, 1, -1)$

20. $x + y \quad\quad = 0$
$\quad y + z \quad = 1$
$\quad\quad z + w = 0$
$x \quad - w \quad = 3$

Matrix form:
$$\begin{bmatrix} 1 & 1 & 0 & 0 \\ 0 & 1 & 1 & 0 \\ 0 & 0 & 1 & 1 \\ 1 & 0 & 0 & -1 \end{bmatrix}\begin{bmatrix} x \\ y \\ z \\ w \end{bmatrix} = \begin{bmatrix} 0 \\ 1 \\ 0 \\ 3 \end{bmatrix}$$

Solving gives
$$\begin{bmatrix} x \\ y \\ z \\ w \end{bmatrix} = \begin{bmatrix} 1 & 1 & 0 & 0 \\ 0 & 1 & 1 & 0 \\ 0 & 0 & 1 & 1 \\ 1 & 0 & 0 & -1 \end{bmatrix}^{-1}\begin{bmatrix} 0 \\ 1 \\ 0 \\ 3 \end{bmatrix}$$
$$= \begin{bmatrix} 1/2 & -1/2 & 1/2 & 1/2 \\ 1/2 & 1/2 & -1/2 & -1/2 \\ -1/2 & 1/2 & 1/2 & 1/2 \\ 1/2 & -1/2 & 1/2 & -1/2 \end{bmatrix}\begin{bmatrix} 0 \\ 1 \\ 0 \\ 3 \end{bmatrix}$$

$$= \begin{bmatrix} 1 \\ -1 \\ 2 \\ -2 \end{bmatrix}$$

Solution: $(x, y, z, w) = (1, -1, 2, -2)$

21. Reduce by dominance: Row 1 dominates row 2, so eliminate row 2:

$$\begin{bmatrix} 2 & 1 & 3 & 2 \\ 2 & 0 & 1 & 3 \end{bmatrix}$$

Now column 2 dominates all the other columns, so eliminate all but column 2:

$$\begin{bmatrix} 1 \\ 0 \end{bmatrix}$$

Row 1 dominates row 2, so we eliminate row 2, leaving just the single entry 1. This means that the optimal strategies are pure strategies, corresponding to the first row and second column: $R = [1, 0, 0]$, $C = [0, 1, 0, 0]^T$. The corresponding expected value is the entry in the first row, second column, $e = 1$.

22. We begin by trying to reduce by dominance. However, no row dominates any other, and no column dominates any other, so we cannot reduce this game. Since it is larger than 2×2, we look to see if it is strictly determined. We circle the row minima and box the column maxima:

$$\begin{bmatrix} \boxed{3} & \boxed{\ominus3} & -2 \\ \ominus1 & \boxed{3} & 0 \\ 2 & 2 & \boxed{\textcircled{1}} \end{bmatrix}$$

We do have a saddle point: the 1 in the lower right. So, the game is strictly determined and the optimal strategies are $R = [0, 0, 1]$, $C = [0, 0, 1]^T$, with the expected value equal to the saddle point: $e = 1$.

23. We begin by reducing by dominance. The second row dominates the first, so eliminate the first:

$$\begin{bmatrix} -1 & 3 & 0 \\ 3 & 3 & -1 \end{bmatrix}$$

Now, the first column dominates the second, so eliminate the second:

$$\begin{bmatrix} -1 & 0 \\ 3 & -1 \end{bmatrix}$$

We can't reduce any further, so now we look for the players' optimal mixed strategies. We begin with the row player. Take $R = [x \ \ 1-x]$ and $C = [1 \ \ 0]^T$. $e = RPC = -4x + 3$. If we take $R = [x \ \ 1-x]$ and $C = [0 \ \ 1]^T$, we get $f = RPC = x - 1$. Here are the graphs of e and f:

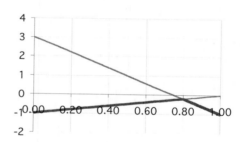

The lower (heavy) portion has its highest point at the intersection of the two lines:

$$-4x + 3 = x - 1$$
$$-5x = -4$$
$$x = 4/5 = 0.8$$

So the optimal row strategy is $R = [0.8 \ \ 0.2]$ for the 2×2 game, which corresponds to $[0 \ \ 0.8 \ \ 0.2]$ for the original game.

For the column player we take $R = [1 \ \ 0]$ and $C = [x \ \ 1-x]$, getting $e = RPC = -x$. Taking $R = [0 \ \ 1]$ and $C = [x \ \ 1-x]$ we get $f = RPC = 4x - 1$. Here are the graphs of e and f:

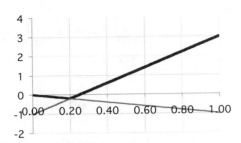

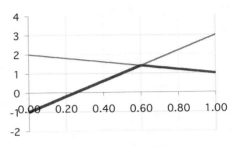

The upper (heavy) portion has its lowest point at the intersection of the two lines:

$$-x = 4x - 1$$
$$-5x = -1$$
$$x = 1/5 = 0.2$$

So, the optimal column strategy is $C = [0.2 \quad 0.8]$ for the 2×2 game, which corresponds to $[0.2 \quad 0 \quad 0.8]$ for the original game.

The expected value is the 2nd coordinate of either intersection point above, or the product RPC with the optimal strategies. In any case, it is $e = -0.2$.

24. We begin by reducing by dominance. The first row dominates the second, so eliminate the second:

$$\begin{bmatrix} 1 & 4 & 3 & 3 \\ 2 & 0 & -1 & 2 \end{bmatrix}$$

Now, the third column dominates the second and the fourth, so we can eliminate those two columns:

$$\begin{bmatrix} 1 & 3 \\ 2 & -1 \end{bmatrix}$$

We can't reduce any further, so now we look for the players' optimal mixed strategies. We begin with the row player. Take $R = [x \quad 1-x]$ and $C = [1 \quad 0]^T$. $e = RPC = -x + 2$. If we take $R = [x \quad 1-x]$ and $C = [0 \quad 1]^T$, we get $f = RPC = 4x - 1$. Here are the graphs of e and f:

The lower (heavy) portion has its highest point at the intersection of the two lines:

$$-x + 2 = 4x - 1$$
$$-5x = -3$$
$$x = 3/5 = 0.6$$

So the optimal row strategy is $R = [0.6 \quad 0.4]$ for the 2×2 game, which corresponds to $[0.6 \quad 0 \quad 0.4]$ for the original game.

For the column player we take $R = [1 \quad 0]$ and $C = [x \quad 1-x]$, getting $e = RPC = -2x + 3$. Taking $R = [0 \quad 1]$ and $C = [x \quad 1-x]$ we get $f = RPC = 3x - 1$. Here are the graphs of e and f:

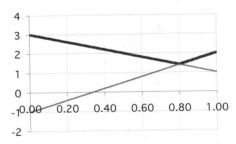

The upper (heavy) portion has its lowest point at the intersection of the two lines:

$$-2x + 3 = 3x - 1$$
$$-5x = -4$$
$$x = 4/5 = 0.8$$

So, the optimal column strategy is $C = [0.8 \quad 0.2]$ for the 2×2 game, which corresponds to $[0.8 \quad 0 \quad 0.2 \quad 0]$ for the original game.

The expected value is the 2nd coordinate of either intersection point above, or the product

RPC with the optimal strategies. In any case, it is *e* = 1.4.

25. $(I - A)^{-1} = \begin{bmatrix} 0.7 & -0.1 \\ 0 & 0.7 \end{bmatrix}^{-1} = \begin{bmatrix} 10/7 & 10/49 \\ 0 & 10/7 \end{bmatrix}$

$X = (I - A)^{-1}D = \begin{bmatrix} 1100 \\ 700 \end{bmatrix}$

26. $(I - A)^{-1} = \begin{bmatrix} 0.3 & -0.1 \\ -0.1 & 0.3 \end{bmatrix}^{-1}$

$= \begin{bmatrix} 15/4 & 5/4 \\ 5/4 & 15/4 \end{bmatrix}$

$X = (I - A)^{-1}D = \begin{bmatrix} 6250 \\ 8750 \end{bmatrix}$

27. $(I - A)^{-1} = \begin{bmatrix} 0.8 & -0.2 & -0.2 \\ 0 & 0.8 & -0.2 \\ 0 & 0 & 0.8 \end{bmatrix}^{-1}$

$= \begin{bmatrix} 5/4 & 5/16 & 25/64 \\ 0 & 5/4 & 5/16 \\ 0 & 0 & 5/4 \end{bmatrix}$

$X = (I - A)^{-1}D = \begin{bmatrix} 48{,}125 \\ 22{,}500 \\ 10{,}000 \end{bmatrix}$

28. $(I - A)^{-1} = \begin{bmatrix} 0.5 & -0.1 & 0 \\ -0.1 & 0.5 & -0.1 \\ 0 & -0.1 & 0.5 \end{bmatrix}^{-1}$

$= \begin{bmatrix} 48/23 & 10/23 & 2/23 \\ 10/23 & 50/23 & 10/23 \\ 2/23 & 10/23 & 48/23 \end{bmatrix}$

$X = (I - A)^{-1}D = \begin{bmatrix} 70{,}000 \\ 120{,}000 \\ 70{,}000 \end{bmatrix}$

29. Inventory $= \begin{bmatrix} 2500 & 4000 & 3000 \\ 1500 & 3000 & 1000 \end{bmatrix}$

Sales $= \begin{bmatrix} 300 & 500 & 100 \\ 100 & 450 & 200 \end{bmatrix}$

To obtain the inventory in the warehouses at the end of June, we subtract the June sales from the June 1 stock (inventory):

Inventory – Sales

$= \begin{bmatrix} 2500 & 4000 & 3000 \\ 1500 & 3000 & 1000 \end{bmatrix} - \begin{bmatrix} 300 & 500 & 100 \\ 100 & 450 & 200 \end{bmatrix}$

$= \begin{bmatrix} 2200 & 3500 & 2900 \\ 1400 & 2550 & 800 \end{bmatrix}$

30. Monthly Sales $= \begin{bmatrix} 280 & 550 & 100 \\ 50 & 500 & 120 \end{bmatrix}$

Assuming no restock, the inventory x months after July 1 is obtained by subtracting x times the monthly sales:

$N = \begin{bmatrix} 2200 & 3500 & 2900 \\ 1400 & 2550 & 800 \end{bmatrix}$

$\quad - x \begin{bmatrix} 280 & 550 & 100 \\ 50 & 500 & 120 \end{bmatrix}$

31. Projected July sales figures given in Question 4(c) are $\begin{bmatrix} 280 & 550 & 100 \\ 50 & 500 & 120 \end{bmatrix}$.

Revenue = Quantity × Price

$= \begin{bmatrix} 280 & 550 & 100 \\ 50 & 500 & 120 \end{bmatrix} \begin{bmatrix} 5 \\ 6 \\ 5.5 \end{bmatrix}$

$= \begin{bmatrix} 5250 \\ 3910 \end{bmatrix} \begin{matrix} \text{Texas} \\ \text{Nevada} \end{matrix}$

32. Profit = Revenue – Cost

$\quad$ = Quantity (Selling Price – Cost Price)

$= \begin{bmatrix} 280 & 550 & 100 \\ 50 & 500 & 120 \end{bmatrix} \left(\begin{bmatrix} 5 \\ 6 \\ 5.5 \end{bmatrix} - \begin{bmatrix} 2 \\ 3.5 \\ 1.5 \end{bmatrix} \right)$

$= \begin{bmatrix} 2615 \\ 1880 \end{bmatrix} \begin{matrix} \text{Texas} \\ \text{Nevada} \end{matrix}$

33. July 1 customers = [2000 4000 4000]

Customers at the end of July

$= [2000 \quad 4000 \quad 4000] \begin{bmatrix} 0.8 & 0.1 & 0.1 \\ 0.4 & 0.6 & 0 \\ 0.2 & 0 & 0.8 \end{bmatrix}$

$= [4000 \quad 2600 \quad 3400]$

34. From Exercise 33, August 1 customers

$$= [2000 \quad 4000 \quad 4000] \begin{bmatrix} 0.8 & 0.1 & 0.1 \\ 0.4 & 0.6 & 0 \\ 0.2 & 0 & 0.8 \end{bmatrix}$$

$$= [4000 \quad 2600 \quad 3400]$$

Customers at the end of August

$$= 4000 \quad 2600 \quad 3400] \begin{bmatrix} 0.8 & 0.1 & 0.1 \\ 0.4 & 0.6 & 0 \\ 0.2 & 0 & 0.8 \end{bmatrix}$$

$$= [4920 \quad 1960 \quad 3120]$$

35. The entries in the matrix give the percentages of customers who switched from one company to another. Such a model does not take the following into account (here are three): (1) It is possible for someone to be a customer at two different enterprises. (2) Some customers may stop using all three of the companies. (3) New customers can enter the field

36. Unknowns

x = Number of shares purchased on July 1

y = Number of shares purchased on August 1

z = Number of shares purchased on September 1

Given information:

Total of 5000 shares purchased:

$x + y + z = 5000$

Total of \$50,000 invested:

$20x + 10y + 5z = 50,000$

Total dividends of \$300 on shares held as of August 15:

$0.10(x + y) = 300$

$0.10x + 0.10y = 300$

We have three equations in three unknowns:

$x + y + z = 5000$

$20x + 10y + 5z = 50,000$

$0.10x + 0.10y = 300$

Matrix form:

$$\begin{bmatrix} 1 & 1 & 1 \\ 20 & 10 & 5 \\ 0.1 & 0.1 & 0 \end{bmatrix} \begin{bmatrix} x \\ y \\ z \end{bmatrix} = \begin{bmatrix} 5000 \\ 50,000 \\ 5 \end{bmatrix}$$

Solving gives

$$\begin{bmatrix} x \\ y \\ z \end{bmatrix} = \begin{bmatrix} 1 & 1 & 1 \\ 20 & 10 & 5 \\ 0.1 & 0.1 & 0 \end{bmatrix}^{-1} \begin{bmatrix} 5000 \\ 50,000 \\ 5 \end{bmatrix}$$

$$= \begin{bmatrix} -.5 & 0.1 & -5 \\ 0.5 & -.1 & 15 \\ 1 & 0 & -10 \end{bmatrix} \begin{bmatrix} 5000 \\ 50,000 \\ 5 \end{bmatrix}$$

$$= \begin{bmatrix} 1000 \\ 2000 \\ 2000 \end{bmatrix}$$

Solution: The company made the following investments: July 1: 1000 shares, August 1: 2000 shares, September 1: 2000 shares

37. We first need to solve #36:

Unknowns

x = Number of shares purchased on July 1

y = Number of shares purchased on August 1

z = Number of shares purchased on September 1

Given information:

Total of 5000 shares purchased:

$x + y + z = 5000$

Total of \$50,000 invested:

$20x + 10y + 5z = 50,000$

Total dividends of \$300 on shares held as of August 15:

$0.10(x + y) = 300$

$0.10x + 0.10y = 300$

We have three equations in three unknowns:

$x + y + z = 5000$

$20x + 10y + 5z = 50,000$

$0.10x + 0.10y = 300$

Matrix form:

$$\begin{bmatrix} 1 & 1 & 1 \\ 20 & 10 & 5 \\ 0.1 & 0.1 & 0 \end{bmatrix} \begin{bmatrix} x \\ y \\ z \end{bmatrix} = \begin{bmatrix} 5000 \\ 50,000 \\ 5 \end{bmatrix}$$

Solving gives

$$\begin{bmatrix} x \\ y \\ z \end{bmatrix} = \begin{bmatrix} 1 & 1 & 1 \\ 20 & 10 & 5 \\ 0.1 & 0.1 & 0 \end{bmatrix}^{-1} \begin{bmatrix} 5000 \\ 50,000 \\ 5 \end{bmatrix}$$

$$= \begin{bmatrix} -.5 & 0.1 & -5 \\ 0.5 & -.1 & 15 \\ 1 & 0 & -10 \end{bmatrix} \begin{bmatrix} 5000 \\ 50,000 \\ 5 \end{bmatrix}$$

$$= \begin{bmatrix} 1000 \\ 2000 \\ 2000 \end{bmatrix}$$

Ta answer the current question, let us write the number of shares purchased (calculated in Exercise 36) as a row matrix:

Shares Purchased = [1000   2000   2000]

Then we can calculated the loss by computing the total purchase cost and subtracting the total proceeds (dividends plus selling price):

Loss = Number of shares × (Purchase price – Dividends – Selling price)

$$= [1000 \ \ 2000 \ \ 2000] \left(\begin{bmatrix} 20 \\ 10 \\ 5 \end{bmatrix} - \begin{bmatrix} 0.10 \\ 0.10 \\ 0 \end{bmatrix} - \begin{bmatrix} 3 \\ 1 \\ 1 \end{bmatrix} \right)$$

$$= [42,700]$$

38. Here is the payoff matrix with the row minima circled and column maxima boxed:

$$P = \begin{bmatrix} 0 & \boxed{\text{–60}} & -40 \\ \boxed{30} & \boxed{20} & \boxed{10} \\ 20 & \boxed{0} & \boxed{15} \end{bmatrix}$$

Since no payoff is both circled and boxed, there are no saddle points, so the game is not strictly determined.

39. $P = \begin{bmatrix} 0 & -60 & -40 \\ 30 & 20 & 10 \\ 20 & 0 & 15 \end{bmatrix}$

$$R = [x \ \ y \ \ z], C = \begin{bmatrix} 0.4 \\ 0.2 \\ 0.4 \end{bmatrix},$$

$e = RPC$

$$= [x \ \ y \ \ z] \begin{bmatrix} 0 & -60 & -40 \\ 30 & 20 & 10 \\ 20 & 0 & 15 \end{bmatrix} \begin{bmatrix} 0.4 \\ 0.2 \\ 0.4 \end{bmatrix}$$

$$= [x \ \ y \ \ z] \begin{bmatrix} -28 \\ 20 \\ 14 \end{bmatrix}$$

$$= -28 + 20y + 14z$$

The largest coefficient is the coefficient of y, so we take

$R = [0 \ 1 \ 0]$

meaning that you should go with the "3 for 1" promotion. The resulting effect on your customer base is then

$e = -28(0) + 20(1) + 14(0) = 20,$

so you will gain 20,000 customers from JungleBooks.

40. Since JungleBooks now has full information about the game, the fundamental principle of game theory comes into effect and you must solve the game to find your minimax optimal strategy. You start by reducing by dominance, which leads to throwing out the "no promotion" options for both players, leaving the following 2×2 game:

$$P = \begin{bmatrix} 20 & 10 \\ 0 & 15 \end{bmatrix}$$

We need only find your optimal strategy as row player. Take $R = [x \ \ 1-x]$ and $C = [1 \ \ 0]^T$. $e = RPC = 20x$. If we take $R = [x \ \ 1-x]$ and $C = [0 \ \ 1]^T$, we get $f = RPC = -5x + 15$. Here are the graphs of e and f:

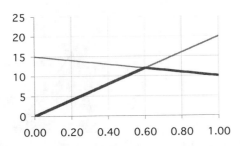

The lower (heavy) portion has its highest point at the intersection of the two lines:

$$20x = -5x + 15$$
$$25x = 15$$
$$x = 3/5 = 0.6$$

So the optimal row strategy is $R = [0.6 \quad 0.4]$ for the 2×2 game, which corresponds to $[0 \quad 0.6 \quad 0.4]$ for the original game. The expected value is the height of the point of intersection, $e = 20(0.6) = 12$, corresponding to a gain of 12,000 customers.

41. We obtain the technology matrix A by dividing each entry in the input-output table by the total in the last row:

$$A = \begin{bmatrix} 0.1 & 0.5 \\ 0.01 & 0.05 \end{bmatrix}$$

42. $(I{-}A)^{-1} = \left(\begin{bmatrix} 1 & 0 \\ 0 & 1 \end{bmatrix} - \begin{bmatrix} 0.1 & 0.5 \\ 0.01 & 0.05 \end{bmatrix} \right)^{-1}$

$$= \begin{bmatrix} 0.9 & -.5 \\ -0.01 & 0.95 \end{bmatrix}^{-1} = \begin{bmatrix} 19/17 & 10/17 \\ 1/85 & 18/17 \end{bmatrix}$$

The (1, 2) entry is $10/17 \approx 0.588$. This gives the number of units of Sector 1 that must be produced to meet a one-unit demand for Sector 2 products. In other words, $0.588 worth of paper must be produced in order to meet a $1 increase in the demand for books.

43. First, compute

$$(I{-}A)^{-1} = \left(\begin{bmatrix} 1 & 0 \\ 0 & 1 \end{bmatrix} - \begin{bmatrix} 0.1 & 0.5 \\ 0.01 & 0.05 \end{bmatrix} \right)^{-1}$$

$$= \begin{bmatrix} 0.9 & -.5 \\ -0.01 & 0.95 \end{bmatrix}^{-1} = \begin{bmatrix} 19/17 & 10/17 \\ 1/85 & 18/17 \end{bmatrix}$$

We are told that the total (external) demand for Bruno Mills' products is

$$\text{Demand} = \begin{bmatrix} 170 \\ 1700 \end{bmatrix}$$

$$\text{Production} = (I{-}A)^{-1} \times \text{Demand}$$

$$= \begin{bmatrix} 19/17 & 10/17 \\ 1/85 & 18/17 \end{bmatrix} \begin{bmatrix} 170 \\ 1700 \end{bmatrix} = \begin{bmatrix} 1190 \\ 1802 \end{bmatrix}$$

Thus, $1190 worth of paper and $1802 worth of books must be produced.

44. We are given the Production vector (capacity) and want to compute the associated demand D:

$$D = X - AX = (I{-}A)X$$

$$= \begin{bmatrix} 0.9 & -.5 \\ -0.01 & 0.95 \end{bmatrix} \begin{bmatrix} 500,000 \\ 200,000 \end{bmatrix}$$

$$= \begin{bmatrix} 350,000 \\ 185,000 \end{bmatrix}$$

Solution: The external demand is $350,000 of paper, $185,000 of books.

Chapter 4
4.1

1. $2x + y \leq 10$

First sketch the graph of

$2x + y = 10$

(graph on the left).

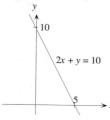 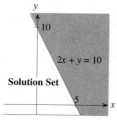

Choose $(0, 0)$ as a test point:

$2(0) + (0) \leq 10$ ✓

Since $(0, 0)$ is in the solution set, we block out the region on the other side of the line as shown above on the right. Since the solution set is not completely enclosed, it is unbounded.

3. $-x - 2y \leq 8$

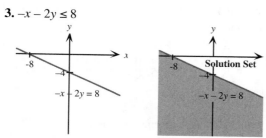

Choose $(0, 0)$ as a test point:

$-0 - 2(0) \leq 8$ ✓

Since $(0, 0)$ is in the solution set, we block out the region on the other side of the line as shown above on the right. Since the solution set is not completely enclosed, it is unbounded.

5. $3x + 2y \geq 5$

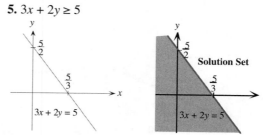

Choose $(0, 0)$ as a test point:

$3(0) + 2(0) \geq 5$ ✗

Since $(0, 0)$ is not in the solution set, we block out the region on the same side of the line as shown above on the right. Since the solution set is not completely enclosed, it is unbounded.

7. $x \leq 3y$

To sketch $x = 3y$, solve for y to obtain $y = \frac{1}{3} x$. This is a line of slope $\frac{1}{3}$ passing through the origin.

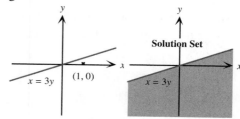

Since $(0, 0)$ is on the line, we choose another test point—say, $(1, 0)$:

$1 \leq 3(0)$ ✗

Since $(1, 0)$ is not in the solution set, we block out the region on the same side of the line as shown above on the right. Since the solution set is not completely enclosed, it is unbounded.

9. $\dfrac{3x}{4} - \dfrac{y}{4} \leq 1$

To sketch the associated line, replace the inequality by equality and solve for y solve for y:

$$\frac{3x}{4} - \frac{y}{4} = 1 \Rightarrow \frac{y}{4} = \frac{3x}{4} - 1$$

$$\Rightarrow y = 3x - 4$$

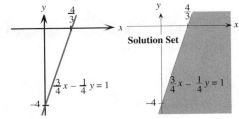

Choose $(0, 0)$ as a test point:

$$\frac{3(0)}{4} - \frac{0}{4} \le 1 \checkmark$$

Since $(0, 0)$ is in the solution set, we block out the region on the other side of the line as shown above on the right. Since the solution set is not completely enclosed, it is unbounded.

11. $x \ge -5$

The line $x = -5$ is a vertical line as shown on the left:

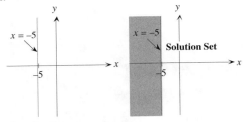

Choose $(0, 0)$ as a test point:

$$0 \ge -5 \checkmark$$

Since $(0, 0)$ is in the solution set, we block out the region on the other side of the line as shown above on the right. Since the solution set is not completely enclosed, it is unbounded.

13. $4x - y \le 8$

$\quad x + 2y \le 2$

The two associated lines are shown below on the left:

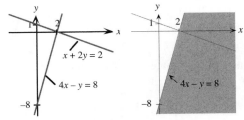

Choose $(0, 0)$ as a test point for the region

$\quad 4x - y \le 8$:

$\quad 4(0) - 0 \le 8 \checkmark$

Therefore we shade to the right of the line $4x - y = 8$ as shown above on the right.

Choose $(0, 0)$ as a test point for the region

$\quad x + 2y \le 2$:

$\quad 0 + 2(0) \le 2 \checkmark$

Therefore we shade above the line $x + 2y = 2$ as shown below.

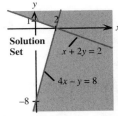

The white region shown is the solution set, which, being not entirely enclosed, is unbounded.

For the corner point, we solve the system

$\quad 4x - y = 8$

$\quad x + 2y = 2$

Multiplying the first equation by 2 and adding gives

$\quad 9x = 18$, so $x = 2$

Substituting $x = 2$ in the first equation gives

$\quad 8 - y = 8$, so $y = 0$

Therefore, the corner point is $(2, 0)$.

15. $3x + 2y \geq 6$

$3x - 2y \leq 6$

$x \geq 0$

The three associated lines are shown below on the left:

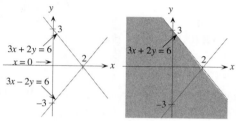

Choose $(0, 0)$ as a test point for the region

$3x + 2y \geq 6$:

$3(0) + 2(0) \geq 6$ ✗

Therefore we shade to the left of the line $3x + 2y = 6$ as shown above on the right.

Now choose $(0, 0)$ as a test point for the region

$3x - 2y \leq 6$:

$3(0) - 2(0) \leq 6$ ✓

Therefore we shade below the line $3x - 2y = 6$ as shown below on the left:

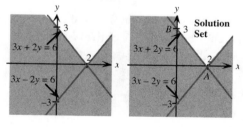

Now choose $(1, 0)$ as a test point for the region $x \geq 0$:

$1 \geq 0$ ✓

Therefore we shade to the left of $x = 0$ as shown above on the right, leaving the solution set as the unshaded area. Again, the solution set is unbounded, since it is no entirely enclosed.

Corner points:

Point	Lines through point	Coordinates
A	$3x + 2y = 6$ $3x - 2y = 6$	$(2, 0)$
B	$x = 0$ $3x + 2y = 6$	$(0, 3)$

17. $x + y \geq 5$

$x \leq 10$

$y \leq 8$

$x \geq 0, y \geq 0$

The last two inequalities $x \geq 0$, $y \geq 0$ tell us that the solution set is in the first quadrant, so we block out everything else, as shown on the left below:

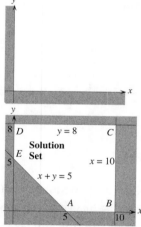

We then add the lines $x+y = 5$, $x = 10$, and $y = 8$ and then shade the appropriate regions, as shown on the right above. The solution set is bounded, since it is entirely enclosed.

Corner points: We can easily read these off the graph:

A: $(5, 0)$, B: $(10, 0)$, C: $(10. 8)$, D: $(0. 8)$, E: $(0, 5)$

19. $20x + 10y \le 100$

$10x + 20y \le 100$

$10x + 10y \le 60$

$x \ge 0, \ y \ge 0$

The last two inequalities $x \ge 0$, $y \ge 0$ tell us that the solution set is in the first quadrant, so we bock out everything else. We then add the lines $20x + 10y = 100$, $10x + 20y = 100$, and $10x + 10y = 60$ and then shade the appropriate regions, as shown below.

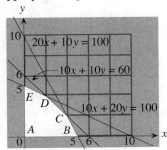

The solution set is the bounded white region.

Corner points:

Point	Lines through point	Coordinates
A	$x = 0, y = 0$	$(0, 0)$
B	$y = 0$ $20x + 10y = 100$	$(5, 0)$
C	$20x + 10y = 100$ $10x + 10y = 60$	$(4, 2)$
D	$10x + 10y = 60$ $10x + 20y = 100$	$(2, 4)$
E	$x = 0$ $10x + 20y = 100$	$(0, 5)$

21. $20x + 10y \ge 100$

$10x + 20y \ge 100$

$10x + 10y \ge 80$

$x \ge 0, \ y \ge 0$

Proceeding as before, we obtain the unbounded solution set shown below:

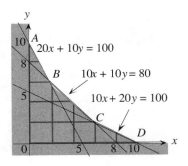

Corner points:

Point	Lines through point	Coordinates
A	$x = 0$ $20x + 10y = 100$	$(0, 10)$
B	$20x + 10y = 100$ $10x + 10y = 80$	$(2, 6)$
C	$10x + 10y = 80$ $10x + 20y = 100$	$(6, 2)$
D	$10x + 20y = 100$ $y = 0$	$(10, 0)$

23. $-3x + 2y \le 5$

$3x - 2y \le 6$

$x \le 2y$

$x \ge 0, y \ge 0$

Solution set (unbounded):

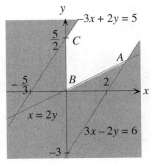

Corner points: We can read two of the corner points, B: $(0, 0)$ and C: $(0, \frac{5}{2})$ directly from the graph. The remaining corner point is A, at the

147

intersection of the lines $x = 2y$ and $3x-2y = 6$. Solving this system of 2 equations gives the point A as $A: (3, \frac{3}{2})$.

25. $2x - y \geq 0$

$\quad x - 3y \leq 0$

$\quad x \geq 0, y \geq 0$

To sketch the lines, solve for y in each case:

$\quad 2x - y = 0 \Rightarrow y = 2x$

$\quad x - 3y = 0 \Rightarrow y = \frac{1}{3}x$

Solution set (unbounded):

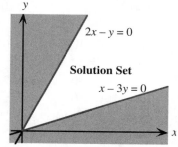

The only corner point is the origin: $(0, 0)$

27. $2.1x - 4.3y \geq 9.7$

To draw this region using technology, solve the associated equation for y:

$\quad 4.3y = 2.1x - 9.7$

$\quad y = \dfrac{2.1}{4.3}x - \dfrac{8.7}{4.3}$

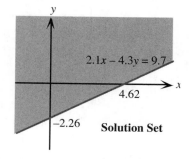

29. $-0.2x + 0.7y \geq 3.3$

$\quad 1.1x + 3.4y \geq 0$

To draw this region using technology, solve the associated equations for y:

$$y = \frac{0.2}{0.7}x + \frac{3.3}{0.7} = \frac{2}{7}x + \frac{33}{7}$$

$$y = -\frac{1.1}{3.4}x = -\frac{11}{34}x$$

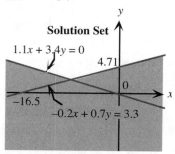

To obtain the coordinates of (the only) corner point, zoom in to it until you can read off the coordinates to two decimal places: $(-7.74, 2.50)$

31. $4.1x - 4.3y \leq 4.4$

$\quad 7.5x - 4.4y \leq 5.7$

$\quad 4.3x + 8.5y \leq 10$

To obtain the associated lines, solve for y:

$$y = \frac{4.1}{4.3}x - \frac{4.4}{4.3}$$

$$y = \frac{7.5}{4.4}x - \frac{5.7}{4.4}$$

$$= -\frac{4.3}{8.5}x + \frac{10}{4.3}$$

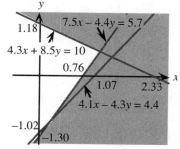

To obtain the coordinates of the corner points, zoom in to it until you can read off the coordinates to two decimal places:

$A: (0.36, -0.68)$, $B: (1.12, 0.61)$

33. Unknowns:

x = # quarts of Creamy Vanilla

y = # quarts of Continental Mocha

Arrange the given information in a table with unknowns across the top:

	Vanilla (x)	Mocha (y)	Available
Eggs	2	1	500
Cream	3	3	900

We can now set up an inequality for each of the items listed on the left:

Eggs: $2x + y \leq 500$

Cream: $3x + 3y \leq 900$

  or $x + y \leq 300$

Since the factory cannot manufacture negative amounts, we also have

  $x \geq 0, y \geq 0$

The solution set is the unshaded region shown below:

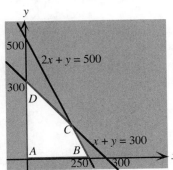

Corner points:

Point	Lines through point	Coordinates
A	$x = 0$ $y = 0$	$(0, 0)$
B	$y = 0$ $2x + y = 500$	$(250, 0)$
C	$2x + y = 500$ $x + y = 300$	$(200, 100)$
D	$x + y = 300$ $x = 0$	$(0, 300)$

35. Unknowns:

x = # ounces of chicken

y = # ounces of grain

Arrange the given information in a table with unknowns across the top:

	Chicken (x)	Grain (y)	Required
Protein	10	2	200
Fat	5	2	150

We can now set up an inequality for each of the items listed on the left (Note that "at least" is represented by "$\geq$"):

Protein: $10x + 2y \geq 200$,

  or $5x + y \geq 100$

Fat: $5x + 2y \geq 150$

The amounts of ingredients cannot be negative:

  $x \geq 0, y \geq 0$

The solution set is the unshaded region shown below:

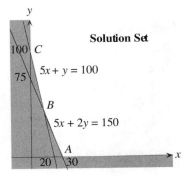

Corner points:

Point	Lines through point	Coordinates
A	$y = 0$ $5x + 2y = 150$	$(30, 0)$
B	$5x + 2y = 150$ $5x + y = 100$	$(10, 50)$
C	$5x + y = 100$ $x = 0$	$(0, 100)$

37. Unknowns:

x = # servings of Mixed Cereal

y = # servings of Mango Tropical Fruit

Arrange the given information in a table with unknowns across the top:

	Cereal (x)	Mango (y)	Required
Calories	60	80	140
Carbs.	11	21	32

We can now set up an inequality for each of the items listed on the left (Note that "at least" is represented by "≥"):

Calories: $60x + 80y \geq 140$,

or $3x + 4y \geq 7$

Carbs: $11x + 21y \geq 32$

The values of the unknowns cannot be negative:

$x \geq 0, y \geq 0$

The solution set is the unshaded region shown below:

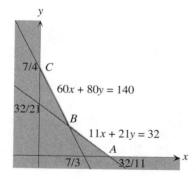

Corner points:

Point	Lines through point	Coordinates
A	$y = 0$ $11x + 21y = 32$	(32/11, 0)
B	$11x + 21y = 32$ $3x + 4y = 7$	(1, 1)
C	$3x + 4y = 7$ $x = 0$	(0, 7/4)

39. Unknowns:

x = # dollars in PNF

y = # dollars in FDMMX

Total invested is up to \$80,000:

$x + y \leq 80,000$

Interest earned is at least \$4200:

$0.06x + 0.05y \geq 4200$,

or $6x + 5y \geq 420,000$

Unknowns nonnegative:

$x \geq 0, y \geq 0$

The solution set is the thin unshaded triangle shown below:

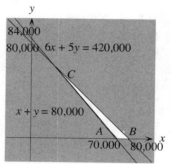

Corner points:

Point	Lines through point	Coordinates
A	$y = 0$ $6x + 5y = 420,000$	(70,000, 0)
B	$y = 0$ $x + y = 80,000$	(80,000, 0)
C	$x + y = 80,000$ $6x + 5y = 420,000$	(20,000, 60,000)

41. Unknowns:

x = # shares in MO

y = # shares in RAI

Total invested is up to \$12,100:

$50x + 55y \leq 12,100$

At least \$550 in dividends

$0.045(50x) + 0.05(55y) \geq 550$,

or $2.25x + 2.75y \geq 550$

Unknowns nonnegative:

$x \geq 0, y \geq 0$

The solution set is the thin unshaded triangle shown below:

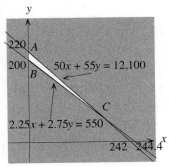

Corner points:

Point	Lines through point	Coordinates
A	$x = 0$ $50x + 55y = 12{,}100$	$(0, 220)$
B	$x = 0$ $2.25x + 2.75y = 550$	$(0, 200)$
C	$50x + 55y = 12{,}100$ $2.25x + 2.75y = 550$	$(220, 20)$

43. Unknowns:

$x = $ # full-page ads in Sports Illustrated

$y = $ # full-page ads of ads in GQ

Readership: $0.65x + 0.15y \geq 3$

At least 3 full-page ads in each magazine:

$\quad x \geq 3, \; y \geq 3$

The solution set is the unshaded region shown below:

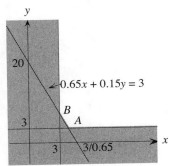

Corner points (coordinates rounded to the nearest whole number):

Point	Lines through point	Coordinates
A	$y = 3$ $0.65x + 0.15y = 3$	$\left(\dfrac{2.55}{0.65}, 3\right)$ $\approx (4, 3)$ (rounded)
B	$x = 3$ $0.65x + 0.15y = 3$	$\left(3, \dfrac{1.05}{0.15}\right)$ $= (3, 7)$

45. Many of the systems of inequalities in the earlier exercises have unbounded solution sets. Another example is: $x \geq 0, \; y \geq 0, \; x+y \geq 1$.

47. The given triangle is the region enclosed by the lines $x = 0, \; y = 0,$ and $x + 2y = 2$ (see figure).

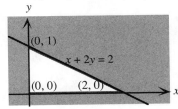

Thus, the region can be described as the solution set of the system $x \geq 0, \; y \geq 0, \; x + 2y \leq 2$.

49. (Answers may vary.) One limitation is that the method is only suitable for situations with two unknown quantities. Another limitation is accuracy, which is never perfect when graphing.

51. There should be at least 3 more grams of ingredient A than ingredient B. Rephrasing this statement gives:

The number of grams of ingredient A exceeds the grams of ingredient B by 3:

$\quad x - y \geq 3$

Choice (C)

53. There should be at least 3 parts (by weight) of ingredient A to 2 parts of ingredient B. That is, 3/2 = 1.5 parts of ingredient A to 1 part of ingredient B. Rephrasing this statement gives: The number of grams of ingredient A is 1.5 times the number of grams of ingredient B:

$x = 1.5y$

$2x = 3y$

$2x - 3y = 0$

Choice (B)

55. There are no feasible solutions; that is, it is impossible to satisfy all the constraints.

4.2

1. Maximize $p = x + y$ subject to

$$x + 2y \le 9$$
$$2x + y \le 9$$
$$x \ge 0, \, y \ge 0$$

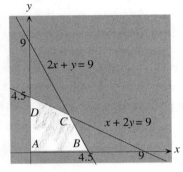

Point	Lines	Coords.	$p = x + y$
A	$x = 0, \, y = 0$	$(0, 0)$	0
B	$y = 0$ $2x + y = 9$	$(4.5, 0)$	4.5
C	$2x + y = 9$ $x + 2y = 9$	$(3, 3)$	**6**
D	$x + 2y = 9$ $x = 0$	$(0, 4.5)$	4.5

Maximum value occurs at point C: $p = 6$ for $x = 3$, $y = 3$.

3. Minimize $c = x + y$ subject to

$$x + 2y \ge 6$$
$$2x + y \ge 6$$
$$x \ge 0, \, y \ge 0$$

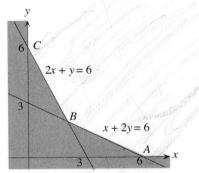

Although the feasible region is unbounded, there is no need to add a bounding rectangle since, by the *FAQ* at the end of the section in the text:

If you are minimizing $c = ax + by$ with a and b nonnegative, $x \ge 0$, and $y \ge 0$, then optimal solutions always exist.

Point	Lines	Coords.	$c = x + y$
A	$y = 0$ $x + 2y = 6$	$(6, 0)$	6
B	$x + 2y = 6$ $2x + y = 6$	$(2, 2)$	**4**
C	$2x + y = 6$ $x = 0$	$(0, 6)$	6

Minimum value occurs at point B: $c = 4$, $x = 2$, $y = 2$.

5. Maximize $p = 3x + y$ subject to

$$3x - 7y \le 0$$
$$7x - 3y \ge 0$$
$$x + y \le 10$$
$$x \ge 0, \, y \ge 0$$

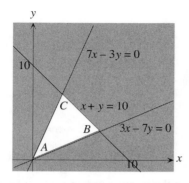

Point	Lines	Coords.	$p = 3x + y$
A	$x = 0, y = 0$	$(0, 0)$	0
B	$3x - 7y = 0$ $x + y = 10$	$(7, 3)$	**24**
C	$7x - 3y = 0$ $x + y = 10$	$(3, 7)$	16

Maximum value occurs at point B: $p = 24$, $x = 7$, $y = 3$.

7. Maximize $p = 3x + 2y$ subject to
$$0.2x + 0.1y \le 1$$
$$0.15x + 0.3y \le 1.5$$
$$10x + 10y \le 60$$
$$x \ge 0, \quad y \ge 0$$

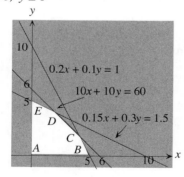

Point	Lines	Coords.	$p = 3x+2y$
A	$x = 0, y = 0$	$(0, 0)$	0
B	$y = 0$ $0.2x+0.1y = 1$	$(5, 0)$	15
C	$0.2x+0.1y = 1$ $10x+10y = 60$	$(4, 2)$	**16**
D	$10x+10y = 60$ $0.15x+0.3y = 1.5$	$(2, 4)$	14
E	$0.15x+0.3y = 1.5$ $x = 0$	$(0, 5)$	10

Maximum value occurs at point C: $p = 16$, $x = 4$, $y = 2$.

9. Minimize $c = 0.2x + 0.3y$ subject to
$$0.2x + 0.1y \ge 1$$
$$0.15x + 0.3y \ge 1.5$$
$$10x + 10y \ge 80$$
$$x \ge 0, \quad y \ge 0$$

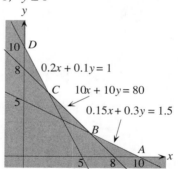

Although the feasible region is unbounded, there is no need to add a bounding rectangle since the coefficients of c are nonnegative.

Point	Lines	Coords	$c = 0.2x+0.3y$
A	$y = 0$ $0.15x+0.3y = 1.5$	$(10, 0)$	2
B	$0.15x+0.3y = 1.5$ $10x+10y = 80$	$(6, 2)$	**1.8**
C	$10x+10y = 80$ $0.2x+0.1y = 1$	$(2, 6)$	2.2

D	$0.2x+0.1y = 1$ $x = 0$	$(0, 10)$	3

Minimum value occurs at point B: $c = 1.8$, $x = 6$, $y = 2$.

11. Maximize and minimize $p = x + 2y$ subject to

$x + y \geq 2$
$x + y \leq 10$
$x - y \leq 2$
$x - y \geq -2$

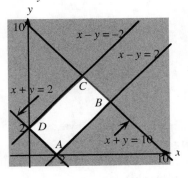

Point	Lines	Coords.	$p = x + 2y$
A	$x + y = 2$ $x - y = 2$	$(2, 0)$	**2** **Min**
B	$x - y = 2$ $x + y = 10$	$(6, 4)$	14
C	$x + y = 10$ $x - y = -2$	$(4, 6)$	**16** **Max**
D	$x - y = -2$ $x + y = 2$	$(0, 2)$	4

Maximum value occurs at point C: $p = 16$, $x = 4$, $y = 6$; minimum value occurs at point A: $p = 2$, $x = 2$, $y = 0$.

13. Maximize $p = 2x + 3y$ subject to

$0.1x + 0.2y \geq 1$
$2x + y \geq 10$
$x \geq 0, y \geq 0$

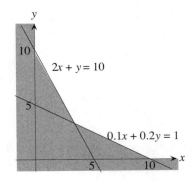

This is an unbounded region, and we wish to maximize $p = 2x + 3y$. According to the *FAQ* at the end of the Section 4.2:

If you are maximizing $p = ax + by$ with a and b nonnegative, then there is no optimal solution unless the feasible region is bounded.

Thus, there is no optimal solution.

15. Minimize $c = 2x + 4y$ subject to

$0.1x + 0.1y \geq 1$
$x + 2y \geq 14$
$x \geq 0, y \geq 0$

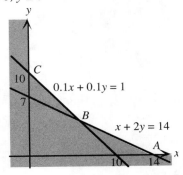

Although the feasible region is unbounded, there is no need to add a bounding rectangle since the coefficients of c are nonnegative

155

Point	Lines	Coords.	$c = 2x+4y$
A	$y = 0$ $x + 2y = 14$	$(14, 0)$	**28**
B	$x + 2y = 14$ $0.1x+0.1y = 1$	$(6, 4)$	**28**
C	$0.1x+0.1y = 1$ $x = 0$	$(0, 10)$	40

Minimum value occurs at points A and B: $c = 28$; $(x, y) = (14, 0)$ and $(6, 4)$, and the line connecting them.

17. Minimize $c = 3x - 3y$ subject to
$$\frac{x}{4} \le y, \quad y \le \frac{2x}{3}$$
$$x + y \ge 5$$
$$x + 2y \le 10$$
$$x \ge 0, y \ge 0$$

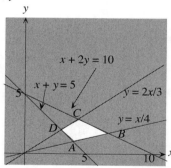

Point	Lines	Coords.	$c = 3x–3y$
A	$y = x/4$ $x + y = 5$	$(4, 1)$	9
B	$y = x/4$ $x + 2y = 10$	$(20/3, 5/3)$	15
C	$x + 2y = 10$ $y = 2x/3$	$(30/7, 20/7)$	30/7
D	$y = 2x/3$ $x + y = 5$	$(3, 2)$	**3**

Minimum value occurs at point D: $c = 3$, $x = 3$, $y = 2$.

19. Maximize $p = x + y$ subject to
$$x + 2y \ge 10$$
$$2x + 2y \le 10$$
$$2x + y \ge 10$$
$$x \ge 0, \; y \ge 0$$

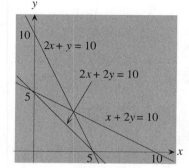

Feasible region is empty—no solutions.

21. (See # 33 in Section 4.1) Unknowns:
x = # quarts of Creamy Vanilla
y = # quarts of Continental Mocha
Maximize $p = 3x + 2y$ subject to
$$2x + y \le 500$$
$$3x + 3y \le 900 \text{ or } x + y \le 300$$
$$x \ge 0, y \ge 0.$$

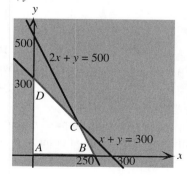

Corner points:

Point	Lines	Coords.	$p = 3x+2y$
A	$x = 0$ $y = 0$	$(0, 0)$	0
B	$y = 0$ $2x + y = 500$	$(250, 0)$	750

C	$2x + y = 500$ $x + y = 300$	(200, 100)	**800**
D	$x + y = 300$ $x = 0$	(0, 300)	600

Maximum value occurs at the point C: $p = 800$; $x = 200$, $y = 100$.

Solution: You should make 200 quarts of vanilla and 100 quarts of mocha.

23. (See # 35 in Section 4.1) Unknowns:

x = # ounces of chicken

y = # ounces of grain

Minimize $c = 10x + y$ subject to

$$10x + 2y \geq 200$$
$$5x + 2y \geq 150$$
$$x \geq 0, y \geq 0.$$

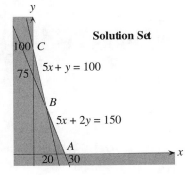

Corner points:

Point	Lines	Coords.	$c = 10x+y$
A	$y = 0$ $5x+2y = 150$	(30, 0)	300
B	$5x+2y = 150$ $5x+y = 100$	(10, 50)	150
C	$5x+y = 100$ $x = 0$	(0, 100)	**100**

Minimum value occurs at the point C: $c = 800$; $x = 0$, $y = 100$.

Solution: Ruff, Inc., should use 100 oz of grain and no chicken.

25. (See # 37 in Section 4.1.) Unknowns:

x = # servings of Mixed Cereal

y = # servings of Mango Tropical Fruit

Minimize $c = 30x + 50y$ subject to

$$11x + 21y \geq 32$$
$$3x + 4y \geq 7$$
$$x \geq 0, y \geq 0.$$

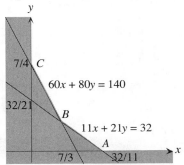

Corner points:

Point	Lines	Coords.	$c = 30x+50y$
A	$y = 0$ $11x+21y = 32$	(32/11, 0)	87.3
B	$11x+21y = 32$ $3x+4y = 7$	(1, 1)	**80**
C	$3x+4y = 7$ $x = 0$	(0, 7/4)	87.5

Minimum value occurs at the point B: $c = 80$; $x = 1$, $y = 1$.

Solution: Feed your child 1 serving of cereal and 1 serving of dessert.

27. Unknowns:

x = # compact fluorescent light bulbs

y = # square ft of insulation

Maximize $p = 2x + 0.2y$ subject to

$$4x + y \le 1200$$
$$x \le 60$$
$$y \le 1100$$
$$x \ge 0, y \ge 0.$$

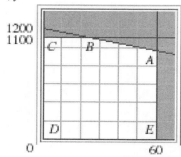

Corner points:

Point	Lines	Coords.	$p = 2x + 0.2y$
A	$4x + y = 1200$ $x = 60$	(60,960)	**312**
B	$y = 1100$ $4x + y = 1200$	(25,1100)	270
C	$y = 1100$ $x = 0$	(0, 1100)	220
D	$y = 0$ $x = 0$	(0, 0)	0
E	$x = 60$ $y = 0$	(60, 0)	120

Maximum value occurs at the point A: $p = 312$; $x = 60$, $y = 960$.

Solution: Purchase 60 compact fluorescent lightbulbs and 960 square feet of insulation for a total saving of $312 per year in energy costs.

29. Unknowns:

x = # of servings of Cell-Tech

y = # servings of Riboforce HP.

Minimize $c = 2.20x + 1.60y$ subject to

$$10x + 5y \ge 80$$
$$2x + y \ge 10$$
$$75x + 15y \le 750$$
$$200x \le 1000$$
$$x \ge 0, y \ge 0$$

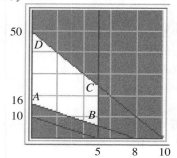

Corner points:

Point	Lines	Coords.	$c = 2.20x+1.60y$
A	$10x+5y = 80$ $x = 0$	(0, 16)	25.6
B	$10x+5y = 80$ $200x = 1000$	(5, 6)	**20.6**
C	$75x+15y = 750$ $200x = 1000$	(5, 25)	51
D	$75x+15y = 750$ $x = 0$	(0, 50)	80

Minimum value occurs at the point B: $c = 20.6$; $x = 5$, $y = 6$.

Solution: Mix 5 servings of Cell-Tech and 6 servings of Riboforce HP for a cost of $20.60.

31. Unknowns:

x = # Dracula salamis

y = # Frankenstein sausages

Maximize $p = x + 3y$ subject to

$x + 2y \leq 1000$

$3x + 2y \leq 2400$

$y \leq 2x$

$x \geq 0, y \geq 0$

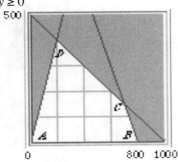

Corner points:

Point	Lines	Coords.	$p = x+3y$
A	$x = 0$ $y = 0$	$(0, 0)$	0
B	$3x+2y = 2400$ $y = 0$	$(800, 0)$	800
C	$x+2y = 1000$ $3x+2y = 2400$	$(700,150)$	1150
D	$x+2y = 1000$ $y-2x = 0$	$(200,400)$	**1400**

Maximum value occurs at the point D: $p = 1400$; $x = 200$, $y = 400$.

Solution: You should make 200 Dracula Salamis and 400 Frankenstein Sausages, for a profit of $1400.

33. Unknowns:

x = # IBM shares

y = # HPQ shares

Maximize $p = 5x + 1.2y$ subject to

$90x + 20y \leq 10,000$

or $9x + 2y \leq 1000$

and $(0.01)(90x) + (0.02)(20y) \geq 120$

or $0.9x + 0.4y \geq 120$

or $9x + 4y \geq 1200$

$x \geq 0, y \geq 0$

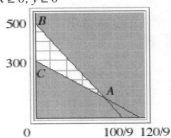

Corner points:

Point	Lines	Coords.	$p = 5x + 1.2y$
A	$9x+2y = 1000$ $9x+4y = 1200$	$(800/9,100)$	$5080/9$
B	$9x+2y = 1000$ $x = 0$	$(0, 500)$	**600**
C	$9x+4y = 1200$ $x = 0$	$(0, 300)$	360

Maximum value occurs at the point B: $p = 600$; $x = 0$, $y = 500$.

Solution: Buy no shares of IBM and 500 shares of HPQ for maximum company earnings of $600.

35. (See # 31 in Section 4.1.) Unknowns:

x = # shares in MO

y = # shares in RAI

Minimize $c = 2.0x + 3.0y$ subject to

$$50x + 55y \leq 12{,}100$$
$$2.25x + 2.75y \geq 550$$
$$x \geq 0, y \geq 0$$

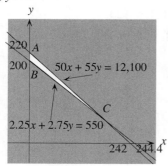

Corner points:

Point	Lines	Coords.	$c =$ 2.0x+3.0y
A	$x = 0$ $50x\ 55y = 12{,}100$	$(0, 220)$	660
B	$x = 0$ $2.25x + 2.75y = 550$	$(0, 200)$	600
C	$50x + 55y = 12{,}100$ $2.25x + 2.75y = 550$	$(220, 20)$	**500**

Minimum value occurs at the point C: $c = 500$; $x = 220$, $y = 20$.

Solution: Buy 220 shares of MO and 20 shares of RAI. The minimum total risk index is $c = 500$.

37. Unknowns:

x = # spots on "Becker"

y = # spots on "The Simpsons"

Maximize $V = 8.3x + 7.5y$, subject to

$$x + y \geq 30$$
$$2x + 1.5y \leq 70$$
$$-x + y \leq 0$$
$$x \geq 0, y \geq 0$$

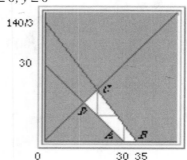

Corner points:

Point	Lines	Coords.	$V =$ 8.3x+7.5y
A	$x + y = 30$ $y = 0$	$(30, 0)$	249
B	$2x + 1.5y = 70$ $y = 0$	$(35, 0)$	290.5
C	$2x + 1.5y = 70$ $-x + y = 0$	$(20, 20)$	**316**
D	$x + y = 30$ $-x + y = 0$	$(15, 15)$	237

Maximum value occurs at the point C: $V = 316$; $x = 20$, $y = 20$.

Solution: You should purchase 20 spots on "Becker" and 20 spots on "The Simpsons."

39. Unknowns:

x = # hours spent in battle instruction per week

y = # hours spent per week in diplomacy instruction

Maximize $p = 50x + 40y$ subject to

$\quad x + y \leq 50$

$\quad x \geq 2y$

$\quad y \geq 10$

$\quad 10x + 5y \geq 400$

$\quad x \geq 0, \, y \geq 0$

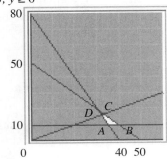

Corner points:

Point	Lines	Coords.	$p = 50x+40y$
A	$y = 10$ $10x+5y = 400$	(35, 10)	2150
B	$x+y = 50$ $y = 10$	(40, 10)	**2400**
C	$x+y = 50$ $x–2y = 0$	(100/3, 50/3)	7000/3
D	$x–2y = 0$ $10x+5y = 400$	(32, 16)	2240

Maximum value occurs at the point B: $p = 2400$; $x = 40$, $y = 10$.

Solution: He should instruct in diplomacy for 10 hours per week and in battle for 40 hours per week, giving a weekly profit of 2400 ducats.

41. Unknowns:

x = # sleep spells

y = # shock spells

Minimize $c = 500x + 750y$ subject to

$\quad 2x+3y \geq 1440$

$\quad 3x+2y \geq 1200$

$\quad x–3y \leq 0$

$\quad -x + 2y \leq 0$

$\quad x \geq 0, \, y \geq 0$

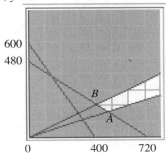

Corner points:

Point	Lines	Coords.	$c = 500x + 750y$
A	$2x+3y = 1440$ $x–3y = 0$	(480, 160)	**360,000**
B	$2x+3y = 1440$ $-x+2y = 0$	(2880/7, 1440/7)	**360,000**

Minimum value occurs at points A and B: $c = 360,000$; $x = 480$, $y = 160$ and $x = 2880/7$, $y = 1440/7$.

Solution: Gillian could expend a minimum of 360,000 pico-shirleys of energy by using 480 sleep spells and 160 shock spells. (There is actually a whole line of solutions joining the one above with $x = 2880/7$, $y = 1440/7$.)

43. Unknowns:

x = # hours for new customers

y = # hours for old customers

Maximize $p = 10x + 30y$ subject to

$$10x + 30y \geq 1200$$
$$x + y \leq 160$$
$$x \geq 100$$
$$x \geq 0, y \geq 0$$

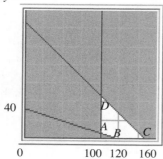

Corner points:

Point	Lines	Coords.	$p =$ $10x+30y$
A	$10x+30y = 1200$ $x = 100$	$(100, 20/3)$	1200
B	$10x+30y = 1200$ $y = 0$	$(120, 0)$	1200
C	$x+y = 160$ $y = 0$	$(160, 0)$	1600
D	$x+y = 160$ $x = 100$	$(100, 60)$	**2800**

Maximum value occurs at the point D: $p = 2800$; $x = 100$, $y = 60$.

Solution: Allocate 100 hours per week for new customers and 60 hours per week for old customers.

45. By the Fundamental Theorem of linear programming, linear programming problems with bounded, non-empty feasible regions always have optimal solutions (Choice (A)).

47. Every point along the line connecting them is also an optimal solution.

49. Here are two simple examples:
(1) (Empty feasible region) Maximize $p = x + y$ subject to $x + y \leq 10$; $x + y \geq 11$, $x \geq 0$, $y \geq 0$.
(2) (Unbounded feasible region and no optimal solutions) Minimize $c = x - y$ subject to $x + y \geq 10$, $x \geq 0$, $y \geq 0$.

53. A simple example is the following: Maximize profit $= p = x + y$ subject to $x \geq 0$, $y \geq 0$. Then p can be made as large as we like by choosing large values of x and/or y. Thus there is no optimal solution to the problem.

55. Mathematically, this means that there are infinitely many possible solutions: one for each point along the line joining the two corner points in question. In practice, select those points with integer solutions (since x and y must be whole numbers in this problem) that are in the feasible region and close to this line, and choose the one that gives the largest profit.

57. The corner points are shown in the following figure:

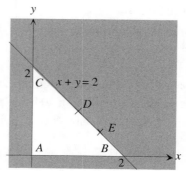

The corner point have coordinates A: $(0, 0)$, B: $(2, 0)$, C:$(0, 2)$, and so $p = xy$ is zero at each of these points. However, at infinitely many points along the diagonal such as D: $(1, 1)$ and E: $(1.5, 0.5)$, the value of p is positive. (For example, at D, $p = (1)(1) = 1$, and at E, $p = (1.5)(0.5) = 0.75$.)

4.3

1. Introduce slack variables and rewrite the constraints and objective function in standard form, and set up the first tableau.

$$x + 2y + s = 6$$
$$-x + y + t = 4$$
$$x + y + u = 4$$
$$-2x - y + p = 0$$

	x	y	s	t	u	p	
s	1	2	1	0	0	0	6
t	−1	1	0	1	0	0	4
u	1	1	0	0	1	0	4
p	−2	−1	0	0	0	1	0

We search for the most negative number in the bottom row (if any). The most negative entry in the bottom row is the −2 in the x-column. So we use this column as the pivot column. The test ratios are: s: 6/1, u: 4/1. The smallest test ratio is u: 4/1. Thus, we pivot on the 1 in the u-row.

	x	y	s	t	u	p		
s	1	2	1	0	0	0	6	$R_1 - R_3$
t	−1	1	0	1	0	0	4	$R_2 + R_3$
u	[1]	1	0	0	1	0	4	
p	−2	−1	0	0	0	1	0	$R_4 + 2R_3$

	x	y	s	t	u	p	
s	0	1	1	0	−1	0	2
t	0	2	0	1	1	0	8
x	1	1	0	0	1	0	4
p	0	1	0	0	2	1	8

Since there are no more negative numbers in the bottom row, we are done, and read off the solution.

Optimal Solution: p = 8/1 = 8; x = 4/1 = 4, y = 0.

3. Introduce slack variables and rewrite the constraints and objective function in standard form, and set up the first tableau.

$$5x - 5y + s = 20$$
$$2x - 10y + t = 40$$
$$-x + y + p = 0$$

	x	y	s	t	p	
s	5	−5	1	0	0	20
t	2	−10	0	1	0	40
p	−1	1	0	0	1	0

We search for the most negative number in the bottom row (if any). The most negative entry in the bottom row is the −1 in the x–column. So we use this column as the pivot column. The test ratios are: s: 20/5, t: 40/2. The smallest test ratio is s: 20/5. Thus, we pivot on the 5 in the s–row.

	x	y	s	t	p		
s	5	−5	1	0	0	20	
t	2	−10	0	1	0	40	$5R_2 - 2R_1$
p	−1	1	0	0	1	0	$5R_3 + R_1$

	x	y	s	t	p	
x	5	−5	1	0	0	20
t	0	−40	−2	5	0	160
p	0	0	1	0	5	20

Since there are no more negative numbers in the bottom row, we are done, and read off the solution:
Optimal Solution: $p = 20/5 = 4$; $x = 20/5 = 4$, $y = 0$.

5. Introduce slack variables and rewrite the constraints and objective function in standard form, and set up the first tableau.

$$5x + 5z + s = 100$$
$$5y - 5z + t = 50$$
$$5x - 5y + u = 50$$
$$-5x + 4y - 3z + p = 0$$

	x	y	z	s	t	u	p	
s	5	0	5	1	0	0	0	100
t	0	5	−5	0	1	0	0	50
u	5	−5	0	0	0	1	0	50
p	−5	4	−3	0	0	0	1	0

The most negative entry in the bottom row is the −5 in the x–column. So we use this column as the pivot column. The test ratios are: s: 100/5, u: 50/5. The smallest test ratio is u: 50/5. Thus, we pivot on the 5 in the u–row.

	x	y	z	s	t	u	p		
s	5	0	5	1	0	0	0	100	$R_1 - R_3$
t	0	5	−5	0	1	0	0	50	
u	[5]	−5	0	0	0	1	0	50	
p	−5	4	−3	0	0	0	1	0	$R_4 + R_3$

	x	y	z	s	t	u	p	
s	0	5	5	1	0	−1	0	50
t	0	5	−5	0	1	0	0	50
x	5	−5	0	0	0	1	0	50
p	0	−1	−3	0	0	1	1	50

The most negative entry in the bottom row is the −3 in the z–column. So we use this column as the pivot column. The only positive entry in this column is the 5 in the s–row, so we pivot on this entry.

	x	y	z	s	t	u	p		
s	0	5	[5]	1	0	−1	0	50	
t	0	5	−5	0	1	0	0	50	$R_2 + R_1$
x	5	−5	0	0	0	1	0	50	
p	0	−1	−3	0	0	1	1	50	$5R_4 + 3R_1$

	x	y	z	s	t	u	p	
z	0	5	5	1	0	−1	0	50
t	0	10	0	1	1	−1	0	100
x	5	−5	0	0	0	1	0	50
p	0	10	0	3	0	2	5	400

Since there are no more negative numbers in the bottom row, we are done, and read off the solution:

Optimal Solution: $p = 400/5 = 80$; $x = 50/5 = 10$, $y = 0$, $z = 50/5 = 10$.

7. Introduce slack variables and rewrite the constraints and objective function in standard form, and set up the first tableau.

$$x + y - z + s = 3$$
$$x + 2y + z + t = 8$$
$$x + y + u = 5$$
$$-7x - 5y - 6z + p = 0$$

	x	y	z	s	t	u	p	
s	1	1	−1	1	0	0	0	3
t	1	2	1	0	1	0	0	8
u	1	1	0	0	0	1	0	5
p	−7	−5	−6	0	0	0	1	0

The most negative entry in the bottom row is the −7 in the x–column. So we use this column as the pivot column. The test ratios are: s: 3/1, t: 8/1, u: 5/1. The smallest test ratio is s: 3/1. Thus, we pivot on the 1 in the s–row.

	x	y	z	s	t	u	p		
s	1	1	−1	1	0	0	0	3	
t	1	2	1	0	1	0	0	8	$R_2 - R_1$
u	1	1	0	0	0	1	0	5	$R_3 - R_1$
p	−7	−5	−6	0	0	0	1	0	$R_4 + 7R_1$

	x	y	z	s	t	u	p	
x	1	1	−1	1	0	0	0	3
t	0	1	2	−1	1	0	0	5
u	0	0	1	−1	0	1	0	2
p	0	2	−13	7	0	0	1	21

The most negative entry in the bottom row is the −13 in the z–column. So we use this column as the pivot column. The test ratios are: t: 5/2, u: 2/1. The smallest test ratio is u: 2/1. Thus, we pivot on the 1 in the u–row.

	x	y	z	s	t	u	p		
x	1	1	−1	1	0	0	0	3	$R_1 + R_3$
t	0	1	2	−1	1	0	0	5	$R_2 - 2R_3$
u	0	0	1	−1	0	1	0	2	
p	0	2	−13	7	0	0	1	21	$R_4 + 13R_3$

	x	y	z	s	t	u	p	
x	1	1	0	0	0	1	0	5
t	0	1	0	1	1	−2	0	1
z	0	0	1	−1	0	1	0	2
p	0	2	0	−6	0	13	1	47

The most negative entry in the bottom row is the −6 in the s–column. So we use this column as the pivot column. The only positive entry in this column is the 1 in the t–row, so we pivot on this entry.

	x	y	z	s	t	u	p		
x	1	1	0	0	0	1	0	5	
t	0	1	0	1	1	−2	0	1	
z	0	0	1	−1	0	1	0	2	$R_3 + R_2$
p	0	2	0	−6	0	13	1	47	$R_4 + 6R_2$

	x	y	z	s	t	u	p	
x	1	1	0	0	0	1	0	5
s	0	1	0	1	1	−2	0	1
z	0	1	1	0	1	−1	0	3
p	0	8	0	0	6	1	1	53

Since there are no more negative numbers in the bottom row, we are done, and read off the solution:
Optimal Solution: $p = 53/1 = 53$; $x = 5/1 = 5$, $y = 0$, $z = 3/1 = 3$.

9. Introduce slack variables and rewrite the constraints and objective function in standard form, and set up the first tableau.

$$5x_1 - x_2 + x_3 + s = 1500$$
$$2x_1 + 2x_2 + x_3 + t = 2500$$
$$4x_1 + 2x_2 + x_3 + u = 2000$$
$$-3x_1 - 7x_2 - 8x_3 + z = 0$$

	x_1	x_2	x_3	s	t	u	z	
s	5	−1	1	1	0	0	0	1500
t	2	2	1	0	1	0	0	2500
u	4	2	1	0	0	1	0	2000
z	−3	−7	−8	0	0	0	1	0

The most negative entry in the bottom row is the −8 in the x_3–column. So we use this column as the pivot column. The test ratios are: s: 1500/1, t: 2500/1, u: 2000/1. The smallest test ratio is s: 1500/1. Thus, we pivot on the 1 in the s–row.

	x_1	x_2	x_3	s	t	u	z		
s	5	−1	$\boxed{1}$	1	0	0	0	1500	
t	2	2	1	0	1	0	0	2500	$R_2 - R_1$
u	4	2	1	0	0	1	0	2000	$R_3 - R_1$
z	−3	−7	−8	0	0	0	1	0	$R_4 + 8R_1$

	x_1	x_2	x_3	s	t	u	z	
x_3	5	−1	1	1	0	0	0	1500
t	−3	3	0	−1	1	0	0	1000
u	−1	3	0	−1	0	1	0	500
z	37	−15	0	8	0	0	1	12000

The most negative entry in the bottom row is the −15 in the x_2–column. So we use this column as the pivot column. The test ratios are: t: 1000/3, u: 500/3. The smallest test ratio is u: 500/3. Thus, we pivot on the 3 in the u–row.

	x_1	x_2	x_3	s	t	u	z		
x_3	5	−1	1	1	0	0	0	1500	$3R_1 + R_3$
t	−3	3	0	−1	1	0	0	1000	$R_2 - R_3$
u	−1	$\boxed{3}$	0	−1	0	1	0	500	
z	37	−15	0	8	0	0	1	12000	$R_4 + 5R_3$

Section 4.3

	x_1	x_2	x_3	s	t	u	z	
x_3	14	0	3	2	0	1	0	5000
t	−2	0	0	0	1	−1	0	500
x_2	−1	3	0	−1	0	1	0	500
z	32	0	0	3	0	5	1	14500

Since there are no more negative numbers in the bottom row, we are done, and read off the solution:
Optimal Solution: $z = 14500/1 = 14500$; $x_1 = 0$, $x_2 = 500/3$, $x_3 = 5000/3$.

11. Introduce slack variables and rewrite the constraints and objective function in standard form, and set up the first tableau.

$$x + y + z + s = 3$$
$$y + z + w + t = 4$$
$$x + z + w + u = 5$$
$$x + y + w + v = 6$$
$$-x +-y - z - w = 0$$

	x	y	z	w	s	t	u	v	p	
s	1	1	1	0	1	0	0	0	0	3
t	0	1	1	1	0	1	0	0	0	4
u	1	0	1	1	0	0	1	0	0	5
v	1	1	0	1	0	0	0	1	0	6
p	−1	−1	−1	−1	0	0	0	0	1	0

The (first) most negative entry in the bottom row is the −1 in the x–column. So we use this column as the pivot column. The test ratios are: s: 3/1, u: 5/1, v: 6/1. The smallest test ratio is s: 3/1. Thus, we pivot on the 1 in the s–row.

	x	y	z	w	s	t	u	v	p		
s	1	1	1	0	1	0	0	0	0	3	
t	0	1	1	1	0	1	0	0	0	4	
u	1	0	1	1	0	0	1	0	0	5	$R_3 - R_1$
v	1	1	0	1	0	0	0	1	0	6	$R_4 - R_1$
p	−1	−1	−1	−1	0	0	0	0	1	0	$R_5 + R_1$

170

	x	y	z	w	s	t	u	v	p	
x	1	1	1	0	1	0	0	0	0	3
t	0	1	1	1	0	1	0	0	0	4
u	0	−1	0	1	−1	0	1	0	0	2
v	0	0	−1	1	−1	0	0	1	0	3
p	0	0	0	−1	1	0	0	0	1	3

The most negative entry in the bottom row is the −1 in the w–column. So we use this column as the pivot column. The test ratios are: t: 4/1, u: 2/1, v: 3/1. The smallest test ratio is u: 2/1. Thus, we pivot on the 1 in the u–row.

	x	y	z	w	s	t	u	v	p		
x	1	1	1	0	1	0	0	0	0	3	
t	0	1	1	1	0	1	0	0	0	4	$R_2 - R_3$
u	0	−1	0	[1]	−1	0	1	0	0	2	
v	0	0	−1	1	−1	0	0	1	0	3	$R_4 - R_3$
p	0	0	0	−1	1	0	0	0	1	3	$R_5 + R_3$

	x	y	z	w	s	t	u	v	p	
x	1	1	1	0	1	0	0	0	0	3
t	0	2	1	0	1	1	−1	0	0	2
w	0	−1	0	1	−1	0	1	0	0	2
v	0	1	−1	0	0	0	−1	1	0	1
p	0	−1	0	0	0	0	1	0	1	5

The most negative entry in the bottom row is the −1 in the y–column. So we use this column as the pivot column. The test ratios are: x: 3/1, t: 2/2, v: 1/1. The smallest test ratio is v: 1/1. Thus, we pivot on the 1 in the v–row.

	x	y	z	w	s	t	u	v	p		
x	1	1	1	0	1	0	0	0	0	3	$R_1 - R_4$
t	0	2	1	0	1	1	−1	0	0	2	$R_2 - 2R_4$
w	0	−1	0	1	−1	0	1	0	0	2	$R_3 + R_4$
v	0	[1]	−1	0	0	0	−1	1	0	1	
p	0	−1	0	0	0	0	1	0	1	5	$R_5 + R_4$

	x	y	z	w	s	t	u	v	p	
x	1	0	2	0	1	0	1	−1	0	2
t	0	0	3	0	1	1	1	−2	0	0
w	0	0	−1	1	−1	0	0	1	0	3
y	0	1	−1	0	0	0	−1	1	0	1
p	0	0	−1	0	0	0	0	1	1	6

The most negative entry in the bottom row is the −1 in the z–column. So we use this column as the pivot column. The test ratios are: x: 2/2, t: 0/3. The smallest test ratio is t: 0/3. Thus, we pivot on the 3 in the t–row.

	x	y	z	w	s	t	u	v	p		
x	1	0	2	0	1	0	1	−1	0	2	$3R_1 - 2R_2$
t	0	0	3	0	1	1	1	−2	0	0	
w	0	0	−1	1	−1	0	0	1	0	3	$3R_3 + R_2$
y	0	1	−1	0	0	0	−1	1	0	1	$3R_4 + R_2$
p	0	0	−1	0	0	0	0	1	1	6	$3R_5 + R_2$

	x	y	z	w	s	t	u	v	p	
x	3	0	0	0	1	−2	1	1	0	6
z	0	0	3	0	1	1	1	−2	0	0
w	0	0	0	3	−2	1	1	1	0	9
y	0	3	0	0	1	1	−2	1	0	3
p	0	0	0	0	1	1	1	1	3	18

Since there are no more negative numbers in the bottom row, we are done, and read off the solution: Optimal Solution: $p = 6$; $x = 6/3 = 2$, $y = 3/3 = 1$, $z = 0/3 = 0$, $w = 9/3 = 3$.

13. Introduce slack variables and rewrite the constraints and objective function in standard form, and set up the first tableau.

$x + y + s = 1$

$y + z + t = 2$

$z + w + u = 3$

$w + v + r = 4$

$-x - y - z - w - v + p = 0$

	x	y	z	w	v	s	t	u	r	p		
s	[1]	1	0	0	0	1	0	0	0	0	1	
t	0	1	1	0	0	0	1	0	0	0	2	
u	0	0	1	1	0	0	0	1	0	0	3	
r	0	0	0	1	1	0	0	0	1	0	4	
p	−1	−1	−1	−1	−1	0	0	0	0	1	0	$R_5 + R_1$

The most negative entry in the bottom row is the −1 in the x–column. So we use this column as the pivot column. The only positive entry in this column is the 1 in the s–row, so we pivot on it and obtain the second tableau:

	x	y	z	w	v	s	t	u	r	p		
x	1	1	0	0	0	1	0	0	0	0	1	
t	0	1	[1]	0	0	0	1	0	0	0	2	
u	0	0	1	1	0	0	0	1	0	0	3	$R_3 - R_2$
r	0	0	0	1	1	0	0	0	1	0	4	
p	0	0	−1	−1	−1	1	0	0	0	1	1	$R_5 + R_2$

The most negative entry in the bottom row is the −1 in the z–column. So we use this column as the pivot column. The test ratios are: t: 2/1, u: 3/1. The smallest test ratio is t: 2/1. Thus, we pivot on the 1 in the t–row, and obtain the third tableau:

	x	y	z	w	v	s	t	u	r	p		
x	1	1	0	0	0	1	0	0	0	0	1	
z	0	1	1	0	0	0	1	0	0	0	2	
u	0	−1	0	[1]	0	0	−1	1	0	0	1	
r	0	0	0	1	1	0	0	0	1	0	4	$R_4 - R_3$
p	0	1	0	−1	−1	1	1	0	0	1	3	$R_5 + R_3$

The most negative entry in the bottom row is the −1 in the w–column. So we use this column as the pivot column. The test ratios are: u: 1/1, r: 4/1. The smallest test ratio is u: 1/1. Thus, we pivot on the 1 in the u–row, and obtain the fourth tableau:

	x	y	z	w	v	s	t	u	r	p	
x	1	1	0	0	0	1	0	0	0	0	1
z	0	1	1	0	0	0	1	0	0	0	2
w	0	-1	0	1	0	0	-1	1	0	0	1
r	0	1	0	0	[1]	0	1	-1	1	0	3
p	0	0	0	0	-1	1	0	1	0	1	4

$R_5 + R_4$

The most negative entry in the bottom row is the -1 in the v–column. So we use this column as the pivot column. The only positive entry in this column is the 1 in the r–row, so we picot on this entry and obtain the fifth tableau:

	x	y	z	w	v	s	t	u	r	p	
x	1	1	0	0	0	1	0	0	0	0	1
z	0	1	1	0	0	0	1	0	0	0	2
w	0	-1	0	1	0	0	-1	1	0	0	1
v	0	1	0	0	1	0	1	-1	1	0	3
p	0	1	0	0	0	1	1	0	1	1	7

Since there are no more negative numbers in the bottom row, we are done, and read off the solution:
Optimal Solution: $p = 7$; $x = 1/1 = 1$, $y = 0$, $z = 2/1 = 2$, $w = 1/1 = 1$, $v = 3/1 = 3$

Note: Making different choices for the pivot columns results in the following alternate solution: $p = 7$; $x = 1$, $y = 0$, $z = 2$, $w = 0$, $v = 4$

Note: In Exercises 15–20, we used the online Simplex Method Tool. When entering the problems, there is no need to enter the inequalities $x \geq 0$, $y \geq 0$, etc.

15. Maximize $p = 2.5x + 4.2y + 2z$ subject to
$$0.1x + y - 2.2z \leq 4.5$$
$$2.1x + y + z \leq 8$$
$$x + 2.2y \leq 5$$
$$x \geq 0, y \geq 0, z \geq 0$$

Tableau #1

x	y	z	s1	s2	s3	p	
0.1	1	-2.2	1	0	0	0	4.5
2.1	1	1	0	1	0	0	8
1	2.2	0	0	0	1	0	5
-2.5	-4.2	-2	0	0	0	1	0

Tableau #2

x	y	z	s1	s2	s3	p	
-0.355	0	-2.2	1	0	-0.455	0	2.23
1.65	0	1	0	1	-0.455	0	5.73
0.455	1	0	0	0	0.455	0	2.27
-0.591	0	-2	0	0	1.91	1	9.55

Tableau #3

x	y	z	s1	s2	s3	p	
3.27	0	0	1	2.2	-1.45	0	14.8
1.65	0	1	0	1	-0.455	0	5.73
0.455	1	0	0	0	0.455	0	2.27
2.7	0	0	0	2	1	1	21

Optimal Solution: p = 21; x = 0, y = 2.27, z = 5.73

17. Maximize $p = x + 2y + 3z + w$ subject to

$$x + 2y + 3z \le 3$$
$$y + z + 2.2w \le 4$$
$$x + z + 2.2w \le 5$$
$$x + y + 2.2w \le 6$$
$$x \ge 0, y \ge 0, z \ge 0, w \ge 0$$

Tableau #1

x	y	z	w	s1	s2	s3	s4	p	
1	2	3	0	1	0	0	0	0	3
0	1	1	2.2	0	1	0	0	0	4
1	0	1	2.2	0	0	1	0	0	5
1	1	0	2.2	0	0	0	1	0	6
-1	-2	-3	-1	0	0	0	0	1	0

Tableau #2

x	y	z	w	s1	s2	s3	s4	p	
0.333	0.667	1	0	0.333	0	0	0	0	1
-0.333	0.333	0	2.2	-0.333	1	0	0	0	3
0.667	-0.667	0	2.2	-0.333	0	1	0	0	4
1	1	0	2.2	0	0	0	1	0	6
0	0	0	-1	1	0	0	0	1	3

Tableau #3

x	y	z	w	s1	s2	s3	s4	p	
0.333	0.667	1	0	0.333	0	0	0	0	1
-0.152	0.152	0	1	-0.152	0.455	0	0	0	1.36
1	-1	0	0	0	-1	1	0	0	1
1.33	0.667	0	0	0.333	-1	0	1	0	3
-0.152	0.152	0	0	0.848	0.455	0	0	1	4.36

Tableau #4

x	y	z	w	s1	s2	s3	s4	p	
0	1	1	0	0.333	0.333	-0.333	0	0	0.667
0	0	0	1	-0.152	0.303	0.152	0	0	1.52
1	-1	0	0	0	-1	1	0	0	1
0	2	0	0	0.333	0.333	-1.33	1	0	1.67
0	0	0	0	0.848	0.303	0.152	0	1	4.52

Optimal Solution: p = 4.52; x = 1, y = 0, z = 0.67, w = 1.52

19. Maximize $p = x - y + z - w + v$ subject to

$x + y \le 1.1$

$y + z \le 2.2$

$z + w \le 3.3$

$w + v \le 4.4$

$x \ge 0, y \ge 0, z \ge 0, w \ge 0, v \ge 0$

Tableau #1

x	y	z	w	v	s1	s2	s3	s4	p	
1	1	0	0	0	1	0	0	0	0	1.1
0	1	1	0	0	0	1	0	0	0	2.2
0	0	1	1	0	0	0	1	0	0	3.3
0	0	0	1	1	0	0	0	1	0	4.4
-1	1	-1	1	-1	0	0	0	0	1	0

Tableau #2

x	y	z	w	v	s1	s2	s3	s4	p	
1	1	0	0	0	1	0	0	0	0	1.1
0	1	1	0	0	0	1	0	0	0	2.2
0	0	1	1	0	0	0	1	0	0	3.3
0	0	0	1	1	0	0	0	1	0	4.4
0	2	-1	1	-1	1	0	0	0	1	1.1

Tableau #3

x	y	z	w	v	s1	s2	s3	s4	p	
1	1	0	0	0	1	0	0	0	0	1.1
0	1	1	0	0	0	1	0	0	0	2.2
0	-1	0	1	0	0	-1	1	0	0	1.1
0	0	0	1	1	0	0	0	1	0	4.4
0	3	0	1	-1	1	1	0	0	1	3.3

Tableau #4

x	y	z	w	v	s1	s2	s3	s4	p	
1	1	0	0	0	1	0	0	0	0	1.1
0	1	1	0	0	0	1	0	0	0	2.2
0	-1	0	1	0	0	-1	1	0	0	1.1
0	0	0	1	1	0	0	0	1	0	4.4
0	3	0	2	0	1	1	0	1	1	7.7

Optimal Solution: $p = 7.7$; $x = 1.1$, $y = 0$, $z = 2.2$, $w = 0$, $v = 4.4$

21. Unknowns:

x = # calculus texts

y = # history texts

z = # marketing texts

Maximize $p = 10x + 4y + 8z$ subject to

$$x + y + z \le 650$$
$$2x + y + 3z \le 1000$$
$$x \ge 0, y \ge 0, z \ge 0.$$

	x	y	z	s	t	p		
s	1	1	1	1	0	0	650	$2R_1 - R_2$
t	2	1	3	0	1	0	1000	
p	−10	−4	−8	0	0	1	0	$R_3 + 5R_2$

	x	y	z	s	t	p	
s	0	1	−1	2	−1	0	300
x	2	1	3	0	1	0	1000
p	0	1	7	0	5	1	5000

Optimal Solution: $p = 5000$; $x = 1000/2 = 500$, $y = 0$, $z = 0$. You should purchase 500 calculus texts, no history texts and no marketing texts. The maximum profit is $5000 per semester.

23. Unknowns:

x = # gallons of PineOrange

y = # gallons of PineKiwi

z = # gallons of OrangeKiwi

Maximize $p = x + 2y + z$ subject to

$$2x + 3y \le 800$$
$$2x + 3z \le 650$$
$$y + z \le 350$$
$$x \ge 0, y \ge 0, z \ge 0.$$

	x	y	z	s	t	u	p		
s	2	3	0	1	0	0	0	800	
t	2	0	3	0	1	0	0	650	
u	0	1	1	0	0	1	0	350	$3R_3 - R_1$
p	−1	−2	−1	0	0	0	1	0	$3R_4 + 2R_1$

	x	y	z	s	t	u	p		
y	2	3	0	1	0	0	0	800	
t	2	0	3	0	1	0	0	650	$R_2 - R_3$
u	−2	0	[3]	−1	0	3	0	250	
p	1	0	−3	2	0	0	3	1600	$R_4 + R_3$

	x	y	z	s	t	u	p		
y	2	3	0	1	0	0	0	800	$2R_1 - R_2$
t	[4]	0	0	1	1	−3	0	400	
z	−2	0	3	−1	0	3	0	250	$2R_3 + R_2$
p	−1	0	0	1	0	3	3	1850	$4R_4 + R_2$

	x	y	z	s	t	u	p	
y	0	6	0	1	−1	3	0	1200
x	4	0	0	1	1	−3	0	400
z	0	0	6	−1	1	3	0	900
p	0	0	0	5	1	9	12	7800

Optimal Solution: $p = 650$; $x = 400/4 = 100$, $y = 1200/6 = 200$, $z = 900/6 = 150$. The company makes a maximum profit of \$650 by making 100 gallons of PineOrange, 200 gallons of PineKiwi, and 150 gallons of OrangeKiwi.

25. Unknowns:

x = # sections of Ancient History

y = # sections of Medieval History

z = # sections of Modern History

Maximize $p = x + 2y + 3z$ (in tens of thousands of dollars) subject to

$$x + y + z \le 45$$
$$100x + 50y + 200z \le 5000$$
$$x + y + 2z \le 60$$
$$x \ge 0, y \ge 0, z \ge 0.$$

	x	y	z	s	t	u	p		
s	1	1	1	1	0	0	0	45	$200R_1 - R_2$
t	100	50	200	0	1	0	0	5000	
u	1	1	2	0	0	1	0	60	$100R_3 - R_2$
p	−1	−2	−3	0	0	0	1	0	$200R_4 + 3R_2$

	x	y	z	s	t	u	p		
s	100	150	0	200	−1	0	0	4000	$R_1 - 3R_3$
z	100	50	200	0	1	0	0	5000	$R_2 - R_3$
u	0	50	0	0	−1	100	0	1000	
p	100	−250	0	0	3	0	200	15000	$R_4 + 5R_3$

	x	y	z	s	t	u	p		
s	100	0	0	200	2	−300	0	1000	
z	100	0	200	0	2	−100	0	4000	$R_2 - R_1$
y	0	50	0	0	−1	100	0	1000	$2R_3 + R_1$
p	100	0	0	0	−2	500	200	20000	$R_4 + R_1$

	x	y	z	s	t	u	p	
t	100	0	0	200	2	−300	0	1000
z	0	0	200	−200	0	200	0	3000
y	100	100	0	200	0	−100	0	3000
p	200	0	0	200	0	200	200	21000

Optimal Solution: $p = 105$; $x = 0$, $y = 3000/100 = 30$, $z = 3000/200 = 15$. The department should offer no Ancient History, 30 sections of Medieval History, and 15 sections of Modern History, for a profit of $\$105 \times 10{,}000 = \$1{,}050{,}000$.

Answers to additional question:

The values of the slack variables are:

$t = 1000/2 = 500$, meaning that there will be 500 students without classes.

$s = u = 0$, meaning that all sections and professors are used.

27. Unknowns:

x = # acres of tomatoes

y = # acres of lettuce

z = # acres of carrots

Maximize $p = 20x + 15y + 5z$ (in hundreds of dollars) subject to

$$x + y + z \le 100$$
$$5x + 4y + 2z \le 400$$
$$4x + 2y + 2z \le 500$$
$$x \ge 0, y \ge 0, z \ge 0.$$

	x	y	z	s	t	u	p		
s	1	1	1	1	0	0	0	100	$5R_1 - R_2$
t	5	4	2	0	1	0	0	400	
u	4	2	2	0	0	1	0	500	$5R_3 - 4R_2$
p	−20	−15	−5	0	0	0	1	0	$R_4 + 4R_2$

	x	y	z	s	t	u	p	
s	0	1	3	5	−1	0	0	100
x	5	4	2	0	1	0	0	400
u	0	−6	2	0	−4	5	0	900
p	0	1	3	0	4	0	1	1600

Optimal Solution: $p = 1600$; $x = 400/5 = 80$, $y = 0$, $z = 0$. Plant 80 acres of tomatoes and no lettuce or carrots.

The slack variable corresponding to the number of acres available is s. The value of s is $s = 1200/5 = 20$, meaning that you will leave 20 acres unplanted.

29. Unknowns:

x = # servings of granola

y = # servings of nutty granola

z = # servings of nuttiest granola

Maximize $p = 6x + 8y + 3z$ subject to

$$x + y + 5z \le 1500$$
$$4x + 8y + 8z \le 10{,}000$$
$$2x + 4y + 8z \le 4000$$
$$x \ge 0, y \ge 0, z \ge 0.$$

	x	y	z	s	t	u	p		
s	1	1	5	1	0	0	0	1500	$4R_1 - R_3$
t	4	8	8	0	1	0	0	10000	$R_2 - 2R_3$
u	2	[4]	8	0	0	1	0	4000	
p	−6	−8	−3	0	0	0	1	0	$R_4 + 2R_3$

	x	y	z	s	t	u	p		
s	[2]	0	12	4	0	−1	0	2000	
t	0	0	−8	0	1	−2	0	2000	
y	2	4	8	0	0	1	0	4000	$R_3 - R_1$
p	−2	0	13	0	0	2	1	8000	$R_4 + R_1$

	x	y	z	s	t	u	p	
x	2	0	12	4	0	−1	0	2000
t	0	0	−8	0	1	−2	0	2000
y	0	4	−4	−4	0	2	0	2000
p	0	0	25	4	0	1	1	10,000

Optimal Solution: $p = 10,000$; $x = 2000/2 = 1000$, $y = 2000/4 = 500$, $z = 0$. The Choral Society can make a profit of \$10,000 by selling 1000 servings of Granola, 500 servings of Nutty Granola and no Nuttiest Granola.

To obtain the ingredients left over, look at the values of the slack variables in the final tableau:

$s = 0$, so there are no toasted oats left over.

$t = 2000/1 = 2000$, so there are 2000 oz. of almonds left over.

$u = 0$, so there are no raisins left over.

31. Unknowns:

$x = $ # oil allocated to process A

$y = $ # oil allocated to process B

$z = $ # oil allocated to process C

Maximize $p = 4x + 4y + 4z$ subject to

$x + y + z \leq 50$

$15x + 10y + 5z \leq 300$, or $3x + 2y + z \leq 60$

$60x + 55y + 50z \leq 5000$, or $12x + 11y + 10z \leq 1000$

$x \geq 0$, $y \geq 0$. $z \geq 0$.

	x	y	z	s	t	u	p		
s	1	1	1	1	0	0	0	50	$3R_1 - R_2$
t	[3]	2	1	0	1	0	0	60	
u	12	11	10	0	0	1	0	1000	$R_3 - 4R_2$
p	−4	−4	−4	0	0	0	1	0	$3R_4 + 4R_2$

	x	y	z	s	t	u	p		
s	0	1	[2]	3	−1	0	0	90	
x	3	2	1	0	1	0	0	60	$2R_2 - R_1$
u	0	3	6	0	−4	1	0	760	$R_3 - 3R_1$
p	0	−4	−8	0	4	0	3	240	$R_4 + 4R_1$

	x	y	z	s	t	u	p	
z	0	1	2	3	−1	0	0	90
x	6	3	0	−3	3	0	0	30
u	0	0	0	−9	−1	1	0	490
p	0	0	0	12	0	0	3	600

Optimal Solution: $p = 200$; $x = 30/6 = 5$, $y = 0$, $z = 90/2 = 45$. Allocate 5 million gals to process A and 45 million gals. to process C.

Another solution (obtained by making different choices for pivoting): Allocate 10 million gals to process B and 40 million gals. to process C.

33. Unknowns:

x = # of servings of Cell-Tech

y = # servings of Riboforce HP

z = # servings of Creatine Transport.

Maximize $p = 10x + 5y + 5z$ subject to

$200x + 100z \le 1000$, or $2x + z \le 10$

$75x + 15y + 35z \le 225$, or $15x + 3y + 7z \le 45$

$x \ge 0, y \ge 0, z \ge 0$.

	x	y	z	s	t	p		
s	2	0	1	1	0	0	10	$15R_1 - 2R_2$
t	[15]	3	7	0	1	0	45	
p	−10	−5	−5	0	0	1	0	$3R_3 + 2R_2$

	x	y	z	s	t	p		
s	0	−6	1	15	−2	0	60	$R_1 + 2R_2$
x	15	[3]	7	0	1	0	45	
p	0	−9	−1	0	2	3	90	$R_3 + 3R_2$

	x	y	z	s	t	p	
s	30	0	15	15	0	0	150
y	15	3	7	0	1	0	45
p	45	0	20	0	5	3	225

Optimal Solution: $p = 75$; $x = 0$, $y = 45/3 = 15$, $z = 0$. Use 15 servings of RiboForce HP and none of the others for a maximum of 75g creatine.

35. x = # IBM shares

y = # HPQ shares

z = # AAPL shares

Maximize $p = 5x + y + z$ subject to $80x + 20y + 70z \le 10{,}000$,

$(0.01)(80x) + (0.02)(20y) \le 200$, or $8x + 4y \le 2000$

	x	y	z	s	t	p	
s	80	20	70	1	0	0	10000
t	8	4	0	0	1	0	2000
p	−5	−1	−1	0	0	1	0

The most negative entry in the bottom row is the −5 in the x-column. So we use this column as the pivot column. The test ratios are: s: 10000/80, t: 2000/9. The smallest test ratio is s: 10000/80. Thus, we pivot on the 80 in the s-row:

	x	y	z	s	t	p	
s	80	20	70	1	0	0	10000
t	8	4	0	0	1	0	2000
p	−5	−1	−1	0	0	1	0

$10R_2 - R_1$

$16R_3 + R_1$

	x	y	z	s	t	p	
x	80	20	70	1	0	0	10000
t	0	20	−70	−1	10	0	10000
p	0	4	54	1	0	16	10000

Since there are no more negative numbers in the bottom row, we are done, and read off the solution:

Optimal Solution: $p = 10000/16 = 625$; $x = 10000/80 = 125$, $y = 0$, $z = 0$

Buy 125 shares of IBM, and no others. The broker is wrong.

37. Unknowns:

x = amount allocated to automobile loans (in $ millions)

y = amount allocated to furniture loans (in $ millions)

z = amount allocated to signature loans (in $ millions)

w = amount allocated to other secured loans

Maximize $p = 8x + 10y + 12z + 10w$ (100 times the return) subject to

$x + y + z + w \le 5$

$-x - y + 9z - w \le 0$

$-x + y + w \le 0$

$-2x + w \le 0$

$x \ge 0, y \ge 0, z \ge 0, w \ge 0.$

	x	y	z	w	s	t	u	v	p	
s	1	1	1	1	1	0	0	0	0	5
t	−1	−1	9	−1	0	1	0	0	0	0
u	−1	1	0	1	0	0	1	0	0	0
v	−2	0	0	1	0	0	0	1	0	0
p	−8	−10	−12	−10	0	0	0	0	1	0

$9R_1 - R_2$

$9R_5 + 12R_2$

	x	y	z	w	s	t	u	v	p		
s	10	10	0	10	9	−1	0	0	0	45	$R_1 - 10R_3$
z	−1	−1	9	−1	0	1	0	0	0	0	$R_2 + R_3$
u	−1	[1]	0	1	0	0	1	0	0	0	
v	−2	0	0	1	0	0	0	1	0	0	
p	−84	−102	0	−102	0	12	0	0	9	0	$R_5 + 102R_3$

	x	y	z	w	s	t	u	v	p		
s	[20]	0	0	0	9	−1	−10	0	0	45	
z	−2	0	9	0	0	1	1	0	0	0	$10R_2 + R_1$
y	−1	1	0	1	0	0	1	0	0	0	$20R_3 + R_1$
v	−2	0	0	1	0	0	0	1	0	0	$10R_4 + R_1$
p	−186	0	0	0	0	12	102	0	9	0	$20R_5 + 186R_1$

	x	y	z	w	s	t	u	v	p	
x	20	0	0	0	9	−1	−10	0	0	45
z	0	0	90	0	9	9	0	0	0	45
y	0	20	0	20	9	−1	10	0	0	45
v	0	0	0	10	9	−1	−10	10	0	45
p	0	0	0	0	1674	54	180	0	180	8370

Optimal Solution: $p = 8370/180 = 46.5$; $x = 45/20 = 2.25$, $y = 45/20 = 2.25$, $z = 45/90 = 0.5$, $w = 0$. Allocate $2,250,000 to automobile loans, $2,250,000 to furniture loans, and $500,000 to signature loans. Another optimal solution (pivot in the w-column): $p = 8370/180 = 46.5$; $x = 45/20 = 2.25$, $y = 0$, $z = 45/90 = 0.5$, $w = 45/20 = 2.25$. Allocate $2,250,000 to automobile loans, $500,000 to signature loans, and $2,250,000 to other secured loans.

In general, Allocate $2,250,000 to automobile loans, $500,000 to signature loans, and $2,250,000 to any combination of furniture loans and other secured loans.

39. Unknowns:

x = amount invested in Warner

y = amount invested in Universal

z = amount invested in Sony

w = amount invested in EMI

Maximize $p = 8x + 20y + 10z + 15w$ subject to

$0.12x + 0.20y + 0.20z + 0.15w \le 15{,}000$, or $12x + 20y + 20z + 15w \le 1{,}500{,}000$

$x - 4y + z + w \le 0$

$x + y + z + w \le 100{,}000$

$x \ge 0,\ y \ge 0,\ z \ge 0,\ w \ge 0.$

	x	y	z	w	s	t	u	p		
s	12	20	20	15	1	0	0	0	1500000	
t	1	-4	1	1	0	1	0	0	0	$5R_2 + R_1$
u	1	1	1	1	0	0	1	0	100000	$20R_3 - R_1$
p	-8	-20	-10	-15	0	0	0	1	0	$R_4 + R_1$

	x	y	z	w	s	t	u	p	
y	12	20	20	15	1	0	0	0	1500000
t	17	0	25	20	1	5	0	0	1500000
u	8	0	0	5	-1	0	20	0	500000
p	4	0	10	0	1	0	0	1	1500000

Optimal Solution: $p = 1{,}500{,}000$; $x = 0$, $y = 1{,}500{,}000/20 = 75{,}000$, $z = 0$, $w = 0$. Invest \$75,000 in Universal, none in the rest. Another optimal solution (pivot in the w-column) is: Invest \$18,750 in Universal, and \$75,000 in EMI.

41. Unknowns:

$x = $ # boards from Tucson to Honolulu

$y = $ # boards from Tucson to Venice Beach

$z = $ # boards from Toronto to Honolulu

$w = $ # boards from Toronto to Venice Beach

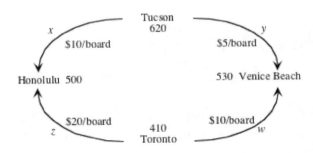

Maximize $p = x + y + z + w$ subject to

$x + y \le 620$

$z + w \le 410$

$x + z \le 500$

$y + w \le 530$

$10x + 5y + 20z + 10w \le 6550$

$x \ge 0,\ y \ge 0,\ z \ge 0.$

	x	y	z	w	s	t	u	v	r	p		
s	1	1	0	0	1	0	0	0	0	0	620	$R_1 - R_3$
t	0	0	1	1	0	1	0	0	0	0	410	
u	[1]	0	1	0	0	0	1	0	0	0	500	
v	0	1	0	1	0	0	0	1	0	0	530	
r	10	5	20	10	0	0	0	0	1	0	6550	$R_5 - 10R_3$
p	-1	-1	-1	-1	0	0	0	0	0	1	0	$R_6 + R_3$

	x	y	z	w	s	t	u	v	r	p		
s	0	[1]	-1	0	1	0	-1	0	0	0	120	
t	0	0	1	1	0	1	0	0	0	0	410	
x	1	0	1	0	0	0	1	0	0	0	500	
v	0	1	0	1	0	0	0	1	0	0	530	$R_4 - R_1$
r	0	5	10	10	0	0	-10	0	1	0	1550	$R_5 - 5R_1$
p	0	-1	0	-1	0	0	1	0	0	1	500	$R_6 + R_1$

	x	y	z	w	s	t	u	v	r	p		
y	0	1	-1	0	1	0	-1	0	0	0	120	$15R_1 + R_5$
t	0	0	1	1	0	1	0	0	0	0	410	$15R_2 - R_5$
x	1	0	1	0	0	0	1	0	0	0	500	$15R_3 - R_5$
v	0	0	1	1	-1	0	1	1	0	0	410	$15R_4 - R_5$
r	0	0	[15]	10	-5	0	-5	0	1	0	950	
p	0	0	-1	-1	1	0	0	0	0	1	620	$15R_6 + R_5$

	x	y	z	w	s	t	u	v	r	p		
y	0	15	0	10	10	0	-20	0	1	0	2750	$R_1 - R_5$
t	0	0	0	5	5	15	5	0	-1	0	5200	$2R_2 - R_5$
x	15	0	0	-10	5	0	20	0	-1	0	6550	$R_3 + R_5$
v	0	0	0	5	-10	0	20	15	-1	0	5200	$2R_4 - R_5$
z	0	0	15	10	-5	0	-5	0	1	0	950	
p	0	0	0	-5	10	0	-5	0	1	15	10250	$2R_6 + R_5$

	x	y	z	w	s	t	u	v	r	p		
y	0	15	–15	0	15	0	–15	0	0	0	1800	$3R_1 + R_4$
t	0	0	–15	0	15	30	15	0	–3	0	9450	$3R_2 - R_4$
x	15	0	15	0	0	0	15	0	0	0	7500	$3R_3 - R_4$
v	0	0	–15	0	–15	0	45	30	–3	0	9450	
w	0	0	15	10	–5	0	–5	0	1	0	950	$9R_5 + R_4$
p	0	0	15	0	15	0	–15	0	3	30	21450	$3R_6 + R_4$

	x	y	z	w	s	t	u	v	r	p	
y	0	45	–60	0	30	0	0	30	–3	0	14850
t	0	0	–30	0	60	90	0	–30	–6	0	18900
x	45	0	60	0	15	0	0	–30	3	0	13050
u	0	0	–15	0	–15	0	45	30	–3	0	9450
w	0	0	120	90	–60	0	0	30	6	0	18000
p	0	0	30	0	30	0	0	30	6	90	73800

Optimal Solution: $p = 73{,}800/90 = 820$; $x = 13{,}050/45 = 290$, $y = 14{,}850/45 = 330$, $z = 0$, $w = 18{,}000/90 = 200$. Make the following shipments: Tucson to Honolulu: 290 boards; Tucson to Venice Beach: 330 boards; Toronto to Honolulu: 0 boards; Toronto to Venice Beach: 200 boards, giving 820 boards shipped.

43. Unknowns:

x = # people you fly from Chicago to Los Angeles

y = # people you fly from Chicago to New York

z = # people you fly from Denver to Los Angeles

w = # people you fly from Denver to New York

Maximize $p = x + y + z + w$ subject to

$195x + 182y + 395z + 166w \leq 4520$

$x + y \leq 20$

$z + w \leq 10$

$x + z \leq 10$

$y + w \leq 15$

$x \geq 0, y \geq 0, z \geq 0, w \geq 0$.

To solve, we used the online Simplex Method Tool. When entering the problems, there is no need to enter the inequalities $x \geq 0$, $y \geq 0$, etc.

Tableau #1

x	y	z	w	s1	s2	s3	s4	s5	p	
195	182	395	166	1	0	0	0	0	0	4520
1	1	0	0	0	1	0	0	0	0	20
0	0	1	1	0	0	1	0	0	0	10
1	0	1	0	0	0	0	1	0	0	10
0	1	0	1	0	0	0	0	1	0	15
−1	−1	−1	−1	0	0	0	0	0	1	0

Tableau #2

x	y	z	w	s1	s2	s3	s4	s5	p	
0	182	200	166	1	0	0	−195	0	0	2570
0	1	−1	0	0	1	0	−1	0	0	10
0	0	1	1	0	0	1	0	0	0	10
1	0	1	0	0	0	0	1	0	0	10
0	1	0	1	0	0	0	0	1	0	15
0	−1	0	−1	0	0	0	1	0	1	10

Tableau #3

x	y	z	w	s1	s2	s3	s4	s5	p	
0	0	382	166	1	−182	0	−13	0	0	750
0	1	−1	0	0	1	0	−1	0	0	10
0	0	1	1	0	0	1	0	0	0	10
1	0	1	0	0	0	0	1	0	0	10
0	0	1	1	0	−1	0	1	1	0	5
0	0	−1	−1	0	1	0	0	0	1	20

Tableau #4

x	y	z	w	s1	s2	s3	s4	s5	p	
0	0	1	0.43	0	−0.48	0	−0.03	0	0	1.96
0	1	0	0.43	0	0.52	0	−1.03	0	0	11.96
0	0	0	0.57	0	0.48	1	0.03	0	0	8.04
1	0	0	−0.43	0	0.48	0	1.03	0	0	8.04
0	0	0	0.57	0	−0.52	0	1.03	1	0	3.04
0	0	0	−0.57	0	0.52	0	−0.03	0	1	21.96

Tableau #5

x	y	z	w	s1	s2	s3	s4	s5	p	
0	0	2.3	1	0.01	−1.1	0	−0.08	0	0	4.52
0	1	−1	0	0	1	0	−1	0	0	10
0	0	−1.3	0	−0.01	1.1	1	0.08	0	0	5.48
1	0	1	0	0	0	0	1	0	0	10
0	0	−1.3	0	−0.01	0.1	0	1.08	1	0	0.48
0	0	1.3	0	0.01	−0.1	0	−0.08	0	1	24.52

Tableau #6

x	y	z	w	s1	s2	s3	s4	s5	p	
0	0	1	1	0	0	1	0	0	0	10
0	1	0.19	0	0.01	0	−0.91	−1.07	0	0	5
0	0	−1.19	0	−0.01	1	0.91	0.07	0	0	5
1	0	1	0	0	0	0	1	0	0	10
0	0	−1.19	0	−0.01	0	−0.09	1.07	1	0	0
0	0	1.19	0	0.01	0	0.09	−0.07	0	1	25

Tableau #7

x	y	z	w	s1	s2	s3	s4	s5	p	
0	0	1	1	0	0	1	0	0	0	10
0	1	−1	0	0	0	−1	0	1	0	5
0	0	−1.11	0	−0.01	1	0.92	0	−0.07	0	5
1	0	2.11	0	0.01	0	0.08	0	−0.93	0	10
0	0	−1.11	0	−0.01	0	−0.08	1	0.93	0	0
0	0	1.11	0	0.01	0	0.08	0	0.07		25

Optimal Solution: $p = 25$; $x = 10$, $y = 5$, $z = 0$, $w = 10$. Fly 10 people from Chicago to Los Angeles, 5 people from Chicago to New York, and 10 people from Denver to New York.

45. The only difficulty with the stated problem is the "≥" in the first constraint: $x - y + z \geq 0$. We can reverse the inequality by multiplying both sides by -1: $-x + y - z \leq 0$

Thus, the given problem can be stated as a standard LP problem:

Maximize $p = 3x - 2y$ subject to

$$-x + y - z \leq 0$$
$$x - y - z \leq 6$$

47. The graphical method applies only to LP problems in two unknowns, whereas the simplex method can be used to solve LP problems with any number of unknowns.

49. She is correct. Since there are only two constraints, there can only be two active variables, giving two or fewer non-zero values for the unknowns at each stage.

51. A basic solution to a system of linear equations is a solution in which all the non-pivotal variables are taken to be zero; that is, all variables whose value is arbitrary are assigned the value zero. To obtain a basic solution for a given system of linear equations, one can row reduce the associated augmented matrix, write down the general solution, and then set all the parameters (variables with "arbitrary" values) equal to zero.

53. No. Let us assume for the sake of simplicity that all the pivots are 1's. (They may certainly be changed to 1's without affecting the value of any of the variables.) Since the entry at the bottom of the pivot column is negative, the bottom row gets replaced by itself plus a positive multiple of the pivot row. The value of the objective function (bottom-right entry) is thus replaced by itself plus a positive multiple of the nonnegative rightmost entry of the pivot row. Therefore, it cannot decrease.

4.4

1. Introduce slack and surplus variables, rewrite the constraints and objective function in standard form, set up the first tableau, and star rows coming from "≥" constraints.

$$x + 2y - s = 6$$
$$-x + y + t = 4$$
$$2x + y + u = 8$$
$$-x - y + p = 0$$

	x	y	s	t	u	p	
*s	1	2	−1	0	0	0	6
t	−1	1	0	1	0	0	4
u	2	1	0	0	1	0	8
p	−1	−1	0	0	0	1	0

The first (only) starred row is the s–row, and its largest positive entry is the 2 in the y–column. So we use this column as the pivot column. The test ratios are: s: 6/2, t: 4/1, u: 8/1. The smallest test ratio is s: 6/2. Thus, we pivot on the 2 in the s–row.

	x	y	s	t	u	p		
*s	1	2	−1	0	0	0	6	
t	−1	1	0	1	0	0	4	$2R_2 - R_1$
u	2	1	0	0	1	0	8	$2R_3 - R_1$
p	−1	−1	0	0	0	1	0	$2R_4 + R_1$

	x	y	s	t	u	p	
y	1	2	−1	0	0	0	6
t	−3	0	1	2	0	0	2
u	3	0	1	0	2	0	10
p	−1	0	−1	0	0	2	6

Since there are no more starred rows, we go to Phase 2, and do the standard simplex method:

	x	y	s	t	u	p		
y	1	2	−1	0	0	0	6	$3R_1 - R_3$
t	−3	0	1	2	0	0	2	$R_2 + R_3$
u	[3]	0	1	0	2	0	10	
p	−1	0	−1	0	0	2	6	$3R_4 + R_3$

	x	y	s	t	u	p		
y	0	6	−4	0	−2	0	8	$R_1 + 2R_2$
t	0	0	[2]	2	2	0	12	
x	3	0	1	0	2	0	10	$2R_3 - R_2$
p	0	0	−2	0	2	6	28	$R_4 + R_2$

	x	y	s	t	u	p	
y	0	6	0	4	2	0	32
s	0	0	2	2	2	0	12
x	6	0	0	−2	2	0	8
p	0	0	0	2	4	6	40

Optimal Solution: $p = 40/6 = 20/3$; $x = 8/6 = 4/3$, $y = 32/6 = 16/3$.

3. Introduce slack and surplus variables, rewrite the constraints and objective function in standard form, set up the first tableau, and star rows coming from "≥" constraints.

$$x + y + s = 25$$
$$x \qquad - t = 10$$
$$-x + 2y - u = 0$$
$$-12x - 10y + p = 0$$

	x	y	s	t	u	p		
s	1	1	1	0	0	0	25	$R_1 - R_2$
*t	[1]	0	0	−1	0	0	10	
*u	−1	2	0	0	−1	0	0	$R_3 + R_2$
p	−12	−10	0	0	0	1	0	$R_4 + 12R_2$

The first starred row above is the t-row, and its largest positive entry is the 1 in the x-column. So we use this column as the pivot column. The test ratios are: s: 25/1, t: 10/1. The smallest test ratio is t: 10/1. Thus, we pivot on the 1 in the t-row.

	x	y	s	t	u	p		
s	0	1	1	1	0	0	15	$2R_1 - R_3$
x	1	0	0	−1	0	0	10	
*u	0	2	0	−1	−1	0	10	
p	0	−10	0	−12	0	1	120	$R_4 + 5R_3$

The only remaining starred row above is the u–row, and its largest positive entry is the 2 in the y–column. So we use this column as the pivot column. The test ratios are: s: 15/1, u: 10/2. The smallest test ratio is u: 10/2. Thus, we pivot on the 2 in the u–row.

	x	y	s	t	u	p	
s	0	0	2	3	1	0	20
x	1	0	0	−1	0	0	10
y	0	2	0	−1	−1	0	10
p	0	0	0	−17	−5	1	170

Since there are no more starred rows, we go to Phase 2, and do the usual simplex method:

	x	y	s	t	u	p		
s	0	0	2	3	1	0	20	
x	1	0	0	−1	0	0	10	$3R_2 + R_1$
y	0	2	0	−1	−1	0	10	$3R_3 + R_1$
p	0	0	0	−17	−5	1	170	$3R_4 + 17R_1$

	x	y	s	t	u	p	
t	0	0	2	3	1	0	20
x	3	0	2	0	1	0	50
y	0	6	2	0	−2	0	50
p	0	0	34	0	2	3	850

Optimal Solution: $p = 850/3$; $x = 50/3 = 50/3$, $y = 50/6 = 25/3$.

5. Introduce slack and surplus variables, rewrite the constraints and objective function in standard form, and set up the first tableau.

$$x + y + z + s = 150$$
$$x + y + z \geq -t = 100$$
$$-2x - 5y - 3z + p = 0$$

	x	y	z	s	t	p		
s	1	1	1	1	0	0	150	$R_1 - R_2$
*t	1	1	1	0	−1	0	100	
p	−2	−5	−3	0	0	1	0	$R_3 + 2R_2$

The first starred row above is the t–row, and its largest positive entry is the 1 in the x–column. So we use this column as the pivot column. The test ratios are: s: 150/1, t: 100/1. The smallest test ratio is t: 100/1. Thus, we pivot on the 1 in the t–row.

	x	y	z	s	t	p	
s	0	0	0	1	1	0	50
x	1	1	1	0	−1	0	100
p	0	−3	−1	0	−2	1	200

Since there are no more starred rows, we go to Phase 2.

	x	y	z	s	t	p		
s	0	0	0	1	1	0	50	
x	1	1	1	0	−1	0	100	
p	0	−3	−1	0	−2	1	200	$R_3 + 3R_2$

	x	y	z	s	t	p		
s	0	0	0	1	1	0	50	
y	1	1	1	0	−1	0	100	$R_2 + R_1$
p	3	0	2	0	−5	1	500	$R_3 + 5R_1$

196

	x	y	z	s	t	p	
t	0	0	0	1	1	0	50
y	1	1	1	1	0	0	150
p	3	0	2	5	0	1	750

Optimal Solution: $p = 750$; $x = 0$, $y = 150/1 = 150$, $z = 0$.

7.
$$x + y + z + w + s = 40$$
$$2x + y - z - w - t = 10$$
$$x + y + z + w - u = 10$$
$$-2x - 3y - z - 4w + p = 0$$

	x	y	z	w	s	t	u	p		
s	1	1	1	1	1	0	0	0	40	$2R_1 - R_2$
*t	[2]	1	−1	−1	0	−1	0	0	10	
*u	1	1	1	1	0	0	−1	0	10	$2R_3 - R_2$
p	−2	−3	−1	−4	0	0	0	1	0	$R_4 + R_2$

The first starred row above is the t–row, and its largest positive entry is the 2 in the x–column. So we use this column as the pivot column. The test ratios are: s: 40/1, t: 10/2, u: 10/1. The smallest test ratio is t: 10/2. Thus, we pivot on the 2 in the t–row.

	x	y	z	w	s	t	u	p		
s	0	1	3	3	2	1	0	0	70	$R_1 - R_3$
x	2	1	−1	−1	0	−1	0	0	10	$3R_2 + R_3$
*u	0	1	3	[3]	0	1	−2	0	10	
p	0	−2	−2	−5	0	−1	0	1	10	$3R_4 + 2R_3$

The first starred row above is the u–row, and its largest positive entry is the 3 in the z–column. So we use this column as the pivot column. The test ratios are: s: 70/3, u: 10/3. The smallest test ratio is u: 10/3. Thus, we pivot on the 3 in the u–row. This will take us to Phase 2:

	x	y	z	w	s	t	u	p		
s	0	0	0	0	2	0	2	0	60	
x	6	4	0	0	0	-2	-2	0	40	
z	0	1	3	[3]	0	1	-2	0	10	
p	0	-4	0	-9	0	-1	-4	3	50	$R_4 + 3R_3$

	x	y	z	w	s	t	u	p		
s	0	0	0	0	2	0	[2]	0	60	
x	6	4	0	0	0	-2	-2	0	40	$R_2 + R_1$
w	0	1	3	3	0	1	-2	0	10	$R_3 + R_1$
p	0	-1	9	0	0	2	-10	3	80	$R_4 + 5R_1$

	x	y	z	w	s	t	u	p		
u	0	0	0	0	2	0	2	0	60	
x	6	[4]	0	0	2	-2	0	0	100	
w	0	1	3	3	2	1	0	0	70	$4R_3 - R_2$
p	0	-1	9	0	10	2	0	3	380	$4R_4 + R_2$

	x	y	z	w	s	t	u	p	
u	0	0	0	0	2	0	2	0	60
y	6	4	0	0	2	-2	0	0	100
w	-6	0	12	12	6	6	0	0	180
p	6	0	36	0	42	6	0	12	1620

Optimal Solution: $p = 1620/12 = 135$; $x = 0$, $y = 100/4 = 25$, $z = 0$, $w = 180/12 = 15$.

9. Convert the given problem into a maximization problem:

Maximize $p = -6x - 6y$ subject to $x + 2y \geq 20$, $2x + y \geq 20$, $x \geq 0$, $y \geq 0$.

Introduce slack and surplus variables, rewrite the constraints and objective function in standard form, and set up the first tableau.

$x + 2y - s = 20$

$2x + y - t = 20$

$6x + 6y + p = 0$

	x	y	s	t	p		
*s	1	2	–1	0	0	20	
*t	2	1	0	–1	0	20	$2R_2 - R_1$
p	6	6	0	0	1	0	$R_3 - 3R_1$

The first starred row above is the s–row, and its largest positive entry is the 2 in the y–column. So we use this column as the pivot column. The test ratios are: s: 20/2, t: 20/1. The smallest test ratio is s: 20/2. Thus, we pivot on the 2 in the s–row.

	x	y	s	t	p		
y	1	2	–1	0	0	20	$3R_1 - R_2$
*t	3	0	1	–2	0	20	
p	3	0	3	0	1	–60	$R_3 - R_2$

The first starred row above is the t–row, and its largest positive entry is the 3 in the x–column. So we use this column as the pivot column. The test ratios are: y: 20/1, t: 20/3. The smallest test ratio is t: 20/3. Thus, we pivot on the 3 in the t–row.

	x	y	s	t	p	
y	0	6	–4	2	0	40
x	3	0	1	–2	0	20
p	0	0	2	2	1	–80

Since there are no more starred rows, we go to Phase 2, and search for the most negative entry in the bottom row (if any). Since there are no negative entries in the bottom row, we are done, and read off the optimal solution: $p = -80$; $x = 20/3$, $y = 40/6 = 20/3$. Since $c = -p$, the minimum value of c is $c = 80$.

11. Convert the given problem into a maximization problem:

Maximize $p = -2x - y - 3z$ subject to $x + y + z \geq 100$, $2x + y \geq 50$, $y + z \geq 50$, $x \geq 0$, $y \geq 0$, $z \geq 0$.

Introduce slack and surplus variables, rewrite the constraints and objective function in standard form, and solve this LP problem:.

$$x + y + z - s = 100$$
$$2x + y = t - 50$$
$$y + z - u = 50$$

	x	y	z	s	t	u	p		
*s	1	1	1	-1	0	0	0	100	$2R_1 - R_2$
*t	[2]	1	0	0	-1	0	0	50	
*u	0	1	1	0	0	-1	0	50	
p	2	1	3	0	0	0	1	0	$R_4 - R_2$

	x	y	z	s	t	u	p		
*s	0	1	2	-2	1	0	0	150	$R_1 - 2R_3$
x	2	1	0	0	-1	0	0	50	
*u	0	1	[1]	0	0	-1	0	50	
p	0	0	3	0	1	0	1	-50	$R_4 - 3R_3$

	x	y	z	s	t	u	p		
*s	0	-1	0	-2	1	[2]	0	50	
x	2	1	0	0	-1	0	0	50	
z	0	1	1	0	0	-1	0	50	$2R_3 + R_1$
p	0	-3	0	0	1	3	1	-200	$2R_4 - 3R_1$

Phase 2:

	x	y	z	s	t	u	p		
u	0	-1	0	-2	1	2	0	50	$R_1 + R_2$
x	2	[1]	0	0	-1	0	0	50	
z	0	1	2	-2	1	0	0	150	$R_3 - R_2$
p	0	-3	0	6	-1	0	2	-550	$R_4 + 3R_2$

	x	y	z	s	t	u	p		
u	2	0	0	-2	0	2	0	100	
y	2	1	0	0	-1	0	0	50	$2R_2 + R_3$
z	-2	0	2	-2	[2]	0	0	100	
p	6	0	0	6	-4	0	2	-400	$R_4 + 2R_3$

	x	y	z	s	t	u	p	
u	2	0	0	−2	0	2	0	100
y	2	2	2	−2	0	0	0	200
t	−2	0	2	−2	2	0	0	100
p	2	0	4	2	0	0	2	−200

Optimal Solution: $c = -p = 200/2 = 100$; $x = 0$, $y = 200/2 = 100$, $z = 0$.

13. Convert the given problem into a maximization problem: Maximize $p = -50x - 50y - 11z$ subject to $2x + z \geq 3$, $2x + y - z \geq 2$, $3x + y - z \leq 3$, $x \geq 0$, $y \geq 0$, $z \geq 0$

	x	y	z	s	t	u	p		
*s	2	0	1	−1	0	0	0	3	$3R_1 - 2R_3$
*t	2	1	−1	0	−1	0	0	2	$3R_2 - 2R_3$
u	$\boxed{3}$	1	−1	0	0	1	0	3	
p	50	50	11	0	0	0	1	0	$3R_4 - 50R_3$

	x	y	z	s	t	u	p		
*s	0	−2	$\boxed{5}$	−3	0	−2	0	3	
*t	0	1	−1	0	−3	−2	0	0	$5R_2 + R_1$
x	3	1	−1	0	0	1	0	3	$5R_3 + R_1$
p	0	100	83	0	0	−50	3	−150	$5R_4 - 83R_1$

	x	y	z	s	t	u	p		
z	0	−2	5	−3	0	−2	0	3	$3R_1 + 2R_2$
*t	0	$\boxed{3}$	0	−3	−15	−12	0	3	
x	15	3	0	−3	0	3	0	18	$R_3 - R_2$
p	0	666	0	249	0	−84	15	−999	$R_4 - 222R_2$

	x	y	z	s	t	u	p	
z	0	0	15	−15	−30	−30	0	15
y	0	3	0	−3	−15	−12	0	3
x	15	0	0	0	15	15	0	15
p	0	0	0	915	3330	2580	15	−1665

Optimal Solution: $c = -p = 1665/15 = 111$; $x = 15/15 = 1$, $y = 3/3 = 1$, $z = 15/15 = 1$.

15. Convert the given problem into a maximization problem: Maximize $p = -x - y - z - w$ subject to $5x - y + w \geq 1000$, $z + w \leq 2000$, $x + y \leq 500$, $x \geq 0$, $y \geq 0$, $z \geq 0$, $w \geq 0$.

	x	y	z	w	s	t	u	p	
*s	$\boxed{5}$	-1	0	1	-1	0	0	0	1000
t	0	0	1	1	0	1	0	0	2000
u	1	1	0	0	0	0	1	0	500
p	1	1	1	1	0	0	0	1	0

$5R_3 - R_1$

$5R_4 - R_1$

	x	y	z	w	s	t	u	p	
x	5	-1	0	1	-1	0	0	0	1000
t	0	0	1	1	0	1	0	0	2000
u	0	6	0	-1	1	0	5	0	1500
p	0	6	5	4	1	0	0	5	-1000

Optimal Solution: $p = -c = 1000/5 = 200$; $x = 1000/5 = 200$, $y = 0$, $z = 0$, $w = 0$.

Note: In Exercises 17–22, we used the online Simplex Method Tool. When entering the problems, there is no need to enter the inequalities $x \geq 0$, $y \geq 0$, etc.

17. Maximize $p = 2x + 3y + 1.1z + 4w$ subject to
$1.2x + y + z + w \leq 40.5$
$2.2x + y - z - w \geq 10$
$1.2x + y + z + 1.2w \geq 10.5$
$x \geq 0$, $y \geq 0$, $z \geq 0$, $w \geq 0$.

Tableau #1

x	y	z	w	s1	s2	s3	p	
1.2	1	1	1	1	0	0	0	40.5
2.2	1	-1	-1	0	-1	0	0	10
1.2	1	1	1.2	0	0	-1	0	10.5
-2	-3	-1.1	-4	0	0	0	1	0

Tableau #2

x	y	z	w	s1	s2	s3	p	
0	0.45	1.55	1.55	1	0.55	0	0	35.05
1	0.45	-0.45	-0.45	0	-0.45	0	0	4.55
0	0.45	1.55	1.75	0	0.55	-1	0	5.05
0	-2.09	-2.01	-4.91	0	-0.91	0	1	9.09

Tableau #3

x	y	z	w	s1	s2	s3	p	
0	0.05	0.18	0	1	0.06	0.89	0	30.58
1	0.57	−0.05	0	0	−0.31	−0.26	0	5.86
0	0.26	0.89	1	0	0.31	−0.57	0	2.89
0	−0.81	2.34	0	0	0.63	−2.81	1	23.28

Tableau #4

x	y	z	w	s1	s2	s3	p	
0	0.06	0.2	0	1.13	0.07	1	0	34.54
1	0.59	0	0	0.29	−0.29	0	0	14.85
0	0.29	1	1	0.65	0.35	0	0	22.68
0	−0.65	2.9	0	3.18	0.82	0	1	120.41

Tableau #5

x	y	z	w	s1	s2	s3	p	
−0.1	0	0.2	0	1.1	0.1	1	0	33.05
1.7	1	0	0	0.5	−0.5	0	0	25.25
−0.5	0	1	1	0.5	0.5	0	0	15.25
1.1	0	2.9	0	3.5	0.5	0	1	136.75

Optimal Solution: $p = 136.75$; $x = 0$, $y = 25.25$, $z = 0$, $w = 15.25$.

19. Minimize $c = 2.2x + y + 3.3z$ subject to

$x + 1.5y + 1.2z \geq 100$

$2x + 1.5y \geq 50$

$1.5y + 1.1z \geq 50$

$x \geq 0, y \geq 0, z \geq 0$.

Tableau #1

x	y	z	s1	s2	s3	−c	
1	1.5	1.2	−1	0	0	0	100
2	1.5	0	0	−1	0	0	50
0	1.5	1.1	0	0	−1	0	50
2.2	1	3.3	0	0	0	1	0

Tableau #2

x	y	z	s1	s2	s3	-c	
1	0	0.1	-1	0	1	0	50
2	0	-1.1	0	-1	1	0	0
0	1	0.73	0	0	-0.67	0	33.33
2.2	0	2.57	0	0	0.67	1	-33.33

Tableau #3

x	y	z	s1	s2	s3	-c	
0	0	0.65	-1	0.5	0.5	0	50
1	0	-0.55	0	-0.5	0.5	0	0
0	1	0.73	0	0	-0.67	0	33.33
0	0	3.78	0	1.1	-0.43	1	-33.33

Tableau #4

x	y	z	s1	s2	s3	-c	
0	-0.89	0	-1	0.5	1.09	0	20.45
1	0.75	0	0	-0.5	0	0	25
0	1.36	1	0	0	-0.91	0	45.45
0	-5.15	0	0	1.1	3	1	-205

Tableau #5

x	y	z	s1	s2	s3	-c	
0	-0.81	0	-0.92	0.46	1	0	18.75
1	0.75	0	0	-0.5	0	0	25
0	0.63	1	-0.83	0.42	0	0	62.5
0	-2.71	0	2.75	-0.27	0	1	-261.25

Tableau #6

x	y	z	s1	s2	s3	-c	
1.08	0	0	-0.92	-0.08	1	0	45.83
1.33	1	0	0	-0.67	0	0	33.33
-0.83	0	1	-0.83	0.83	0	0	41.67
3.62	0	0	2.75	-2.08	0	1	-170.83

Tableau #7

x	y	z	s1	s2	s3	-c	
1	0	0.1	−1	0	1	0	50
0.67	1	0.8	−0.67	0	0	0	66.67
−1	0	1.2	−1	1	0	0	50
1.53	0	2.5	0.67	0	0	1	−66.67

Optimal Solution: $c = 66.67$; $x = 0$, $y = 66.67$, $z = 0$.

21. Minimize $c = 1.1x + y + 1.5z - w$ subject to
$$5.12x - y + w \leq 1000$$
$$z + w \geq 2000$$
$$1.22x + y \leq 500$$
$$x \geq 0, y \geq 0, z \geq 0, w \geq 0.$$

Tableau #1

x	y	z	w	s1	s2	s3	-c	
5.12	−1	0	1	1	0	0	0	1000
0	0	1	1	0	−1	0	0	2000
1.22	1	0	0	0	0	1	0	500
1.1	1	1.5	−1	0	0	0	1	0

Tableau #2

x	y	z	w	s1	s2	s3	-c	
5.12	−1	0	1	1	0	0	0	1000
0	0	1	1	0	−1	0	0	2000
1.22	1	0	0	0	0	1	0	500
1.1	1	0	−2.5	0	1.5	0	1	−3000

Tableau #3

x	y	z	w	s1	s2	s3	-c	
5.12	−1	0	1	1	0	0	0	1000
−5.12	1	1	0	−1	−1	0	0	1000
1.22	1	0	0	0	0	1	0	500
13.9	−1.5	0	0	2.5	1.5	0	1	−500

```
Tableau #4
x        y    z    w    s1    s2    s3    -c
6.34     0    0    1    1     0     1     0     1500
-6.34    0    1    0    -1    -1    -1    0     500
1.22     1    0    0    0     0     1     0     500
15.73    0    0    0    2.5   1.5   1.5   1     250
```

Optimal Solution: $c = -250$; $x = 0$, $y = 500$, $z = 500$, $w = 1500$.

23. Unknowns:

x = # acres of tomatoes

y = # acres of lettuce

z = # acres of carrots

Maximize $p = 20x + 15y + 5z$ (in hundreds of dollars) subject to

$$x + y + z \le 100$$
$$5x + 4y + 2z \ge 400$$
$$4x + 2y + 2z \le 500$$
$$x \ge 0, y \ge 0, z \ge 0.$$

	x	y	z	s	t	u	p		
s	1	1	1	1	0	0	0	100	$5R_1 - R_2$
*t	5	4	2	0	-1	0	0	400	
u	4	2	2	0	0	1	0	500	$5R_3 - 4R_2$
p	-20	-15	-5	0	0	0	1	0	$R_4 + 4R_2$

	x	y	z	s	t	u	p		
s	0	1	3	5	1	0	0	100	
x	5	4	2	0	-1	0	0	400	$R_2 + R_1$
u	0	-6	2	0	4	5	0	900	$R_3 - 4R_1$
p	0	1	3	0	-4	0	1	1600	$R_4 + 4R_1$

	x	y	z	s	t	u	p	
t	0	1	3	5	1	0	0	100
x	5	5	5	5	0	0	0	500
u	0	-10	-10	-20	0	5	0	500
p	0	5	15	20	0	0	1	2000

Optimal Solution: $p = 2000/1 = 2000$; $x = 500/5 = 100$, $y = 0$, $z = 0$

Plant 100 acres of tomatoes and no other crops. This will give you a profit of $200,000. (You will be using all 100 acres of your farm.)

25. Unknowns: x = # mailings to the East Coast, y = # mailings to the Midwest, z = # mailings to the West Coast

Minimize $c = 40x + 60y + 50z$ subject to

$$100x + 100y + 50z \geq 1500$$
$$50x + 100y + 100z \geq 1500$$
$$x \geq 0, y \geq 0, z \geq 0.$$

Associated maximization problem: Maximize $p = -40x - 60y - 50z$ subject to $100x + 100y + 50z \geq 1500$, $50x + 100y + 100z \geq 1500$, $x \geq 0, y \geq 0, z \geq 0$.

	x	y	z	s	t	p		
*s	100	100	50	−1	0	0	1500	
*t	50	100	100	0	−1	0	1500	$2R_2 - R_1$
p	40	60	50	0	0	1	0	$5R_3 - 2R_1$

	x	y	z	s	t	p		
x	100	100	50	−1	0	0	1500	$3R_1 - R_2$
*t	0	100	150	1	−2	0	1500	
p	0	100	150	2	0	5	−3000	$R_3 - R_2$

	x	y	z	s	t	p	
x	300	200	0	−4	2	0	3000
z	0	100	150	1	−2	0	1500
p	0	0	0	1	2	5	−4500

Optimal Solution: $p = -c = 4500/5 = 900$; $x = 3000/300 = 10$, $y = 0$, $z = 1500/150 = 10$. Send 10 mailings to the East coast, none to the Midwest, 10 to the West Coast. Cost: $900. Another Solution resulting in the same cost (pivot on the y-column in the last tableau above) is no mailings to the East coast, 15 to the Midwest, none to the West Coast.

27. Unknowns: x = # quarts orange juice, y = # quarts orange concentrate

Minimize $c = 0.5x + 2y$ subject to

$$x \geq 10{,}000$$
$$y \geq 1000$$
$$10x + 50y \geq 200{,}000$$
$$x \geq 0, y \geq 0.$$

Convert the given problem into a maximization problem: Maximize $p = -0.5x - 2y$ subject to $x \geq 10{,}000$, $y \geq 1000$, $10x + 50y \geq 200{,}000$, $x \geq 0$, $y \geq 0$.

	x	y	s	t	u	p		
*s	1	0	−1	0	0	0	10,000	
*t	0	1	0	−1	0	0	1000	
*u	10	50	0	0	−1	0	200,000	$R_3 - 10R_1$
p	1	4	0	0	0	2	0	$R_4 - R_1$

(Note that we cleared fractions in the last row of the first tableau before starting.)

	x	y	s	t	u	p		
x	1	0	−1	0	0	0	10,000	
*t	0	1	0	−1	0	0	1000	
*u	0	50	10	0	−1	0	100,000	$R_3 - 50R_2$
p	0	4	1	0	0	2	−10,000	$R_4 - 4R_2$

	x	y	s	t	u	p		
x	1	0	−1	0	0	0	10,000	
y	0	1	0	−1	0	0	1000	$50R_2 + R_3$
*u	0	0	10	50	−1	0	50,000	
p	0	0	1	4	0	2	−14,000	$50R_4 - 4R_3$

	x	y	s	t	u	p	
x	1	0	−1	0	0	0	10,000
y	0	50	10	0	−1	0	100,000
t	0	0	10	50	−1	0	50,000
p	0	0	10	0	4	100	−900,000

Optimal Solution: $c = -p = 900{,}000/100 = 9000$; $x = 10{,}000/1 = 10{,}000$, $y = 100{,}000/50 = 2000$. Succulent Citrus should produce 10,000 quarts of orange juice and 2000 quarts of orange concentrate.

29. Unknowns: x = # rock music CDs, y = # rap CDs, and z = # classical CDs.

Maximize $p = 12x + 15y + 12z$ subject to $x - 2y \geq 0$, $x + y + z \leq 20000$, $z \geq 5000$, $x \geq 0$, $y \geq 0$, $z \geq 0$.

	x	y	z	s	t	u	p		
$*s$	$\boxed{1}$	−2	0	−1	0	0	0	0	
t	1	1	1	0	1	0	0	20,000	$R_2 - R_1$
$*u$	0	0	1	0	0	−1	0	5000	
p	−12	−15	−12	0	0	0	1	0	$R_4 + 12R_1$

	x	y	z	s	t	u	p		
x	1	−2	0	−1	0	0	0	0	
t	0	3	1	1	1	0	0	20,000	$R_2 - R_3$
$*u$	0	0	$\boxed{1}$	0	0	−1	0	5000	
p	0	−39	−12	−12	0	0	1	0	$R_4 + 12R_3$

	x	y	z	s	t	u	p		
x	1	−2	0	−1	0	0	0	0	$3R_1 + 2R_2$
t	0	$\boxed{3}$	0	1	1	1	0	15,000	
z	0	0	1	0	0	−1	0	5000	
p	0	−39	0	−12	0	−12	1	60,000	$R_4 + 13R_2$

	x	y	z	s	t	u	p	
x	3	0	0	−1	2	2	0	30,000
y	0	3	0	1	1	1	0	15,000
z	0	0	1	0	0	−1	0	5000
p	0	0	0	1	13	1	1	255,000

Optimal Solution: $p = 255{,}000$; $x = 30{,}000/3 = 10{,}000$, $y = 15{,}000/3 = 5000$, $z = 5000/1 = 5000$. Stock 10,000 rock CDs, 5000 rap CDs and 5000 classical CDs for a maximum retail value of $255,000.

31. Unknowns: x = # servings of cereal, y = # servings of dessert, z = # servings of juice.

Minimize $c = 10x + 53y + 27z$ subject to

$60x + 80y + 60z \geq 120$, or $3x + 4y + 3z \geq 6$

$45y + 120z \geq 120$, or $3y + 8z \geq 8$

$x \geq 0$, $y \geq 0$, $z \geq 0$.

Associated maximization problem: Maximize $p = -10x - 53y - 27z$ subject to $3x + 4y + 3z \geq 6$, $3y + 8z \geq 8$, $x \geq 0$, $y \geq 0$, $z \geq 0$.

	x	y	z	s	t	p		
*s	3	$\boxed{4}$	3	−1	0	0	6	
*t	0	3	8	0	−1	0	8	$4R_2 - 3R_1$
p	10	53	27	0	0	1	0	$4R_3 - 53R_1$

	x	y	z	s	t	p		
y	3	4	3	−1	0	0	6	$23R_1 - 3R_2$
*t	−9	0	$\boxed{23}$	3	−4	0	14	
p	−119	0	−51	53	0	4	−318	$23R_3 + 51R_2$

	x	y	z	s	t	p		
y	$\boxed{96}$	92	0	−32	12	0	96	
z	−9	0	23	3	−4	0	14	$96R_2 + 9R_1$
p	−3196	0	0	1372	−204	92	−6600	$96R_3 + 3196R_1$

	x	y	z	s	t	p	
x	96	92	0	−32	12	0	96
z	0	828	2208	0	−276	0	2208
p	0	294032	0	29440	18768	8832	−326784

Optimal Solution: $c = -p = 326784/8832 = 37$; $x = 96/96 = 1$, $y = 0$, $z = 2208/2208 = 1$. Provide one serving of cereal, one serving of juice, and no dessert!

33. Unknowns: x = # bundles from Nadir, y = # bundles from Blunt, z = # bundles from Sonny.
Minimize $c = 3x + 4y + 5z$ (in thousands of dollars) subject to $5x + 10y + 15z \geq 150$, $10x + 10y + 10z \geq 200$, $15x + 10y + 10z \geq 150$, $x \geq 0$, $y \geq 0$, $z \geq 0$.
Associated maximization problem: Maximize $p = -3x - 4y - 5z$ subject to $5x + 10y + 15z \geq 150$, $10x + 10y + 10z \geq 200$, $15x + 10y + 10z \geq 150$, $x \geq 0$, $y \geq 0$, $z \geq 0$.

	x	y	z	s	t	u	p		
*s	5	10	[15]	–1	0	0	0	150	
*t	10	10	10	0	–1	0	0	200	$3R_2 - 2R_1$
*u	15	10	10	0	0	–1	0	150	$3R_3 - 2R_1$
p	3	4	5	0	0	0	1	0	$3R_4 - R_1$

	x	y	z	s	t	u	p		
z	5	10	15	–1	0	0	0	150	$7R_1 - R_3$
*t	20	10	0	2	–3	0	0	300	$3.5R_2 - 2R_3$
*u	[35]	10	0	2	0	–3	0	150	
p	4	2	0	1	0	0	3	–150	$35R_4 - 4R_3$

	x	y	z	s	t	u	p		
z	0	60	105	–9	0	3	0	900	$R_1 - 6R_3$
*t	0	30	0	6	–21	12	0	1500	$2R_2 - 3R_3$
x	35	[10]	0	2	0	–3	0	150	
p	0	30	0	27	0	12	105	–5850	$R_4 - 3R_3$

	x	y	z	s	t	u	p		
z	–210	0	105	–21	0	[21]	0	0	
*t	–105	0	0	0	–21	21	0	1050	$R_2 - R_1$
y	35	10	0	2	0	–3	0	150	$7R_3 + R_1$
p	–105	0	0	21	0	21	105	–6300	$R_4 - R_1$

	x	y	z	s	t	u	p		
u	–210	0	105	–21	0	21	0	0	$R_1 + 2R_2$
*t	[105]	0	–105	21	–21	0	0	1050	
y	35	70	105	–7	0	0	0	1050	$3R_3 - R_2$
p	105	0	–105	42	0	0	105	–6300	$R_4 - R_2$

	x	y	z	s	t	u	p	
u	0	0	−105	21	−42	21	0	2100
x	105	0	−105	21	−21	0	0	1050
y	0	210	420	−42	21	0	0	2100
p	0	0	0	21	21	0	105	−7350

Optimal Solution: $p = -c = 7350/105 = 70$; $x = 1050/105 = 10$, $y = 2100/210 = 10$, $z = 0$. Buy 10 bundles from Nadir, 10 from Blunt, and none from Sony. Cost: $70,000. Another solution resulting in the same cost (pivot on the z-column in the last tableau) is 15 bundles from Nadir, none from Blunt, and 5 from Sonny.

Note: In the Exercises marked for technology, we used the online Simplex Method Tool. When entering the problems, there is no need to enter the inequalities $x \geq 0$, $y \geq 0$, etc.

35. Unknowns: $x = \#$ servings of Cell–Tech, $y = \#$ servings of Riboforce HP, $z = \#$ servings of Creatine Transport.
Minimize $c = 2.20x + 1.60y + 0.60z$ subject to $10x + 5y + 5z \geq 80$, $2x + y + z \geq 10$, $75x + 15y + 35z \leq 750$, $200x + 100z \leq 1000$, $x \geq 0$, $y \geq 0$, $z \geq 0$.
Associated maximization problem: Maximize $p = -2.20x - 1.60y - 0.60z$ subject to $10x + 5y + 5z \geq 80$, $2x + y + z \geq 10$, $75x + 15y + 35z \leq 750$, $200x + 100z \leq 1000$, $x \geq 0$, $y \geq 0$, $z \geq 0$.

```
Tableau #1
x        y        z        s1       s2       s3       s4       p
10       5        5        -1       0        0        0        0        80
2        1        1        0        -1       0        0        0        10
75       15       35       0        0        1        0        0        750
200      0        100      0        0        0        1        0        1000
2.2      1.6      0.6      0        0        0        0        1        0

Tableau #2
x        y        z        s1       s2       s3       s4       p
0        5        0        -1       0        0        -0.05    0        30
0        1        0        0        -1       0        -0.01    0        0
0        15       -2.5     0        0        1        -0.37    0        375
1        0        0.5      0        0        0        0.01     0        5
0        1.6      -0.5     0        0        0        -0.01    1        -11
```

Tableau #3

x	y	z	s1	s2	s3	s4	p	
0	0	0	-1	5	0	0	0	30
0	1	0	0	-1	0	-0.01	0	0
0	0	-2.5	0	15	1	-0.22	0	375
1	0	0.5	0	0	0	0.01	0	5
0	0	-0.5	0	1.6	0	0.01	1	-11

Tableau #4

x	y	z	s1	s2	s3	s4	p	
0	0	0	-0.2	1	0	0	0	6
0	1	0	-0.2	0	0	-0.01	0	6
0	0	-2.5	3	0	1	-0.22	0	285
1	0	0.5	0	0	0	0.01	0	5
0	0	-0.5	0.32	0	0	0.01	1	-20.6

Tableau #5

x	y	z	s1	s2	s3	s4	p	
0	0	0	-0.2	1	0	0	0	6
0	1	0	-0.2	0	0	-0.01	0	6
5	0	0	3	0	1	-0.2	0	310
2	0	1	0	0	0	0.01	0	10
1	0	0	0.32	0	0	0.01	1	-15.6

Optimal Solution: $c = p = 15.6$; $x = 0$, $y = 6$, $z = 10$. Mix 6 servings of Riboforce HP and 10 servings of Creatine Transport for a cost of $15.60.

37. Unknowns: x = # convention–style hotels, y = # vacation–style hotels, and z = # motels.
(a) Minimize $C = 100x + 20y + 4z$ subject to $500x + 200y + 50z \geq 1400$, $x \geq 1$, $z \leq 2$, $x \geq 0$, $y \geq 0$, $z \geq 0$.
Associated maximization problem: Maximize $p = -100x - 20y - 4z$ subject to $500x + 200y + 50z \geq 1400$, $x \geq 1$, $z \leq 2$, $x \geq 0$, $y \geq 0$, $z \geq 0$.

	x	y	z	s	t	u	p		
*s	500	200	50	-1	0	0	0	1400	$R_1 - 500R_2$
*t	1	0	0	0	-1	0	0	1	
u	0	0	1	0	0	1	0	2	
p	100	20	4	0	0	0	1	0	$R_4 - 100R_2$

	x	y	z	s	t	u	p		
*s	0	200	50	-1	500	0	0	900	
x	1	0	0	0	-1	0	0	1	$500R_2 + R_1$
u	0	0	1	0	0	1	0	2	
p	0	20	4	0	100	0	1	-100	$5R_4 - R_1$

	x	y	z	s	t	u	p		
t	0	200	50	-1	500	0	0	900	
x	500	200	50	-1	0	0	0	1400	$R_2 - R_1$
u	0	0	1	0	0	1	0	2	
p	0	-100	-30	1	0	0	5	-1400	$2R_4 + R_1$

	x	y	z	s	t	u	p		
y	0	200	50	-1	500	0	0	900	$R_1 - 50R_3$
x	500	0	0	0	-500	0	0	500	
u	0	0	1	0	0	1	0	2	
p	0	0	-10	1	500	0	10	-1900	$R_4 + 10R_3$

	x	y	z	s	t	u	p	
y	0	200	0	-1	500	-50	0	800
x	500	0	0	0	-500	0	0	500
z	0	0	1	0	0	1	0	2
p	0	0	0	1	500	10	10	-1880

Optimal Solution: $c = -p = 1880/10 = 188$; $x = 500/500 = 1$, $y = 800/200 = 4$, $z = 2/1 = 2$. Build 1 convention–style hotel, 4 vacation style hotels and 2 small motels. The total cost will amount to $188 million.

(b) Since 20% of the $188 million cost is $37.6 million, you will still be covered by the $50 million subsidy.

39. x = # boards sent from Tucson to Honolulu, y = # boards sent from Tucson to Venice Beach, z = # boards sent from Toronto to Honolulu, w = # boards sent from Toronto to Venice Beach.

Minimize $c = 10x + 5y + 20z + 10w$ subject to

$x + y \le 620$

$z + w \le 410$

$x + z \ge 500$

$y + w \ge 530$

$x \ge 0, y \ge 0, z \ge 0, w \ge 0.$

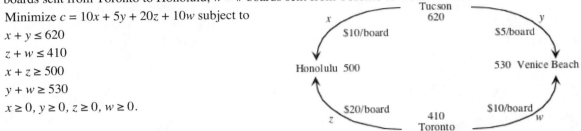

Convert to a minimization problem and solve:

	x	y	z	w	s	t	u	v	p		
s	1	1	0	0	1	0	0	0	0	620	$R_1 - R_3$
t	0	0	1	1	0	1	0	0	0	410	
*u	$\boxed{1}$	0	1	0	0	0	−1	0	0	500	
*v	0	1	0	1	0	0	0	−1	0	530	
p	10	5	20	10	0	0	0	0	1	0	$R_5 - 10R_3$

	x	y	z	w	s	t	u	v	p		
s	0	$\boxed{1}$	−1	0	1	0	1	0	0	120	
t	0	0	1	1	0	1	0	0	0	410	
x	1	0	1	0	0	0	−1	0	0	500	
*v	0	1	0	1	0	0	0	−1	0	530	$R_4 - R_1$
p	0	5	10	10	0	0	10	0	1	−5000	$R_5 - 5R_1$

	x	y	z	w	s	t	u	v	p		
y	0	1	−1	0	1	0	1	0	0	120	$R_1 + R_4$
t	0	0	1	1	0	1	0	0	0	410	$R_2 - R_4$
x	1	0	1	0	0	0	−1	0	0	500	$R_3 - R_4$
*v	0	0	$\boxed{1}$	1	−1	0	−1	−1	0	410	
p	0	0	15	10	−5	0	5	0	1	−5600	$R_5 - 15R_4$

	x	y	z	w	s	t	u	v	p		
y	0	1	0	1	0	0	0	−1	0	530	$R_1 - R_4$
t	0	0	0	0	1	1	1	1	0	0	
x	1	0	0	−1	1	0	0	1	0	90	$R_3 + R_4$
z	0	0	1	$\boxed{1}$	−1	0	−1	−1	0	410	
p	0	0	0	−5	10	0	20	15	1	−11750	$R_5 + 5R_4$

	x	y	z	w	s	t	u	v	p	
y	0	1	−1	0	1	0	1	0	0	120
t	0	0	0	0	1	1	1	1	0	0
x	1	0	1	0	0	0	−1	0	0	500
w	0	0	1	1	−1	0	−1	−1	0	410
p	0	0	5	0	5	0	15	10	1	−9700

Optimal Solution: $c = -p = 9700/1 = 9700$; $x = 500/1 = 500$, $y = 120/1 = 120$, $z = 0$, $w = 410/1 = 410$. Make the following shipments: Tucson to Honolulu: 500 boards per week; Tucson to Venice Beach: 120 boards per week; Toronto to Honolulu: 0 boards per week; Toronto to Venice Beach: 410 boards per week. Minimum weekly cost is $9,700.

41. Unknowns: x = amount overdrawn from Congressional Integrity Bank, y = amount from Citizens' Trust, z = amount from Checks R Us.

Maximize $p = x$ subject to $x + y + z \le 10,000$, $x + y \le \frac{1}{4}(x + y + z)$, or $3x + 3y - z \le 0$, $x \ge 2500$, $x \ge 0$, $y \ge 0$, $z \ge 0$.

	x	y	z	s	t	u	p		
s	1	1	1	1	0	0	0	10,000	$3R_1 - R_2$
t	$\boxed{3}$	3	−1	0	1	0	0	0	
*u	1	0	0	0	0	−1	0	2500	$3R_3 - R_2$
p	−1	0	0	0	0	0	1	0	$3R_4 + R_2$

	x	y	z	s	t	u	p		
s	0	0	4	3	−1	0	0	30,000	$R_1 - 4R_3$
x	3	3	−1	0	1	0	0	0	$R_2 + R_3$
*u	0	−3	$\boxed{1}$	0	−1	−3	0	7500	
p	0	3	−1	0	1	0	3	0	$R_4 + R_3$

216

	x	y	z	s	t	u	p	
s	0	12	0	3	3	12	0	0
x	3	0	0	0	0	−3	0	7500
z	0	−3	1	0	−1	−3	0	7500
p	0	0	0	0	0	−3	3	7500

$4R_2 + R_1$
$4R_3 + R_1$
$4R_4 + R_1$

	x	y	z	s	t	u	p	
u	0	12	0	3	3	12	0	0
x	12	12	0	3	3	0	0	30,000
z	0	0	4	3	−1	0	0	30,000
p	0	12	0	3	3	0	12	30,000

Optimal Solution: $p = 30,000/12 = 2500$; $x = 30,000/12 = 2500$, $y = 0$, $z = 30,000/4 = 7500$. Withdraw $2500 from Congressional Integrity Bank, $0 from Citizens' Trust, $7500 from Checks R Us.

43. Unknowns: x = # people you fly from Chicago to Los Angeles, y = # people you fly from Chicago to New York, z = # people you fly from Denver to Los Angeles, w = # people you fly from Denver to New York

Minimize $c = 195x + 182y + 395z + 166w$ subject to $x + y \leq 20$, $z + w \leq 10$, $x + z \geq 10$, $y + w \geq 15$ $x \geq 0$, $y \geq 0$, $z \geq 0$, $w \geq 0$.

Tableau #1

x	y	z	w	s1	s2	s3	s4	−c	
1	1	0	0	1	0	0	0	0	20
0	0	1	1	0	1	0	0	0	10
1	0	1	0	0	0	−1	0	0	10
0	1	0	1	0	0	0	−1	0	15
195	182	395	166	0	0	0	0	1	0

Tableau #2

x	y	z	w	s1	s2	s3	s4	−c	
0	1	−1	0	1	0	1	0	0	10
0	0	1	1	0	1	0	0	0	10
1	0	1	0	0	0	−1	0	0	10
0	1	0	1	0	0	0	−1	0	15
0	182	200	166	0	0	195	0	1	−1950

217

Tableau #3

x	y	z	w	s1	s2	s3	s4	-c	
0	1	-1	0	1	0	1	0	0	10
0	0	1	1	0	1	0	0	0	10
1	0	1	0	0	0	-1	0	0	10
0	0	1	1	-1	0	-1	-1	0	5
0	0	382	166	-182	0	13	0	1	-3770

Tableau #4

x	y	z	w	s1	s2	s3	s4	-c	
0	1	0	1	0	0	0	-1	0	15
0	0	0	0	1	1	1	1	0	5
1	0	0	-1	1	0	0	1	0	5
0	0	1	1	-1	0	-1	-1	0	5
0	0	0	-216	200	0	395	382	1	-5680

Tableau #5

x	y	z	w	s1	s2	s3	s4	-c	
0	1	-1	0	1	0	1	0	0	10
0	0	0	0	1	1	1	1	0	5
1	0	1	0	0	0	-1	0	0	10
0	0	1	1	-1	0	-1	-1	0	5
0	0	216	0	-16	0	179	166	1	-4600

Tableau #6

x	y	z	w	s1	s2	s3	s4	-c	
0	1	-1	0	0	-1	0	-1	0	5
0	0	0	0	1	1	1	1	0	5
1	0	1	0	0	0	-1	0	0	10
0	0	1	1	0	1	0	0	0	10
0	0	216	0	0	16	195	182	1	-4520

Optimal Solution: $c = 4520$; $x = 10$, $y = 5$, $z = 0$, $w = 10$. Fly 10 people from Chicago to LA, 5 from Chicago to New York, none from Denver to LA, 10 from Denver to NY at a total cost of $4520.

45. Unknowns: x = # cardiologists hired, y = # rehabilitation specialists hired, z = # infectious disease specialists hired.

Maximize $r = 12x + 19y + 14z$ ($10,000) subject to $x + y + 3z \geq 27$, $10x + 10y + 10z \leq 200-30 = 170$, $x \geq 0$, $y \geq 0$, $z \geq 0$.

Tableau #1

x	y	z	s1	s2	r	
1	1	3	-1	0	0	27
1	1	1	0	1	0	17
-12	-19	-14	0	0	1	0

Tableau #2

x	y	z	s1	s2	r	
1/3	1/3	1	-1/3	0	0	9
2/3	2/3	0	1/3	1	0	8
-22/3	-43/3	0	-14/3	0	1	126

Tableau #3

x	y	z	s1	s2	r	
0	0	1	-1/2	-1/2	0	5
1	1	0	1/2	3/2	0	12
7	0	0	5/2	43/2	1	298

Optimal Solution: $r = 298$; $x = 0$, $y = 12$, $z = 5$. Hire no more cardiologists, 12 rehabilitation specialists, and 5 infectious disease specialists.

47. In a general linear programming problem, the solution $x = 0$, $y = 0$, ...) represented by the initial tableau may not be feasible (feasible solutions must satisfy all the constraints)., and this shows up as a basic solution with some negative values for surplus variables. In order for a basic solution to be feasible, none of the basic variables (including surplus and slack variables) can be negative. In phase I we use pivoting to arrive at a basic solution where no basic variables are negative

49. The basic solution corresponding to the initial tableau has all the unknowns equal to zero, and this is not a feasible solution, since it does not satisfy the given inequality.

51. We can rewrite the given problem as a standard LP problem by multiplying both sides of the first constraint by –1: Maximize $p = x + y$ subject to $-x + 2y \leq 0$, $2x + y \leq 10$, $x \geq 0$, $y \geq 0$. Therefore, this problem can be solved using the techniques of either section (choice (C)).

53. Examples are Exercises 1 and 2.

55. A simple example is: Maximize $p = x + y$ subject to $x + y \le 10$, $x + y \ge 20$, $x \ge 0$, $y \ge 0$.

	x	*y*	*s*	*t*	*p*		
s	1	1	1	0	0	10	
*t	1	1	0	−1	0	20	$R_2 - R_1$
p	−1	−1	0	0	1	0	$R_3 + R_1$

	x	*y*	*s*	*t*	*p*	
x	1	1	1	0	0	10
*t	0	0	−1	−1	0	10
p	0	0	1	0	1	10

We now find it impossible to find a feasible solution, since there are no positive entries in the starred row. This problem has an empty feasible region, and hence no optimal solution.

4.5

1. Maximize $p = 2x+y$ subject to

$x + 2y \le 6$

$-x + y \le 2$

$x \ge 0, y \ge 0$

$$\begin{bmatrix} 1 & 2 & 6 \\ -1 & 1 & 2 \\ 2 & 1 & 0 \end{bmatrix} \rightarrow \begin{bmatrix} 1 & -1 & 2 \\ 2 & 1 & 1 \\ 6 & 2 & 0 \end{bmatrix}$$

Minimize $c = 6s+2t$ subject to

$s - t \ge 2$

$2s + t \ge 1$

$s \ge 0, t \ge 0$

3. Minimize $c = 2s+t+3u$ subject to

$s + t + u \ge 100$

$2s + t \ge 50$

$s \ge 0, t \ge 0, u \ge 0$

$$\begin{bmatrix} 1 & 1 & 1 & 100 \\ 2 & 1 & 0 & 50 \\ 2 & 1 & 3 & 0 \end{bmatrix} \rightarrow \begin{bmatrix} 1 & 2 & 2 \\ 1 & 1 & 1 \\ 1 & 0 & 3 \\ 100 & 50 & 0 \end{bmatrix}$$

Maximize $p = 100x+50y$ subject to

$x + 2y \le 2$

$x + y \le 1$

$x \le 3$

$x \ge 0, y \ge 0$

5. Maximize $p = x+y+z+w$ subject to

$x + y + z \le 3$

$y + z + w \le 4$

$x + z + w \le 5$

$x + y + w \le 6$

$x \ge 0, y \ge 0, z \ge 0, w \ge 0$

$$\begin{bmatrix} 1 & 1 & 1 & 0 & 3 \\ 0 & 1 & 1 & 1 & 4 \\ 1 & 0 & 1 & 1 & 5 \\ 1 & 1 & 0 & 1 & 6 \\ 1 & 1 & 1 & 1 & 0 \end{bmatrix} \rightarrow \begin{bmatrix} 1 & 0 & 1 & 1 & 1 \\ 1 & 1 & 0 & 1 & 1 \\ 1 & 1 & 1 & 0 & 1 \\ 0 & 1 & 1 & 1 & 1 \\ 3 & 4 & 5 & 6 & 0 \end{bmatrix}$$

Minimize $c = 3s+4t+5u+6v$ subject to

$s + u + v \ge 1$

$s + t + v \ge 1$

$s + t + u \ge 1$

$t + u + v \ge 1$

$s \ge 0, t \ge 0, u \ge 0, v \ge 0$

7. Minimize $c = s+3t+u$ subject to

$5s - t + v \ge 1000$

$u - v \ge 2000$

$s + t \ge 500$

$s \ge 0, t \ge 0, u \ge 0, v \ge 0$

$$\begin{bmatrix} 5 & -1 & 0 & 1 & 1000 \\ 0 & 0 & 1 & -1 & 2000 \\ 1 & 1 & 0 & 0 & 500 \\ 1 & 3 & 1 & 0 & 0 \end{bmatrix} \rightarrow$$

$$\begin{bmatrix} 5 & 0 & 1 & 1 \\ -1 & 0 & 1 & 3 \\ 0 & 1 & 0 & 1 \\ 1 & -1 & 0 & 0 \\ 1000 & 2000 & 500 & 0 \end{bmatrix}$$

Maximize

$p = 1000x+2000y+500z$ subject to

$5x + z \le 1$

$-x + z \le 3$

$y \le 1$

$x - y \le 0$

$x \ge 0, y \ge 0, z \ge 0$

9. Minimize $c = s + t$ subject to

$s + 2t \ge 6$

$2s + t \ge 6$

$s \ge 0, \ t \ge 0$

$$\begin{bmatrix} 1 & 2 & 6 \\ 2 & 1 & 6 \\ 1 & 1 & 0 \end{bmatrix} \rightarrow \begin{bmatrix} 1 & 2 & 1 \\ 2 & 1 & 1 \\ 6 & 6 & 0 \end{bmatrix}$$

Maximize $p = 6x + 6y$ subject to

$x + 2y \le 1$

$2x + y \le 1$

$x \ge 0, y \ge 0$

Solve the dual problem using the standard simplex method:

	x	y	s	t	p		
s	1	2	1	0	0	1	$2R_1 - R_2$
t	[2]	1	0	1	0	1	
p	−6	−6	0	0	1	0	$R_3 + 3R_2$

	x	y	s	t	p		
s	0	[3]	2	−1	0	1	
x	2	1	0	1	0	1	$3R_2 - R_1$
p	0	−3	0	3	1	3	$R_3 + R_1$

	x	y	s	t	p	
y	0	3	2	−1	0	1
x	6	0	−2	4	0	2
p	0	0	2	2	1	4

The solution to the primal problem is $c = 4/1 = 4$; $s = 2/1 = 2$, $t = 2/1 = 2$.

11. Minimize $c = 6s+6t$
subject to
$$s + 2t \geq 20$$
$$2s + t \geq 20$$
$$s \geq 0, t \geq 0$$

$$\begin{bmatrix} 1 & 2 & 20 \\ 2 & 1 & 20 \\ 6 & 6 & 0 \end{bmatrix} \rightarrow \begin{bmatrix} 1 & 2 & 6 \\ 2 & 1 & 6 \\ 20 & 20 & 0 \end{bmatrix}$$

Maximize $p = 20x+20y$
subject to
$$x + 2y \leq 6$$
$$2x + y \leq 6$$
$$x \geq 0, y \geq 0$$

Solve the dual problem using the standard simplex method:

	x	y	s	t	p		
s	1	2	1	0	0	6	$2R_1 - R_2$
t	[2]	1	0	1	0	6	
p	−20	−20	0	0	1	0	$R_3 + 10R_2$

	x	y	s	t	p		
s	0	[3]	2	−1	0	6	
x	2	1	0	1	0	6	$3R_2 - R_1$
p	0	−10	0	10	1	60	$3R_3 + 10R_1$

	x	y	s	t	p	
y	0	3	2	−1	0	6
x	6	0	−2	4	0	12
p	0	0	20	20	3	240

The solution to the primal problem is $c = 240/3 = 80$; $s = 20/3$, $t = 20/3$.

13. First, rewrite the given problem with no decimals in the constraints:

Minimize $c = 0.2s + 0.3t$

subject to

$2s + t \geq 10$

$s + 2t \geq 10$

$s + t \geq 8$

$s \geq 0, t \geq 0$

$$\begin{bmatrix} 2 & 1 & 10 \\ 1 & 2 & 10 \\ 1 & 1 & 8 \\ 0.2 & 0.3 & 0 \end{bmatrix} \rightarrow$$

$$\begin{bmatrix} 2 & 1 & 1 & 0.2 \\ 1 & 2 & 1 & 0.3 \\ 10 & 10 & 8 & 0 \end{bmatrix}$$

Maximize $p = 10x + 10y + 8z$

subject to

$2x + y + z \leq 0.2$

$x + 2y + z \leq 0.3$

$x \geq 0, y \geq 0, z \geq 0$

Solve the dual problem using the standard simplex method (Note: do not rewrite the constraints without decimals at this stage — clear them in the first step of the simplex method.)

	x	y	z	s	t	p		
s	2	1	1	1	0	0	0.2	$10R_1$
t	1	2	1	0	1	0	0.3	$10R_1$
p	−10	−10	−8	0	0	1	0	

	x	y	z	s	t	p		
s	20	10	10	10	0	0	2	
t	10	20	10	0	10	0	3	$2R_2 - R_1$
p	−10	−10	−8	0	0	1	0	$2R_3 + R_1$

	x	y	z	s	t	p		
x	20	10	10	10	0	0	2	$3R_1 - R_2$
t	0	30	10	−10	20	0	4	
p	0	−10	−6	10	0	2	2	$3R_3 + R_2$

223

	x	y	z	s	t	p		
x	60	0	[20]	40	−20	0	2	
y	0	30	10	−10	20	0	4	$2R_2 - R_1$
p	0	0	−8	20	20	6	10	$5R_3 + 2R_1$

	x	y	z	s	t	p	
z	60	0	20	40	−20	0	2
y	−60	60	0	−60	60	0	6
p	120	0	0	**180**	**60**	**30**	**54**

The solution to the primal problem is $c = 54/30 = 1.8$; $s = 180/30 = 6$, $t = 60/30 = 2$.

15. Minimize $c = 2s+t$ subject to

$3s + t \geq 30$

$s + t \geq 20$

$s + 3t \geq 30$

$s \geq 0, t \geq 0$

$$\begin{bmatrix} 3 & 1 & 30 \\ 1 & 1 & 20 \\ 1 & 3 & 30 \\ 2 & 1 & 0 \end{bmatrix} \rightarrow \begin{bmatrix} 3 & 1 & 1 & 2 \\ 1 & 1 & 3 & 1 \\ 30 & 20 & 30 & 0 \end{bmatrix}$$

Maximize $p = 30x+20y+30z$ subject to

$3x + y + z \leq 2$

$x + y + 3z \leq 1$

$x \geq 0, y \geq 0, z \geq 0$

Solve the dual problem using the standard simplex method:

	x	y	z	s	t	p		
s	[3]	1	1	1	0	0	2	
t	1	1	3	0	1	0	1	$3R_2 - R_1$
p	−30	−20	−30	0	0	1	0	$R_3 + 10R_1$

	x	y	z	s	t	p		
x	3	1	1	1	0	0	2	$8R_1 - R_2$
t	0	2	[8]	−1	3	0	1	
p	0	−10	−20	10	0	1	20	$2R_3 + 5R_2$

	x	y	z	s	t	p		
x	24	6	0	9	−3	0	15	$R_1 - 3R_2$
z	0	[2]	8	−1	3	0	1	
p	0	−10	0	15	15	2	45	$R_3 + 5R_2$

	x	y	z	s	t	p	
x	24	0	−24	12	−12	0	12
y	0	2	8	−1	3	0	1
p	0	0	40	**10**	**30**	2	**50**

The solution to the primal problem is $c = 50/2 = 25$; $s = 10/2 = 5$, $t = 30/2 = 15$.

17. Minimize $c = s+2t+3u$

    subject to

    $3s + 2t + u \geq 60$

    $2s + t + 3u \geq 60$

    $s \geq 0,\, t \geq 0,\, u \geq 0$

$$\begin{bmatrix} 3 & 2 & 1 & 60 \\ 2 & 1 & 3 & 60 \\ 1 & 2 & 3 & 0 \end{bmatrix} \rightarrow \begin{bmatrix} 3 & 2 & 1 \\ 2 & 1 & 2 \\ 1 & 3 & 3 \\ 60 & 60 & 0 \end{bmatrix}$$

Maximize $p = 60x+60y$

    subject to

    $3x + 2y \leq 1$

    $2x - y \leq 2$

    $x + 3y \leq 3$

    $x \geq 0,\, y \geq 0$

Solve the dual problem using the standard simplex method:

	x	y	s	t	u	p		
s	[3]	2	1	0	0	0	1	
t	2	−1	0	1	0	0	2	$3R_2 - 2R_1$
u	1	3	0	0	1	0	3	$3R_3 - R_1$
p	−60	−60	0	0	0	1	0	$R_4 + 20R_1$

	x	y	s	t	u	p		
x	3	[2]	1	0	0	0	1	
t	0	−7	−2	3	0	0	4	$2R_2 + 7R_1$
u	0	7	−1	0	3	0	8	$2R_3 - 7R_1$
p	0	−20	20	0	0	1	20	$R_4 + 10R_1$

	x	y	s	t	u	p	
y	3	2	1	0	0	0	1
t	21	0	3	6	0	0	15
u	−21	0	−9	0	6	0	9
p	30	0	**30**	**0**	**0**	1	30

The solution to the primal problem is $c = 30/1 = 30$; $s = 30/1 = 30$, $t = 0$, $u = 0$.

19. Minimize $c = 2s+t+3u$

subject to

$s + t + u \geq 100$

$2s + t \geq 50$

$t + u \geq 50$

$s \geq 0, t \geq 0, u \geq 0$

$$\begin{bmatrix} 1 & 1 & 1 & 100 \\ 2 & 1 & 0 & 50 \\ 0 & 1 & 1 & 50 \\ 2 & 1 & 3 & 0 \end{bmatrix} \rightarrow \begin{bmatrix} 1 & 2 & 0 & 2 \\ 1 & 1 & 1 & 1 \\ 1 & 0 & 1 & 3 \\ 100 & 50 & 50 & 0 \end{bmatrix}$$

Maximize $p = 100x+50y+50z$

subject to

$x + 2y + z \leq 2$

$x + y + z \leq 1$

$x + z \leq 3$

$x \geq 0, y \geq 0, z \geq 0$

Solve the dual problem using the standard simplex method:

	x	y	z	s	t	u	p		
s	1	2	1	1	0	0	0	2	$R_1 - R_2$
t	[1]	1	1	0	1	0	0	1	
u	1	0	1	0	0	1	0	3	$R_3 - R_2$
p	−100	−50	−50	0	0	0	1	0	$R_4 + 100R_2$

	x	y	z	s	t	u	p	
s	0	1	0	1	−1	0	0	1
x	1	1	1	0	1	0	0	1
u	0	−1	0	0	−1	1	0	2
p	0	50	50	0	100	0	1	100

The solution to the primal problem is $c = 100/1 = 100$; $s = 0$, $t = 100/1 = 100$, $u = 0$.

21. Minimize $c = s+t+u$

subject to

$3s + 2t + u \geq 60$

$2s + t + 3u \geq 60$

$s + 3t + 2u \geq 60$

$s \geq 0, t \geq 0, u \geq 0$

$$\begin{bmatrix} 3 & 2 & 1 & 60 \\ 2 & 1 & 3 & 60 \\ 1 & 3 & 2 & 60 \\ 1 & 1 & 1 & 0 \end{bmatrix} \rightarrow$$

$$\begin{bmatrix} 3 & 2 & 1 & 1 \\ 2 & 1 & 3 & 1 \\ 1 & 3 & 2 & 1 \\ 60 & 60 & 60 & 0 \end{bmatrix}$$

Maximize $p = 60x+60y+60z$

subject to

$3x + 2y + z \leq 1$

$2x + 1y + 3z \leq 1$

$1x + 3y + 2z \leq 1$

$x \geq 0, y \geq 0, z \geq 0$

Solve the dual problem using the standard simplex method:

	x	y	z	s	t	u	p		
s	[3]	2	1	1	0	0	0	1	
t	2	1	3	0	1	0	0	1	$3R_2 - 2R_1$
u	1	3	2	0	0	1	0	1	$3R_3 - R_1$
p	−60	−60	−60	0	0	0	1	0	$R_4 + 20R_1$

	x	y	z	s	t	u	p		
x	3	2	1	1	0	0	0	1	$7R_1 - R_2$
t	0	−1	$\boxed{7}$	−2	3	0	0	1	
u	0	7	5	−1	0	3	0	2	$7R_3 - 5R_2$
p	0	−20	−40	20	0	0	1	20	$7R_4 + 40R_2$

	x	y	z	s	t	u	p		
x	21	15	0	9	−3	0	0	6	$\frac{1}{3}R_1$
z	0	−1	7	−2	3	0	0	1	
u	0	$\boxed{54}$	0	3	−15	21	0	9	$\frac{1}{3}R_1$
p	0	−180	0	60	120	0	7	180	

	x	y	z	s	t	u	p		
x	7	5	0	3	−1	0	0	2	$18R_1 - 5R_3$
z	0	−1	7	−2	3	0	0	1	$18R_2 + R_3$
u	0	$\boxed{18}$	0	1	−5	7	0	3	
p	0	−180	0	60	120	0	7	180	$R_4 + 10R_3$

	x	y	z	s	t	u	p	
x	126	0	0	49	7	−35	0	21
z	0	0	126	−35	49	7	0	21
y	0	54	0	3	−15	21	0	9
p	0	0	0	70	70	70	7	210

The solution to the primal problem is $c = 210/7 = 30$; $s = 70/7 = 10$, $t = 70/7 = 10$, $u = 70/7 = 10$.

23. Add $k = 2$ to each entry and put 1s to the right and below:

$$\begin{bmatrix} 1 & 3 & 4 & 1 \\ 4 & 1 & 0 & 1 \\ 1 & 1 & 1 & 0 \end{bmatrix}$$

This gives the following LP problem:

Maximize $p = x + y + z$ subject to

$x + 3y + 4z \le 1$

$4x + y \qquad \le 1$

$x \ge 0, \, y \ge 0, \, z \ge 0$

Here are the tableaux:

	x	y	z	s	t	p		
s	1	3	4	1	0	0	1	$4R_1 - R_2$
t	[4]	1	0	0	1	0	1	
p	−1	−1	−1	0	0	1	0	$4R_3 + R_2$

	x	y	z	s	t	p	
s	0	11	[16]	4	−1	0	3
x	4	1	0	0	1	0	1
p	0	−3	−4	0	1	4	1	$4R_3 + R_1$

	x	y	z	s	t	p		
z	0	[11]	16	4	−1	0	3	
x	4	1	0	0	1	0	1	$11R_2 - R_1$
p	0	−1	0	4	3	16	7	$11R_3 + R_1$

	x	y	z	s	t	p	
y	0	11	16	4	−1	0	3
x	44	0	−16	−4	12	0	8
p	0	0	16	48	32	176	80

The solution to the primal problem is $p = 80/176 = 5/11$; $x = 8/44 = 2/11$, $y = 3/11$, $z = 0$. So, the column player's optimal strategy is $C = [2/5 \ 3/5 \ 0]^T$ and the value of the game is $e = 11/5 - 2 = 1/5$. The solution to the dual problem is $s = 48/176 = 3/11$, $t = 32/176 = 2/11$, so the row player's optimal strategy is $R = [3/5 \ 2/5]$.

25. Add $k = 2$ to each entry and put 1s to the right and below:

$$\begin{bmatrix} 1 & 3 & 4 & 1 \\ 4 & 1 & 0 & 1 \\ 3 & 4 & 2 & 1 \\ 1 & 1 & 1 & 0 \end{bmatrix}$$

This gives the following LP problem:

Maximize $p = x + y + z$ subject to
$$x + 3y + 4z \leq 1$$
$$4x + y \leq 1$$
$$3x + 4y + 2z \leq 1$$
$$x \geq 0, \; y \geq 0, \; z \geq 0$$

Here are the tableaux:

	x	y	z	s	t	u	p		
s	1	3	4	1	0	0	0	1	$4R_1 - R_2$
t	4	1	0	0	1	0	0	1	
u	3	4	2	0	0	1	0	1	$4R_3 - 3R_2$
p	-1	-1	-1	0	0	0	1	0	$4R_4 + R_2$

	x	y	z	s	t	u	p		
s	0	11	16	4	-1	0	0	3	$R_1 - 2R_3$
x	4	1	0	0	1	0	0	1	
u	0	13	8	0	-3	4	0	1	
p	0	-3	-4	0	1	0	4	1	$2R_4 + R_3$

	x	y	z	s	t	u	p		
s	0	-15	0	4	5	-8	0	1	
x	4	1	0	0	1	0	0	1	$5R_2 - R_1$
z	0	13	8	0	-3	4	0	1	$5R_3 + 3R_1$
p	0	7	0	0	-1	4	8	3	$5R_4 + R_1$

	x	y	z	s	t	u	p	
t	0	-15	0	4	5	-8	0	1
x	20	20	0	-4	0	8	0	4
z	0	20	40	12	0	-4	0	8
p	0	20	0	4	0	12	40	16

The solution to the primal problem is $p = 16/40 = 2/5$; $x = 4/20 = 1/5$, $y = 0$, $z = 8/40 = 1/5$. So, the column player's optimal strategy is $C = [1/2 \; 0 \; 1/2]^T$ and the value of the game is $e = 5/2 - 2 = 1/2$.

The solution to the dual problem is $s = 4/40 = 1/10$, $t = 0$, $u = 12/40 = 3/10$, so the row player's optimal strategy is $R = [1/4 \; 0 \; 3/4]$.

27. The first row is dominated by the last, so we can remove it. That is as far as we can reduce by dominance. Add $k = 3$ to each entry and put 1s to the right and below:

$$\begin{bmatrix} 5 & 2 & 1 & 0 & 1 \\ 4 & 5 & 3 & 4 & 1 \\ 3 & 5 & 6 & 6 & 1 \\ 1 & 1 & 1 & 1 & 0 \end{bmatrix}$$

This gives the following LP problem:

Maximize $p = x + y + z + w$ subject to

$$5x + 2y + z \quad\quad \leq 1$$
$$4x + 5y + 3z + 4w \leq 1$$
$$3x + 5y + 6z + 6w \leq 1$$
$$x \geq 0,\ y \geq 0,\ z \geq 0$$

Here are the tableaux:

	x	y	z	w	s	t	u	p		
s	**5**	2	1	0	1	0	0	0	1	
t	4	5	3	4	0	1	0	0	1	$5R_2 - 4R_1$
u	3	5	6	6	0	0	1	0	1	$5R_3 - 3R_1$
p	−1	−1	−1	−1	0	0	0	1	0	$5R_4 + R_1$

	x	y	z	w	s	t	u	p		
x	5	2	1	0	1	0	0	0	1	
t	0	17	11	**20**	−4	5	0	0	1	
u	0	19	27	30	−3	0	5	0	2	$2R_3 - 3R_2$
p	0	−3	−4	−5	1	0	0	5	1	$4R_4 + R_2$

	x	y	z	w	s	t	u	p		
x	5	2	1	0	1	0	0	0	1	$21R_1 - R_3$
w	0	17	11	20	−4	5	0	0	1	$21R_2 - 11R_3$
u	0	−13	**21**	0	6	−15	10	0	1	
p	0	5	−5	0	0	5	0	20	5	$21R_4 + 5R_3$

	x	y	z	w	s	t	u	p	
x	105	55	0	0	15	15	−10	0	20
w	0	500	0	420	−150	270	−110	0	10
z	0	−13	21	0	6	−15	10	0	1
p	0	40	0	0	30	30	50	420	110

The solution to the primal problem is $p = 110/420 = 11/42$; $x = 20/105 = 4/21$, $y = 0$, $z = 1/21$, $w = 10/420 = 1/42$. So, the column player's optimal strategy is $C = [8/11\ 0\ 2/11\ 1/11]^T$ and the value of the game is $e = 42/11 - 3 = 9/11$.

The solution to the dual problem is $s = 30/420 = 1/14$, $t = 30/420 = 1/14$, $u = 50/420 = 5/42$, so the row player's optimal strategy is $R = [0\ 3/11\ 3/11\ 5/11]$. (Note that we put in 0 for the first row that we removed at the beginning.)

29. Unknowns: s = # ounces of fish, t = # ounces of cornmeal

Minimize $c = 5s + 5t$ subject to $8t + 4t \geq 48$, $4s + 8t \geq 48$, $s \geq 0$, $t \geq 0$.

Dualize: $\begin{bmatrix} 8 & 4 & 48 \\ 4 & 8 & 48 \\ 5 & 5 & 0 \end{bmatrix} \rightarrow \begin{bmatrix} 8 & 4 & 5 \\ 4 & 8 & 5 \\ 48 & 48 & 0 \end{bmatrix}$

Dual problem: Maximize $p = 48x + 48y$ subject to $8x + 4y \leq 5$, $4x + 8y \leq 5$, $x \geq 0$, $y \geq 0$.

Solve the dual problem:

	x	y	s	t	p		
s	8	4	1	0	0	5	
t	4	8	0	1	0	5	$2R_2 - R_1$
p	−48	−48	0	0	1	0	$R_3 + 6R_1$

	x	y	s	t	p		
x	8	4	1	0	0	5	$3R_1 - R_2$
t	0	12	−1	2	0	5	
p	0	−24	6	0	1	30	$R_3 + 2R_2$

	x	y	s	t	p	
x	24	0	4	−2	0	10
y	0	12	−1	2	0	5
p	0	0	4	4	1	40

The solution to the primal problem is $c = 40/1 = 40$, $s = 4/1 = 4$, $t = 4/1 = 4$. Meow should use 4 ounces each of fish and cornmeal, for a total cost of 40¢ per can. The shadow costs are the values of x and y in the final tableau: $x = 10/24 = 5/12$¢ per gram of protein, $y = 5/12$¢ per gram of fat.

31. $s = $ # ounces of chicken, $t = $ # ounces of grain

Minimize $c = 10s + t$ subject to $10s + 2t \geq 200$, $5s + 2t \geq 150$, $s \geq 0$, $t \geq 0$.

Dualize: $\begin{bmatrix} 10 & 2 & 200 \\ 5 & 2 & 150 \\ 10 & 1 & 0 \end{bmatrix} \rightarrow \begin{bmatrix} 10 & 5 & 10 \\ 2 & 2 & 1 \\ 200 & 150 & 0 \end{bmatrix}$

Dual problem: Maximize $p = 200x + 150y$ subject to $10x + 5y \leq 10$, $2x + 2y \leq 1$, $x \geq 0$, $y \geq 0$.
Solve the dual problem:

	x	y	s	t	p		
s	10	5	1	0	0	10	$R_1 - 5R_2$
t	2	2	0	1	0	1	
p	−200	−150	0	0	1	0	$R_3 + 100R_2$

	x	y	s	t	p	
s	0	−5	1	−5	0	5
x	2	2	0	1	0	1
p	0	50	0	100	1	100

The solution to the primal problem is $c = 100/1 = 100$; $s = 0$, $t = 100/1 = 100$. Use no chicken and 100 oz of grain for a total cost of \$1. The shadow costs are the values of x and y in the final tableau: $x = 1/2$¢ per g of protein, $y = 0$¢ per gram of fat.

33. Unknowns: $s = $ # servings of cereal, $t = $ # servings of dessert, $u = $ # servings of juice.

Minimize $c = 10s + 53t + 27u$ subject to $60s + 80t + 60u \geq 120$, $45s + 120u \geq 120$, $s \geq 0$, $t \geq 0$, $u \geq 0$.

Dualize: $\begin{bmatrix} 60 & 80 & 60 & 120 \\ 0 & 45 & 120 & 120 \\ 10 & 53 & 27 & 0 \end{bmatrix} \rightarrow \begin{bmatrix} 60 & 0 & 10 \\ 80 & 45 & 53 \\ 60 & 120 & 27 \\ 120 & 120 & 0 \end{bmatrix}$

Dual problem: Maximize $p = 120x + 120y$ subject to $60x \leq 10$, $80x + 45y \leq 53$, $60x + 120y \leq 27$, $x \geq 0$, $y \geq 0$.
Solve the dual problem:

	x	y	s	t	u	p		
s	60	0	1	0	0	0	10	
t	80	45	0	1	0	0	53	$12R_2 - 16R_1$
u	60	120	0	0	1	0	27	$R_3 - R_1$
p	−120	−120	0	0	0	1	0	$R_4 + 2R_1$

	x	y	s	t	u	p		
x	60	0	1	0	0	0	10	
t	0	540	−16	12	0	0	476	$2R_2 - 9R_3$
u	0	120	−1	0	1	0	17	
p	0	−120	2	0	0	1	20	$R_4 + R_3$

	x	y	s	t	u	p	
x	60	0	1	0	0	0	10
t	0	0	−23	24	−9	0	799
y	0	120	−1	0	1	0	17
p	0	0	1	0	1	1	37

The solution to the primal problem is $c = 37/1 = 37$; $s = 1/1 = 1$, $t = 0$, $u = 1$. Prepare one serving of cereal, one serving of juice, and no dessert! for a total cost of 37¢. The shadow costs are the values of x and y in the final tableau: $x = 10/60 = 1/6$¢ per calorie and $y = 17/120$¢ per % U.S. RDA of Vitamin C.

35. Unknowns: s = # mailings to the East Coast, t = # mailings to the Midwest, u = # mailings to the West Coast

Minimize $c = 40s + 60y + 50z$ subject to $100s + 100t + 50u \geq 1500$, $50s + 100t + 100u \geq 1500$, $s \geq 0$, $t \geq 0$, $u \geq 0$.

Dualize:
$$\begin{bmatrix} 100 & 100 & 50 & 1500 \\ 50 & 100 & 100 & 1500 \\ 40 & 60 & 50 & 0 \end{bmatrix} \rightarrow \begin{bmatrix} 100 & 50 & 40 \\ 100 & 100 & 60 \\ 50 & 100 & 50 \\ 1500 & 1500 & 0 \end{bmatrix}$$

Dual problem: Maximize $p = 1500x + 1500y$ subject to $100x + 50y \leq 40$, $100x + 100y \leq 60$, $50x + 100y \leq 50$, $x \geq 0$, $y \geq 0$.

Solve the dual problem:

	x	y	s	t	u	p		
s	100	50	1	0	0	0	40	
t	100	100	0	1	0	0	60	$R_2 - R_1$
u	50	100	0	0	1	0	50	$2R_3 - R_1$
p	−1500	−1500	0	0	0	1	0	$R_4 + 15R_1$

	x	y	s	t	u	p		
x	100	50	1	0	0	0	40	$3R_1 - R_3$
t	0	50	−1	1	0	0	20	$3R_2 - R_3$
u	0	150	−1	0	2	0	60	
p	0	−750	15	0	0	1	600	$R_4 + 5R_3$

	x	y	s	t	u	p	
x	300	0	4	0	−2	0	60
t	0	0	−2	3	−2	0	0
y	0	150	−1	0	2	0	60
p	0	0	10	0	10	1	900

The solution to the primal problem is $c = 900/1 = 900$; $s = 10/1 = 10$, $t = 0$, $u = 10/1 = 10$. Send 10 mailings to the East coast, none to the Midwest, 10 to the West Coast. Cost: \$900. The shadow costs are the values of x and y in the final tableau: $x = 60/300 = 1/5 = 20$¢ per Democrat and $y = 60/150 = 2/5 = 40$¢ per Republican.

Another possible solution: 15 mailings to the Midwest and no mailing to the coasts. Cost: \$900. Shadow costs: 20¢ per Democrat and 40¢ per Republican.

37. Unknowns: $s = $ # sleep spells, $t = $ # shock spells

Minimize $c = 500s + 750t$ subject to $2s + 3t \geq 1440$, $3s + 2t \geq 1200$, $-s + 3t \geq 0$, $s - 2t \geq 0$, $s \geq 0$, $t \geq 0$.

Dualize:
$$\begin{bmatrix} 2 & 3 & 1440 \\ 3 & 2 & 1200 \\ -1 & 3 & 0 \\ 1 & -2 & 0 \\ 500 & 750 & 0 \end{bmatrix} \rightarrow \begin{bmatrix} 2 & 3 & -1 & 1 & 500 \\ 3 & 2 & 3 & -2 & 750 \\ 1440 & 1200 & 0 & 0 & 0 \end{bmatrix}$$

Dual problem: Maximize $p = 1440x + 1200y$ subject to $2x + 3y - z + w \leq 500$, $3x + 2y + 3z - 2w \leq 750$, $x \geq 0$, $y \geq 0$, $z \geq 0$, $w \geq 0$.

Solve the dual problem:

	x	y	z	w	s	t	p		
s	2	3	–1	1	1	0	0	500	
t	3	2	3	–2	0	1	0	750	$2R_1 - 3R_2$
p	–1440	–1200	0	0	0	0	1	0	$R_3 + 480R_2$

	x	y	z	w	s	t	p		
x	2	3	–1	1	1	0	0	500	$9R_2 + R_1$
t	0	–5	9	–7	–3	2	0	0	
p	0	960	–720	720	720	0	1	360000	$R_3 + 80R_1$

	x	y	z	w	s	t	p	
x	18	22	0	2	6	2	0	4500
z	0	–5	9	–7	–3	2	0	0
p	0	560	0	160	**480**	**160**	1	**360,000**

The solution to the primal problem $360,000/1 = 360,000$ 480 160. Gillian should use 480 sleep spells and 160 shock spells, costing 360,000 pico-shirleys of energy. Another solution (pivot on the 3 in the first step): 2880/7 sleep spells and 1440/7 shock spells.

39. We observe that the first column dominates the last, so we can eliminate the last. In the matrix that remains, the second row dominates the third, so we eliminate the third. We cannot reduce any further. Add $k = 500$ to each entry and put 1s to the right and below:

$$\begin{bmatrix} 300 & 200 & 1 \\ 0 & 1000 & 1 \\ 1 & 1 & 0 \end{bmatrix}$$

This gives the following LP problem:

Maximize $p = x + y$ subject to

$$300x + 200y \le 1$$
$$1000y \le 1$$
$$x \ge 0, y \ge 0$$

Here are the tableaux:

	x	y	s	t	p		
s	300	200	1	0	0	1	
t	0	1000	0	1	0	1	
p	–1	–1	0	0	1	0	$300R_3 + R_1$

	x	y	s	t	p		
x	300	200	1	0	0	1	$5R_1 - R_2$
t	0	1000	0	1	0	1	
p	0	−100	1	0	300	1	$10R_3 + R_2$

	x	y	s	t	p	
x	1500	0	5	−1	0	4
y	0	1000	0	1	0	1
p	0	0	10	1	3000	11

The solution to the primal problem is $p = 11/3000$; $x = 4/1500 = 1/375$, $y = 1/1000$. So, T. N. Spend's optimal strategy is $C = [8/11\ \ 3/11\ \ 0]^T$ and the value of the game is $e = 3000/11 - 500 = -2500/11 \approx 227$.

The solution to the dual problem is $s = 10/3000$, $t = 1/3000$, so T. L. Down's optimal strategy is $R = [10/11\ \ 1/11\ \ 0]$.

T. N. Spend should spend about 73% of the days in Littleville, 27% in Metropolis, and skip Urbantown. T. L. Down should spend about 91% of the days in Littleville, 9% in Metropolis, and skip Urbantown. The expected outcome is that T. L. Down will lose about 227 votes per day of campaigning.

41. With A the row player and B the column player, the payoff matrix is

$$P = \begin{bmatrix} 2 & -1 & 0 \\ -1 & 4 & -1 \\ 0 & -1 & 6 \end{bmatrix}$$

This game does not reduce. Add $k = 1$ to each entry and put 1s to the right and below:

$$\begin{bmatrix} 3 & 0 & 1 & 1 \\ 0 & 5 & 0 & 1 \\ 1 & 0 & 7 & 1 \\ 1 & 1 & 1 & 0 \end{bmatrix}$$

This gives the following LP problem:

Maximize $p = x + y + z$ subject to

$$3x \qquad + z \le 1$$
$$\quad\ 5y \quad\ \ \le 1$$
$$x \qquad +7z \le 1$$
$$x \ge 0,\ y \ge 0,\ z \ge 0$$

Here are the tableaux:

	x	y	z	s	t	u	p		
s	3	0	1	1	0	0	0	1	
t	0	5	0	0	1	0	0	1	
u	1	0	7	0	0	1	0	1	$3R_3 - R_1$
p	−1	−1	−1	0	0	0	1	0	$3R_4 + R_1$

	x	y	z	s	t	u	p		
x	3	0	1	1	0	0	0	1	
t	0	5	0	0	1	0	0	1	
u	0	0	20	−1	0	3	0	2	
p	0	−3	−2	1	0	0	3	1	$5R_4 + 3R_2$

	x	y	z	s	t	u	p		
x	3	0	1	1	0	0	0	1	$20R_1 - R_3$
y	0	5	0	0	1	0	0	1	
u	0	0	20	−1	0	3	0	2	
p	0	0	−10	5	3	0	15	8	$2R_4 + R_3$

	x	y	z	s	t	u	p		
x	60	0	0	21	0	−3	0	18	
y	0	5	0	0	1	0	0	1	
z	0	0	20	−1	0	3	0	2	
p	0	0	0	9	6	3	30	18	

The solution to the primal problem is $p = 18/30 = 3/5$; $x = 18/60 = 3/10$, $y = 1/5$, $z = 2/20 = 1/10$. So, player B's optimal strategy is $C = [1/2 \ 1/3 \ 1/6]^T$ and the value of the game is $e = 5/3 - 1 = 2/3$.

The solution to the dual problem is $s = 9/30 = 3/10$, $t = 6/30 = 1/5$, $u = 3/30 = 1/10$, so player A's optimal strategy is $R = [1/2 \ 1/3 \ 1/6]$.

Each player should show one finger with probability 1/2, two fingers with probability 1/3, and three fingers with probability 1/6. The expected outcome is that player A will with 2/3 point per round, on average.

Section 4.5

43. Let Colonel Blotto be the row player and Captain Kije the column player. Label possible moves by the number of regiments sent to the first of the two locations; the remaining regiments are sent to the other location. The payoff matrix is then the following:

	Captain Kije			
	0	1	2	3
0	4	2	1	0
1	1	3	0	−1
2	−2	2	2	−2
3	−1	0	3	1
4	0	1	2	4

Colonel Blotto labels the rows 0, 1, 2, 3, 4.

We cannot reduce this game. Add $k = 2$ to each entry and put 1s to the right and below:

$$\begin{bmatrix} 6 & 4 & 3 & 2 & 1 \\ 3 & 5 & 2 & 1 & 1 \\ 0 & 4 & 4 & 0 & 1 \\ 1 & 2 & 5 & 3 & 1 \\ 2 & 3 & 4 & 6 & 1 \\ 1 & 1 & 1 & 1 & 0 \end{bmatrix}$$

This gives the following LP problem:

Maximize $p = x + y + z + w$ subject to

$$6x + 4y + 3z + 2w \le 1$$
$$3x + 5y + 2z + w \le 1$$
$$4y + 4z \le 1$$
$$x + 2y + 5z + 3w \le 1$$
$$2x + 3y + 4z + 6w \le 1$$
$$x \ge 0, y \ge 0, z \ge 0, w \ge 0$$

Here are the tableaux:

	x	y	z	w	s	t	u	v	r	p		
s	6	4	3	2	1	0	0	0	0	0	1	
t	3	5	2	1	0	1	0	0	0	0	1	$2R_2 - R_1$
u	0	4	4	0	0	0	1	0	0	0	1	
v	1	2	5	3	0	0	0	1	0	0	1	$6R_4 - R_1$
r	2	3	4	6	0	0	0	0	1	0	1	$3R_5 - R_1$
p	−1	−1	−1	−1	0	0	0	0	0	1	0	$6R_6 + R_1$

238

	x	y	z	w	s	t	u	v	r	p		
x	6	4	3	2	1	0	0	0	0	0	1	$8R_1 - R_5$
t	0	6	1	0	-1	2	0	0	0	0	1	
u	0	4	4	0	0	0	1	0	0	0	1	
v	0	8	27	16	-1	0	0	6	0	0	5	$R_4 - R_5$
r	0	5	9	[16]	-1	0	0	0	3	0	2	
p	0	-2	-3	-4	1	0	0	0	0	6	1	$4R_6 + R_5$

	x	y	z	w	s	t	u	v	r	p		
x	48	27	15	0	9	0	0	0	-3	0	6	$2R_1 - 9R_2$
t	0	[6]	1	0	-1	2	0	0	0	0	1	
u	0	4	4	0	0	0	1	0	0	0	1	$3R_3 - 2R_2$
v	0	3	18	0	0	0	0	6	-3	0	3	$2R_4 - R_2$
w	0	5	9	16	-1	0	0	0	3	0	2	$6R_5 - 5R_2$
p	0	-3	-3	0	3	0	0	0	3	24	6	$2R_6 + R_2$

	x	y	z	w	s	t	u	v	r	p		
x	96	0	21	0	27	-18	0	0	-6	0	3	$10R_1 - 21R_3$
y	0	6	1	0	-1	2	0	0	0	0	1	$10R_2 - R_3$
u	0	0	[10]	0	2	-4	3	0	0	0	1	
v	0	0	35	0	1	-2	0	12	-6	0	5	$2R_4 - 7R_3$
w	0	0	49	96	-1	-10	0	0	18	0	7	$10R_5 - 49R_3$
p	0	0	-5	0	5	2	0	0	6	48	13	$2R_6 + R_3$

	x	y	z	w	s	t	u	v	r	p	
x	960	0	0	0	228	-96	-63	0	-60	0	9
y	0	60	0	0	-12	24	-3	0	0	0	9
z	0	0	20	0	4	-8	6	0	0	0	2
v	0	0	0	0	-12	24	-21	24	-12	0	3
w	0	0	0	960	-108	96	-147	0	180	0	21
p	0	0	0	0	12	0	3	0	12	96	27

The solution to the primal problem is $p = 27/96 = 9/32$; $x = 9/960 = 3/320$, $y = 9/60 = 3/20$, $z = 2/20 = 1/10$, $w = 21/960 = 7/320$. So, Captain Kije's optimal strategy is $C = [1/30 \ 8/15 \ 16/45 \ 7/90]^T$ and the value of the game is $e = 32/9 - 2 = 14/9$. In fact, this is only one of several optimal strategies for Captain Kije; which you find depends on the choices of pivots you make in the simplex method. Other optimal strategies are the one given above, in reverse order, and $[1/18 \ 4/9 \ 4/9 \ 1/18]^T$.

The solution to the dual problem (which is unique) is $s = 12/96 = 1/8$, $t = 0$, $u = 3/96 = 1/32$, $v = 0$, $r = 12/96 = 1/8$, so Colonel Blotto's optimal strategy is $R = [4/9 \ 0 \ 1/9 \ 0 \ 4/9]$.

Write moves as (x, y) where x represents the number of regiments sent to the first location and y represents the number sent to the second location. Colonel Blotto should play $(0, 4)$ with probability 4/9, $(2, 2)$ with probability 1/9, and $(4, 0)$ with probability 4/9. Captain Kije has several optimal strategies, one of which is to play $(0, 3)$ with probability 1/30, $(1, 2)$ with probability 8/15, $(2, 1)$ with probability 16/45, and $(3, 0)$ with probability 7/90. The expected outcome is that Colonel Blotto will win 14/9 points on average.

45. The dual of a standard minimization problem satisfying the nonnegative objective condition is a standard maximization problem, which can be solved using the standard simplex algorithm, thus avoiding the need to do Phase I.

47. Answers will vary, We use a minimization problem that does not satisfy the nonnegative objective condition. An example is: Minimize $c = x - y$ subject to $x - y \geq 100$, $x + y \geq 200$, $x \geq 0$, $y \geq 0$. this problem can be solved using the techniques in Section 4.4.

49. Unknowns: $s = \#$ convention–style hotels, $t = \#$ vacation–style hotels, and $u = \#$ motels. Minimize $C = 100s + 20t + 4u$ subject to $500s + 200t + 50u \geq 1400$, $s \geq 1$, $u \leq 2$, $s \geq 0$, $t \geq 0$, $u \geq 0$. Rewrite the constraint $z \leq 2$ using "$\geq$":

Minimize $C = 100s + 20t + 4u$ subject to $500x + 200t + 50u \geq 1400$, $s \geq 1$, $-u \leq -2$, $s \geq 0$, $t \geq 0$, $u \geq 0$.

Dualize:
$$\begin{bmatrix} 500 & 200 & 50 & 1400 \\ 1 & 0 & 0 & 1 \\ 0 & 0 & -1 & -2 \\ 100 & 20 & 4 & 0 \end{bmatrix} \rightarrow \begin{bmatrix} 500 & 1 & 0 & 100 \\ 200 & 0 & 0 & 20 \\ 50 & 0 & -1 & 4 \\ 1400 & 1 & -2 & 0 \end{bmatrix}$$

Dual problem: Maximize $P = 1400x + y - 2z$ subject to $500x + y \leq 100$, $200x \leq 20$, $50x - z \leq 4$, $x \geq 0$, $y \geq 0$, $z \geq 0$.

Solve the dual problem:

	x	y	z	s	t	u	p		
s	500	1	0	1	0	0	0	100	$R_1 - 10R_3$
t	200	0	0	0	1	0	0	20	$R_2 - 4R_3$
u	50	0	-1	0	0	1	0	4	
p	-1400	-1	2	0	0	0	1	0	$R_4 + 28R_3$

	x	y	z	s	t	u	p		
s	0	1	10	1	0	−10	0	60	$4R_1 - 10R_2$
t	0	0	[4]	0	1	−4	0	4	
x	50	0	−1	0	0	1	0	4	$4R_3 + R_2$
p	0	−1	−26	0	0	28	1	112	$4R_4 + 26R_2$

	x	y	z	s	t	u	p	
s	0	[4]	0	4	−10	0	0	200
z	0	0	4	0	1	−4	0	4
x	200	0	0	0	1	0	0	20
p	0	−4	0	0	26	8	4	552	$R_4 + R_1$

	x	y	z	s	t	u	p	
y	0	4	0	4	−10	0	0	200
z	0	0	4	0	1	−4	0	4
x	200	0	0	0	1	0	0	20
p	0	0	0	4	16	8	4	752

The solution to the primal problem is $c = 752/4 = 188$; $s = 4/4 = 1$, $t = 16/4 = 4$, $u = 8/4 = 2$. Build 1 convention-style hotel, 4 vacation style hotels and 2 small motels.

Chapter 4 Review Exercises

1. $2x - 3y \le 12$

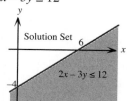

Unbounded

2. $x \le 2y$

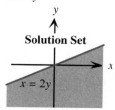

Unbounded

3. $x + 2y \le 20$
$3x + 2y \le 30$
$x \ge 0, y \ge 0$

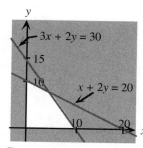

Bounded
Corner points: $(0, 0)$, $(0, 10)$,
$(5, 15/2)$, $(10, 0)$

4. $3x + 2y \ge 6$
$2x - 3y \le 6$
$3x - 2y \ge 0$
$x \ge 0, y \ge 0$

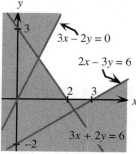

Unbounded
Corner points: $(2, 0)$, $(3, 0)$ $(1, 3/2)$

5. Maximize $p = 2x + y$ subject to
$3x + y \le 30$
$x + y \le 12$
$x + 3y \le 30$
$x \ge 0, y \ge 0$

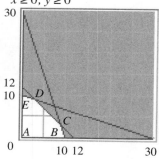

Point	Lines	Coords.	$p = 2x+y$
A	$x = 0$ $y = 0$	$(0, 0)$	0
B	$3x+y = 30$ $y = 0$	$(10, 0)$	20
C	$3x+y = 30$ $x+y = 12$	$(9, 3)$	**21**
D	$x+y = 12$ $x+3y = 30$	$(3, 9)$	15
E	$x+3y = 30$ $x = 0$	$(0, 10)$	10

Maximum value occurs at C:
$p = 21; x = 9, y = 3$

6. Maximize $p = 2x + 3y$ subject to

$$x + y \geq 10$$
$$2x + y \geq 12$$
$$x + y \leq 20$$
$$x \geq 0, y \geq 0$$

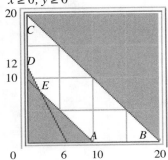

Point	Lines	Coords.	$p = 2x+3y$
A	$y = 0$ $x+y = 10$	(10, 0)	20
B	$y = 0$ $x+y = 20$	(20, 0)	40
C	$x = 0$ $x+y = 20$	(0, 20)	**60**
D	$2x+y = 12$ $x = 0$	(0, 12)	36
E	$x+y = 10$ $2x+y = 12$	(2, 8)	28

Maximum value occurs at C:

$p = 60; x = 0, y = 20$

7. Minimize $c = 2x + y$ subject to

$$3x + y \geq 30$$
$$x + 2y \geq 20$$
$$2x - y \geq 0$$
$$x \geq 0, y \geq 0$$

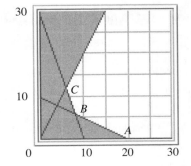

Point	Lines	Coords.	$c = 2x+y$
A	$x+2y = 20$ $y = 0$	(20, 0)	40
B	$3x+y = 30$ $x+2y = 20$	(8, 6)	**22**
C	$3x+y = 30$ $2x-y = 0$	(6, 12)	24

Minimum value occurs at B:

$c = 22; x = 8, y = 6$

8. Minimize $c = 3x + y$ subject to

$$3x + 2y \geq 6$$
$$2x - 3y \leq 0$$
$$3x - 2y \geq 0$$
$$x \geq 0, y \geq 0$$

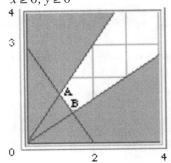

Point	Lines	Coords.	$p = 2x+3y$
A	$3x+2y = 6$	$(1, 3/2)$	**9/2**
	$3x-2y = 0$		
B	$3x+2y = 6$	$(18/13, 12/13)$	$66/13$
	$2x-3y = 0$		

Minimum value occurs at A:
$c = 9/2$; $x = 1$, $y = 3/2$

9. Introduce slack variables and set up the initial tableau:

$$x + 2y + 2z + s = 60$$
$$2x + y + 3z + t = 60$$
$$-x - y - 2z + p = 0$$

	x	y	z	s	t	p		
s	1	2	2	1	0	0	60	$3R_1 - 2R_2$
t	2	1	$\boxed{3}$	0	1	0	60	
p	−1	−1	−2	0	0	1	0	$3R_3 + 2R_2$

The most negative entry in the bottom row in the above tableau is the −2 in the z–column. So we use this column as the pivot column. The test ratios are: s: 60/2, t: 60/3. The smallest test ratio is t: 60/3. Thus, we pivot on the 3 in the t–row.

	x	y	z	s	t	p		
s	−1	$\boxed{4}$	0	3	−2	0	60	
z	2	1	3	0	1	0	60	$4R_2 - R_1$
p	1	−1	0	0	2	3	120	$4R_3 + R_1$

The most negative entry in the bottom row in the above tableau is the −1 in the y–column. So we use this column as the pivot column. The test ratios are: s: 60/4, z: 60/1. The smallest test ratio is s: 60/4. Thus, we pivot on the 4 in the s–row.

	x	y	z	s	t	p	
y	−1	4	0	3	−2	0	60
z	9	0	12	−3	6	0	180
p	3	0	0	3	6	12	540

Optimal Solution: $p = 45$; $x = 0$, $y = 60/4 = 15$, $z = 180/12 = 15$.

10. Introduce slack variables and set up the initial tableau:

$$x + 2y + 2z + s = 60$$
$$2x + y + 3z + t = 60$$
$$x + 3y + 6z + u = 60$$
$$-x - y - 2z + p = 0$$

	x	y	z	s	t	u	p		
s	1	2	2	1	0	0	0	60	$3R_1 - R_3$
t	2	1	3	0	1	0	0	60	$2R_2 - R_3$
u	1	3	6	0	0	1	0	60	
p	−1	−1	−2	0	0	0	1	0	$3R_4 + R_3$

The most negative entry in the bottom row in the above tableau is the −2 in the z–column. So we use this column as the pivot column. The test ratios are: s: 60/2, t: 60/3, u: 60/6. The smallest test ratio is u: 60/6. Thus, we pivot on the 6 in the u–row.

	x	y	z	s	t	u	p		
s	2	3	0	3	0	−1	0	120	$3R_1 - 2R_2$
t	3	−1	0	0	2	−1	0	60	
z	1	3	6	0	0	1	0	60	$3R_3 - R_2$
p	−2	0	0	0	0	1	3	60	$3R_4 + 2R_2$

The most negative entry in the bottom row in the above tableau is the −2 in the x–column. So we use this column as the pivot column. The test ratios are: s: 120/2, t: 60/3, z: 60/1. The smallest test ratio is t: 60/3. Thus, we pivot on the 3 in the t–row.

	x	y	z	s	t	u	p		
s	0	11	0	9	−4	−1	0	240	$10R_1 - 11R_3$
x	3	−1	0	0	2	−1	0	60	$10R_2 + R_3$
z	0	10	18	0	−2	4	0	120	
p	0	−2	0	0	4	1	9	300	$5R_4 + R_3$

The most negative entry in the bottom row in the above tableau is the −2 in the y–column. So we use this column as the pivot column. The test ratios are: s: 240/11, z: 120/10. The smallest test ratio is z: 120/10. Thus, we pivot on the 10 in the z–row.

	x	y	z	s	t	u	p	
s	0	0	−198	90	−18	−54	0	1080
x	30	0	18	0	18	−6	0	720
y	0	10	18	0	−2	4	0	120
p	0	0	18	0	18	9	45	1620

Optimal Solution: $p = 36$; $x = 720/30 = 24$, $y = 120/10 = 12$, $z = 0$.

11. Introduce slack and surplus variables and set up the initial tableau:

$x + y + z - s = 100$

$y + z + t = 80$

$x + z + u = 80$

$-x - y - 3z + p = 0$

	x	y	z	s	t	u	p		
*s	1	1	1	−1	0	0	0	100	$R_1 - R_3$
t	0	1	1	0	1	0	0	80	
u	1	0	1	0	0	1	0	80	
p	−1	−1	−3	0	0	0	1	0	$R_4 + R_3$

The first starred row in the above tableau is the s-row, and its (first) largest positive entry is the 1 in the x-column. So we use this column as the pivot column. The test ratios are: s: 100/1, u: 80/1. The smallest test ratio is u: 80/1. Thus, we pivot on the 1 in the u-row.

	x	y	z	s	t	u	p		
*s	0	1	0	−1	0	−1	0	20	
t	0	1	1	0	1	0	0	80	$R_2 - R_1$
x	1	0	1	0	0	1	0	80	
p	0	−1	−2	0	0	1	1	80	$R_4 + R_1$

The first starred row in the above tableau remains the s–row, and its largest positive entry is the 1 in the y–column. So we use this column as the pivot column. The test ratios are: s: 20/1, t: 80/1. The smallest test ratio is s: 20/1. Thus, we pivot on the 1 in the s–row.

	x	y	z	s	t	u	p		
y	0	1	0	−1	0	−1	0	20	
t	0	0	1	1	1	1	0	60	
x	1	0	1	0	0	1	0	80	$R_3 - R_2$
p	0	0	−2	−1	0	0	1	100	$R_4 + 2R_2$

Since there are no more starred rows in the above tableau, we are in Phase 2, and choose the most negative number in the bottom row to give us the pivot column:

	x	y	z	s	t	u	p	
y	0	1	0	−1	0	−1	0	20
z	0	0	1	1	1	1	0	60
x	1	0	0	−1	−1	0	0	20
p	0	0	0	1	2	2	1	220

Optimal Solution: $p = 220$; $x = 20/1 = 20$, $y = 20/1 = 20$, $z = 60/1 = 60$.

12.

	x	y	s	t	u	p		
*s	1	2	−1	0	0	0	12	$R_1 - 2R_3$
t	2	1	0	1	0	0	12	$R_2 - R_3$
u	1	1	0	0	1	0	5	
p	−2	−1	0	0	0	1	0	$R_4 + R_3$

	x	y	s	t	u	p	
*s	−1	0	−1	0	−2	0	2
t	1	0	0	1	−1	0	7
y	1	1	0	0	1	0	5
p	−1	0	0	0	1	1	5

There is no longer any positive entry in the starred row, meaning that the feasible region is empty, so there is no solution.

13. Associated maximization problem:

Maximize $p = -x - 2y - 3z$ subject to $3x + 2y + z \geq 60$, $2x + y + 3z \geq 60$, $x \geq 0$, $y \geq 0$, $z \geq 0$.

Phase 1:

	x	y	z	s	t	p		
*s	3	2	1	−1	0	0	60	
*t	2	1	3	0	−1	0	60	$3R_2 - 2R_1$
p	1	2	3	0	0	1	0	$3R_3 - R_1$

	x	y	z	s	t	p		
x	3	2	1	−1	0	0	60	$7R_1 - R_2$
*t	0	−1	7	2	−3	0	60	
p	0	4	8	1	0	3	−60	$7R_3 - 8R_2$

Phase 2:

	x	y	z	s	t	p		
x	21	15	0	−9	3	0	360	$2R_1 + 9R_2$
z	0	−1	7	2	−3	0	60	
p	0	36	0	−9	24	21	−900	$2R_3 + 9R_2$

	x	y	z	s	t	p		
x	42	21	63	0	−21	0	1260	
s	0	−1	7	2	−3	0	60	
p	0	63	63	0	21	42	−1260	

Optimal Solution: $p = -c = 1260/42 = 30$; $x = 1260/42 = 30$, $y = 0$, $z = 0$.

14. Since technology is indicated, we use the online Simplex Method Tool. When entering the problems, there is no need to enter the inequalities $x \geq 0$, $y \geq 0$, etc.

```
Tableau #1

x        y        z        s1       s2       s3       -c
3        2        1        -1       0        0        0
2        1        3        0        -1       0        0        60
1        3        2        0        0        -1       0        60
1        1        -1       0        0        0        1        0
```

. . .

Tableau #6

x	y	z	s1	s2	s3	-c	
5	1	0	-2	0	1	0	60
3	2	1	-1	0	0	0	60
7	5	0	-3	1	0	0	120
4	3	0	-1	0	0	1	60

Looking at the sixth tableau, we find that there is no possible pivot above the most negative entry in the bottom row, indicating that the feasible region is unbounded (and no optimal solution exists).

15.

	x	y	z	w	s	t	u	v	$-c$	
*s	1	1	0	0	-1	0	0	0	0	30
*t	1	0	1	0	0	-1	0	0	0	20
u	1	1	0	-1	0	0	1	0	0	10
v	0	1	1	-1	0	0	0	1	0	10
$-c$	1	1	1	1	0	0	0	0	1	0

The first starred row is the s-row, and its (first) largest positive entry is the 1 in the x-column. So we use this column as the pivot column. The test ratios are: s: 30/1, t: 20/1, u: 10/1. The smallest test ratio is u: 10/1. Thus, we pivot on the 1 in the u-row.

	x	y	z	w	s	t	u	v	$-c$		
*s	1	1	0	0	-1	0	0	0	0	30	$R_1 - R_3$
*t	1	0	1	0	0	-1	0	0	0	20	$R_2 - R_3$
u	[1]	1	0	-1	0	0	1	0	0	10	
v	0	1	1	-1	0	0	0	1	0	10	
$-c$	1	1	1	1	0	0	0	0	1	0	$R_5 - R_3$

	x	y	z	w	s	t	u	v	$-c$	
*s	0	0	0	1	-1	0	-1	0	0	20
*t	0	-1	1	1	0	-1	-1	0	0	10
x	1	1	0	-1	0	0	1	0	0	10
v	0	1	1	-1	0	0	0	1	0	10
$-c$	0	0	1	2	0	0	-1	0	1	-10

The first starred row is the s-row, and its largest positive entry is the 1 in the w-column. So we use this column as the pivot column. The test ratios are: s: 20/1, t: 10/1. The smallest test ratio is t: 10/1. Thus, we pivot on the 1 in the t-row.

	x	y	z	w	s	t	u	v	$-c$		
*s	0	0	0	1	−1	0	−1	0	0	20	$R_1 - R_2$
*t	0	−1	1	$\boxed{1}$	0	−1	−1	0	0	10	
x	1	1	0	−1	0	0	1	0	0	10	$R_3 + R_2$
v	0	1	1	−1	0	0	0	1	0	10	$R_4 + R_2$
$-c$	0	0	1	2	0	0	−1	0	1	−10	$R_5 - 2R_2$

	x	y	z	w	s	t	u	v	$-c$	
*s	0	1	−1	0	−1	1	0	0	0	10
w	0	−1	1	1	0	−1	−1	0	0	10
x	1	0	1	0	0	−1	0	0	0	20
v	0	0	2	0	0	−1	−1	1	0	20
$-c$	0	2	−1	0	0	2	1	0	1	−30

The first starred row is the s-row, and its (first) largest positive entry is the 1 in the y-column. So we use this column as the pivot column. There is only one positive ratio, in the s-row. Thus, we pivot on the 1 in the s-row.

	x	y	z	w	s	t	u	v	$-c$		
*s	0	$\boxed{1}$	−1	0	−1	1	0	0	0	10	
w	0	−1	1	1	0	−1	−1	0	0	10	$R_2 + R_1$
x	1	0	1	0	0	−1	0	0	0	20	
v	0	0	2	0	0	−1	−1	1	0	20	
$-c$	0	2	−1	0	0	2	1	0	1	−30	$R_5 - 2R_1$

	x	y	z	w	s	t	u	v	$-c$	
y	0	1	-1	0	-1	1	0	0	0	10
w	0	0	0	1	-1	0	-1	0	0	20
x	1	0	1	0	0	-1	0	0	0	20
v	0	0	2	0	0	-1	-1	1	0	20
$-c$	0	0	1	0	2	0	1	0	1	-50

Since there are no more starred rows, we go to Phase 2, and search for the most negative entry in the bottom row (if any). Since there are no negative entries in the bottom row, we are done.

Optimal Solution: $c = -(-50/1) = 50$; $x = 20/1 = 20$, $y = 10/1 = 10$, $z = 0$, $w = 20/1 = 20$

16.

	x	y	z	w	s	t	u	$-c$	
*s	1	1	0	0	-1	0	0	0	30
t	0	1	-1	0	0	1	0	0	20
u	0	0	1	-1	0	0	1	0	10
$-c$	4	1	1	1	0	0	0	1	0

The first starred row is the s-row, and its (first) largest positive entry is the 1 in the x-column. So we use this column as the pivot column. The only positive test ration is s: 30/1. Thus, we pivot on the 1 in the s-row.

	x	y	z	w	s	t	u	$-c$		
*s	1	1	0	0	-1	0	0	0	30	
t	0	1	-1	0	0	1	0	0	20	
u	0	0	1	-1	0	0	1	0	10	
$-c$	4	1	1	1	0	0	0	1	0	$R_4 - 4R_1$

	x	y	z	w	s	t	u	$-c$	
x	1	1	0	0	-1	0	0	0	30
t	0	1	-1	0	0	1	0	0	20
u	0	0	1	-1	0	0	1	0	10
$-c$	0	-3	1	1	4	0	0	1	-120

Since there are no more starred rows, we go to Phase 2, and search for the most negative entry in the bottom row (if any). The most negative entry in the bottom row is the -3 in the y-column. So we use this column as the pivot column. The test ratios are: x: 30/1, t: 20/1. The smallest test ratio is t: 20/1. Thus, we pivot on the 1 in the t-row.

	x	y	z	w	s	t	u	$-c$		
x	1	1	0	0	-1	0	0	0	30	$R_1 - R_2$
t	0	$\boxed{1}$	-1	0	0	1	0	0	20	
u	0	0	1	-1	0	0	1	0	10	
$-c$	0	-3	1	1	4	0	0	1	-120	$R_4 + 3R_2$

	x	y	z	w	s	t	u	$-c$	
x	1	0	1	0	-1	-1	0	0	10
y	0	1	-1	0	0	1	0	0	20
u	0	0	1	-1	0	0	1	0	10
$-c$	0	0	-2	1	4	3	0	1	-60

The most negative entry in the bottom row is the -2 in the z-column. So we use this column as the pivot column. The test ratios are: x: 10/1, u: 10/1. The smallest test ratio is u: 10/1. Thus, we pivot on the 1 in the u-row.

	x	y	z	w	s	t	u	$-c$		
x	1	0	1	0	-1	-1	0	0	10	$R_1 - R_3$
y	0	1	-1	0	0	1	0	0	20	$R_2 + R_3$
u	0	0	$\boxed{1}$	-1	0	0	1	0	10	
$-c$	0	0	-2	1	4	3	0	1	-60	$R_4 + 2R_3$

	x	y	z	w	s	t	u	$-c$	
x	1	0	0	1	-1	-1	-1	0	0
y	0	1	0	-1	0	1	1	0	30
z	0	0	1	-1	0	0	1	0	10
$-c$	0	0	0	-1	4	3	2	1	-40

The most negative entry in the bottom row is the -1 in the w-column. So we use this column as the pivot column. There is only one non-negative test ratio, in the x-row. Thus, we pivot on the 1 in the x-row.

	x	y	z	w	s	t	u	$-c$		
x	1	0	0	$\boxed{1}$	-1	-1	-1	0	0	
y	0	1	0	-1	0	1	1	0	30	$R_2 + R_1$
z	0	0	1	-1	0	0	1	0	10	$R_3 + R_1$
$-c$	0	0	0	-1	4	3	2	1	-40	$R_4 + R_1$

	x	y	z	w	s	t	u	$-c$	
w	1	0	0	1	-1	-1	-1	0	0
y	1	1	0	0	-1	0	0	0	30
z	1	0	1	0	-1	-1	0	0	10
$-c$	1	0	0	0	3	2	1	1	-40

Since there are no more negative numbers in the bottom row, we are done, and read off the solution:
Optimal Solution: $c = -(-40/1) = 40$; $x = 0$, $y = 30/1 = 30$, $z = 10/1 = 10$, $w = 0/1 = 0$.

17. Minimize $c = 2s + t$ subject to $3s + 2t \geq 60$, $2s + t \geq 60$, $x + 3y \geq 60$, $s \geq 0$, $t \geq 0$.

Dualize: $\begin{bmatrix} 3 & 2 & 60 \\ 2 & 1 & 60 \\ 1 & 3 & 60 \\ 2 & 1 & 0 \end{bmatrix} \rightarrow \begin{bmatrix} 3 & 2 & 1 & 2 \\ 2 & 1 & 3 & 1 \\ 60 & 60 & 60 & 0 \end{bmatrix}$

Dual problem: Maximize $p = 60x + 60y + 60z$ subject to $3x + 2y + z \leq 2$, $2x + y + 43z \leq 1$, $x \geq 0$, $y \geq 0$, $z \geq 0$.

Solve the dual problem:

	x	y	z	s	t	p		
s	3	2	1	1	0	0	2	$2R_1 - 3R_2$
t	$\boxed{2}$	1	43	0	1	0	1	
p	-60	-60	-60	0	0	1	0	$R_3 + 30R_2$

	x	y	z	s	t	p		
s	0	1	-127	2	-3	0	1	$R_1 - R_2$
x	2	$\boxed{1}$	43	0	1	0	1	
p	0	-30	1230	0	30	1	30	$R_3 + 30R_2$

	x	y	z	s	t	p	
s	-2	0	-170	2	-4	0	0
y	2	1	43	0	1	0	1
p	60	0	2520	0	60	1	60

Solution to the primal problem: $c = 60$; $s = 0$, $t = 60$. Another possible solution: $c = 60$; $s = 24$, $t = 12$. Since the original unknowns were x and y, we write these optimal solutions as: $c = 60$; $x = 0$, $y = 60$ and $c = 60$; $x = 24$, $y = 12$.

18. Minimize $c = 2s + t + 2u$ subject to $3s + 2t + u \geq 100$, $2s + t + 3u \geq 200$, $t \geq 0$, $t \geq 0$, $u \geq 0$.

Dualize: $\begin{bmatrix} 3 & 2 & 1 & 100 \\ 2 & 1 & 3 & 200 \\ 2 & 1 & 2 & 0 \end{bmatrix} \rightarrow \begin{bmatrix} 3 & 2 & 2 \\ 2 & 1 & 1 \\ 1 & 3 & 2 \\ 100 & 200 & 0 \end{bmatrix}$

Dual problem: Maximize $p = 100x + 200y$ subject to $3x + 2y \leq 2$, $2x + y \leq 1$, $x + 3y \leq 2$.
Solve the dual problem:

	x	y	s	t	u	p		
s	3	2	1	0	0	0	2	$3R_1 - 2R_3$
t	2	1	0	1	0	0	1	$3R_2 - R_3$
u	1	$\boxed{3}$	0	0	1	0	2	
p	-100	-200	0	0	0	1	0	$3R_4 + 200R_3$

	x	y	s	t	u	p		
s	7	0	3	0	-2	0	2	$5R_1 - 7R_2$
t	$\boxed{5}$	0	0	3	-1	0	1	
y	1	3	0	0	1	0	2	$5R_3 - R_2$
p	-100	0	0	0	200	3	400	$R_4 + 20R_2$

	x	y	s	t	u	p	
s	0	0	15	-21	-3	0	3
x	5	0	0	3	-1	0	1
y	0	15	0	-3	6	0	9
p	0	0	0	60	180	3	420

Solution to the primal problem: $c = 420/3 = 140$; $s = 0$, $t = 60/3 = 20$, $u = 180/3 = 60$. Since the original unknowns were x, y, and z, we write the optimal solution as: $c = 140$; $x = 0$, $y = 20$, $z = 60$.

19. First rewrite the given problem as a standard minimization problem with all the constraints using"$\geq$":
Minimize $c = 2s + t$ subject to $3s + 2t \geq 10$, $-2s + t \geq -30$, $s + 3t \geq 60$, $s \geq 0$, $t \geq 0$.

Dualize: $\begin{bmatrix} 3 & 2 & 10 \\ -2 & 1 & -30 \\ 1 & 3 & 60 \\ 2 & 1 & 0 \end{bmatrix} \rightarrow \begin{bmatrix} 3 & -2 & 1 & 2 \\ 2 & 1 & 3 & 1 \\ 10 & -30 & 60 & 0 \end{bmatrix}$

Dual problem: Maximize $p = 10x - 30y + 60z$ subject to $3x - 2y + z \leq 2$, $2x + y + 3z \leq 1$, $x \geq 0$, $y \geq 0$, $z \geq 0$.
Solve the dual problem:

	x	y	z	s	t	p		
s	3	−2	1	1	0	0	2	$3R_1 - R_2$
t	2	1	③	0	1	0	1	
p	−10	30	−60	0	0	1	0	$R_3 + 20R_2$

	x	y	z	s	t	p	
s	7	−7	0	3	−1	0	5
z	2	1	3	0	1	0	1
p	30	50	0	0	20	1	20

Solution to the primal problem: $c = 20/1 = 20$; $s = 0$, $t = 20/1 = 20$. Since the original unknowns were x and y, we write the optimal solution as: $c = 20$; $x = 0$, $y = 20$.

20. First rewrite the given problem as a standard minimization problem with all the constraints using"$\geq$":
Minimize $c = 2s + t + 2u$ subject to $3s - 2t + u \geq 100$, $-2s - t + 3u \geq -200$, $s \geq 0$, $t \geq 0$, $u \geq 0$.

Dualize: $\begin{bmatrix} 3 & -2 & 1 & 100 \\ -2 & -1 & 3 & -100 \\ 2 & 1 & 2 & 0 \end{bmatrix} \rightarrow \begin{bmatrix} 3 & -2 & 2 \\ -2 & -1 & 1 \\ 1 & 3 & 2 \\ 100 & -100 & 0 \end{bmatrix}$

Maximize $p = 100x - 100y$ subject to $3x - 2y \leq 2$, $-2x - 2y \leq 1$, $x + 3y \leq 2$, $x \geq 0$, $y \geq 0$.
Solve the dual problem:

	x	y	s	t	u	p		
s	③	−2	1	0	0	0	2	
t	−2	−2	0	1	0	0	1	$3R_2 + 2R_1$
u	1	3	0	0	1	0	2	$3R_3 - R_1$
p	−100	100	0	0	0	1	0	$3R_4 + 100R_1$

	x	y	s	t	u	p	
x	3	−2	1	0	0	0	2
t	0	−10	2	3	0	0	7
u	0	11	−1	0	3	0	4
p	0	100	100	0	0	3	200

Solution to the primal problem: $c = 200/3$; $s = 100/3$, $t = 0$, $u = 0$. Since the original unknowns were x, y, and z, we write the optimal solution as: $c = 200/3$; $x = 100/3$, $y = 0$, $z = 0$.

21. The first column is dominated by the last, so we can eliminate it. We cannot reduce any further. Add $k = 2$ to each entry and put 1s to the right and below:

$$\begin{bmatrix} 4 & 1 & 1 \\ 0 & 3 & 1 \\ 1 & 2 & 1 \\ 1 & 1 & 0 \end{bmatrix}$$

This gives the following LP problem:

Maximize $p = x + y$ subject to

$4x + y \le 1$

$3y \le 1$

$x + 2y \le 1$

$x \ge 0,\ y \ge 0$

Here are the tableaux:

	x	y	s	t	u	p		
s	[4]	1	1	0	0	0	1	
t	0	3	0	1	0	0	1	
u	1	2	0	0	1	0	1	$4R_3 - R_1$
p	−1	−1	0	0	0	1	0	$4R_4 + R_1$

	x	y	s	t	u	p		
x	4	1	1	0	0	0	1	$3R_1 - R_2$
t	0	[3]	0	1	0	0	1	
u	0	7	−1	0	4	0	3	$3R_3 - 7R_2$
p	0	−3	1	0	0	4	1	$R_4 + R_2$

	x	y	s	t	u	p	
x	12	0	3	−1	0	0	2
y	0	3	0	1	0	0	1
u	0	0	−3	−7	12	0	2
p	0	0	1	1	0	4	2

The solution to the primal problem is $p = 2/4 = 1/2$; $x = 2/12 = 1/6$, $y = 1/3$. So, the column player's optimal strategy is $C = [0\ 1/3\ 2/3]^T$ and the value of the game is $e = 2 - 2 = 0$.

The solution to the dual problem is $s = 1/4$, $t = 1/4$, $u = 0$, so the row player's optimal strategy is $R = [1/2\ 1/2\ 0]$.

22. The first row dominates the second, so remove the second. We cannot reduce any further. Add $k = 3$ to each entry and put 1s to the right and below:

$$\begin{bmatrix} 0 & 3 & 4 & 1 \\ 3 & 2 & 1 & 1 \\ 1 & 1 & 1 & 0 \end{bmatrix}$$

This gives the following LP problem:

Maximize $p = x + y + z$ subject to

$$3y + 4z \le 1$$
$$3x + 2y + z \le 1$$
$$x \ge 0,\ y \ge 0,\ z \ge 0$$

Here are the tableaux:

	x	y	z	s	t	p	
s	0	3	4	1	0	0	1
t	[3]	2	1	0	1	0	1
p	−1	−1	−1	0	0	1	0

$3R_3 + R_2$

	x	y	z	s	t	p		
s	0	3	[4]	1	0	0	1	
x	3	2	1	0	1	0	1	$4R_2 - R_1$
p	0	−1	−2	0	1	3	1	$2R_3 + R_1$

	x	y	z	s	t	p	
z	0	3	4	1	0	0	1
x	12	5	0	−1	4	0	3
p	0	1	0	1	2	6	3

The solution to the primal problem is $p = 3/6 = 1/2$; $x = 3/12 = 1/4$, $y = 0$, $z = 1/4$. So, the column player's optimal strategy is $C = [1/2\ 0\ 1/2]^T$ and the value of the game is $e = 2 - 3 = -1$.

The solution to the dual problem is $s = 1/6$, $t = 2/6 = 1/3$, so the row player's optimal strategy is $R = [1/3\ 0\ 2/3]$.

23. This game cannot be reduced. Add $k = 3$ to each entry and put 1s to the right and below:

$$\begin{bmatrix} 0 & 1 & 6 & 1 \\ 4 & 3 & 3 & 1 \\ 1 & 5 & 4 & 1 \\ 1 & 1 & 1 & 0 \end{bmatrix}$$

This gives the following LP problem:

Maximize $p = x + y + z$ subject to

$$y + 6z \le 1$$
$$4x + 3y + 3z \le 1$$
$$x + 5y + 4z \le 1$$
$$x \ge 0,\ y \ge 0,\ z \ge 0$$

Here are the tableaux:

	x	y	z	s	t	u	p		
s	0	1	6	1	0	0	0	1	
t	4	3	3	0	1	0	0	1	
u	1	5	4	0	0	1	0	1	$4R_3 - R_2$
p	−1	−1	−1	0	0	0	1	0	$4R_4 + R_2$

	x	y	z	s	t	u	p		
s	0	1	6	1	0	0	0	1	$17R_1 - R_3$
x	4	3	3	0	1	0	0	1	$17R_2 - 3R_3$
u	0	17	13	0	−1	4	0	3	
p	0	−1	−1	0	1	0	4	1	$17R_4 + R_3$

	x	y	z	s	t	u	p		
s	0	0	89	17	1	−4	0	14	
x	68	0	12	0	20	−12	0	8	$89R_2 - 12R_1$
y	0	17	13	0	−1	4	0	3	$89R_3 - 13R_1$
p	0	0	−4	0	16	4	68	20	$89R_4 + 4R_1$

	x	y	z	s	t	u	p	
z	0	0	89	17	1	−4	0	14
x	6052	0	0	−204	1768	−1020	0	544
y	0	1513	0	−221	−102	408	0	85
p	0	0	0	68	1428	340	6052	1836

The solution to the primal problem is $p = 1836/6052 = 27/89$; $x = 544/6052 = 8/89$, $y = 85/1513 = 5/89$, $z = 14/89$. So, the column player's optimal strategy is $C = [8/27 \ 5/27 \ 14/27]^T$ and the value of the game is $e = 89/27 - 3 = 8/27$.

The solution to the dual problem is $s = 68/6052 = 1/89$, $t = 1428/6052 = 21/89$, $u = 340/6052 = 5/89$, so the row player's optimal strategy is $R = [1/27 \ 21/27 \ 5/27] = [1/27 \ 7/9 \ 5/27]$.

24. This game cannot be reduced. Add $k = 4$ to each entry and put 1s to the right and below:

$$\begin{bmatrix} 0 & 2 & 1 & 1 \\ 5 & 1 & 2 & 1 \\ 1 & 5 & 0 & 1 \\ 1 & 1 & 1 & 0 \end{bmatrix}$$

This gives the following LP problem:

Maximize $p = x + y + z$ subject to
$$2y + z \le 1$$
$$5x + y + 2z \le 1$$
$$x + 5y \le 1$$
$$x \ge 0, \ y \ge 0, \ z \ge 0$$

Here are the tableaux:

	x	y	z	s	t	u	p		
s	0	2	1	1	0	0	0	1	
t	[5]	1	2	0	1	0	0	1	
u	1	5	0	0	0	1	0	1	$5R_3 - R_2$
p	−1	−1	−1	0	0	0	1	0	$5R_4 + R_2$

	x	y	z	s	t	u	p		
s	0	2	1	1	0	0	0	1	$12R_1 - R_3$
x	5	1	2	0	1	0	0	1	$24R_2 - R_3$
u	0	[24]	−2	0	−1	5	0	4	
p	0	−4	−3	0	1	0	5	1	$6R_4 + R_3$

	x	y	z	s	t	u	p		
s	0	0	14	12	1	−5	0	8	$25R_1 - 7R_2$
x	120	0	[50]	0	25	−5	0	20	
y	0	24	−2	0	−1	5	0	4	$25R_3 + R_2$
p	0	0	−20	0	5	5	30	10	$5R_4 + 2R_2$

	x	y	z	s	t	u	p	
s	−840	0	0	300	−150	−90	0	60
z	120	0	50	0	25	−5	0	20
y	120	600	0	0	0	120	0	120
p	240	0	0	0	75	15	150	90

The solution to the primal problem is $p = 90/150 = 3/5$; $x = 0$, $y = 120/600 = 1/5$, $z = 20/50 = 2/5$. So, the column player's optimal strategy is $C = [0\ 1/3\ 2/3]^T$ and the value of the game is $e = 5/3 - 4 = -7/3$. The solution to the dual problem is $s = 0$, $t = 75/150 = 1/2$, $u = 15/150 = 1/10$, so the row player's optimal strategy is $R = [0\ 5/6\ 1/6]$.

25. Apply the simplex method to the given problem:

	x	y	s	t	p		
*s	−2	[1]	−1	0	0	1	
*t	1	−2	0	−1	0	1	$R_2 + 2R_1$
c	1	2	0	0	1	0	$R_3 - 2R_1$

	x	y	s	t	p	
y	−2	1	−1	0	0	1
*t	−3	0	−2	−1	0	3
c	5	0	2	0	1	−2

Since there are no positive entries in the t-row (other than the right-hand side) we cannot go to the next step. Thus, it is impossible to get into the feasible region by completing Phase 1. In other words, there are no feasible solutions (choice A).

Another way of seeing this would be to graph the constraints. This would show an empty feasible region.

26. Apply the simplex method to the given problem:

	x	y	s	t	p		
s	-2	1	1	0	0	1	$R_1 + 2R_2$
t	$\boxed{1}$	-2	0	1	0	2	
p	-1	-1	0	0	1	0	$R_3 + R_2$

	x	y	s	t	p	
s	0	-3	1	2	0	5
x	1	-2	0	1	0	2
p	0	-3	0	1	1	2

There are now no positive entries to use as pivot in the y-column. This indicates that the objective function is unbounded (choice B).

Another way of seeing this would be to set up the problem graphically. This would show an unbounded feasible region with nonnegative coefficients of p. Thus, p can be made as large as we like.

27. First note that the objective function, $Z = x_1 + 4x_2 + 2x_3 - 10$, is not linear because of the "-10". However, maximizing $p = x_1 + 4x_2 + 2x_3$ will also maximize Z, provided we remember to subtract 10 from the optimal value of the objective function when we are done. Thus, we solve the problem using $p = x_1 + 4x_2 + 2x_3$ as our objective function:

	x_1	x_2	x_3	s	t	p		
s	4	1	1	1	0	0	45	$R_1 - R_2$
t	-1	$\boxed{1}$	2	0	1	0	0	
p	-1	-4	-2	0	0	1	0	$R_3 + 4R_2$

	x_1	x_2	x_3	s	t	p		
s	$\boxed{5}$	0	-1	1	-1	0	45	
x_2	-1	1	2	0	1	0	0	$5R_2 + R_1$
p	-5	0	6	0	4	1	0	$R_3 + R_1$

	x_1	x_2	x_3	s	t	p	
x_1	5	0	-1	1	-1	0	45
x_2	0	5	9	1	4	0	45
p	0	0	5	1	3	1	45

Optimal Solution: $p = 45$; $x_1 = 45/5 = 9$, $x_2 = 45/5 = 9$, $x_3 = 0$. Since the maximum value of p is 45, the maximum value of Z is $45 - 10 - 35$.

28. Unknowns: $x = \#$ packages from Duffin House, $y = \#$ packages from Higgins Press. Minimize $c = 50x + 80y$ subject to $5x + 5y \geq 4000$, $5x + 10y \geq 6000$, $x \geq 0$, $y \geq 0$.

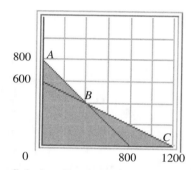

Point	Lines	Coords.	$c = 50x+80y$
A	$5x+5y = 4000$ $x = 0$	$(0, 800)$	64,000
B	$5x+5y = 4000$ $5x+10y = 6000$	$(400, 400)$	**52,000**
C	$5x+10y = 6000$ $y = 0$	$(1200, 0)$	60,000

Solution: Purchase 400 packages from each for a minimum cost of $52,000.

29. Since the lowest cost is given by ordering the same number of packages from each supplier (point B in the graph above), changing that by ordering at least 20% more packages from Duffin as from Higgins will result in a higher cost, since the solution will no longer be point B but another point in the feasible region (choice B), assuming it is possible to meet all the conditions. On the other hand, it is conceivable that the extra constraint will result in an empty feasible region (choice D). Thus, the correct answers are (B) and (D).

30. Unknowns: as in Exercise 28. Minimize $c = 50x + 80y$ subject to $5x + 5y \geq 4000$, $5x + 10y \geq 6000$, $x \geq 1.2y$, or $x - 1.2y \geq 0$, $x \geq 0$, $y \geq 0$.

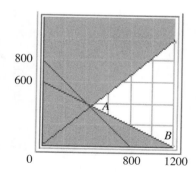

Point	Lines	Coords.	$c = 50x+80y$
A	$5x+10y = 6000$ $x-1.2y = 0$	$(450, 375)$	**52,500**
B	$5x+10y = 6000$ $y = 0$	$(1200, 0)$	60,000

Solution: Purchase 450 packages from Duffin House, 375 from Higgins Press for a minimum cost of $52,500.

31. Unknowns: as in Exercise 28. Minimize $c = 50x + 80y$ subject to $5x + 5y \geq 4000$, $5x + 10y \geq 6000$, $x - 1.2y \geq 0$, $50x \leq 80y$, or $50x - 80y \leq 0$ $x \geq 0$, $y \geq 0$.

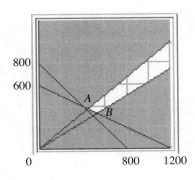

Point	Lines	Coords.	$c = 50x+80y$
A	$5x+10y = 6000$ $x-1.2y = 0$	(450, 375)	**52,500**
B	$5x+10y = 6000$ $50x-80y = 0$	(1600/3, 1000/3)	160,000/3

Solution: Purchase 450 packages from Duffin House, 375 from Higgins Press for a minimum cost of $52,500.

32. Unknowns: $x = $ # paperbacks, $y = $ # quality paperbacks, $z = $ # hardcovers

Maximize $p = x + 2y + 3z$ subject to $3x + 2y + z \leq 6000$, $2x + y + 3z \leq 6000$, $x + y + z \leq 2200$, $x \geq 0$, $y \geq 0$, $z \geq 0$.

	x	y	z	s	t	u	p		
s	3	2	1	1	0	0	0	6000	$3R_1 - R_2$
t	2	1	$\boxed{3}$	0	1	0	0	6000	
u	1	1	1	0	0	1	0	2200	$3R_3 - R_2$
p	-1	-2	-3	0	0	0	1	0	$R_4 + R_2$

	x	y	z	s	t	u	p		
s	7	5	0	3	-1	0	0	12,000	$2R_1 - 5R_3$
z	2	1	3	0	1	0	0	6000	$2R_2 - R_3$
u	1	$\boxed{2}$	0	0	-1	3	0	600	
p	1	-1	0	0	1	0	1	6000	$2R_4 + R_3$

	x	y	z	s	t	u	p	
s	9	0	0	6	3	-15	0	21,000
z	3	0	6	0	3	-3	0	11,400
y	1	2	0	0	-1	3	0	600
p	3	0	0	0	1	3	2	12600

Optimal Solution: $p = 12{,}600/2 = 6300$; $x = 0$, $y = 600/2 = 300$, $z = 11{,}400/6 = 1900$. Duffin should print no paperbacks, 300 quality paperbacks and 1900 hardcovers for a total daily profit of $6300.

33. Convert the given problem to a maximization problem:

Maximize $p = -50x - 150y - 100z$ subject to $5x + 10y + 5z \geq 4000$, $2x + 10y + 5z \geq 6000$, $x - y + z \leq 0$, $x \geq 0$, $y \geq 0$, $z \geq 0$.

	x	y	z	s	t	u	p		
*s	5	10	5	−1	0	0	0	4000	
*t	2	10	5	0	−1	0	0	6000	$R_2 - R_1$
u	1	−1	1	0	0	1	0	0	$10R_3 + R_1$
p	50	150	100	0	0	0	1	0	$R_4 - 15R_1$

	x	y	z	s	t	u	p		
y	5	10	5	−1	0	0	0	4000	$R_1 + R_2$
*t	−3	0	0	1	−1	0	0	2000	
u	15	0	15	−1	0	10	0	4000	$R_3 + R_2$
p	−25	0	25	15	0	0	1	−60000	$R_4 - 15R_2$

	x	y	z	s	t	u	p	
y	2	10	5	0	−1	0	0	6000
s	−3	0	0	1	−1	0	0	2000
u	12	0	15	0	−1	10	0	6000
p	20	0	25	0	15	0	1	−90000

Optimal Solution: $c = -p = 90000$; $x = 0$, $y = 6000/10 = 600$, $z = 0$.

34. Unknowns: $x = $ # science credits, $y = $ # fine arts credits, $z = $ # liberal arts credits, $w = $ # math credits. Minimize $C = 300x + 300y + 200z + 200w$ subject to $x + y + z + w \geq 120$, $x - y \geq 0$, $-2x + w \leq 0$, $-y + 3z - 3w \leq 0$, $x \geq 0$, $y \geq 0$, $z \geq 0$, $w \geq 0$.

35. Since technology is indicated, we use the online Simplex Method Tool. When entering the problems, there is no need to enter the inequalities $x \geq 0$, $y \geq 0$, etc.

```
Tableau #1
x      y      z      w      s1     s2     s3     s4     -c
1      1      1      1      -1     0      0      0      0      120
1      -1     0      0      0      -1     0      0      0      0
-2     0      0      1      0      0      1      0      0      0
0      -1     3      -3     0      0      0      1      0      0
300    300    200    200    0      0      0      0      1      0
```

Chapter 4 Review Exercises

Tableau #2

x	y	z	w	s1	s2	s3	s4	−c	
0	2	1	1	−1	1	0	0	0	120
1	−1	0	0	0	−1	0	0	0	0
0	−2	0	1	0	−2	1	0	0	0
0	−1	3	−3	0	0	0	1	0	0
0	600	200	200	0	300	0	0	1	0

Tableau #3

x	y	z	w	s1	s2	s3	s4	−c	
0	2	1	1	−1	1	0	0	0	120
2	0	1	1	−1	−1	0	0	0	120
0	0	1	2	−1	−1	1	0	0	120
0	0	7	−5	−1	1	0	2	0	120
0	0	−100	−100	300	0	0	0	1	−36000

Tableau #4

x	y	z	w	s1	s2	s3	s4	−c	
0	7	0	6	−3	3	0	−1	0	360
7	0	0	6	−3	−4	0	−1	0	360
0	0	0	19	−6	−8	7	−2	0	720
0	0	7	−5	−1	1	0	2	0	120
0	0	0	−1200	2000	100	0	200	7	−240000

Tableau #5

x	y	z	w	s1	s2	s3	s4	−c	
0	19	0	0	−3	15	−6	−1	0	360
19	0	0	0	−3	−4	−6	−1	0	360
0	0	0	19	−6	−8	7	−2	0	720
0	0	19	0	−7	−3	5	4	0	840
0	0	0	0	4400	−1100	1200	200	19	−528000

Tableau #6

x	y	z	w	s1	s2	s3	s4	−c	
0	19	0	0	−3	15	−6	−1	0	360
15	4	0	0	−3	0	−6	−1	0	360
0	8	0	15	−6	0	3	−2	0	720
0	1	5	0	−2	0	1	1	0	240
0	220	0	0	660	0	120	20	3	−79200

`Optimal Solution: c = 26400; x = 24, y = 0, z = 48, w = 48.`
Billy Sean should take the following combination: Sciences: 24 credits, Fine Arts: no credits, Liberal Arts: 48 credits, Mathematics: 48 credits, for a total cost of $26,400.

36. Unknowns: x = # packages from New York to O'Hagan.com, y = # packages form New York to FantasyBooks.com, z = # packages from Illinois to O'Hagan.com,, w = # packages form Illinois to FantasyBooks.com.

Minimize $c = 20x + 50y + 30z + 40w$ subject to $x + y \le 600$, $y + z \le 300$, $x + z \ge 600$, $y + w \ge 200$, $x \ge 0$, $y \ge 0$, $z \ge 0$, $w \ge 0$.

Associated maximization problem: Maximize $p = -20x - 50y - 30z - 40w$ subject to $x + y \le 600$, $y + z \le 300$, $x + z \ge 600$, $y + w \ge 200$, $x \ge 0$, $y \ge 0$, $z \ge 0$, $w \ge 0$.

	x	y	z	w	s	t	u	v	p		
s	1	1	0	0	1	0	0	0	0	600	$R_1 - R_3$
t	0	1	1	0	0	1	0	0	0	300	
*u	[1]	0	1	0	0	0	-1	0	0	600	
*v	0	1	0	1	0	0	0	-1	0	200	
p	20	50	30	40	0	0	0	0	1	0	$R_5 - 20R_3$

	x	y	z	w	s	t	u	v	p		
s	0	[1]	-1	0	1	0	1	0	0	0	
t	0	1	1	0	0	1	0	0	0	300	$R_2 - R_1$
x	1	0	1	0	0	0	-1	0	0	600	
*v	0	1	0	1	0	0	0	-1	0	200	$R_4 - R_1$
p	0	50	10	40	0	0	20	0	1	-12,000	$R_5 - 50R_1$

	x	y	z	w	s	t	u	v	p		
y	0	1	-1	0	1	0	1	0	0	0	$2R_1 + R_2$
t	0	0	[2]	0	-1	1	-1	0	0	300	
x	1	0	1	0	0	0	-1	0	0	600	$2R_3 - R_2$
*v	0	0	1	1	-1	0	-1	-1	0	200	$2R_4 - R_2$
p	0	0	60	40	-50	0	-30	0	1	-12,000	$R_5 - 30R_2$

	x	y	z	w	s	t	u	v	p		
y	0	2	0	0	1	1	1	0	0	300	
z	0	0	2	0	-1	1	-1	0	0	300	
x	2	0	0	0	1	-1	-1	0	0	900	
$*v$	0	0	0	[2]	-1	-1	-1	-2	0	100	
p	0	0	0	40	-20	-30	0	0	1	-21,000	$R_5 - 20R_4$

	x	y	z	w	s	t	u	v	p		
y	0	2	0	0	1	1	1	0	0	300	$R_1 - R_2$
z	0	0	2	0	-1	[1]	-1	0	0	300	
x	2	0	0	0	1	-1	-1	0	0	900	$R_3 + R_2$
w	0	0	0	2	-1	-1	-1	-2	0	100	$R_4 + R_2$
p	0	0	0	0	0	-10	20	40	1	-23,000	$R_5 + 10R_2$

	x	y	z	w	s	t	u	v	p		
y	0	2	-2	0	[2]	0	2	0	0	0	
t	0	0	2	0	-1	1	-1	0	0	300	$2R_2 + R_1$
x	2	0	2	0	0	0	-2	0	0	1200	
w	0	0	2	2	-2	0	-2	-2	0	400	$R_4 + R_1$
p	0	0	20	0	-10	0	10	40	1	-20,000	$R_5 + 5R_1$

	x	y	z	w	s	t	u	v	p	
s	0	2	-2	0	2	0	2	0	0	0
t	0	2	2	0	0	2	0	0	0	600
x	2	0	2	0	0	0	-2	0	0	1200
w	0	2	0	2	0	0	0	-2	0	400
p	0	10	10	0	0	0	20	40	1	-20,000

Optimal Solution: $c = -p = 20{,}000$; $x = 1200/2 = 600$, $y = 0$, $z = 0$, $w = 400/2 = 200$. The lowest cost is $20,000; New York to O'Hagan.com: 600 packages, New York to FantasyBooks.com: 0 packages, Illinois to O'Hagan.com: 0 packages, Illinois to FantasyBooks.com: 200 packages.

37. First reduce the game: We can eliminate O'Hagan's option of offering no promotion, and then we can eliminate FantasyBook's no promotion option. The entries in the remaining payoff matrix are nonnegative, so we put 1s to the right and below:

$$\begin{bmatrix} 20 & 10 & 15 & 1 \\ 0 & 15 & 10 & 1 \\ 1 & 1 & 1 & 0 \end{bmatrix}$$

This gives the following LP problem:

Maximize $p = x + y + z$ subject to

$$20x + 10y + 15z \leq 1$$
$$15y + 10z \leq 1$$
$$x \geq 0,\ y \geq 0,\ z \geq 0$$

Here are the tableaux:

	x	y	z	s	t	p	
s	20	10	15	1	0	0	1
t	0	15	10	0	1	0	1
p	−1	−1	−1	0	0	1	0

$20R_3 + R_1$

	x	y	z	s	t	p	
x	20	10	15	1	0	0	1
t	0	15	10	0	1	0	1
p	0	−10	−5	1	0	20	1

$3R_1 - 2R_2$

$3R_3 + 2R_2$

	x	y	z	s	t	p	
x	60	0	25	3	−2	0	1
y	0	15	10	0	1	0	1
p	0	0	5	3	2	60	5

The solution to the primal problem is $p = 5/60 = 1/12$; $x = 1/60$, $y = 1/15$, $z = 0$. So, FantasyBook's optimal strategy is $C = [0\ 1/5\ 4/5\ 0]^T$ and the value of the game is $e = 12$.

The solution to the dual problem is $s = 3/60 = 1/20$, $t = 2/60 = 1/30$, so O'HaganBook's optimal strategy is $R = [0\ 3/5\ 2/5]$.

Chapter 5
5.1

1. $PV = 2000$, $r = 0.06$, $t = 1$

$$INT = PVrt$$
$$= 2000(0.06)(1) = \$120$$
$$FV = PV + INT$$
$$= 2000 + 120 = \$2120$$

3. $PV = 20{,}200$, $r = 0.05$, $t = \frac{1}{2}$

$$INT = PVrt$$
$$= 20{,}200(0.05)(\tfrac{1}{2}) = \$505$$
$$FV = PV + INT$$
$$= 20{,}200 + 505 = \$20{,}705$$

5. $PV = 10{,}000$, $r = 0.03$, $t = 10/12$

$$INT = PVrt$$
$$= 10{,}000(0.03)(10/12) = \$250$$
$$FV = PV + INT$$
$$= 10{,}000 + 250 = \$10{,}250$$

7. $PV = 10{,}000/(1 + 0.02 \times 5) = \$9090.91$

9. $PV = 1000/(1 + 0.07 \times 0.5) = \$966.18$

11. $PV = 15{,}000/(1 + 0.03 \times 15/12) = \$14{,}457.83$

13. $FV = 5000(1 + 0.08 \times 0.5) = \$5200$

15. $FV = \$1000$, $r = 0.045$, $t = 6$

$$FV = PV(1 + rt)$$
$$1000 = PV(1 + 0.045 \times 6)$$
$$1000 = PV(1.27)$$
$$PV = \frac{1000}{1.27} \approx \$787.40$$

17. We are given the interest and asked to compute r.

$$INT = 250, \ PV = 1000, \ t = 5$$

$$INT = PVrt$$
$$250 = 1000 \times r \times 5 = 5000r$$
$$r = \frac{250}{5000} = 0.05, \text{ or } 5\%$$

19. We are given present and future values, and want to compute t.

$$FV = 4640, \ PV = 4000, \ r = 0.08$$
$$FV = PV(1 + rt)$$
$$4640 = 4000(1 + 0.08t)$$
$$1 + 0.08t = \frac{4640}{4000} = 1.160$$
$$0.08t = 0.160$$
$$t = \frac{0.160}{0.08} = 2 \text{ years.}$$

21. To say that the discount rate is 3.705% means that its selling price (PV) is 3.705% lower than its maturity value (FV). To simplify the calculation, let us use a T bill with a maturity value of $10,000:

$$PV = 10{,}000 - 10{,}000(0.03705/2)$$
$$= 9814.75. \ 10{,}000$$
$$= 9814.75(1 + 0.5r)$$
$$= 9814.75 + 4907.375r$$
$$r = (10{,}000 - 9814.75)/4907.375$$
$$= 0.03775, \text{ or } 3.775\%$$

23. $1050 = 1000(1 + 4r/52) = 1000 + 4000r/52$; $r = 50 \times 52/4000 = 0.65$, which is 65% interest.

25. You will pay $5000 \times 0.09 \times 2 = \$900$ in interest on the loan. Adding the $100 fee, you pay the bank a total of $1000 over the two years. To find the effective rate, we solve $6000 = 5000(1 + 2r) = 5000 + 10{,}000r$; $r = 1000/10{,}000 = 0.1$, so the rate is 10%.

27. $PV = 3.28$, $FV = 45.74$,

$t = 92$ months $= 92/12$ years

$45.74 = 3.28(1 + 92r/12) = 3.28 + 25.1467r$

$r = (45.74 - 3.28)/25.1467 = 1.6885 = 168.85\%$

29. The only dates on which would you would have gotten an increase rather than a decrease were November 2004, February 2005, and August 2005. Calculating the annual returns as in the preceding exercises, we get the following figures:

Jan. 2002–Nov. 2004: 60.45%

Jan. 2002–Feb. 2005: 85.28%

Jan. 2002–Aug. 2005: 75.37%

The largest annual return was 85.28% if you sold in February 2005.

31. No. Simple interest increase is linear. The graph is visibly not linear in that time period. Further, we can compare slopes between marked points to see if the slope remained roughly constant: From December 1997 to August 1999 the slope was $(16.31 - 3.28)/(20/12) = 7.818$ while, from August 1999 to March 2000 the slope was $(33.95 - 16.31)/(7/12) = 30.24$. These slopes are quite different.

33. 1950 population: $PV = 500,000$

2000 population: $FV = 2,800,000$

$INT = FV - PV$

$\qquad = 2,800,000 - 500,000$

$\qquad = 2,300,000$

$INT = PVrt$

$2,300,000 = 500,000r(50) = 25,000,000r$

$r = \dfrac{2,300,000}{25,000,000} \approx 0.092$ or 9.2%

35. We are given:

$PV = 1950$ population $= 500,000$

$INT = 0.092$ (from Exercise 23)

$t = 60$ (years to 2010)

$FV = PV(1 + rt)$

$\qquad = 500,000(1 + 0.092 \times 60)$

$\qquad = 3,260,000$

37. $PV = 1950$ population $= 500,000$

$\qquad INT = 0.092$ (from Exercise 33)

After t years, the population will be

$\qquad FV = PV(1 + rt)$

$\qquad\quad = 500,000(1 + 0.092t)$

$\qquad\quad = 500,000 + 46t$

($t =$ time in years since 1950).

Graph:

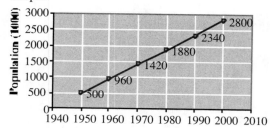

39. (A) is the only possible choice because the equation $FV = PV(1+rt) = PV + PVrt$ gives the future value as a linear function of time.

41. Wrong. In simple interest growth, the change each year is a fixed percentage of the *starting* value, and not the preceding year's value. (Also see the next exercise.)

43. Simple interest is always calculated on a constant amount, PV. If interest is paid into your account, then the amount on which interest is calculated increases, and hence does not remain constant.

5.2

1. $PV = \$10{,}000$, $r = 0.03$, $m = 1$, $t = 10$

$$FV = PV\left(1 + \frac{r}{m}\right)^{mt}$$
$$= 10{,}000(1 + 0.03)^{10}$$
$$= 10{,}000(1.03)^{10} \approx \$13{,}439.16$$

Technology: `10000*(1+0.03)^10`

3. $PV = \$10{,}000$, $r = 0.025$, $m = 4$, $t = 5$

$$FV = PV\left(1 + \frac{r}{m}\right)^{mt}$$
$$= 10{,}000\left(1 + \frac{0.025}{4}\right)^{(4)(5)}$$
$$= 10{,}000(1.00625)^{20} \approx \$11{,}327.08$$

Technology: `10000*(1+0.025/4)^(4*5)`

5. $PV = \$10{,}000$, $r = 0.065$, $m = 365$, $t = 10$

$$FV = PV\left(1 + \frac{r}{m}\right)^{mt}$$
$$= 10{,}000\left(1 + \frac{0.065}{365}\right)^{(365)(10)}$$
$$= 10{,}000\left(1 + \frac{0.065}{365}\right)^{3650} \approx \$19{,}154.30$$

Technology:
`10000*(1+0.065/365)^(365*10)`

7. Since the monthly interest rate is given, we use the formula $FV = PV(1 + i)^n$.

$PV = \$10{,}000$, $i = 0.002$, $n = 10 \times 12 = 120$

$$FV = PV(1 + i)^n$$
$$= 10{,}000(1 + 0.002)^{120}$$
$$= 10{,}000(1.002)^{120} \approx \$12{,}709.44$$

Technology: `10000*(1+0.002)^120`

9. $FV = \$1000$, $t = 10$, $r = 0.05$, $m = 1$

$$PV = \frac{FV}{\left(1 + \dfrac{r}{m}\right)^{mt}}$$
$$= \frac{1000}{\left(1 + \dfrac{0.05}{1}\right)^{(1)(10)}}$$
$$= \frac{1000}{1.05^{10}} \approx \$613.91$$

Technology: `1000/(1+0.05)^10`

11. $FV = \$1000$, $t = 5$, $r = 0.042$, $m = 52$

$$PV = \frac{FV}{\left(1 + \dfrac{r}{m}\right)^{mt}}$$
$$= \frac{1000}{\left(1 + \dfrac{0.042}{52}\right)^{(52)(5)}}$$
$$= \frac{1000}{\left(1 + \dfrac{0.042}{52}\right)^{260}} \approx \$810.65$$

Technology:
`1000/(1+0.042/52)^(52*5)`

13. $FV = \$1000$, $n = 4$, $i = -0.05$

$$PV = \frac{FV}{(1 + i)^n}$$
$$= \frac{1000}{(1 - 0.05)^4} = \frac{1000}{0.95^4} \approx \$1{,}227.74$$

Technology: `1000/(1-0.05)^4`

15. $r_{nom} = 0.05$, $m = 4$

$$r_{eff} = \left(1 + \frac{r_{nom}}{m}\right)^m - 1$$
$$= \left(1 + \frac{0.05}{4}\right)^4 - 1$$
$$\approx 0.0509, \text{ or } 5.09\%$$

Technology: `(1+0.05/4)^4-1`

17. $r_{\text{nom}} = 0.10$, $m = 12$

$$r_{\text{eff}} = \left(1 + \frac{r_{\text{nom}}}{m}\right)^m - 1$$

$$= \left(1 + \frac{0.10}{12}\right)^{12} - 1$$

$$\approx 0.1047, \text{ or } 10.47\%$$

Technology: `(1+0.10/12)^12-1`

19. $r_{\text{nom}} = 0.10$, $m = 365 \times 24 = 8760$

$$r_{\text{eff}} = \left(1 + \frac{r_{\text{nom}}}{m}\right)^m - 1$$

$$= \left(1 + \frac{0.10}{8760}\right)^{8760} - 1$$

$$\approx 0.1052, \text{ or } 10.52\%$$

Technology: `(1+0.10/8760)^ 8760-1`

21. $PV = \$1000$, $r = 0.06$, $m = 4$, $t = 4$

$$FV = PV\left(1 + \frac{r}{m}\right)^{mt}$$

$$= 1000\left(1 + \frac{0.06}{4}\right)^{(4)(4)}$$

$$= 1000(1.015)^{16} \approx \$1268.99$$

The deposit will have grown by
$$\$1268.99 - \$1000 = \$268.99$$

Technology: `1000*(1+0.06/4)^(4*4)`

23. $PV = \$3000$, $r = -0.06$, $m = 1$, $t = 3$

$$FV = PV\left(1 + \frac{r}{m}\right)^{mt}$$

$$= 3000\left(1 - \frac{0.06}{1}\right)^{(1)(3)}$$

$$= 3000(0.94)^3 \approx \$2491.75$$

Technology: `3000*(1-0.06)^3`

25. $FV = \$5000$, $r = 0.055$, $t = 10$, $m = 1$

$$PV = \frac{FV}{\left(1 + \frac{r}{m}\right)^{mt}}$$

$$= \frac{5000}{\left(1 + \frac{0.055}{1}\right)^{(1)(10)}}$$

$$= \frac{5000}{1.055^{10}} \approx \$2927.15$$

Technology: `5000/(1+0.055)^10`

27. Gold: $PV = \$5000$, $r = 0.10$, $m = 1$, $t = 10$

$$FV = PV\left(1 + \frac{r}{m}\right)^{mt}$$

$$= 5000\left(1 + \frac{0.10}{1}\right)^{(1)(10)}$$

$$= 5000(1.10)^{10} \approx \$12,968.71$$

CDs: $PV = \$5000$, $r = 0.05$, $m = 2$, $t = 10$

$$FV = PV\left(1 + \frac{r}{m}\right)^{mt}$$

$$= 5000\left(1 + \frac{0.05}{2}\right)^{(2)(10)}$$

$$= 5000(1.025)^{20} \approx \$8193.08$$

Combined value after 10 years
$$= \$12,968.71 + \$8193.08 = \$21,161.79$$

Technology: `5000*(1+0.10)^10+`
`5000*(1+0.05/2)^(2*10)`

29. $PV = \$200,000$, $i = -0.02$, $n = 10$

$$FV = PV(1 + i)^n$$

$$= 200,000(1 - 0.02)^{10}$$

$$= 200,000(0.98)^{10} \approx \$163,414.56$$

Technology: `200000*(1-0.02)^10`

31. $FV = \$100,000$, $r = 0.04$, $m = 1$, $t = 15$

$$PV = \frac{FV}{\left(1 + \frac{r}{m}\right)^{mt}}$$

$$= \frac{100,000}{\left(1 + \dfrac{0.04}{1}\right)^{(1)(15)}}$$

$$= \frac{100,000}{1.04^{15}} \approx \$55,526.45 \text{ per year}$$

Technology: `100000/(1+0.04)^15`

33. $FV = \$1,000,000$, $r = 0.06$, $m = 1$, $t = 30$

$$PV = \frac{FV}{\left(1 + \dfrac{r}{m}\right)^{mt}}$$

$$= \frac{1,000,000}{\left(1 + \dfrac{0.06}{1}\right)^{(1)(30)}}$$

$$= \frac{1,000,000}{1.06^{30}} \approx \$174,110$$

Technology: `1000000/(1+0.06)^30`

35. $FV = \$297.91$, $n = 6 \times 3 = 18$, $i = -0.05$

$$PV = \frac{FV}{(1 + i)^n}$$

$$= \frac{297.91}{(1 - 0.05)^{18}} = \frac{297.91}{0.95^{18}} \approx \$750.00$$

Technology: `297.91/(1-0.05)^18`

37. $FV = \$30,000$, $n = 5$, $i = 0.02$

$$PV = \frac{FV}{(1 + i)^n}$$

$$= \frac{30,000}{(1 + 0.02)^5} = \frac{30,000}{1.02^5} \approx \$27,171.92$$

Technology: `30000/(1+0.02)^5`

39. $FV = \$200,000$, $n = 10$, $i = 0.06$

$$PV = \frac{FV}{(1 + i)^n}$$

$$= \frac{200,000}{(1 + 0.06)^{10}} = \frac{200,000}{1.06^{10}} \approx \$111,678.96$$

Technology: `200000/(1+0.06)^10`

41. Step 1: Calculate the future value of the investment:

$$PV = \$1000, \ i = 0.05, \ t = 2$$
$$FV = PV(1 + i)^n$$
$$= 1000(1 + 0.05)^2$$
$$= 1000(1.05)^2 \approx \$1102.50$$

Technology: `1000*(1+0.05)^2`

Step 2: Discount this value using inflation:

$$FV = \$1102.50, \ i = 0.03, \ t = 2$$
$$PV = \frac{FV}{(1 + i)^n}$$
$$= \frac{1102.50}{(1 + 0.03)^2} = \frac{1102.50}{1.03^2} \approx \$1039.21$$

Technology: `1102.50/(1+0.03)^2`

43. Compare the effective yields of the two investments:

First Investment:

$$r_{\text{nom}} = 0.12, \ m = 1$$
$$r_{\text{eff}} = r_{\text{nom}} = 0.12, \text{ or } 12\%$$

Second Option:

$$r_{\text{nom}} = 0.119, \ m = 12$$
$$r_{\text{eff}} = \left(1 + \frac{r_{\text{nom}}}{m}\right)^m - 1$$
$$= \left(1 + \frac{0.119}{12}\right)^{12} - 1$$
$$\approx 0.1257, \text{ or } 12.57\%$$

This is the better investment

Technology: `(1+0.119/12)^12-1`

45. $PV = \$24$, $i = 0.062$, $n = 2001 - 1626 = 375$

$$FV = PV(1 + i)^n$$
$$= 24(1 + 0.062)^{375}$$
$$= 24(1.062)^{375} \approx \$150{,}281 \text{ million}$$

This is more than the 2001 estimated real estate value of \$136,106 million. Thus, the Manhattan Indians could have bought the island back in 2001.

Technology: $24*(1+0.062)^{375}$

47. $PV = 100$ reals, $i = 0.08$, $n = 5$

$$FV = PV(1 + i)^n$$
$$= 100(1 + 0.008)^5$$
$$= 100(1.08)^5 \approx 147 \text{ reals}$$

Technology: $100*(1+0.08)^5$

49. $FV = 1000$ pesos, $i = 0.03$, $n = 10$

$$PV = \frac{FV}{(1 + i)^n}$$
$$= \frac{1000}{(1 + 0.03)^{10}} = \frac{1000}{1.03^{10}} \approx 744 \text{ pesos}$$

Technology: $1000/(1+0.03)^{10}$

51. Step 1: Calculate the future value of the investment:

$PV = 1000$ pesos, $r = 0.08$, $m = 2$, $t = 10$

$$FV = PV\left(1 + \frac{r}{m}\right)^{mt}$$
$$= 1000\left(1 + \frac{0.08}{2}\right)^{(2)(10)}$$
$$= 1000(1.04)^{20} \approx 2191.12 \text{ pesos}$$

Technology: $1000*(1+0.08/2)^{20}$

Step 2: Discount this amount using the inflation rate:

$FV = 2191.12$, $i = 0.06$, $n = 10$

$$PV = \frac{FV}{(1 + i)^n}$$
$$= \frac{2191.12}{(1 + 0.06)^{10}} = \frac{2191.12}{1.06^{10}} \approx 1224 \text{ pesos}$$

Technology: $2191.12/(1+0.06)^{10}$

Note: If we first round the answer of Step 1 to the nearest peso, in Step 2 we get

$$PV = \frac{2191}{1.06^{10}} \approx 1223 \text{ pesos}$$

53. We compare the future values of one unit of the currency for a one-year period:

Chile: $PV = 1$ peso, $r = 0.043$, $m = 1$, $t = 1$

$$FV = PV\left(1 + \frac{r}{m}\right)^{mt}$$
$$= 1\left(1 + \frac{0.043}{1}\right)^{(1)(1)}$$
$$= 1.043 \text{ pesos}$$

Now discount this using inflation:

$FV = 1.043$, $i = 0.03$, $n = 1$

$$PV = \frac{FV}{(1 + i)^n}$$
$$= \frac{1.043}{(1 + 0.03)^1} = \frac{1.043}{1.03} \approx 1.0126 \text{ pesos}$$

Ecuador: $PV = 1$ sucre, $r = 0.162$, $m = 2$, $t = 1$

$$FV = PV\left(1 + \frac{r}{m}\right)^{mt}$$
$$= 1\left(1 + \frac{0.162}{2}\right)^{(2)(1)}$$
$$= 1.081^2 \approx 1.1686 \text{ sucres}$$

Now discount this using inflation:

$FV = 1.1686$, $i = 0.15$, $n = 1$

$$PV = \frac{FV}{(1 + i)^n}$$
$$\approx \frac{1.1686}{(1 + 0.15)^1} = \frac{1.1686}{1.15} \approx 1.0162 \text{ sucres}$$

The investment in Ecuador is better.

Note A more accurate answer for Ecuador is 1.0161 pesos, which we get by using more decimal places in the calculation of 1.081^2.

55. $PV = 3.28$, $FV = 45.74$, $t = 92$ months = 92/12 years

$$FV = PV\left(1 + \frac{r}{m}\right)^{mt}$$

$$45.74 = 3.28(1 + r)^{92/12}$$

Solve for r:

$$r = (45.74/3.28)^{12/92} - 1 = 0.4102 = 41.02\%$$

57. The only dates on which would you would have gotten an increase rather than a decrease were November 2004, February 2005, and August 2005. Calculating the annual returns as in the preceding exercises, we get the following figures:

Jan. 2002–Nov. 2004: 42.22%

Jan. 2002–Feb. 2005: 51.90%

Jan. 2002–Aug. 2005: 41.76%

The largest annual return was 51.90% if you sold in February 2005.

59. No. Compound interest increase is exponential. The graph looks roughly exponential in that period, but to really tell we can compare interest rates between marked points to see if the rate remained roughly constant: From December 1997 to August 1999 the rate was $(16.31/3.28)^{12/20} - 1 = 1.6179$ or 161.79%, while from August 1999 to March 2000 the rate was $(33.95/16.31)^{12/7} - 1 = 2.5140$ or 251.40%. These rates are quite different.

61. My investment:

$PV = \$5000$, $r = 0.054$, $m = 2$

$$FV = PV\left(1 + \frac{r}{m}\right)^{mt}$$

$$= 5000\left(1 + \frac{0.054}{2}\right)^{2t} = 5000(1.027)^{2t}$$

Friend's investment:

$PV = \$6000$, $r = 0.048$, $m = 2$

$$FV = PV\left(1 + \frac{r}{m}\right)^{mt}$$

$$= 6000\left(1 + \frac{0.048}{2}\right)^{2t} = 6000(1.024)^{2t}$$

Technology Formulas:

`Y₁ = 5000*1.027^(2*x)`

`Y₂ = 6000*1.024^(2*x)`

Graph:

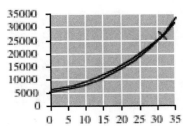

Zoomed-in View:

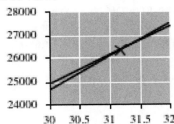

The graphs cross around $t \approx 31$ years. The value of the investment is

$$5000(1.027)^{2(31)} \approx \$26100 \quad \text{(rounded to 3 significant digits)}$$

63. $PV = 40,000$ $i = 1.0$ (a 100% increase per period)

$$FV = PV(1 + i)^n = 40,000(1+1.0)^n$$

$$= 40,000(2)^n$$

`Y₁ = 40000*2^x`

`Y₂ = 1000000`

Graph:

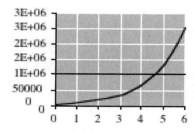

Zoomed-in View:

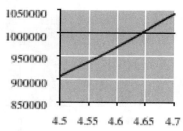

$n \approx 4.65$. Since n measures 6-month periods, this corresponds to $4.65/2 \approx 2.3$ years.

65. (a) The amount you paid for the bond is its present value.

$FV = \$100,000, i = 0.15, t = 30$

$PV = \dfrac{FV}{(1 + i)^n}$

$\approx \dfrac{100,000}{(1 + 0.15)^{30}} = \dfrac{100,000}{1.15^{30}} \approx \$1510.31$

(b) The value of the bond at any time is its present value at that time, given the prevailing interest rate. Since it will pay $100,000 in 13 years' time, we have:

$FV = \$100,000, i = 0.0475, t = 13$

$PV = \dfrac{FV}{(1 + i)^n}$

$\approx \dfrac{100,000}{(1 + 0.0.475)^{13}} = \dfrac{100,000}{1.0475^{13}} \approx \$54,701.29$

(c) By part (a) the bond cost you $1510.31 and, by part (b), was worth $54,701.29 after 17 years.

$PV = \$1510.31, FV = \$54,701.29$

276

$FV = PV(1 + i)^n$

$54,701.29 = 1510.31(1 + i1)^{17}$

$\dfrac{54,701.29}{1510.31} = (1 + i1)^{17}$

$1 + i = \left(\dfrac{54,701.29}{1510.31}\right)^{1/17}$

$i = \left(\dfrac{54,701.29}{1510.31}\right)^{1/17} - 1 \approx 0.2351, \text{ or } 23.51\%$

Technology:

```
(54701.29 /1510.31)^(1/17)-1
```

67. The function $y = P(1 + {}^r/_m)^{mx}$ is not a linear function of x, but an exponential function. Thus, its graph is not a straight line.

69. Wrong. Its growth can be modeled by
$0.01(1 + 0.10)^t = 0.01(1.10)^t$.
This is an exponential function of t; not a linear one.

71. A compound-interest investment behaves as though it was being compounded once a year at the effective rate. Thus, if two equal investments have the same effective rates, they grow at the same rate.

73. The effective rate exceeds the nominal rate when the interest is compounded more than once a year, since then interest is being paid on interest accumulated during each year, resulting in a larger effective rate. Conversely, if the interest is compounded less often than once a year, the effective rate is less than the nominal rate.

75. Compare their future values in constant dollars. That is, future compute their future values, and then discount each for inflation. The investment with the larger future value is the better investment.

77. $PV = 100$, $r = 0.10$, $m = 1, 10, 100, \ldots$

```
Y₁ = 100*(1.10)^x
Y₂ = 100*(1+ 0.10/10)^(10*x)
Y₃ = 100*(1+ 0.10/100)^(100*x)
```

...

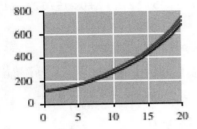

The graphs are approaching a particular curve (shown darker) as m gets larger, approximately the curve given by the largest value of m.

5.3

1. $FV = PMT\dfrac{(1 + i)^n - 1}{i}$

$PMT = \$100,$

$i = $ interest paid each period $= 0.05/12$

$n = $ total number of periods $= 12 \times 10 = 120$

$FV = 100\dfrac{(1 + 0.05/12)^{120} - 1}{(0.05/12)} \approx \$15,528.23$

Technology:

`100*((1+0.05/12)^120-1)/(0.05/12)`

3. $FV = PMT\dfrac{(1 + i)^n - 1}{i}$

$PMT = \$1000,$

$i = $ interest paid each period $= 0.07/4$

$n = $ total number of periods $= 4 \times 20 = 80$

$FV = 1000\dfrac{(1 + 0.07/4)^{80} - 1}{(0.07/4)} \approx \$171,793.82$

Technology:

`1000*((1+0.07/4)^80-1)/(0.07/4)`

5. We need to calculate the sum of
$FV = PV(1 + i)^n$
and
$FV = PMT\dfrac{(1 + i)^n - 1}{i}$
where

$PV = \$5000, PMT = \$100,$

$i = $ interest paid each period $= 0.05/12$

$n = $ total number of periods $= 12 \times 10 = 120$

$FV = 5000(1 + 0.05/12)^{120}$

$\quad + 100\dfrac{(1 + 0.05/12)^{120} - 1}{(0.05/12)} \approx \$23,763.28$

Technology:

`5000*(1+0.05/12)^120+`
`100*((1+0.05/12)^120-1)/(0.05/12)`

7. $PMT = FV\dfrac{i}{(1 + i)^n - 1}$

$FV = \$10,000, i = 0.05/12, n = 12 \times 5 = 60$

$PMT = 10,000\dfrac{(0.05/12)}{(1 + 0.05/12)^{60} - 1}$

$\quad \approx \$147.05$

Technology:

`10000*0.05/12/((1+0.05/12)^60-1)`

9. $PMT = FV\dfrac{i}{(1 + i)^n - 1}$

$FV = \$75,000, i = 0.06/4, n = 4 \times 20 = 80$

$PMT = 75,000\dfrac{(0.06/4)}{(1 + 0.06/4)^{80} - 1}$

$\quad \approx \$491.12$

Technology:

`75000*0.06/4/((1+0.06/4)^80-1)`

11. We first account for the future value of the $10,000 already in the account:

$PV = \$10,000, i = 0.05/12, n = 12 \times 5 = 60$

$FV = PV(1 + i)^n = 10,000(1 + 0.05/12)^{60}$

We subtract this from $20,000 to get the future value of the payments, so:

$PMT = FV\dfrac{i}{(1 + i)^n - 1}$ where

$FV = \$20,000 - 10,000(1 + 0.05/12)^{60}$:

PMT
$= \dfrac{[20,000 - 10,000(1 + 0.05/12)^{60}]\,(0.05/12)}{(1 + 0.05/12)^{60} - 1}$

$\quad \approx \$105.38$

Technology:

`(20000-10000*(1+0.05/12)^60)*`
`0.05/12/((1+0.05/12)^60-1)`

13. $PMT = \$500, i = 0.03/12, n = 12 \times 20 = 240$

$$PV = PMT\frac{1 - (1 + i)^{-n}}{i}$$

$$= 500\frac{1 - (1 + 0.03/12)^{-240}}{0.03/12}$$

$$\approx \$90,155.46$$

Technology:

```
500*(1-(1+0.03/12)^(-240))/
(0.03/12)
```

15. $PMT = \$1500, i = 0.06/4, n = 4 \times 20 = 80$

$$PV = PMT\frac{1 - (1 + i)^{-n}}{i}$$

$$= 1500\frac{1 - (1 + 0.06/4)^{-80}}{0.06/4}$$

$$\approx \$69,610.99$$

Technology:

```
1500*(1-(1+0.06/4)^(-80))/(0.06/4)
```

17. $FV = \$10,000, PMT = \$500, i = 0.03/12,$
$n = 12 \times 20 = 240$

We need to fund both the future value and the payments, so the present value is the sum

$$PV = FV(1 + i)^{-n} + PMT\frac{1 - (1 + i)^{-n}}{i}$$

$$= 10,000(1 + 0.03/12)^{-240}$$

$$+ \ 500\frac{1 - (1 + 0.03/12)^{-240}}{0.03/12}$$

$$\approx \$95,647.68$$

Technology:

```
10000*(1+0.03/12)^(-240)+
500*(1-(1+0.03/12)^(-240))/
(0.03/12)
```

19. $PV = \$100,000, i = 0.03/12,$
$n = 12 \times 20 = 240$

$$PMT = PV\frac{i}{1 - (1 + i)^{-n}}$$

$$= 100,000\frac{0.03/12}{1 - (1 + 0.03/12)^{-240}}$$

$$\approx \$554.60$$

Technology:

```
100000*(0.03/12)/
(1-(1+0.03/12)^(-240))
```

21. $PV = \$75,000, i = 0.04/4 = 0.01,$
$n = 4 \times 20 = 80$

$$PMT = PV\frac{i}{1 - (1 + i)^{-n}}$$

$$= 75,000\frac{0.01}{1 - (1 + 0.01)^{-80}}$$

$$\approx \$1366.41$$

Technology:

```
75000*0.01/(1-(1+0.01)^(-80))
```

23. Part of the present value has to fund the future value of \$10,000:

$FV = \$10,000, i = 0.03/12, n = 12 \times 20 = 240$

so

$$FV(1 + i)^{-n} = 10,000(1 + 0.03/12)^{-240}$$

is required; we subtract this from the present value and use

$$PV = 100,000 - 10,000(1 + 0.03/12)^{-240}$$

in the payment formula.

$$PMT = PV\frac{i}{1 - (1 + i)^{-n}}$$

$$= \frac{(0.03/12)[100,000 - 10,000(1 + 0.03/12)^{-240}]}{1 - (1 + 0.03/12)^{-240}}$$

$$\approx \$524.14$$

Technology:

```
(0.03/12)*(100000-
10000*(1+0.03/12)^(-240))/
(1-(1+0.03/12)^(-240))
```

25. $PV = \$10,000$, $i = 0.09/12 = 0.0075$, $n = 4 \times 12 = 48$

$$PMT = PV\frac{i}{1 - (1 + i)^{-n}}$$

$$= 10,000\frac{0.0075}{1 - (1 + 0.0075)^{-48}}$$

$$\approx \$248.85$$

Technology:

```
10000*0.0075/(1-(1+0.0075)^(-48))
```

27. $PV = \$100,000$, $i = 0.05/4$, $n = 4 \times 20 = 80$

$$PMT = PV\frac{i}{1 - (1 + i)^{-n}}$$

$$= 100,000\frac{0.05/4}{1 - (1 + 0.05/4)^{-80}}$$

$$\approx \$1984.65$$

Technology:

```
100000*(0.05/4)/
(1-(1+0.05/4)^(-80))
```

29. The periodic payments are based on a 4.875% annual payment. For payments twice a year, this is

$$PMT = 1000(0.04875/2) = \$24.375$$

Since the bond yield is 4.880%,

$$i = 0.0488/2 = 0.0244$$

$$n = 2 \times 10 = 20$$

The present value comes from the future value of $1000 and the payments, which we treat like an annuity:

$$PV = FV(1 + i)^{-n} + PMT\frac{1 - (1 + i)^{-n}}{i}$$

$$PV = 1000(1 + 0.0244)^{-20}$$

$$+ \; 24.375\frac{1 - (1 + 0.0244)^{-20}}{0.0244}$$

$$= 1000(1.0244)^{-20} + 24.375\frac{1 - 1.0244^{-20}}{0.0244}$$

$$\approx \$999.61$$

Technology:

```
1000*1.0244^(-20)+24.375*(1-
1.0244^(-20))/0.0244
```

Online Time Value of Money Utility

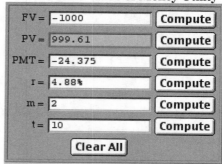

31. The periodic payments are based on a 3.625% annual payment. For payments twice a year, this is

$$PMT = 1000(0.03625/2) = \$18.125$$

Since the bond yield is 3.705%,

$$i = 0.03705/2 = 0.018525$$

$$n = 2 \times 2 = 4$$

The present value comes from the future value of $1000 and the payments, which we treat like an annuity:

$$PV = FV(1 + i)^{-n} + PMT\frac{1 - (1 + i)^{-n}}{i}$$

$$PV = 1000(1 + 0.018525)^{-4}$$

$$+ \; 18.125\frac{1 - (1 + 0.018525)^{-4}}{0.018525}$$

$$= 1000(1.018525^{-4}$$

$$+ \; 18.125\frac{1 - 1.018525^{-4}}{0.018525}$$

$$\approx \$998.47$$

Technology:

```
1000*1.018525^(-4)+
18.125*(1-1.018525^(-
4))/0.018525
```

Online Time Value of Money Utility

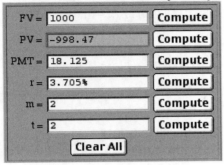

33. FV = maturity value = $1000.

$PMT = 1000(0.035/2) = \$17.50$

PV = Selling price = $994.69

$n = 2 \times 5 = 10$

Using technology, we compute the interest rate to be approximately 3.617%. (The online utility rounds this to two decimal places.)

TI-83/84

Online Time Value of Money Utility

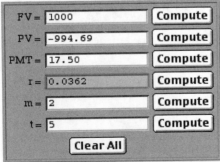

35. FV = maturity value = $1000.

$PMT = 1000(0.03/2) = \$15.00$

PV = Selling price = $998.86

$n = 2 \times 2 = 4$

Using technology, we compute the interest rate to be approximately 3.059%. (The online utility rounds this to two decimal places.)

TI-83/84

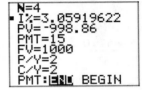

Online Time Value of Money Utility

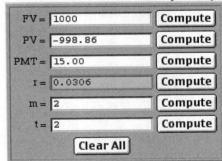

37. This is a two-stage process:

(1) A sinking fund to build the retirement fund.

(2) An annuity depleting the fund during retirement.

For Stage 2, we have

$PMT = \$5000$, $i = 0.03/12 = 0.0025$,

$n = 20 \times 12 = 240$,

and we need to calculate the starting value PV.

$$PV = PMT \frac{1 - (1 + i)^{-n}}{i}$$

$$= 5000 \frac{1 - (1 + 0.0025)^{-240}}{0.0025}$$

$$= 5000 \frac{1 - (1.0025)^{-240}}{0.0025}$$

$$\approx \$901,554.57$$

Technology:
`5000*(1-1.0025^(-240))/0.0025`

Thus, in Stage 1, you need to accumulate $901,554.57 in the sinking fund:

$FV = \$901,554.57$, $i = 0.03/12 = 0.0025$,

$n = 40 \times 12 = 480$,

$$PMT = FV\frac{i}{(1+i)^n - 1}$$
$$= \frac{901,554.57 \times 0.0025}{1.0025^{480} - 1}$$
$$\approx \$973.54 \text{ per month}$$

Technology:
`901554.57*0.0025/(1.0025^480-1)`

39. This is a two-stage process:

(1) A sinking fund to build the retirement fund.

(2) An annuity depleting the fund during retirement.

Stage 1: Building the retirement fund

$PMT = \$1200$, $i = 0.04/4 = 0.01$,

$n = 4 \times 40 = 160$.

$$FV = PMT\frac{(1+i)^n - 1}{i}$$
$$= 1200\frac{1.01^{160} - 1}{0.01}$$
$$\approx \$469,659.16$$

Technology:
`1200*(1.01^160-1)/0.01`

Stage 2: Depleting the fund:

$PV = \$469,659.16$, $i = 0.04/4 = 0.01$,

$n = 4 \times 25 = 100$

$$PMT = PV\frac{i}{1 - (1+i)^{-n}}$$
$$= \frac{469,659.16 \times 0.01}{1 - 1.01^{-100}}$$
$$\approx \$7451.49$$

282

Technology:
`469659.16*0.01/(1-1.01^(-100))`

41. Solid Savings & Loan:

$PV = \$10,000$, $i = 0.09/12 = 0.0075$,

$n = 12 \times 4 = 48$

$$PMT = PV\frac{i}{1 - (1+i)^{-n}}$$
$$= \frac{10,000 \times 0.0075}{1 - 1.0075^{-48}} \approx \$248.85$$

Technology:
`10000*0.0075/(1-1.0075^(-48))`

Fifth Federal Bank & Trust:

$PV = \$10,000$, $i = 0.07/12$, $n = 12 \times 3 = 36$

$$PMT = PV\frac{i}{1 - (1+i)^{-n}}$$
$$= \frac{10,000 \times 0.07/12}{1 - (1+0.07/12)^{-36}} \approx \$308.77$$

Technology:
`10000*(0.07/12)/`
`(1-(1+0.07/12)^(-36))`

Answer: You should take the loan from Solid Savings & Loan: it will have payments of $248.85 per month. The payments on the other loan would be more than $300 per month.

43. We can construct an amortization table using the technique outlined in the Technology Guides. For example, using Excel, we might set it up as follows:

	A	B	C	D	E	F
1	Month	Outstanding Principal	Payment on Principle	Interest Payment		
2	0	$50,000			Rate	8%
3	1	=B2-C3	=F$4-D3	=DOLLAR(B2*F$2/12)	Years	15
4	2				Payment	=DOLLAR(-PMT(F2/12,F3*12,B2))
5	3					
6	4					

Adding the payments on principal (Column C) and interest payments (Column D) for each year will give the following table:

Year	Interest	Payment on Principal
1	$3,934.98	$1,798.98
2	$3,785.69	$1,948.27
3	$3,623.97	$2,109.99
4	$3,448.84	$2,285.12
5	$3,259.19	$2,474.77
6	$3,053.77	$2,680.19
7	$2,831.32	$2,902.64
8	$2,590.39	$3,143.57
9	$2,329.48	$3,404.48
10	$2,046.91	$3,687.05
11	$1,740.88	$3,993.08
12	$1,409.47	$4,324.49
13	$1,050.54	$4,683.42
14	$661.81	$5,072.15
15	$240.84	$5,491.80

45. We use the TI-83/84 Finance features. First, calculate the monthly payments for years 1–5:

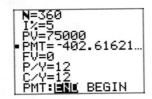

Payments for the first 5 years are therefore $402.62/month.

Now compute the balance at the end of this period (60 months): On the Home Screen, enter

bal(60,2)

to obtain the remaining balance of $68,871.25.

For years 6–30, compute payments on the remaining balance:

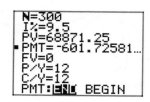

Payments for last 25 years are therefore $601.73/month.

47. We use the TI-83/84 Finance features. First, calculate the monthly payments for the original loan:

The original monthly payments were therefore $824.79.

Now compute the balance at the end of 4 years (48 months): On the Home Screen, enter

```
bal(48,2)
```

to obtain the remaining balance of $93,383.39.

Had you continued with his loan for the full period of 30 years (360 periods) the remaining interest would have been (use the Home Screen again)

```
∑Int(1,360,2)−∑Int(1,48,2)
```

which gives $163,946.49.

You now refinance the outstanding $93,383.39 at 6.875% interest:

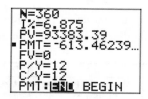

This gives your new monthly payments as $613.46. The total interest you will pay is given by calculating

```
∑Int(1,360,2)
```

giving $127,464.72. The total savings on interest is therefore

$163,946.49 − $127,464.72 = $36,481.77.

49. The payments on the loan, ignoring the fee, are

$$PMT = 5000\frac{0.09/12}{1 - (1 + 0.09/12)^{-24}} = \$228.42$$

Add to each payment 100/24 = $4.17 to get a new payment of $232.59. Using technology or trial-and-error, this monthly payment on a 2-year $5000 loan corresponds to an interest rate of 10.81%.

51.

This gives 153.5/12 ≈ 13 years to retirement.

Online Time Value of Money Utility

FV =	100000	Compute
PV =	0	Compute
PMT =	-500	Compute
I =	4%	Compute
m =	12	Compute
t =	12.7919	Compute

Clear All

53.

TI-83/84

```
■N=55.79763048
  I%=15
  PV=2000
  PMT=-50
  FV=0
  P/Y=12
  C/Y=12
  PMT:END BEGIN
```

This gives $55.798/12 \approx 4.5$ years to repay the debt.

Online Time Value of Money Utility

FV =	0	Compute
PV =	2000	Compute
PMT =	-50	Compute
I =	15%	Compute
m =	12	Compute
t =	4.6498	Compute

Clear All

55. Graph the future value of both accounts using the formula for the future value of a sinking fund.

Your account:

$i = 0.045/12 = 0.00375$, $PMT = \$100$

$$FV = PMT\frac{(1+i)^n - 1}{i}$$

$$= 100\frac{1.00375^n - 1}{0.00375}$$

To graph this, use the technology formula

`100*(1.00375^x-1)/0.00375`

Lucinda's account:

$i = 0.065/12$, $PMT = \$75$.

$$FV = PMT\frac{(1+i)^n - 1}{i}$$

$$= 75\frac{(1+0.065/12)^n - 1}{(0.065/12)}$$

To graph this, use the technology formula

`75*((1+0.065/12)^x-1)/(0.065/12)`

Graph:

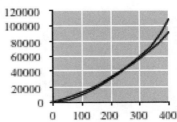

The graphs appear to cross around $n = 300$, so we zoom in there:

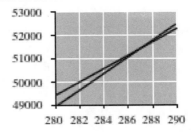

The graphs cross around $n = 287$, so the number of years is approximately $t = 287.5/12 \approx 24$ years.

57. He is wrong because his estimate ignores the interest that will be earned by your annuity—both while it is increasing and while it is decreasing. Your payments will be

285

considerably smaller (depending on the interest earned).

59. He is not correct. For instance, the payments on a $100,000 10-year mortgage 12% are $1434.71, while for a 20-year mortgage at the same rate, they are $1101.09, which is a lot more than half the 10-year mortgage payment.

61. $PV = FV(1 + i)^{-n}$

$$= PMT\frac{(1 + i)^n - 1}{i}(1 + i)^{-n}$$

$$= PMT\frac{1 - (1 + i)^{-n}}{i}$$

Chapter 5 Review Exercises

1. $FV = 6000(1 + 0.0475 \times 5) = \$7425.00$

2. $FV = 10{,}000(1 + 0.0525 \times 2.5) = \$11{,}312.50$

3. $FV = 6000(1 + 0.0475/12)^{60} = \$7604.88$

4. $FV = 10{,}000(1 + 0.0525/2)^5 = \$11{,}383.24$

5. $FV = 100\dfrac{(1 + 0.0475/12)^{60} - 1}{0.0475/12} = \$6757.41$

6. $FV = 2000\dfrac{(1 + 0.0525/2)^5 - 1}{0.0525/2} = \$10{,}538.96$

7. $PV = 6000/(1 + 0.0475 \times 5) = \$4848.48$

8. $PV = 10{,}000/(1 + 0.0525 \times 2.5) = \$8839.78$

9. $PV = 6000(1 + 0.0475/12)^{-60} = \$4733.80$

10. $PV = 10{,}000(1 + 0.0525/2)^{-5} = \$8784.85$

11. $PV = 100\dfrac{1 - (1 + 0.0475/12)^{-60}}{0.0475/12} = \$5331.37$

12. $PV = 2000\dfrac{1 - (1 + 0.0525/2)^{-5}}{0.0525/2} = \$9258.32$

13. $PMT = 12{,}000\dfrac{0.0475/12}{(1 + 0.0475/12)^{60} - 1} = \$177.58$

14. $PMT = 20{,}000\dfrac{0.0525/2}{(1 + 0.0525/2)^5 - 1} = \$3795.44$

15. $PMT = 6000\dfrac{0.0475/12}{1 - (1 + 0.0475/12)^{-60}} = \$112.54$

16. $PMT = 10{,}000\dfrac{0.0525/2}{1 - (1 + 0.0525/2)^{-5}} = \$2160.22$

17. $PMT = 10{,}000\dfrac{0.0475/12}{1 - (1 + 0.0475/12)^{-60}} = \$187.57$

18. $PMT = 15{,}000\dfrac{0.0525/2}{1 - (1 + 0.0525/2)^{-5}} = \$3240.33$

19. Each interest payment is $10{,}000 \times 0.06/2 = \$300$. For an annuity earning 7% and paying $300 every six months for 5 years, the present value is

$$PV = 300\dfrac{1 - (1 + 0.07/2)^{-10}}{0.07/2} = \$2494.98$$

The present value of the $10,000 maturity value is

$$PV = 10{,}000(1 + 0.07/2)^{-10} = \$7089.19$$

The total price is $\$2494.98 + 7089.19 = \$9584.17$

20. Each interest payment is $10{,}000 \times 0.06/2 = \$300$. For an annuity earning 5% and paying $300 every six months for 5 years, the present value is

$$PV = 300\dfrac{1 - (1 + 0.05/2)^{-10}}{0.05/2} = \$2625.62$$

The present value of the $10,000 maturity value is

$$PV = 10{,}000(1 + 0.05/2)^{-10} = \$7811.98$$

The total price is $\$2625.62 + 7811.98 = \$10{,}437.60$

21. 5.346% (using, for example, the TI-83/84 TVM Solver)

22. 4.662% (using, for example, the TI-83/84 TVM Solver)

23. $10,000 = 6000(1 + 0.0475t) = 6000 + 285t$
$t = (10,000 - 6000)/285 = 14.0$ years

24. $15,000 = 10,000(1 + 0.0525t) = 10,000 + 525t$
$t = (15,000 - 10,000)/525 = 9.5$ years

25. $10,000 = 6000(1 + 0.0475/12)^{12t}$
To solve algebraically requires logarithms:
$t = \dfrac{\log(10,000/6000)}{12\log(1 + 0.0475/12)} \approx 10.8$ years
We could also find this using, for example, the TI-83/84 TVM Solver

26. $15,000 = 10,000(1 + 0.0525/2)^{2t}$
To solve algebraically requires logarithms:
$t = \dfrac{\log(15,000/10,000)}{2\log(1 + 0.0525/2)} \approx 7.8$ years
We could also find this using, for example, the TI-83/84 TVM Solver

27. $10,000 = 100\dfrac{(1 + 0.0475/12)^{12t} - 1}{0.0475/12}$
To solve algebraically requires logarithms:
$t = \dfrac{\log[0.0475 \times 10,000/(100 \times 12) + 1]}{12\log(1 + 0.0475/12)}$
≈ 7.0 years
We could also find this using, for example, the TI-83/84 TVM Solver

28. $15,000 = 2000\dfrac{(1 + 0.0525/2)^{2t} - 1}{0.0525/2}$
To solve algebraically requires logarithms:

$t = \dfrac{\log[0.0525 \times 15,000/(2000 \times 2) + 1]}{12\log(1 + 0.0525/2)}$
≈ 3.5 years
We could also find this using, for example, the TI-83/84 TVM Solver

29. Use the compound interest formula:
$FV = PV(1 + t)^n$
where $PV = \$150,000$, $i = 0.20$, $n = 1, 2, 3, 4, 5$
2000: $FV = 150,000(1.20) = \$180,000$
2001: $FV = 150,000(1.20)^2 = \$216,000$
2002: $FV = 150,000(1.20)^3 = \$259,200$
2003: $FV = 150,000(1.20)^4 = \$311,040$
2004: $FV = 150,000(1.20)^5 = \$373,248$

Revenues first surpass \$300,000 in 2003.

30. $PV = \$20,000$, $i = -0.0375$, $n = 7$ (quarters)
$FV = PV(1 + t)^n$
$= 20.000(1 - 0.0375)^7 \approx \$15,305.06$

31. After the first day of trading, the value of each share will be \$6. Thereafter, the shares appreciate in value by 8% per month for 6 months, and you desire a future value of at least \$500,000.
$FV = 500,000$, $i = 0.08$, $n = 6$
$PV = FV(1 + i)^{-n}$
$= 500,000 \times 1.08^{-6} \approx \$315,084.81$
Therefore, since each share will be worth \$6 after the first day, the number of shares you must sell is at least
$\dfrac{315,084.81}{6} \approx 52,514.14$
Since you can offer only a whole number of shares, we must round this up to get the minimum desired future value. Thus, you should offer at least 52,515 shares.

32. Industrial Bank:

$PV = 250,000$, $i = 0.095/12$, $n = 12 \times 10 = 120$

$$PMT = PV\frac{i}{1 - (1 + i)^{-n}}$$

$$= \frac{250,000 \times 0.095/12}{1 - (1+0.095/12)^{-120}} \approx \$3234.94$$

Technology:

```
250000*(0.095/12)/
(1-(1+0.095/12)^(-120))
```

Expansion Loans:

$PV = 250,000$, $i = 0.065/12$, $n = 12 \times 8 = 96$

$$PMT = PV\frac{i}{1 - (1 + i)^{-n}}$$

$$= \frac{250,000 \times 0.065/12}{1 - (1+0.065/12)^{-96}} \approx \$3346.56$$

Technology:

```
250000*(0.065/12)/
(1-(1+0.065/12)^(-96))
```

33. $i = 0.065/12$, $PMT = 3000$, $n = 12 \times 8 = 96$

$$PV = PMT\frac{1 - (1 + i)^{-n}}{i}$$

$$= 3000\frac{1 - (1+0.065/12)^{-96}}{(0.065/12)} \approx \$224,111$$

(to the nearest dollar)

Technology:

```
3000*(1-(1+0.065/12)^(-96))/
(0.065/12)
```

34. $PV = 250,000$, $FV = 0$, $PMT = 3000$, $n = 12 \times 8 = 96$,

and we are seeking the interest rate

TI-83/84

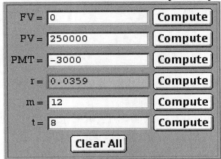

Online Time Value of Money Utility

The interest rate would be 3.59%.

35. $PV = 50,000$, $PMT = 1000 + 800$ (company contribution) $= 1800$, $i = 0.073/12$, $n = 12 \times 10 = 120$. Considering the contribution of the present \$50,000 as well as the payments, we get

$$FV = PV(1 + i)^n + PMT\frac{(1 + i)^n - 1}{i}$$

$$= 50,000(1+0.073/12)^{120}$$

$$+ 1800\frac{(1+0.073/12)^{120}-1}{(0.073/12)}$$

$$\approx \$420,275$$

(to the nearest dollar)

Technology:

```
50000*(1+0.073/12)^120+
1800*((1+0.073/12)^120-1)/
(0.073/12)
```

36. For the company's contribution, take

$PMT = 800$, $i = 0.073/12$, $n = 12\times10 = 120$

$$FV = PMT\frac{(1 + i)^n - 1}{i}$$

$$= 800\frac{(1+0.073/12)^{120}-1}{(0.073/12)} \approx \$140,778$$

(to the nearest dollar)

Technology:

```
800*((1+0.073/12)^120-1)/
(0.073/12)
```

37. We first take out the effect of the initial $50,000:

$PV = 50,000$, $i = 0.073/12$, $n = 12\times10 = 120$

$$FV = PV(1 + i)^n = 50,000(1+0.073/12)^{120}$$

Thus, the payments have to result in a future value of

$$FV = 500,000 - 50,000(1+0.073/12)^{120}$$

$$\approx 396,475.163$$

We can now use the payment formula to determine the necessary payments:

$$PMT = FV\frac{i}{(1 + i)^n - 1}$$

$$= \frac{396,475.163\times0.073/12}{(1+0.073/12)^{120} - 1}$$

$$\approx \$2553.06$$

Since $800 of this is contributed by the company, Callahan's payments should be

$$\$2553.06 - \$800 = \$1453.06$$

38. First compute the amount Callahan needs at the start of retirement:

$PMT = 5000$, $i = 0.087/12$, $n = 12\times30 = 360$

$$PV = PMT\frac{1 - (1 + i)^{-n}}{i}$$

$$= 5000\frac{1 - (1+0.087/12)^{-360}}{(0.087/12)}$$

$$\approx \$638,461.93.$$

Technology:

```
5000*(1-(1+0.087/12)^(-360))/
(0.087/12)
```

In order to accumulate this amount, using the information from Exercise 35, we have:

$PV = 50,000$, $FV = 638,461.93$, $i = 0.073/12$,

$n = 12\times10 = 120$

The initial $50,000 will grow to

$$50,000(1+0.073/12)^{120}$$

so the payments need to result in a future value of only

$$638,461.93 - 50,000(1+0.073/12)^{120}$$

$$\approx 534,937.093$$

Now use the payment formula:

$$PMT = FV\frac{i}{(1 + i)^n - 1}$$

$$= \frac{534,937.093\times0.073/12}{(1+0.073/12)^{120} - 1}$$

$$\approx \$3039.90$$

Since $800 of this is contributed by the company, Callahan's payments should be

$$\$3039.90 - \$800 = \$2239.90$$

39. The bond will pay interest every six months amounting to

$$50,000(0.072/2) = \$1800$$

For someone purchasing this bond after one year, there will be 9 years to maturity, Think of the bond as an investment that will pay the owner $1800 every six months for 9 years, at which time it will pay $50,000. This is exactly the behavior of an annuity paired with an investment with future value $50,000.

$FV = 50,000$, $PMT = 1800$, $i = 0.063/2 = 0.0315$, $n = 2\times9 = 18$

The present value has contributions both from the investment and the annuity:

$$PV = FV(1 + i)^{-n} + PMT\frac{1 - (1 + i)^{-n}}{i}$$

$$= 50{,}000(1.0315)^{-18} + 1800\frac{1 - 1.0315^{-180}}{0.0315}$$

$$\approx \$53{,}055.66$$

Technology:
```
50000*1.0315^(-18)+
1800*(1-1.0315^(-18))/0.0315
```

40. This is similar to Exercise 39, but with $FV = 50{,}000$, $PMT = 1800$, $i = 0.06/2 = 0.03$, $n = 2 \times 8\frac{1}{2} = 17$

$$PV = FV(1 + i)^{-n} + PMT\frac{1 - (1 + i)^{-n}}{i}$$

$$= 50{,}000(1.03)^{-17} + 1800\frac{1 - 1.03^{-170}}{0.03}$$

$$\approx \$53{,}949.84$$

Technology:
```
50000*1.03^(-17)+
1800*(1-1.03^(-17))/0.03
```

41. Here, $FV = 50{,}000$, $PV = 54{,}000$, $PMT = 1800$, $n = 2 \times 8\frac{1}{2} = 17$,

and we are seeking the interest rate.

TI-83/84

Online Time Value of Money Utility

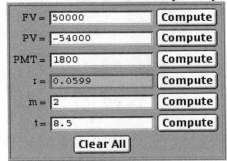

The interest rate would have to be 5.99%.

Chapter 6

6.1

1. The elements of F are the four seasons: spring, summer, fall, winter. Thus,

F = {spring, summer, fall, winter}

3. The elements of I are all the positive integers no greater than 6, namely 1, 2, 3, 4, 5, 6. Thus,

I = {1, 2, 3, 4, 5, 6}

5. $A = \{n \mid n$ is a positive integer and $0 \le n \le 3\}$ = {1, 2, 3} (Note that 0 is not positive, so we exclude it.)

7. $B = \{n \mid n$ is an even positive integer and $0 \le n \le 8\}$ = {2, 4, 6, 8}

9. (a) If the coins are distinguishable,

S = {(H,H), (H,T), (T,H), (T,T)}

(Note that (H, T) and (T, H) are different outcomes, since the first coin is distinguished from the second.)

(b) If the coins are indistinguishable, then (H, T) and (T, H) are the same outcome, and so

S = {(H,H), (H,T), (T,T)}

11. If the dice are distinguishable, then the outcomes can be thought of as ordered pairs. Thus, since the numbers add to 6,

S = {(1,5), (2,4), (3,3), (4,2), (5,1)}

13. If the dice are indistinguishable, then the outcomes are characterized only by the numbers that come up, and not by the order. Thus, since the numbers add to 6,

S = {(1,5), (2,4), (3,3)}

15. Since the numbers facing up can never add to 13 (the largest sum is 12), there are no such outcomes. In other words,

$S = \varnothing$

17.

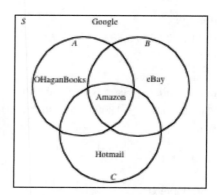

19.

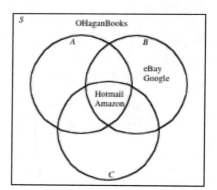

21. $A \cup B$ is the set of all elements that are in A or B (or both): June, Janet, Jill, Justin, Jeffrey, Jello. Thus,

$A \cup B$ = {June, Janet, Jill, Justin, Jeffrey, Jello}
    = A

23. $A \cup \varnothing$ is the set of all elements that are in A or $\varnothing$ (or both). Since $\varnothing$ has no elements,

$A \cup \varnothing = A$

for every set A.

25. $B{\cup}C$ = {Janet, Jello, Justin}${\cup}${Sally, Solly, Molly, Jolly, Jello}

= {Janet, Justin, Sally, Solly, Molly, Jolly, Jello}

Therefore,

$A{\cup}(B{\cup}C)$

= {June, Janet, Jill, Justin, Jeffrey, Jello} ${\cup}$ {Janet, Justin, Sally, Solly, Molly, Jolly, Jello}

= {June, Janet, Jill, Justin, Jeffrey, Jello, Sally, Solly, Molly, Jolly}

27. $C{\cap}B$ is the set of all elements that are common to both C and B. Thus,

$C{\cap}B$ = {Jello}

29. $A{\cap}{\varnothing}$ is the set of all elements that are common to both A and ${\varnothing}$. Since ${\varnothing}$ has no elements,

$A{\cap}{\varnothing} = {\varnothing}$

For every set A.

31. $A{\cap}B$ = {Janet, Jello, Justin}

Therefore,

$(A{\cap}B){\cap}C$

= {Janet, Jello, Justin} ${\cap}$ {Sally, Solly, Molly, Jolly, Jello}

= {Jello}

33. $A{\cap}B$ = {Janet, Jello, Justin}

Therefore,

$(A{\cap}B){\cup}C$

= {Janet, Jello, Justin} ${\cup}$ {Sally, Solly, Molly, Jolly, Jello}

= {Janet, Justin, Jello, Sally, Solly, Molly, Jolly}

35. $A \times C$ is the set of all ordered pairs (a, b) with $a \in A$ and $c \in C$: Therefore,

$A{\times}C$ = {(small, triangle), (small, square), (medium, triangle), (medium, square), (large, triangle), (large, square)}

37. $A \times B$ is the set of all ordered pairs (a, b) with $a \in A$ and $b \in B$: Therefore,

$A{\times}B$ = {(small, blue), (small, green), (medium, blue), (medium, green), (large, blue), (large, green)}

39. To represent $B \times C$ we use the elements of B = {blue, green} for the row headings, and the elements of C = {triangle, square} for the column headings:

	A	B	C
1		Triangle	Square
2	Blue	Blue Triangle	Blue Square
3	Green	Green Triangle	Green Square

41. To represent $A \times B$ we use the elements of A = {small, medium, large} for the row headings, and the elements of B = {blue, green} for the column headings:

	A	B	C
1		Blue	Green
2	Small	Small Blue	Small Green
3	Medium	Medium Blue	Medium Green
4	Large	Large Blue	Large Green

43. If a die is rolled and a coin is tossed, each outcome is a pair (b, a) where b is an outcome when a die is rolled and a is an outcome when a coin is tossed. Thus, the set of outcomes is

$B \times A$ = {1H, 1T, 2H, 2T, 3H, 3T, 4H, 4T, 5H, 5T, 6H, 6T}

45. If a coin is tossed 3 times, each outcome is a triple (a_1, a_2, a_3) where a_1 and a_2 and a_3 are outcomes when a coin is tossed. Thus, the set of outcomes is

$A \times A \times A$ = {HHH, HHT, HTH, HTT, THH, THT, TTH, TTT}

47. *E* is the set of outcomes in which at least one die shows an even number. Thus, *E'* is the set of outcomes in which *both* die show an odd number:

$E' = \{(1,1), (1,3), (1,5), (3,1), (3,3), (3,5), (5,1), (5,3), (5,5)\}$

49. *E∪F* is the set of outcomes in which either at least one die is even, or at least one is odd. Since this includes *all* the outcomes (in every outcome there is either an odd or even outcome), its complement is empty:

$(E∪F)' = \varnothing$

51. By Exercise 47, *E'* is the set of outcomes in which *both* die show an odd number: By Exercise 48, *F'* is the set of outcomes in which both dice show an even number. Thus, *E'∪F'* is the set of outcomes in which either both are even, or both are odd:

$E'∪F' = \{(1,1), (1,3), (1,5), (3,1), (3,3), (3,5),$
$(5,1), (5,3), (5,5), (2,2), (2,4), (2,6), (4,2), (4,4),$
$(4,6), (6,2), (6,4), (6,6)\}$

53. $(A∪B)'$ is the region outside *A∪B*, while $A'∩B'$ is the overlap of the region outside *A* with that outside *B* (the grey region in the diagram below):

$(A∪B)' = A'∩B'$

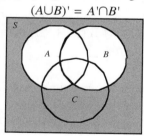

55. $(A∩B)∩C$ is the overlap of *A∩B* and *C*, which is the same as the overlap of all three sets: *A, B, C*. Similarly for *A∩(B∩C)* (the grey region in the diagram below):

$(A∩B)∩C = A∩(B∩C)$

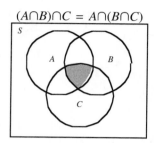

57. *A∪(B∩C)* is the union of *A* with *B∩C*, and so consists of all elements in *A* together with those in both *B* and *C*. This is the same as *(A∪B)∩(A∪C)* (the grey region in the diagram below):

$A∪(B∩C) = (A∪B)∩(A∪C)$

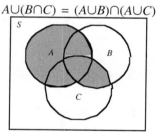

59. $S' = \varnothing$

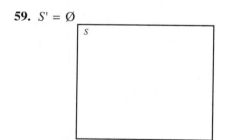

61. The set of clients who owe her money is
$A = \{$Acme, Crafts, Effigy, Global$\}$
The set of clients who have done at least \$10,000 worth of business with her is
$B = \{$Acme, Brothers, Crafts, Dion$\}$
Therefore, the set of clients who owe her money and have done at least \$10,000 worth of business with her is
$A∩B = \{$Acme, Crafts$\}$

63. The set of clients who have done at least $10,000 worth of business with her is

B = {Acme, Brothers, Crafts, Dion}

The set of clients who have employed her in the last year is

C = {Acme, Crafts, Dion, Effigy, Global, Hilbert}

Therefore, the set of clients who have done at least $10,000 worth of business with her or have employed her in the last year is

$B \cup C$ = {Acme, Brothers, Crafts, Dion, Effigy, Global, Hilbert}

65. The set of clients who do not owe her money is

A' = {Brothers, Dion, Floyd, Global, Hilbert}

The set of clients who have employed her in the last year is

C = {Acme, Crafts, Dion, Effigy, Global, Hilbert}

Therefore, the set of clients who do not owe her money and have employed her in the last year is

$A' \cap C$ = {Dion, Hilbert}

67. The clients who owe her money is

A = {Acme, Crafts, Effigy, Global}

The set of clients who have not done at least $10,000 worth of business with her is

B' = {Effigy, Floyd, Global, Hilbert}

The set of clients who have not employed her in the last year is

C' = {Brothers, Floyd}

Therefore, the set of clients who owe her money, have not done at least $10,000 worth of business with her, and have not employed her in the last year is

$A \cap B' \cap C'$ = $\emptyset$

69. One can organize the spreadsheet as follows: For the row headings, use the years 2003, 2004, 2005, and 2006. For the column headings, use Sail Boats, Motor Boats, and Yachts:

◇	A	B	C	D
1		**Sail Boats**	**Motor Boats**	**Yachts**
2	**2003**	(2003 Sail Boats)	(2003 Motor Boats)	(2003 Yachts)
3	**2004**	(2004 Sail Boats)	(2004 Motor Boats)	(2004 Yachts)
4	**2005**	(2005 Sail Boats)	(2005 Motor Boats)	(2005 Yachts)
5	**2006**	(2006 Sail Boats)	(2006 Motor Boats)	(2006 Yachts)

This setup gives a tabular representation of the cartesian product

{2003, 2004, 2005, 2006} × {Sail Boats, Motor Boats, Yachts}

Alternatively, one could use the years for the column headings and {Sail Boats, Motor Boats, Yachts} for the row headings, and obtain a representation of

{Sail Boats, Motor Boats, Yachts} × {2003, 2004, 2005, 2006}

71. The collection of all iPods and jPods combined is the set of all items that are either iPods (that is, in the set I) or jPods (that is, in the set J), and is therefore the union of the two sets: $I \cup J$.

73. Techno music that is neither European nor Dutch:

Techno AND NOT (European OR Dutch)

Techno $\cap$ (European $\cup$ Dutch) $'$

(Choice B).

75. Let A = {1}, B = {2}, and C = {1, 2}. Then

$(A \cap B) \cup C$ = {1, 2}

but

$A \cap (B \cup C)$ = {1}.

In general, $A \cap (B \cup C)$ must be a subset of A, but $(A \cap B) \cup C$ need not be; also, $(A \cap B) \cup C$ must contain C as a subset, but $A \cap (B \cup C)$ need not.

77. A universal set is a set containing all "things" currently under consideration. When discussing sets of positive integers the universe might be the set of all positive integers, or the set of all integers (positive, negative, and 0), or any other set containing the set of all positive integers.

79. $A \cap (B \cup C')$ means A, and either B or not C. Thus, for instance, take A as the set of suppliers who deliver components on time, B as the set of suppliers whose components are known to be of high quality, and C as the set of suppliers who do not promptly replace defective components. Then selecting suppliers in $A \cap (B \cup C')$ means selecting those who deliver components on time and are either companies whose components are of high quality or are suppliers who promptly replace defective components.

81. Let A = movies that are violent, B = movies that are shorter than two hours, C = movies that have a tragic ending, and D = movies that have an unexpected ending. The given sentence can be rewritten as "She prefers movies in $A' \cap B \cap (C \cup D)'$." It can also be rewritten as "She prefers movies in $A' \cap B \cap C' \cap D'$."

6.2

1. $n(A)$ = Number of elements in A = 4, $n(B)$ = Number of elements in B = 5. Therefore,

$n(A) + n(B) = 4 + 5 = 9$.

3. $A \cup B$ = {Dirk, Johan, Frans, Sarie, Tina, Klaas, Henrika}

$n(A \cup B) = 7$

5. $B \cap C$ = {Frans}

$A \cup (B \cap C)$ = {Dirk, Johan, Frans, Sarie}

$n(A \cup (B \cap C)) = 4$

7. From Exercise 3, we know that

$n(A \cup B) = 7$

On the other hand,

$n(A) + n(B) - n(A \cap B) = 4 + 5 - 2 = 7$

Therefore,

$n(A \cup B) = n(A) + n(B) - n(A \cap B)$

9. $n(A \times A) = n(A)n(A) = 2 \times 2 = 4$

11. $n(B \times C) = n(B)n(C) = 6 \times 3 = 18$

13. $n(A \times B \times B) = n(A)n(B \times B) = n(A)n(B)n(B)$
$= 2 \times 6 \times 6 = 72$

15. $n(A \cup B) = n(A) + n(B) - n(A \cap B)$
$= 43 + 20 - 3 = 60$

17. $n(A \cup B) = n(A) + n(B) - n(A \cap B)$
$100 = 60 + 60 - n(A \cap B)$
$n(A \cap B) = 60 + 60 - 100 = 20$

Exercises 19–26 use
S = {BA, MU, SO, SU, LI, MS, WA, RT, DU, LY}
A = {SO, LI, MS, RT}
B = {BA, MU, SO}

19. $n(A') = n(S) - n(A) = 10 - 4 = 6$

21. $n((A \cap B)') = n(S) - n(A \cap B) = 10 - 1 = 9$

23. A' = {BA, MU, SU, WA, DU, LY}
B' = {SU, LI, MS, WA, RT, DU, LY}
$A' \cap B'$ = {SU, WA, DU, LY}
$n(A' \cap B') = 4$

25. $n((A \cap B)') = 9$ from Exercise 21
$n(A') + n(B') - n((A \cup B)') = 6 + 7 - 4 = 9$
Therefore, $n((A \cap B)') = n(A') + n(B') - n((A \cup B)')$.

27. Assign labels to the regions of the diagram where the quantities are unknown:

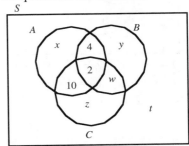

$n(A) = 20 \Rightarrow x + 16 = 20 \Rightarrow \underline{x = 4}$
$n(B \cap C) = 8 \Rightarrow 2 + w = 8 \Rightarrow \underline{w = 6}$
$n(B) = 20 \Rightarrow 6 + y + w = 20$
$\quad \Rightarrow 6 + y + 6 = 20 \Rightarrow \underline{y = 8}$
$n(C) = 28 \Rightarrow 12 + z + w = 28$
$\quad \Rightarrow 12 + z + 6 = 28 \Rightarrow \underline{z = 10}$
$n(S) = 50 \Rightarrow x + y + z + w + t + 16 = 50$
$\quad \Rightarrow 4 + 8 + 10 + 6 + t + 16 = 50 \Rightarrow \underline{t = 6}$

Thus, the completed diagram is

297

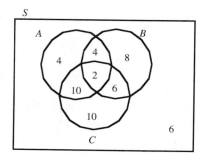

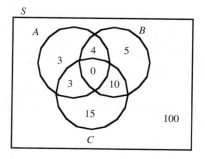

29. Assign labels to the regions of the diagram where the quantities are unknown:

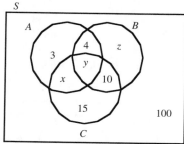

$n(A) = 10 \Rightarrow x + y + 7 = 10 \Rightarrow x + y = 3$
$n(B) = 19 \Rightarrow y + z + 14 = 19 \Rightarrow y + z = 5$
$n(S) = 140 \Rightarrow x + y + z + 132 = 140 \Rightarrow x + y + z = 8$

This is a system of 3 linear equations in 3 unknowns:

$x + y = 3$ (1)
$y + z = 5$ (2)
$x + y + z = 8$ (3)

To solve, we can substitute (1) into (3), giving
$3 + z = 8 \Rightarrow \underline{z = 5}$
(2) now gives $y + 5 = 5 \Rightarrow \underline{y = 0}$
(1) now gives $x + 0 = 3 \Rightarrow \underline{x = 3}$
Thus, the solution is $(x, y, z) = (3, 0, 5)$. The completed diagram is:

31. Let C be the set of web sites containing "costenoble" and let W be the set of web sites containing the word "waner". We are told that
$\quad n(C) = 11,000, \ n(W) = 69,000, \ n(C \cap W) = 4000$
Therefore,
$\quad n(C \cup W) = n(C) + n(W) - n(C \cap W)$
$\quad\quad = 11,000 + 69,000 - 4000 = 76000$

33. Let B be the set of people who had black hair, and let R be the set of people who had a whole row to themselves. We are told that $n(B \cup R) = 37$, $n(B) = 33$, and $n(R) = 6$. We are asked to find $n(B \cap R)$.
$\quad n(B \cup R) = n(B) + n(R) - n(B \cap R)$
$\quad 37 = 33 + 6 - n(B \cap R)$
$\quad n(B \cap R) = 33 + 6 - 37 = 2$

35. $C \cap N$ is the set of authors who are both successful and new.
$C \cup N$ is the set of authors who are either successful or new (or both).
$n(C) = 30$, $n(N) = 20$, $n(C \cap N) = 5$, $n(C \cup N) = 45$
$\quad n(C \cup N) = n(C) + n(N) - n(C \cap N)$
$\quad 45 = 30 + 20 - 5 \ \checkmark$

37. $C \cap N'$ is the set of authors who are successful but not new—in other words, the set of authors who are successful and established.

$n(C \cap N') = 25$

39. Of the 80 established authors, 25 are successful. Thus, the percentage of established authors who are successful is

$\frac{25}{80} = 0.3125$, or 31.25%

Of the 30 successful authors, 25 are established. Thus, the percentage of successful authors who are established is

$\frac{25}{30} \approx 0.8333$, or 83.33%

41. The set of dollars spent on new boats in 2001 is $N \cap C$; $n(N \cap C) = 8$ billion.

43. The set of dollars spent in 2001 excluding new boats is $C \cap N'$.

$n(C \cap N') = 8 + 5 = 13$ billion

45. The set of dollars spent in 1999 on new and used boats is $A \cap (N \cup U)$.

$n(A \cap (N \cup U)) = 7 + 7 = 14$ billion

47. The set of non-Internet stocks that increased is $V \cap I'$.

	P	E	I	Total
V	10	5	15	30
N	30	0	10	40
D	10	5	15	30
Total	50	10	40	100

$n(V \cap I') = 10 + 5 = 15$

49. $n(P' \cup N)$ is the number of stocks that were either not pharmaceutical stocks, or were unchanged in value after a year (or both).

	P	E	I	Total
V	10	5	15	30
N	30	0	10	40
D	10	5	15	30
Total	50	10	40	100

$n(P' \cup N) = 5 + 15 + 30 + 10 + 5 + 15 = 80$

51. $n(V \cap I) = 15$; $n(I) = 40$

	P	E	I	Total
V	10	5	15	30
N	30	0	10	40
D	10	5	15	30
Total	50	10	40	100

$\frac{n(V \cap I)}{n(I)} = \frac{15}{40} = \frac{3}{8}$

This is the fraction of Internet stocks that increased in value.

53. Let S be the set of all children in the study, let R be the set of children who had rickets and let U be the set of urban children. We are told that

$n(S) = 1556$, $n(R) = 1024$, $n(U) = 243$, $n(U \cap R) = 93$.

Represent the given information in a Venn diagram:

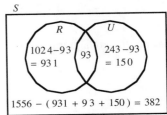

(a) From the diagram, $n(R \cap U') = 931$.
(b) From the diagram, $n(U' \cap R') = 382$.

55. (a) Following is a Venn diagram with most of the unknowns marked with labels:

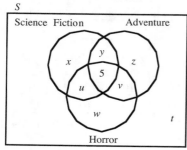

15 had seen a science fiction movie and a horror movie $\Rightarrow 5 + u = 15 \Rightarrow \underline{u = 10}$

5 had seen an adventure movie and a horror movie $\Rightarrow 5 + v = 5 \Rightarrow \underline{v = 0}$

25 had seen a science fiction movie and an adventure movie $\Rightarrow 5 + y = 25 \Rightarrow \underline{y = 20}$

35 had seen a horror movie 35 had seen a horror movie $\Rightarrow 5 + u + v + w = 35$

$\Rightarrow 5 + 10 + 0 + w = 35 \Rightarrow \underline{w = 20}$

55 had seen an adventure movie $\Rightarrow 5 + y + z + v = 55$

$\Rightarrow 5 + 20 + z + 0 = 55 \Rightarrow \underline{z = 30}$

40 had seen a science fiction movie $\Rightarrow 5 + x + y + u = 40$

$\Rightarrow 5 + x + 20 + 10 = 40 \Rightarrow \underline{x = 5}$

Finally, the total number of people in the survey was 100

$\Rightarrow 5 + x + y + z + u + v + w + t = 100$

$\Rightarrow 5 + 5 + 20 + 30 + 10 + 0 + 20 + t = 100$

$\Rightarrow \underline{t = 10}$

Here is the completed diagram:

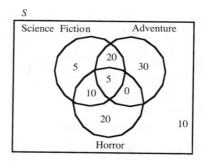

(b) Of the 40 people who had seen science fiction, 15 saw a horror movie. Therefore, the percentage of science fiction movie fans who are also horror movie fans can be estimated as

$\dfrac{15}{40} = 0.375$, or 37.5%.

57. Let R be the set of students who liked rock music, and let C be the set of those who liked classical music.

$n(R) = 22$

$n(R \cap C) = 5$

We are also given other information that we do not need.

Since 5 of the 22 people who liked rock also liked classical, the other 17 did not like classical. Therefore, 17 of those that enjoyed rock did not enjoy classical music

59. The number of elements in the Cartesian product of two finite sets is the product of the number of elements in the two sets.

61. Answers will vary

63. Since

$n(A \cup B) = n(A) + n(B) - n(A \cap B)$,

$n(A \cup B) \neq n(A) + n(B)$ when $n(A \cap B) \neq 0$; that is, when $A \cap B \neq \emptyset$.

65. Since $A \subseteq A \cup B$, the only way they can have the same number of elements is if $A = A \cup B$; that is, when $B \subseteq A$.

67. $n(A \cup B \cup C) =$
$n(A) + n(B) + n(C) - n(A \cap B) - n(B \cap C) - n(A \cap C) + n(A \cap B \cap C)$

6.3

1. Alternative 1: 2 outcomes
  Alternative 2: 3 outcomes
  Alternative 3: 5 outcomes
Total number of outcomes: $2+3+5 = 10$

3. Step 1: 2 outcomes
  Step 2: 3 outcomes
  Step 3: 5 outcomes
Total number of outcomes: $2\times3\times5 = 30$

5. Alternative 1: $1\times2 = 2$ outcomes
  Alternative 2: $2\times2 = 4$ outcomes
Total number of outcomes: $2 + 4 = 6$

7. Step 1: $1+2 = 3$ outcomes
  Step 2: $2+2+1 = 5$ outcomes
Total number of outcomes: $3 \times 5 = 15$

9. Alternative 1: $(3+1)\times2 = 8$ outcomes
  Alternative 2: 5 outcomes
Total number of outcomes: $8 + 5 = 13$

11. Step 1: $(3\times1)+2 = 6$ outcomes
  Step 2: 5 outcomes
Total number of outcomes: $6 \times 5 = 30$

13. Decision algorithm: start with 4 empty slots, and select slots in which to place the letters.
  Step 1: Select a slot to place the b: 4 choices
  Step 2: Places the a's in the remaining slots: 1 choice
Total number of outcomes: $4\times1 = 4$

15. Decision algorithm:
  Step 1: Select a flavor: 31 choices
  Step 2: Select cone, cup, or sundae: 3 choices
Total number of outcomes: $31\times3 = 93$

17. Decision algorithm:
  Step 1: Select the first bit: 2 choices
  Step 2: Select the second bit: 2 choices
  Step 3: Select the third bit: 2 choices
  Step 4: Select the fourth bit: 2 choices
Total number of outcomes: $2\times2\times2\times2 = 16$

19. Decision algorithm: start with 6 empty slots, and select slots in which to place the ternary digits.
  Step 1: Select a slot to place the 1: 6 choices
  Step 2: Select a slot to place the 2: 5 choices
  Step 3: Place 0s in the remaining slots: 1 choice
Total number of outcomes: $6\times5\times1 = 30$

21. Alternative 1 (Gummy candy):
    Step 1: Select a size: 3 choices
    Step 2: Select a shape: 3 choices
  Alternative 2 (Licorice nibs):
    Step 1: Select a size: 2 choices
    Step 2: Select a color: 2 choices
Total number of outcomes: $3\times3 + 2\times2 = 13$

23. Alternative 1: Single disc:
    Step 1: Select a capacity: 2 choices
    Step 2: Select a color: 5 choices
  Alternative 2: Box of discs:
    Step 1: Select box size: 2 choices
    Step 2: Select a capacity: 2 choices
    Step 3: Select a color: 2 choices
Total number of choices: $(2\times5)+(2\times2\times2) = 18$

25. Step 1: Answer 1st t/f question: 2 choices
  Step 2: Answer 2nd t/f question: 2 choices
  . . . .
  Step 10: Answer 10th t/f question: 2 choices
  Step 11: Answer 1st multiple choice question: 5 choices

Step 12: Answer 2nd multiple choice question: 5 choices

Total number of choices: $(2 \times 2 \times ... \times 2) \times (5 \times 5) = 2^{10} \times 5^2 = 25{,}600$

27. Alternative 1: Do Part A:

Steps 1–8: Answer the 8 t/f questions; 2 choices each, giving 2^8 possible choices

Alternative 2: Answer Part B:

Steps 1–5: Answer 5 multiple choice questions with 5 choices each, giving 5^5 possible choices

Total number of choices: $2^8 + 5^5 = 3381$

29.

(a) Step 1: Select a mutual fund: 4 choices

Step 2: Select a muni. bond fund: 3 choices

Step 3: Select a stock: 8 choices

Step 4: Select a precious metal: 3 choices

Total number of choices: $4 \times 3 \times 8 \times 3 = 288$

(b) Step 1: Select 3 mutual funds: 4 choices (select which one to leave out)

Step 2: Select 2 muni bond funds: 3 choices (select which one to leave out)

Step 3: Select one stock: 8 choices

Step 4: Select 2 precious metals 3 choices (select which one to leave out)

Total number of choices: $4 \times 3 \times 8 \times 3 = 288$

31. The number of possible characters that can be represented equals the number of possible bytes:

Steps 1–8: Select 0 or 1 for each bit: $2 \times 2 \times ... \times 2 = 2^8$ choices

Thus, the number of possible characters is $2^8 = 256$.

33. Decision algorithm to obtain a symmetry of the five-pointed star:

Alternative 1: Pure rotation: 5 choices

Alternative 2: Rotation followed by a flip: 5 choices

Total number of symmetries: $5 + 5 = 10$

35. Alternative 1: Letter only: 26 choices

Alternative 2: Letter plus digit

Step 1: Choose a letter: 26 choices

Step 2: Choose a digit: 10 choices

Total number of variables $= 26 + 26 \times 10 = 286$

37. Step 1: Select a winner of the North Carolina-Central Connecticut game: 2 choices

Step 2: Select a winner of the Virginia-Syracuse game: 1 choice (We know that Syracuse is the winner)

Step 3: Select the overall winner: 2 choices

Total number of choices: $2 \times 1 \times 2 = 4$

39.

(a) Step 1: Choose the first digit: 8 choices

Steps 2–7 Choose the remaining 6 digits: 10^6 choices

Total number: $8 \times 10^6 = 8{,}000{,}000$

(b) To count the numbers beginning with 463, 460, or 400, use the following decision algorithm:

Alternative 1: Start with 463

Steps 1–4: Choose the remaining 4 digits: 10^4 choices

Alternative 2: Start with 460

Steps 1–4: Choose the remaining 4 digits: 10^4 choices

Alternative 3: Start with 400

Steps 1–4: Choose the remaining 4 digits: 10^4 choices

Total number of choices: $10^4 + 10^4 + 10^4 = 30{,}000$

(c) Use the following decision algorithm:

Step 1: Select the 1st digit (other than 0 or 1): 8 choices

Step 2: Select a 2nd digit different from the 1st digit: 9 choices

Step 2: Select a 3rd digit different from the 2nd digit: 9 choices

. . .

Step 7: Select a 7th digit different from the 6th digit: 9 choices

Total number of choices: $8 \times 9^6 = 4{,}251{,}528$

41. (a) Steps 1–3: Choose three bases (4 choices each): $4^3 = 64$ choices

(b) Steps 1–3: Choose n bases (4 choices each): 4^n choices

(c) From part (b) with $n - 2.1 \times 10^{10}$, one obtains a total of
$$4^n = 4^{2.1 \times 10^{10}} \text{ possible DNA chains}$$

43. (a) Steps 1—6: Select 6 hexadecimal digits (16 choices per digit): $16^6 = 16{,}777{,}216$ possible colors

(b) Step 1: Choose the 1st repeating pair of digits: 16 choices

Step 2: Choose the 2nd repeating pair of digits: 16 choices

Step 3: Choose the 3rd repeating pair of digits: 16 choices

Total number of colors: $16^3 = 4096$

(c) Step 1: Choose the digit x: 16 choices

Step 2: Choose the digit y: 16 choices

Total number of grayscale shades: $16^2 = 256$

(d) Step 1: Choose the position for the sequence xy: 3 choices

Step 2: Choose the values of x and y: 16^2 choices

Total number of choices: 3×16^2 choices. However, the above decision algorithm leads to the sequence

000000 in three ways. The number of pure colors is therefore $3 \times 16^2 - 2 = 766$.

45. Step 1: Choose a male actor to play Escalus: 10 choices

Step 2: Choose a male actor to play Paris: 9 choices

...

Step 6: Choose a male actor to play Tybalt: 5 choices

Step 7: Choose a male actor to play Friar Lawrence: 4 choices

Step 8: Choose a female actor to play Lady Montague: 8 choices

Step 9: Choose a female actor to play Lady Capulet: 7 choices

Step 10: Choose a female actor to play Juliet: 6 choices

Step 11: Choose a female actor to play Juliet's Nurse: 5 choices

Total number of casts:
$(10 \times 9 \times 8 \times 7 \times 6 \times 5 \times 4) \times (8 \times 7 \times 6 \times 5) =$
$1{,}016{,}064{,}000$

47.

(a) Steps 1–3 Choose 3 letters: 26^3 choices

Steps 4–6: Choose 3 digits: 10^3 choices

Total number of license plates: $26^3 \times 10^3 = 17{,}576{,}000$

(b) Steps 1–2 Choose 2 letters: 26^2 choices

Step 3: Choose a letter other than I, O or Q: 23 choices

Steps 4–6: Choose 3 digits: 10^3 choices

Total number of license plates: $26^2 \times 23 \times 10^3 = 15{,}548{,}000$

(c) A decision algorithm for the number of reserved plates:

Step 1: Choose VET, MDZ, or DPZ: 3 choices

Step 2–4: Choose 3 digits: 10^3 choices

Total number of reserved plates: 3×10^3

From part (b), the total number of possible plates is 15,548,000. Therefore, the number of unreserved plates is

$$15,548,000 - 3 \times 10^3 = 15,545,000$$

49.

(a) Start with 4 empty slots in which to place the letters R and D.

Step 1: Select a slot for the single D: 4 choices

Step 2: Place Rs in the remaining slots: 1 choice

Total number of sequences: $4 \times 1 = 4$

(b) Each route in the maze is a sequence of 4 moves: either right (R) or down (D). Since only one down move is possible, the number of such routes equals the number of four-letter sequences that contain only the letters R and D, with D occurring only once: 4 possibilities, by part (a)

(c) If we allowed left and/or up moves, we could get an unlimited number of routes, such as

RRRD, RLRRRD, RLRLRRRD, ...

51.

(a) Step 1: Choose a 1st cylinder: 6 choices

Step 2: Choose a 2nd cylinder on the opposite side: 3 choices

Step 3: Choose a 3rd cylinder on the same side as the first: 2 choices

Step 4: Chose a 4th cylinder on the same side as the second: 2 choices

Step 5: Chose a 5th cylinder on the same side as the first: 1 choice

Step 6: Chose a 6th cylinder on the same side as the second: 1 choice

Total number of firing sequences:

$$6 \times 3 \times 2 \times 2 \times 1 \times 1 = 72$$

(b) Use the same decision algorithm as for part (a), except that, in Step 1, we only have 3 choices (the cylinders on the left).

Total number of firing sequences:

$$3 \times 3 \times 2 \times 2 \times 1 \times 1 = 36$$

53. Decision algorithm for producing a painting:

Steps 1–5: Select blue or grey for each of the odd-numbered lines: 2^5 choices

Step 6: Select a single color for the remaining lines: 3 choices

Total number of paintings: $2^5 \times 3 = 96$

55. (a) Decision algorithm for incorrect codes that will open the lock:

Step 1: Decide which digit will be the wrong one: 4 choices

Step 2: Choose an incorrect digit in that place: 9 choices (one is correct, leaving 9 incorrect ones)

Total number of incorrect codes that will open the lock: $4 \times 9 = 36$

(b) From part (a), there are 36 incorrect codes that will open the lock. Also, there is only 1 correct code. Therefore, the total number of codes that will open the lock is

$$36 + 1 = 40$$

57. Decision algorithm for creating a calendar:

Step 1: Choose a day of the week for Jan 1: 7 choices

Step 2: Decide whether or not it is a leap year: 2 choices

Total number of possible calendars: $7 \times 2 = 14$

59. Decision algorithm to select a particular iteration:

Step 1: Choose a value for i: 10 choices

Step 2: Choose a value for j: 19 choices
    (19 integers in the range $2-20$)

Step 3: Choose a value for k: 10 choices

Total number of iterations: $10 \times 19 \times 10 = 1900$

61. Decision algorithm to place a single 1×1 block:

Step 1: Choose a position in the left-right direction: m choices

Step 2: Choose a position in the front-back direction: n choices

Step 3: Choose a position in the up-down direction: r choices

Total number of possible solids: mnr

63. The number of possible sequences of length 1 is 2 (either a dot: $\cdot$ or a dash: $-$)

The number of possible sequences of length 2 is $2 \times 2 = 4$ (2 steps, 2 choices per step):

$\cdot\cdot$ $\cdot-$ $-\cdot$ $--$

The number of possible sequences of length 3 is 2^3 = 8 (3 steps, 2 choices per step)

The number of possible sequences of length 4 is 2^4 = 16 (3 steps, 2 choices per step)

So far, using different lengths from 1 to 4, we can encode $1 + 2 + 4 + 8 + 16 = 31$ possible letters, which is enough to include the whole alphabet. (Using lengths 1–3 will not work, sine that will give only $1 + 2 + 4 + 8 = 15$ possible letters). Thus, we need to use sequences of up to 4 dots and dashes.

65. The multiplication principle is based on the cardinality of the <u>Cartesian product</u> of two sets.

67. The decision algorithm produces every pair of shirts twice, first in one order and then in the other. Therefore, it is not valid. The actual number of pairs of shirts is half of the number computed by the algorithm: $90/2 = 45$.

69. Think of placing the five squares in a row of five empty slots. Step1: choose a slot for the blue square 5 choices. Step 2: choose a slot for the green square: 4 choices. Step 3: choose the remaining 3 slots for the yellow squares, 1 choice. Hence there are 20 possible five-square sequences.

6.4

1. $6! = 6 \times 5 \times 4 \times 3 \times 2 \times 1 = 720$

3. $\dfrac{8!}{6!} = \dfrac{8 \times 7 \times \cancel{6 \times 5 \times 4 \times 3 \times 2 \times 1}}{\cancel{6 \times 5 \times 4 \times 3 \times 2 \times 1}} = 56$

5. $P(6, 4) = 6 \times 5 \times 4 \times 3 = 360$

7. $\dfrac{P(6, 4)}{4!} = \dfrac{6 \times 5 \times 4 \times 3}{4 \times 3 \times 2 \times 1} = 15$

9. $C(3, 2) = \dfrac{3 \times 2}{2 \times 1} = 3$

Alternatively,

$\quad C(3, 2) = C(3, 3-2) = C(3, 1) = 3$

11. $C(10, 8) = C(10, 2) \quad 10-8 = 2$

$\qquad = \dfrac{10 \times 9}{2 \times 1} = 45$

13. $C(20, 1) = \dfrac{20}{1} = 20$

15. $C(100, 98) = C(100, 2) \qquad 100-98 = 2$

$\qquad = \dfrac{100 \times 99}{2 \times 1} = 4950$

17. The number of ordered lists of 4 items chosen from 6 is

$\quad P(6, 4) = 6 \times 5 \times 4 \times 3 = 360$

19. The number of unordered sets of 3 objects chosen from 7 is

$\quad C(7, 3) = \dfrac{7 \times 6 \times 5}{3 \times 2 \times 1} = 35$

21. Each 5-letter sequence containing the letters b, o, g, e, y is a list of 5 letters chosen from the above 5 (or a permutation of the 5 letters) and the number of these is

$\quad P(5, 5) = 5! = 5 \times 4 \times 3 \times 2 \times 1 = 120$

23. Each 3-letter sequence is a list of 3 letters chosen from the given 6, and so the number of 3-letter sequences is

$\quad P(6, 3) = 6 \times 5 \times 4 = 120$

25. Each 3-letter unordered set is a list of 3 letters chosen from the given 6, and so the number of 3-letter sets is

$\quad C(6, 3) = \dfrac{6 \times 5 \times 4}{3 \times 2 \times 1} = 20$

27. Since there are repeated letters, we use a decision algorithm to construct such a sequence. Start with 6 empty slots.

Step 1: Choose a slot for the k: 6 choices

Step 2: Choose 3 slots from the remaining 5 for the a's: $C(5, 3)$ choices

Step 3: Choose 2 slots from the remaining 2 for the u's: $C(2, 2)$ slots

Total number of sequences:

$\quad 6 \times C(5,3) \times C(2,2) = 60$

29. There are a total of 10 marbles in the bag. The number of sets of 4 chosen from 10 marbles is

$\quad C(10, 4) = 210$

31. Decision algorithm for assembling a collection of 4 marbles that includes all the red ones:

Step 1: Choose 3 red marbles: $C(3, 3) = 1$ choice

Step 2: Choose 1 non-red marble: $C(7, 1) = 7$ choices

(There are 7 non-red marbles to choose from.)

Total number of sets $= C(3, 3)C(7, 1) = 7$

33. Decision algorithm for assembling a collection of 4 marbles that includes no red ones:

    Step 1: Choose 4 non-red marble: $C(7, 4) = 35$ choices

       (There are 7 non-red marbles to choose from.)

Total number of sets = $C(7, 4) = 35$

35. Decision algorithm for assembling a collection as specified:

    Step 1: Choose 1 red marble: 3 choices

    Step 2: Choose 1 green marble: 2 choices

    Step 3: Choose 1 yellow marble: 2 choices

    Step 4: Choose 1 orange marble: 2 choices

Total number of sets: $3 \times 2 \times 2 \times 2 = 24$

37. A set containing at least 2 red marbles must contain either 2 or 3 red marbles.

    Alternative 1: Decide on 2 red marbles

        Step 1: Choose 2 red marbles: $C(3, 2)$ choices

        Step 2: Choose 3 non-red marbles: $C(7, 3)$ choices

    Alternative 3: Decide on 3 red marbles

        Step 1: Choose 3 red marbles: $C(3, 3)$ choices

        Step 2: Choose 2 non-red marbles: $C(7, 2)$ choices

Total number of choices:

    $C(3, 2)C(7, 3) + C(3, 3)C(7, 2) = 126$

39. A set containing at most 1 yellow marbles must contain either 0 or 1 yellow marbles.

    Alternative 1: Decide on 0 yellow marbles

        Step 1: Choose 5 non-yellow marbles: $C(8, 5)$ choices

    Alternative 2: Decide on 1 yellow marble

        Step 1: Choose 1 yellow marble: $C(2, 1)$ choices

    Step 2: Choose 4 non-yellow marbles: $C(8, 4)$ choices

Total number of choices:

    $C(8,5) + C(2,1)C(8,4) = 196$

41. Decision algorithm for assembling a collection as specified:

    Alternative 1: Use the lavender marble but no yellow ones

        Step 1: Select the lavender marble: $C(1, 1) = 1$ choice

        Step 2: Select 4 non-lavender non-yellow marbles: $C(7, 4)$ choices

    Alternative 2: Use a yellow marble but no lavender ones

        Step 1: Select 1 yellow marble: $C(2, 1)$ choices

        Step 2: Select 4 non-lavender non-yellow marbles: $C(7, 4)$ choices

Total number of sets:

    $C(1, 1)C(7, 4) + C(2, 1)C(7, 4) = 105$

43. Think of the outcome of a sequence of 30 dice throws as a sequence of "words" of length 30 using the "letters" 1, 2, 3, ..., 6. Using this interpretation, a sequence with 5 ones is a 30-letter word containing 5 ones. For the decision algorithm, start with 30 empty slots and choose numbers to fill the slots:

    Step 1: Choose 5 slots for the ones: $C(30, 5)$ choices

    Steps 2–26: Choose one of 2, 3, 4, 5, 6 to fill each of the remaining 25 slots: $5 \times 5 \times ... \times 5 = 5^{25}$ choices

Total number of sequences with 5 ones: $C(30, 5) \times 5^{25}$

Since there are 6^{30} different sequences possible, the fraction of sequences with 5 ones is

$$\frac{C(30, 5) \times 5^{25}}{6^{30}} = 0.192$$

45. For the decision algorithm, start with 30 empty slots and choose numbers to fill the slots:

  Step 1: Choose 15 slots for the even numbers: $C(30, 15)$ choices

  Next 15 steps: Choose 2, 4, or 6 for each of these 15 slots: 3^{15} choices

  Next 15 steps: Choose 1, 3, or 5 for each of the remaining 15 slots: 3^{15} choices

Since there are 6^{30} different sequences possible, the fraction of sequences with exactly 15 even numbers is

$$\frac{C(30, 15) \times 3^{15} \times 3^{15}}{6^{30}} = 0.144$$

47. Each itinerary is a list of 4 venues chosen from 4. Thus, the number of itineraries is

$$P(4, 4) = 4! = 4 \times 3 \times 2 \times 1 = 24$$

49. Decision algorithm for constructing a two pair hand:

  Step 1: Select 2 denominations for the pairs: $C(13, 2)$ choices

  Step 2: Select two cards of the lowest-ranked denomination above: $C(4, 2)$ choices

  Step 3: Select two cards of the other denomination above: $C(4, 2)$ choices

  Step 4: Select a single card that belongs to neither of the two denominations: $C(44, 1) = 44$ choices

  $(52 - 8 = 44)$

Total number of two pair hands:

  $C(13, 2)C(4, 2)C(4, 2)C(44, 1) = 123{,}552$

51. Decision algorithm for constructing a two of a kind hand:

  Step 1: Select a denomination for the two cards: $C(13, 1) = 13$ choices

  Step 2: Select 2 cards of the above denomination: $C(4, 2)$ choices

  Step 3: Select 3 denominations for the singles: $C(12, 3)$ choices

  Step 4: Select 1 card of the lowest-ranked denomination above: $C(4, 1) = 4$ choices

  Step 5: Select 1 card of the next-ranked denomination: $C(4, 1) = 4$ choices

  Step 6: Select 1 card of the highest-ranked: $C(4, 1) = 4$ choices

Total number of two of a kind hands:

  $13 \times C(4, 2)C(12, 3) \times 4 \times 4 \times 4 = 1{,}098{,}240$

53. A straight is a run of 5 cards of consecutive denominations: A, 2, 3, 4, 5 up through 10, J, Q, K, A, but not all of the same suit. If we ignore the restriction about the suits we get the following decision algorithm:

  Step 1: Select a starting denomination for the straight: (A, 2, ..., 10): 10 choices

  Steps 2–6: Select a card of each of the above denominations: $4 \times 4 \times 4 \times 4 \times 4 = 4^{5}$ choices

Total number of runs of 5 cards: 10×4^{5}

However, these runs include hands in which all 5 cards are of the same suit (straight flushes). There are 10×4 of these (Step 1: select a starting card; Step 2: Select a single suit). Excluding the straight flushes gives

Total number of straights:

  $10 \times 4^{5} - 10 \times 4 = 10{,}200$

55. (a) Since a portfolio consists of a set of 5 stocks chosen from the 10, the number of possible portfolios is $C(10, 5) = 252$ portfolios.

(b) Decision algorithm for assembling a portfolio as per part (b):

Step 1: Choose MO and GM: $C(2, 2) = 1$ choice

Step 2: Choose 3 more stocks from 6 (MO, GM, GE, DD excluded): $C(6, 3) = 20$

Total number of choices: $1 \times 20 = 20$ portfolios

(c) Five of the listed stocks have yields above 4%. Decision algorithm for assembling a portfolio as per part (c):

Alternative 1: Exactly 4 stocks have yields above 4%:

Step 1: Choose 4 stocks with yields above 4%: $C(5, 4) = 5$ choices

Step 2: Choose 1 stock with a yield not above 4%: $C(5, 1) = 5$ choices

Total number of choices for Alternative 1: $5 \times 5 = 25$ choices

Alternative 2: All 5 stocks have yields above 4%:

Step 1: Choose 5 stocks with yields above 4%: $C(5, 5) = 1$ choice

Total number of choices for Alternative 2: 1 choice

Total number of portfolios = $25 + 1 = 26$

57. (a) Decision algorithm for assembling a collection of stocks:

Step 1: Choose 3 tech stocks: $C(6, 3) = 20$ choices

Step 2: Choose 2 non-tech stocks: $C(6, 2) = 15$ choices

Total number of choices: $20 \times 15 = 300$ collections

(b) Decision algorithm for assembling a collection of stocks that declined in value (there were 3 stocks in each category that declined):

Step 1: Choose 3 declining tech stocks: $C(3, 3) = 1$ choice

Step 2: Choose 2 declining non-tech stocks: $C(3, 2) = 3$ possible choices

Total number of choices: $1 \times 3 = 3$ collections

(c) Since only 3 collections of out a possible 300 consist entirely of stocks that declined in value, the chances of selecting one of those collections is 3 in 300; that is, 1 in 100, or .01.

59. (a) The number of groups of 4 chosen from 10 is $C(10, 4) = 210$

(b) Decision algorithm for assembling a group of 4 movies that satisfies the Lara twins:

Alternative 1: Exactly one of "Alien Vs. Predator" or "Sky Captain"

Step 1: Choose 1 out of "Alien Vs. Predator" and "Sky Captain" : $C(2, 1) = 2$ choices

Step 2: Choose 3 from the remaining 7 ("Shall we Dance" is excluded): $C(7, 3) = 35$ choices

Total number of groups in Alternative 1 is therefore $2 \times 35 = 70$

Alternative 2: Both "Alien Vs. Predator" and "Sky Captain"

Step 1: Choose 2 out of "Alien Vs. Predator" and "Sky Captain" : $C(2, 1) = 2$ choices

Step 2: Choose 2 from the remaining 7 ("Alien Vs. Predator", "Sky Captain", and "Shall we Dance" excluded) $C(7, 2) = 21$ choices

Total number of groups in Alternative 2 is therefore $2 \times 21 = 42$

Total number of groups that make the Lara twins happy is therefore $35 + 42 = 77$

(c) Since 77 of the 210 groups (less than half) make the Lara twins happy, they are less likely than not to be satisfied with a random selection.

61.

(a) Each itinerary is a permutation of the 23 listed cities: 23! of them altogether

(b) Once the first five stops are determined, a decision algorithm for completing the itinerary is:

Steps 1–18: Select a different city from the remaining 18 for each remaining stop: 18×17×...×2×1 choices

Total number of itineraries: 18×17×...×2×1 = 18!

(c) Decision algorithm for constructing an itinerary of the required type (think of making an itinerary as filling a sequence of 23 empty slots with different cities):

Step 1: Choose a starting point in the itinerary to insert the sequence of 5 named cities: 1 through 19: 19 choices

Next 18 steps: Select a different city from the remaining 18 for each remaining slot: 18×17×...×2×1 choices

Total number of itineraries: 19×18!

63. Start with 11 empty slots. Decision algorithm:

Step 1: Select a slot for the m: $C(11, 1) = 11$ choices

Step 2: Select 4 slots for the i's: $C(10, 4)$ choices

Step 3: Select 6 slots for the s's: $C(6, 4)$ choices

Step 4: Select 2 slots for the p's: $C(2, 2) = 1$ choice

Total number of sequences:

$C(11, 1)C(10, 4)C(6, 4)C(2, 2)$

65. Start with 11 empty slots. Decision algorithm:

Step 1: Select 2 slots for the m's: $C(11, 2)$ choices

Step 2: Select 1 slot for the e: $C(9, 1) = 9$ choices

Step 3: Select 1 slot for the g: $C(8, 1) = 8$ choices

Step 4: Select 3 slots for the a's: $C(7, 3)$ choices

Step 5: Select 1 slot for the l: $C(4, 1) = 4$ choices

Step 6: Select 1 slot for the o: $C(3, 1) = 3$ choices

Step 7: Select 1 slot for the n: $C(2, 1) = 2$ choices

Step 8: Select 1 slot for the i: $C(1, 1) = 1$ choice

Total number of sequences:

$C(11,2)C(9,1)C(8,1)C(7,3)C(4,1)C(3,1)$
$C(2,1)C(1,1)$

67. Start with 10 empty slots. Decision algorithm:

Step 1: Select 2 slots for the c's: $C(10, 2)$ choices

Step 2: Select 4 slots for the a's: $C(8, 4)$ choices

Step 3: Select 1 slot for the s: $C(4, 1) = 4$ choices

Step 4: Select 1 slot for the b: $C(3, 1) = 3$ choices

Step 5: Select 1 slot for the l: $C(2, 1) = 2$ choices

Step 6: Select 1 slot for the n: $C(1, 1) = 1$ choice

Total number of sequences:

$C(10, 2)C(8, 4)C(4, 1)C(3, 1)C(2, 1)C(1, 1)$

69. We wish to compute the number of groups of 4 chosen from 5 contestants (since we are excluding Ben and Ann): $C(5, 4) = 5$ possible groups (Choice A)

71. Each handshake corresponds to a pair of people, so the question is really asking how many pairs of people there are in a group of 10: $C(10, 2) = 45$ (Choice D).

73. (a) Note: In a set of 3 numbers, all three are different. The number of combinations = the number of sets of 3 numbers chosen from 40: $C(40, 3) = 9880$

(b) Decision algorithm for constructing a combination in which a number appears twice:

　Step 1: Select the number that appears twice: 40 choices

　Step 3: Select another number: 39 choices

Total number of new combinations: $40 \times 39 = 1560$

(c) Decision algorithm for constructing a combination in which a number appears 3 times:

　Step 1: Select the number that appears 3 times: 40 choices

This gives 40 new combinations.

Total number of combinations:

　$9880 + 1560 + 40 = 11{,}480$

75. (a) The number of combinations of 2 equations chosen from the 20 constraints is $C(20, 2) = 190$ systems of equations to solve

(b) Replace 20 by n, getting $C(n, 2)$.

77. You should choose the multiplication principle because the multiplication principle can be used to solve all problems that call for the formulas for permutations, as well as others.

79. Urge your friend not to focus on formulas, but instead learn to formulate decision algorithms and use the principles of counting.

81. It is ambiguous on the following point: are the three students to play different characters, or are they to play a group of three, such as "three guards." This should be made clear in the exercise.

Chapter 6 Review Exercises

1. The negative integers greater than or equal to -3 are -3, -2, and -1. Thus,

$$N = \{-3, -2, -1\}$$

2. There are $2 \times 2 \times 2 \times 2 \times 2 = 2^5 = 32$ possible outcomes:

$S = \{$HHHHH, HHHHT, HHHTH, HHHTT, HHTHH, HHTHT, HHTTH, HHTTT, HTHHH, HTHHT, HTHTH, HTHTT, HTTHH, HTTHT, HTTTH, HTTTT, THHHH, THHHT, THHTH, THHTT, THTHH, THTHT, THTTH, THTTT, TTHHH, TTHHT, TTHTH, TTHTT, TTTHH, TTTHT, TTTTH, TTTTT$\}$

3. If the dice are distinguishable, then the outcomes can be thought of as ordered pairs. Thus, since the numbers are different, S is the set of all ordered pairs of distinct numbers in the range 1–6:

$S = \{(1,2), (1,3), (1,4), (1,5), (1,6), (2,1), (2,3),$ $(2,4), (2,5), (2,6), (3,1), (3,2), (3,4), (3,5), (3,6),$ $(4,1), (4,2), (4,3), (4,5), (4,6), (5,1), (5,2), (5,3),$ $(5,4), (5,6), (6,1), (6,2), (6,3), (6,4), (6,5)\}$

4. $A = \{1, 2, 3, 4, 5\}$

$\quad B = \{3, 4, 5\}$

$\quad C = \{1, 2, 5, 6, 7\}$

$A \cap B = \{3, 4, 5\}$

$(A \cap B) \cup C = \{3, 4, 5\} \cup C$

$\qquad\qquad = \{1, 2, 3, 4, 5, 6, 7\}$

$B \cup C = \{1, 2, 3, 4, 5, 6, 7\}$

$A \cap (B \cup C) = A \cap \{1, 2, 3, 4, 5, 6, 7\}$

$\qquad\qquad = \{1, 2, 3, 4, 5\}$

5. $A = \{a, b\}$, $B = \{b, c\}$, $S = \{a, b, c, d\}$

$\quad B' = \{a, d\}$

$\quad A \cup B' = \{a, b\} \cup \{a, d\} = \{a, b, d\}$

$\quad A \times B' = \{(a, a), (a, d), (b, a), (b, d)\}$

6. The set of all customers who owe money but owe less than \$1000 is the set of all customers in A and not in B: $A \cap B'$

7. The set of outcomes when a day in August and a time of that day are selected is the set of pairs (day in August, time). That is, it is the set $A \times B$.

8. The set of outcomes in which both dice show an even number or sum to 7 is the set of all outcomes not in E or in F (or both): $E' \cup F$

9. The set of all integers that are not positive odd perfect squares is the set of integers other than those in P and E' ($E' = $ the odd integers) and Q: $(P \cap E' \cap Q)'$ or $P' \cup E \cup Q'$

10. $n(A \cup B) = n(A) + n(B) - n(A \cap B)$

$\qquad 32 = 24 + 24 - n(A \cap B)$

$\quad n(A \cap B) = 24 + 24 - 32 = 16$

11. Let S be the set of all novels in your home; $n(S) = 400$

Let A be the event that you have read a novel: $n(A) = 150$

Let B be the event that Roslyn has read a novel: $n(A) = 200$

We are also told that $n(A \cap B) = 50$.

To compute $n(A \cup B)$ we use

$\quad n(A \cup B) = n(A) + n(B) - n(A \cap B)$

$\qquad\qquad = 150 + 200 - 50 = 300$

The event that neither of you has read a novel is $(A \cup B)'$, and its cardinality is given by

$\quad n[(A \cup B)'] = n(S) - n(A \cup B)$

$\qquad\qquad = 400 - 300 = 100$

Note that the formula used for the last calculation is

$\quad n(C') = n(S) - n(C)$

12. A choice of a model and color corresponds to a pair (model, color); that is, to an element of $A \times B$, where A is the set of models; $n(A) = 3$, and B is the set of colors; $n(B) = 5$.

$$n(A \times B) = n(A)n(B) = 3 \cdot 5 = 15$$

13. The number of outcomes from rolling the dice is $n(A \times B)$, where A is the set of outcomes of the red dice and B is the set of outcomes of the green one.

$$n(A \times B) = n(A)n(B) = 6 \cdot 6 = 36$$

The set of losing combinations has the form $A \cup B$ where here A is the set of doubles and B is the set of outcomes in which the green die shows an odd number and the red die shows an even number. To compute its cardinality, we use

$$n(A \cup B) = n(A) + n(B) - n(A \cap B)$$
$$= 6 + 3 \times 3 - 0 = 15$$

The set of winning combinations is the complement of the set C of losing combinations, and is given by

$$n(C') = n(S) - n(C) = 36 - 15 = 21.$$

14. Decision algorithm for constructing a full house with either two Kings and three Queens or two Queens and three Kings:

Alternative 1: Two Kings and three Queens
Step 1: Select a pair of suits for the Kings: $C(4, 2)$ choices
Step 2: Select three suite for the Queens: $C(4, 3)$ choices
Total for Alternative 1: $C(4, 2)C(4, 3)$ choices

The second alternative (two Queens and three Kings) gives the same number of choices. Therefore, the total number of such hands is $2C(4, 2)C(4, 3)$

15. Decision algorithm for constructing a two of a kind hand with no Aces:

Step 1: Select a denomination other than Ace for the pair: $C(12, 1)$ choices
Step 2: Select 2 cards of that denomination: $C(4, 2)$ choices
Step 3: Select 3 denominations (other than Ace or the one already selected above) for the remaining singles: $C(11, 3)$ choices
Step 4: Select a suit for the highest-ranked denomination just chosen: $C(4, 1)$ choices
Step 5: Select a suit for the next highest-ranked denomination just chosen: $C(4, 1)$ choices
Step 6: Select a suit for the lowest-ranked denomination just chosen: $C(4, 1)$ choices

Total number of hands:
$C(12, 1)C(4, 2)C(11, 3)C(4, 1)C(4, 1)C(4, 1)$

16. Decision algorithm for constructing a three of a kind hand with no Aces:
Step 1: Select a denomination other than Ace for the triple: $C(12, 1)$ choices
Step 2: Select 3 cards of that denomination: $C(4, 3)$ choices
Step 3: Select 2 denominations (other than Ace or the one already selected above) for the remaining singles: $C(11, 2)$ choices
Step 4: Select a suit for the highest-ranked denomination just chosen: $C(4, 1)$ choices
Step 5: Select a suit for the lowest-ranked denomination just chosen: $C(4, 1)$ choices

Total number of hands:
$C(12, 1)C(4, 3)C(11, 2)C(4, 1)C(4, 1)$

17. Decision algorithm for constructing a straight flush:
Step 1: Select a suit for the flush: $C(4, 1)$ choices

Step 2: Select a starting denomination for the consecutive run: $C(10, 1)$ choices (A, 2, 3, 4, 5 up through 10, J, Q, K, A)

Total number of hands: $C(4, 1)C(10, 1)$

18. The number of sets of 5 marbles chosen from 12 is $C(12, 5) = 792$

19. Decision algorithm for constructing a set of 5 marbles that include all the red ones:
  Step 1: Select 4 red marbles: $C(4, 4)$ choices
  Step 2: Select 1 non-red marble: $C(8, 1)$ choices

Total number of choices: $C(4, 4)C(8, 1) = 8$

20. By the above calculations, of 8 of the 792 possible sets of 5 marbles include all the red ones. Therefore, $792 - 8 = 784$ do not.

21. Decision algorithm for constructing a set of 5 marbles that include at least 2 yellow ones:
  Alternative 1: Use 2 yellow ones:
    Step 1: Select 2 yellow marbles: $C(3, 2)$ choices
    Step 2: Select 3 non-yellow marbles: $C(9, 3)$ choices
  Alternative 2: Use 3 yellow ones:
    Step 1: Select 3 yellow marbles: $C(3, 3)$ choices
    Step 2: Select 2 non-yellow marbles: $C(9, 2)$ choices

Total number of choices:
  $C(3, 2)C(9, 3) + C(3, 3)C(9, 2) = 288$

22. Decision algorithm for constructing a set of 5 marbles that include at most one of the red ones but no yellow ones:

Alternative 1: Use no red ones:
  Step 1: Select 5 non-red, non-yellow marbles: $C(5, 5)$ choices
Alternative 2: Use 1 red one:
  Step 1: Select 1 red marble: $C(4, 1)$ choices
  Step 2: Select 4 non-red, non-yellow marbles: $C(5, 4)$ choices

Total number of choices:
  $C(5, 5) + C(4, 1)C(5, 4) = 21$

23. $S \cup T$ is the set of books that are either sci-fi or stored in Texas (or both)
$$n(S \cup T) = n(S) + n(T) - n(S \cap T)$$
$$= 33{,}000 + 94{,}000 - 15{,}000$$
$$= 112{,}000$$

24. $H \cap C$ is the set of horror books in California
  $n(H \cap C) = 12{,}000$
(Intersection of the Horror column and California row)

25. $C \cup S'$ is the set of books that are either stored in California or not sci-fi. To compute its cardinality, add the corresponding entries (shown below in thousands)

	S	H	R	O	Total
W	10	12	12	30	64
C	8	12	6	16	42
T	15	15	20	44	94
Total	33	39	38	90	200

$n(C \cup S') = 175{,}000$

26. $(R \cap T) \cup H$ is the set of romance books stored in Texas together with all the horror books.

315

	S	H	R	O	Total
W	10	12	12	30	64
C	8	12	6	16	42
T	15	15	20	44	94
Total	33	39	38	90	200

$n((R \cap T) \cup H) = 59,000$

27. $R \cap (T \cup H)$ is the set of romance books that are also horror books or stored in Texas.

	S	H	R	O	Total
W	10	12	12	30	64
C	8	12	6	16	42
T	15	15	20	44	94
Total	33	39	38	90	200

$n(R \cap (T \cup H)) = 20,000$

28. $(S \cap W) \cup (H \cap C')$ is the set of sci-fi books stored in Washington together with the horror books not stored in California.

	S	H	R	O	Total
W	10	12	12	30	64
C	8	12	6	16	42
T	15	15	20	44	94
Total	33	39	38	90	200

$n((S \cap W) \cup (H \cap C')) = 37,000$

In Exercises 29–33, we let H denote the set of OHaganBooks.com customers, J the set of JungleBooks.com customers, and F the set of FarmerBooks.com customers.

29. We use a Venn diagram:

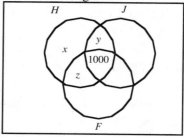

We are told that
OHaganBooks.com had 3500 customers:
$$x + y + z + 1000 = 3500$$
2000 customers were shared with JungleBooks.com:
$$y + 1000 = 2000,$$
so $y = 1000$
1500 customers were shared with FarmerBooks.com:
$$z + 1000 = 1500$$
so $z = 500$
We can now compute x from the first equation:
$$x + 1000 + 500 + 1000 = 3500$$
$$x = 1000$$
We are asked for the number that are exclusive OHaganBooks.com customers:
$$x = 1000$$

30. Let us start with the information from Exercise 29 filled in:

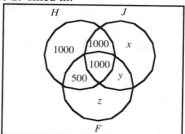

JungleBooks.com has a total of 3600 customers:
$$x + y + 2000 = 3600$$
$$x + y = 1600$$

FarmerBooks.com has 3400 customers:

$$y + z + 1500 = 3400$$

$$y + z = 1900$$

JungleBooks.com and FarmerBooks.com share 1100 customers between them:

$$y + 1000 = 1100$$

so, $y = 100$

giving

$$x + 100 = 1600$$

so, $x = 1500$

$$100 + z = 1900$$

so, $z = 1800$

The customers they share that are not OHaganBooks.com customers constitute

$$x + y + z = 1500 + 100 + 1800 = 3400$$

31. Let us start with the information from Exercise 29 filled in:

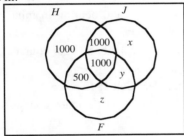

JungleBooks.com has a total of 3600 customers:

$$x + y + 2000 = 3600$$

$$x + y = 1600$$

FarmerBooks.com has 3400 customers:

$$y + z + 1500 = 3400$$

$$y + z = 1900$$

JungleBooks.com and FarmerBooks.com share 1100 customers between them:

$$y + 1000 = 1100$$

so, $y = 100$

giving

$$x + 100 = 1600$$

so, $x = 1500$

$$100 + z = 1900$$

so, $z = 1800$

Thus, OHaganBooks has 1000 exclusive customers, JungleBooks has $x = 1500$ exclusive customers, and FarmerBooks has $z = 1800$ exclusive customers—the most.

32. Here is the complete data from Exercises 29 and 30 above:

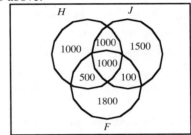

$$n(H \cup J) = 3500 + 1500 + 100 = 5100$$

$$n(H \cup F) = 3500 + 100 + 1800 = 5400$$

Thus, OHaganBooks should merge with FarmerBooks.com for a combined customer base of 5400.

33. Here is the complete data from Exercises 29 and 31 above:

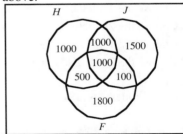

If OHaganBooks merged with JungleBooks, their combined exclusive customer base would be

$$n[(H \cup J) \cap F']$$
$$= 5100 - (1000 + 500 + 100) = 3500$$

If OHaganBooks merged with FarmerBooks, their combined exclusive customer base would be

$$n[(H \cup F) \cap J']$$
$$= 5400 - (1000 + 1000 + 100) = 3300$$

The larger of the two results from a merger with JungleBooks.com — 3500 exclusive customers.

34. Decision algorithm for constructing a 3-letter code:

Step 1: Choose the 1st letter: 26 choices
Step 2: Choose the 2nd letter: 26 choices
Step 3: Choose the 3rd letter: 26 choices

Total number of codes: $26 \times 26 \times 26 = 26^3 = 17{,}576$

35. Decision algorithm for constructing a 3-letter code with 3 different letters:

Step 1: Choose the 1st letter: 26 choices
Step 2: Choose the 2nd letter: 25 choices
Step 3: Choose the 3rd letter: 26 choices

Total number of codes: $26 \times 25 \times 24 = 15{,}600$

36. Decision algorithm for constructing a code:

Step 1: Choose the 1st letter: 26 choices
Step 2: Choose the 2nd letter: 25 choices
Step 3: Choose the 1st digit: 9 choices
Step 4: Choose the 2nd digit: 10 choices

Total number of codes: $26 \times 25 \times 9 \times 10 = 58{,}500$

37. 2 different letters, 3 different digits not starting with zero: $26 \times 25 \times 9 \times 9 \times 8 = 421{,}200$
2 different letters, 4 different digits not starting with zero: $26 \times 25 \times 9 \times 9 \times 8 \times 7 = 2{,}948{,}400$

38. Decision algorithm for constructing a course schedule meeting the minimum requirements:

Step 1: Select 3 liberal arts courses: $C(6, 3)$ choices
Step 2: Select 3 math courses: $C(6, 3)$ choices (Note: At this stage, you have must select a minimum of 4 courses to make up the required 10, and so you must select 2 sciences and 2 fine arts.)
Step 3: Select 2 science courses: $C(5, 2)$ choices
Step 4: Select 2 fine arts courses: $C(6, 2)$ choices

Total number of schedules:
$$C(6, 3)C(6, 3)C(5, 2)C(6, 2)$$
$$= 20?20 \times 10 \times 15 = 60{,}000$$

39. Decision algorithm for constructing a course schedule meeting the minimum requirements, including Calc I, and bearing in mind the fine arts complication:

Step 1: Select 3 liberal arts courses: $C(6, 3)$ choices
Step 2: Select Calc I: 1 choice
Step 3: Select 2 other math courses: $C(5, 2)$ choices
Step 4: Select 2 science courses: $C(5, 2)$ choices
Step 5: Select 2 fine arts courses other than the single pair that cannot be taken in the first year: $C(6, 2) - 1$ choices

Total number of schedules:
$$C(6, 3)C(5, 2)C(5, 2)[C(6, 2) - 1]$$
$$= 20 \times 10 \times 10 \times (15 - 1) = 28{,}000$$

40. Decision algorithm for constructing a course schedule meeting all of the above requirements, and the physics complication:

Alternative 1: Take neither Physics I nor Physics II

  Step 1: Select 3 liberal arts courses: $C(6, 3)$ choices

  Step 2: Select Calc I: 1 choice

  Step 3: Select 2 other math courses: $C(5, 2)$ choices

  Step 4: Select 2 science courses other than Physics I and II: $C(3, 2)$ choices

  Step 5: Select 2 fine arts courses other than the single pair that cannot be taken in the first year: $C(6, 2) - 1$ choices

Total number of schedules for Alternative 1:

$C(6, 3)C(5, 2)C(3, 2)[C(6, 2) - 1]$

  $= 20 \times 10 \times 3 \times (15-1) = 8400$

Alternative 2: Take Physics I but not Physics II

  Step 1: Select 3 liberal arts courses: $C(6, 3)$ choices

  Step 2: Select Calc I: 1 choice

  Step 3: Select 2 other math courses: $C(5, 2)$ choices

  Step 4: Select Physics I: 1 choice

  Step 5: Select 1 science course other than Physics I and II: $C(3, 1)$ choices

Step 6: Select 2 fine arts courses other than the single pair that cannot be taken in the first year: $C(6, 2) - 1$ choices

Total number of schedules for Alternative 1:

$C(6, 3)C(5, 2)C(3, 1)[C(6, 2) - 1]$

  $= 20 \times 10 \times 3 \times (15-1) = 8400$

Alternative 3: Take Physics I and Physics II

  Step 1: Select 3 liberal arts courses: $C(6, 3)$ choices

  Step 2: Select Calc I: 1 choice

  Step 3: Select 2 other math courses: $C(5, 2)$ choices

  Step 4: Select Physics I & II: 1 choice

  Step 5: Select 2 fine arts courses other than the single pair that cannot be taken in the first year: $C(6, 2) - 1$ choices

Total number of schedules for Alternative 1:

$C(6, 3)C(5, 2)[C(6, 2) - 1]$

  $= 20 \times 10 \times (15-1) = 2800$

Total number of course schedules:

  $8400 + 8400 + 2800 = 19,600$

Chapter 7

7.1

1. S = {HH, HT, TH, TT}

E = {HH, HT, TH}

3. S = {HHH, HHT, HTH, HTT, THH,
THT, TTH, TTT}

E = {HTT, THT, TTH, TTT}

5. S = $\begin{cases} (1,1)\ (1,2)\ (1,3)\ (1,4)\ (1,5)\ (1,6) \\ (2,1)\ (2,2)\ (2,3)\ (2,4)\ (2,5)\ (2,6) \\ (3,1)\ (3,2)\ (3,3)\ (3,4)\ (3,5)\ (3,6) \\ (4,1)\ (4,2)\ (4,3)\ (4,4)\ (4,5)\ (4,6) \\ (5,1)\ (5,2)\ (5,3)\ (5,4)\ (5,5)\ (5,6) \\ (6,1)\ (6,2)\ (6,3)\ (6,4)\ (6,5)\ (6,6) \end{cases}$

E = {(1, 4), (2, 3), (3, 2), (4, 1)}

7. S = $\begin{cases} (1,1)\ (1,2)\ (1,3)\ (1,4)\ (1,5)\ (1,6) \\ (2,2)\ (2,3)\ (2,4)\ (2,5)\ (2,6) \\ (3,3)\ (3,4)\ (3,5)\ (3,6) \\ (4,4)\ (4,5)\ (4,6) \\ (5,5)\ (5,6) \\ (6,6) \end{cases}$

(We removed all repeating pairs.)

E = {(1, 3), (2, 2)}

9. S as in Exercise 7

E = {(2, 2), (2, 3), (2, 5), (3, 3), (3, 5), (5, 5)}

11. S = {m, o, z, a, r, t}

E = {o, a}

13. There are $4 \times 3 = 12$ possible sequences of 2 different letters chosen from *sore:*

S = {(s, o), (s, r), (s, e), (o, s), (o, r), (o, e),
(r, s), (r, o), (r, e), (e, s), (e, o), (e, r)}

E = {(o, s), (o, r), (o, e), (e, s), (e, o), (e, r)}

15. There are $5 \times 4 = 20$ possible sequences of 2 different digits chosen from 0–4:

S = {01, 02, 03, 04, 10, 12, 13, 14, 20, 21,
23, 24, 30, 31, 32, 34, 40, 41, 42, 43}

E = {10, 20, 21, 30, 31, 32, 40, 41, 42, 43}

17. S = {domestic car, imported car, van,
antique car, antique truck}

E = {van, antique truck}

19. (a) The sample space is the set of all sets of 4 gummy candies chosen from the packet of 12.

(b) The event that April will get the combination she desires is the set of all sets of 4 gummy candies in which two are strawberry and two are blackcurrant.

21. (a) The sample space is the set of all lists of 14 people chosen from 20.

(b) The event that Colin Powell is the Secretary of State is the set of all lists of 14 people chosen from 20, in which Colin Powell occupies the first (Secretary of State) position.

23. A: the red die shows 1; B: the numbers add to 4

"The red die shows 1 *and* the numbers add to 4" is the event $A \cap B$. *And =* $\cap$

There is only one possible outcome in $A \cap B$:

$A \cap B$ = {(1, 4)}

Therefore, $n(A \cap B) = 1$

25. B: the numbers add to 4

"The numbers do not add to 4" is the event B'.

One has B = {(1, 3), (2, 2), (3, 1)}

$n(B') = n(S) - n(B) = 36 - 3 = 33$

27. "The numbers do not add to 4" is the event B'. "The numbers add to 11" is the event D'.

Therefore, "The numbers do not add to 4 *but* they do add to 11" is the event $B'\cap D'$.
$B'\cap D' = \{(6, 5), (5, 6)\}$
$n(B'\cap D') = 2$

29. "At least one of the numbers is 1" is the event C. "The numbers add to 4" is the event B.

Therefore, "At least one of the numbers is 1 or the numbers add to 4" is the event $C\cup B$
$C\cup B = \{(1, 1), (1, 2), (1, 3), (1, 4), (1, 5), (1, 6), (6, 1), (5, 1), (4, 1), (3, 1), (2, 1), (2, 2)\}$
$n(C\cup B) = 12$

31. W: You will use the web site tonight.
I: Your math grade will improve.

Therefore, "You will use the web site tonight *and* your math grade will improve." is the event $W\cap I$.
And $= \cap$

33. E: You will use the web site every night.
I: Your math grade will improve.

Therefore, "Either you will use the web site every night, *or* your math grade will *not* improve." is the event $E\cup I'$. *Or* $= \cup$

35. I: Your math grade will improve.
W: You will use the web site tonight.
E: You will use the web site every night.

The given statement is: "Either your math grade will improve, *or* you will use the web site tonight *but* you will not use it every night."

The comma after "improve" breaks the statement into two pieces:

Either
your math grade will improve: (I)
Or
you will use the web site tonight *but* you will not use it every night: ($W\cap E'$).

Therefore, the given statement is the event $I\cup(W\cap E')$.

37. I: Your math grade will improve.
W: You will use the web site tonight.
E: You will use the web site every night.

The given statement is: "Either your math grade will improve *or* you will use the web site tonight, *but* you will not use it not every night."

The comma after "tonight" breaks the statement into two pieces:

Either your math grade will improve *or* you will use the web site tonight, ($I\cup W$)
But
you will not use it every night: E'.

Therefore, the given statement is the event $(I\cup W)\cap E'$

39. Looking at the map, we find the following regions that saw an increase in housing prices of 15% or more: New England, Pacific, Middle Atlantic. Therefore,
$E = \{$New England, Pacific, Middle Atlantic$\}$

41. *E*: The region you choose saw an increase in housing prices of 15% or more.

  F: The region you choose is on the east coast.

E∪*F* translates to *E or F:* You choose a region that saw an increase in housing prices of 15% or more *or* is on the east coast.

  E∪*F* = {Pacific, New England, Middle Atlantic, South Atlantic}

E∩*F* translates to *E and F:* You choose a region that saw an increase in housing prices of 15% or more *and* is on the east coast.

E∩*F* = {New England, Middle Atlantic}

43. Two events are mutually exclusive if they have no outcomes in common.

(a) *E*: You choose a region from among the three with the highest percentage increase in housing prices.

  E = { Pacific, New England, Middle Atlantic}

  F: You choose a region that is not on the east or west coast.

  F = { West North Central, West South Central, East North Central, East South Central }

Since *E* and *F* have no regions in common, *E*∩*F* = Ø., so they are mutually exclusive.

(b) *E*: You choose a region from among the three with the highest percentage increase in housing prices.

  E = { Pacific, New England, Middle Atlantic}

  F: You choose a region that is not on the west coast.

F = { West North Central, West South Central, East North Central, East South Central, New England, Middle Atlantic, South Atlantic }

Since *E*∩*F* = {New England, Middle Atlantic}, *E*∩*F* ≠ Ø., so they are not mutually exclusive.

45. *S*: an author is successful.

  N: an author is new.

S∩*N* is the event that an author is successful and new.

S∪*N* is the event that an author is either successful or new

$n(S \cap N) = 5$ (See table)

	N	*E*	**Total**
S	5	25	30
U	15	55	70
Total	20	80	100

$n(S \cup N) = 45$ (See table)

	N	*E*	**Total**
S	5	25	30
U	15	55	70
Total	20	80	100

47. Since *N* and *E* have no outcomes in common (an author cannot be both new and experienced), they are mutually disjoint.

  N and *S* are not mutually disjoint, since
  $n(N \cap S) = 5$ (See table)

  S and *E* are not mutually disjoint, since
  $n(S \cap E) = 25$ (See table)

	N	*E*	**Total**
S	5	25	30
U	15	55	70
Total	20	80	100

49. *S*: an author is successful.

N: an author is new.

S∩*N'* is the event that an author is successful but not a new author.

$n(S \cap N') = 25$ (See table)

	N	*E = N'*	Total
S	5	25	30
U	15	55	70
Total	20	80	100

51. The total number of established authors is 80. Of these, 25 are successful. Thus, the percentage that are successful is

$$\frac{25}{80} = 31.25\%$$

The total number of successful authors is 30. Of these, 25 are successful. Thus, the percentage that are experienced is

$$\frac{25}{30} \approx 83.33\%$$

53. "A stock's value increased *but* it was not an Internet stock." is the event *V*∩*I'*.

$n(V \cap I') = 15$ (See table)

E'

	P	*E*	*I*	Total
V	10	5	15	30
N	30	0	10	40
D	10	5	15	30
Total	50	10	40	100

55. $n(P' \cup N) = 80$. (See table)

	P	*E*	*I*	Total
V	10	5	15	30
N	30	0	10	40
D	10	5	15	30
Total	50	10	40	100

P': The stock was not a pharmaceutical stock.

N: The stock was unchanged in value.

Therefore, $n(P' \cup N)$ is the number of stocks that were either not pharmaceutical stocks, or were unchanged in value after a year (or both).

57. For a pair of events to be mutually exclusive, they must have no outcomes in common. Three kinds can be identified:

(1) Pairs of events corresponding to separate columns: *P* and *E*, *P* and *I*, *E* and *I*.

(2) Pairs of events corresponding to separate rows: *X* and *N*, *V* and *D*, *N* and *D*.

(3) Pairs corresponding to a row and column whose intersection is empty: *N* and *E*.

59. $n(V \cap I) = 15$; $n(I) = 40$

	P	*E*	*I*	Total
V	10	5	15	30
N	30	0	10	40
D	10	5	15	30
Total	50	10	40	100

$$\frac{n(V \cap I)}{n(I)} = \frac{15}{40} = \frac{3}{8}$$

This is the fraction of Internet stocks that increased in value.

61. *E*: The dog's flight drive is strongest.

H: The dog's fight drive is weakest.

G: The dog's fight drive is strongest.

(a) "The dog's flight drive is not strongest *and* its fight drive is weakest" is the event $E' \cap H$.

(b) "The dog's flight drive is strongest *or* its fight drive is weakest" is the event $E \cup H$.

(c) "Neither the dog's flight drive *nor* fight drive are strongest" is the event $E' \cap G' = (E \cup G)'$

63. (a) The dog's fight and flight drives are both strongest.: Intersection of right-most column and bottom row: {9}

(b) The dog's fight drive is strongest, but its flight drive is neither weakest nor strongest: Intersection of right-most column and middle row: {6}

65. (a) {1, 4, 7} This is the left-most column: The dog's fight drive is weakest.

(b) {1, 9}: The dog's fight and flight drives are either both strongest or both weakest.

(c) {3, 6, 7, 8, 9} This is the union of the right-most column and the bottom row: Either the dog's fight drive is strongest, or its flight drive is strongest.

67. S is the set of sets of 4 gummy bears chosen from 6; $n(S) = C(6, 4) = 15$.

Decision algorithm for constructing a set including the raspberry gummy bear:

    Step 1; Select the raspberry one: $C(1, 1) = 1$ choice

    Step 2: Select 3 non-raspberry ones: $C(5, 3) = 10$ choices

Total number of choices: $n(E) = 1 \times 10 = 10$

69. (a) Decision algorithm for constructing a finish (winner, second place and third place) for the race:

    Step 1: Select the winner: 7 choices

    Step 2: Select second place: 6 choices

    Step 3; Select third place: 5 choices

Total number of finishes: $n(S) = 7 \times 6 \times 5 = 210$

(b) E: Electoral College is in second or third place.

    F: Celera is the winner.

Therefore, $E \cap F$ is the event that Celera wins and Electoral College is in second or third place. In other words, it is the set of all lists of three horses in which Celera is first and Electoral College is second or third.

Decision algorithm for constructing a finish in $E \cap F$:

    Alternative 1: Electoral College is second:

        Step 1: Select Celera as the winner: 1 choice

        Step 2: Select Electoral College for second place: 1 choice

        Step 3; Select third place: 5 choices

Total number of finishes for Alternative 1: $1 \times 1 \times 5 = 5$

    Alternative 2: Electoral College is third:

        Step 1: Select Celera as the winner: 1 choice

        Step 2: Select Electoral College for third place: 1 choice

        Step 3; Select second place: 5 choices

Total number of finishes for Alternative 2: $1 \times 1 \times 5 = 5$

Therefore, $n(E \cap F) = 5 + 5 = 10$.

71. The sample space consists of all sets of 3 marbles chosen from 8.

$$n(S) = C(8, 3) = 56$$

73. Using the multiplication principle:

$$n(E) = C(4, 1) \times C(2, 1) \times C(2, 1) = 16$$

75. An event is a <u>subset of the sample space</u>.

77. $(E \cap F)'$ is the complement of the event $E \cap F$. Since $E \cap F$ is the event that both E and F occur,

$(E \cap F)'$ is the event that E and F do not both occur.

79. True; Consider the following experiment: Select an element of the set S at random. Then the sample space is the set of elements of S. In other words, the sample space is S.

81. Answers may vary. Cast a die and record the remainder when the number facing up is divided by 2.

83. Yes. For instance, $E = \{(2, 5), (5, 1)\}$ and $F = \{(4, 3)\}$ are two such events.

7.2

1. $P(E) = \dfrac{fr(E)}{N} = \dfrac{40}{100} = .4$

3. $N = 800$, $fr(E) = 640$

$P(E) = \dfrac{fr(E)}{N} = \dfrac{640}{800} = .8$

5. We obtain the estimated probability distribution by dividing the frequencies by $N = 4000$:

HH	1100	**Probability**
HT	950	$\dfrac{1100}{4000} = .275$
TH	1200	$\dfrac{950}{4000} = .2375$
TT	750	$\dfrac{1200}{4000} = .3$
HH	1100	$\dfrac{750}{4000} = .1875$

7. Refer to the estimated probability distribution in the solution to Exercise 5. The event the that the second coin lands with heads up is

$E = \{HH, TH\}$

$P(E) = .275 + .3 = .575$

9. From the solution to Exercise 7, the second coin comes up heads approximately 58% of the time. Therefore, the second coin *seems* slightly biased in favor of heads, especially in view of the fact that the coin was tossed 4000 times. On the other hand, it is conceivable that the coin is fair and that heads came up 58% of the time purely by chance. Deciding which conclusion is more reasonable requires some knowledge of inferential statistics.

11. Use the following formulas to generate binary digits (0 represents heads, say, and 1 represents tails):

TI-83/84: `randInt(0,1)`
To obtain `randInt`, follow $\boxed{\text{MATH}}$ $\rightarrow$ PRB.

Excel: =RANDBETWEEN(0,1)

Answers will vary. The estimated probability that heads comes up should be around .5.

13. Simulate tossing two coins by generating two random binary digits (0 represents heads, say, and 1 represents tails). If the outcome is one head and one tail, the digits should add to 1. Otherwise, they will add to 0 or 2. The sum of these digits can be obtained as follows:

TI-83/84:
`randInt(0,1)+randInt(0,1)`
To obtain `randInt`, follow $\boxed{\text{MATH}}$ $\rightarrow$ PRB.

Excel:
`=RANDBETWEEN(0,1)+`
`    RANDBETWEEN(0,1)`

Then count how many 1s you get.
Answers will vary. The estimated probability that the outcome is one head and one tail should be around .5.

15. $P(E) = \dfrac{n(E)}{n(S)} = \dfrac{5}{20} = \dfrac{1}{4}$

17. $P(E) = \dfrac{n(E)}{n(S)} = \dfrac{10}{10} = 1$

19. $n(S) = 4$, $n(E) = 3$

$P(E) = \dfrac{n(E)}{n(S)} = \dfrac{3}{4}$

21. $S = \{HH, HT, TH, TT\}$; $n(S) = 4$

$E = \{HH, HT, TH\}$; $n(E) = 3$

$P(E) = \dfrac{n(E)}{n(S)} = \dfrac{3}{4}$

23. $S = \{HHH, HHT, HTH, HTT, THH,$
$THT, TTH, TTT\}$; $n(S) = 8$
$E = \{HTT, THT, TTH, TTT\}$; $n(E) = 4$
$$P(E) = \frac{n(E)}{n(S)} = \frac{4}{8} = \frac{1}{2}$$

25. $n(S) = 36$, $n(E) = 4$
$$P(E) = \frac{n(E)}{n(S)} = \frac{4}{36} = \frac{1}{9}$$

27. $n(S) = 36$, $n(E) = 0$
$$P(E) = \frac{n(E)}{n(S)} = \frac{0}{36} = 0$$

29. $n(S) = 36$,
$E = \{(2, 2), (2, 3), (2, 5), (3, 2), (3, 3),$
$(3, 5), (5, 2), (5, 3), (5, 5)\}$; $n(E) = 9$
$$P(E) = \frac{n(E)}{n(S)} = \frac{9}{36} = \frac{1}{4}$$

31. $E = \{(4, 4), (2, 3)\}$
The outcomes for a pair of indistinguishable dice
are not equally likely, so we cannot use $P(E) = n(E)/n(S)$. Using Example 7 in the textbook,
$$P(4, 4) = \frac{1}{36}$$
$$P(2, 3) = \frac{1}{18}$$
Therefore, $P(E) = \frac{1}{36} + \frac{1}{18} = \frac{1}{12}$

The corresponding event for distinguishable dice is
$\{(4, 4), (2, 3), (3, 2)\}$
(We can also use this to compute $P(E)$:
$$P(E) = \frac{n(E)}{n(S)} = \frac{3}{36} = \frac{1}{12}\,)$$

33. Start with

Outcome	1	2	3	4	5	6
Probability	x	$2x$	x	$2x$	x	$2x$

Since the sum of the probabilities of the outcomes must be 1, we get
$$9x = 1,$$
so $x = \dfrac{1}{9}$
This gives:

Outcome	1	2	3	4	5	6
Probability	$\frac{1}{9}$	$\frac{2}{9}$	$\frac{1}{9}$	$\frac{2}{9}$	$\frac{1}{9}$	$\frac{2}{9}$

$$P(\{1, 2, 3\}) = \frac{1}{9} + \frac{2}{9} + \frac{1}{9} = \frac{4}{9}$$

35. Start with

Outcome	1	2	3	4
Probability	$8x$	$4x$	$2x$	x

Since the sum of the probabilities of the outcomes must be 1, we get
$$15x = 1,$$
so $x = \dfrac{1}{15}$
This gives:

Outcome	1	2	3	4
Probability	$\frac{8}{15}$	$\frac{4}{15}$	$\frac{2}{15}$	$\frac{1}{15}$

37. (a) Take E to be the event that a randomly selected pound of beef is purchased by McDonald's.
$N = 25$ billion, $fr(E) = 1$ billion
$$P(E) = \frac{fr(E)}{N} = \frac{1}{25} = .04$$

(b) Take F to be the event that a randomly selected pound of potatoes is not purchased by McDonald's.
$N = 46$ billion, $fr(F) = 46 - 1 = 45$ billion
$$P(F) = \frac{fr(F)}{N} = \frac{45}{46} \approx .98$$

39.(a) We obtain the estimated probability distribution by dividing the frequencies by $N = 1+4+4+1 = 10$:

Outcome	Frequency	Probability
3	1	$\frac{1}{10} = .1$
2	4	$\frac{4}{10} = .4$
1	4	$\frac{4}{10} = .4$
0	1	$\frac{1}{10} = .1$

(b) Take E to be the event that a randomly selected small SUV will have a crash test rating of "Acceptable" (2) or better.

$E = \{2, 3\}$

$P(E) = .4 + .1 = .5$

41. The number households is $N = 1000$.

(a) Here is a chart showing the frequencies and computation of estimated probabilities:

Event	Frequency	Probability
Dial-up	628	$\frac{628}{1000} \approx .63$
Cable Modem	206	$\frac{206}{1000} \approx .21$
DSL	152	$\frac{152}{1000} \approx .15$
Other	14	$\frac{14}{1000} \approx .01$

(b) $E = \{$ Cable Modem, DSL$\}$

$N = 1000, fr(E) = 206 + 152 = 358$

$P(E) = \frac{fr(E)}{N} = \frac{358}{1000} = .358 \approx .36$

Alternatively,

$P(E) = P(\text{Cable Modem}) + P(\text{DSL})$

$\approx .21 + .15 = .36$

43. If we classify the given data as low (L), middle (M) or high (H), we get:

$$M, L, M, L, L, L, L, M, H, H$$

This gives the following distribution:

Outcome	L	M	H
Frequency	5	3	2
Probability	$\frac{5}{10} = .5$	$\frac{3}{10} = .3$	$\frac{2}{10} = .2$

45. $P(E \cap S) = \frac{fr(E \cap S)}{N} = \frac{25}{100} = .25$

	N	E	Total
S	5	25	30
U	15	55	70
Total	20	80	100

47. $P(N) = \frac{fr(N)}{N} = \frac{20}{100} = .2$

	N	E	Total
S	5	25	30
U	15	55	70
Total	20	80	100

49. $P(U) = \frac{fr(U)}{N} = \frac{70}{100} = .7$

	N	E	Total
S	5	25	30
U	15	55	70
Total	20	80	100

51. Restrict attention to the successful authors only:

$P(\text{Successful author is established}) = \frac{25}{30} = \frac{5}{6}$

	N	E	Total
S	5	25	30

53. Restrict attention to the established authors only:

P(Established author is successful)

$$= \frac{25}{80} = \frac{5}{16}$$

	E
S	25
U	55
Total	80

55. P(Contaminated) $= .8$

Therefore,

$$P(U) = 1 - .8 = .2$$

Of the contaminated chicken, 20% had the strain resistant to antibiotics. Therefore,

$$P(R) = .2 \times .8 = .16 \qquad (20\% \text{ of } 80\%)$$

The other 80% of contaminated chicken is not resistant. Therefore,

$$P(C) = .8 \times .8 = .64 \qquad (80\% \text{ of } 80\%)$$

This gives the following probability distribution:

Outcome	U	C	R
Probability	.2	.64	.16

57. Conventionally grown produce:

The events are:

No pesticide (NP)

Single pesticide (SP)

Multiple pesticides (MP)

We are told that 73% of conventionally grown foods had residues from at least one pesticide. Therefore,

$$P(NP) = 1 - .73 = .27$$

We are also told that conventionally grown foods were 6 times as likely to contain multiple pesticides as organic foods, and that 10% of organic foods had multiple pesticides. Therefore,

$$P(MP) = 6 \times .10 = .60$$

This leaves

$$P(SP) = 1 - (.27 + .60) = .13$$

Probability distribution for conventionally grown produce:

Outcome	NP	SP	MP
Probability	.27	.13	.60

Organic produce:

We are told that

$$P(MP) = .10$$

and that 23% had either single or multiple pesticide residues. Therefore,

$$P(SP) = .23 - .10 = .13$$

This leaves

$$P(NP) = 1 - (.10 + .13) = .77$$

Probability distribution for organic produce:

Outcome	NP	SP	MP
Probability	.77	.13	.10

59. P(False negative) $= \dfrac{10}{400} = .025$

$$P(\text{False positive}) = \frac{10}{200} = .05$$

61. We had the following probability distribution in Exercise 55:

Outcome	U	C	R
Probability	.2	.64	.16

To simulate it, generate integers in the range 1–100 using the following formula:

TI-83/84: `randInt(1,100)`

Excel: `=RANDBETWEEN(1,100)`

Regard integers in the range 1–20 as representing U, integers in the range 21–84 as representing C, and those in the range 85–100 as representing R. Then compute the frequency distribution, which should approximate the probability distribution given above.

63. The total number of items sold was $n(S) = 1030 + 210 + 230 = 1470$ million.
$$P(E) = \frac{n(E)}{n(S)} = \frac{1030}{1470} \approx .70$$

65. $P(E) = \frac{n(E)}{n(S)} = \frac{1030+230}{1470} \approx .86$

67. $P(E) = \frac{n(E)}{n(S)} = \frac{1030+230}{1470} \approx .86$

69. The probability distribution is obtained by converting the percentages to decimals:

Outcome	Probability
Hispanic or Latino	.42
White (not Hispanic)	.37
African American	.09
Asian	.08
Other	.04

P(Neither White nor Asian) $= 1 - (.37+.08) = .55$

71. (a) $S = $ {stock market success, sold to other concern, fail}

(b) P(Stock market success) $= \frac{2}{10} = .2$

P(Sold to other concerns) $= \frac{3}{10} = .3$

P(Fail) $= 1 - (.2 + .3) = .5$

(c) To realize profits for early investors, a start-up venture must be either a stock market success or sold to another concern, so
P(Profit) $= .3 + .2 = .5$

73. The outcomes of the experiment are: SUV, pickup, passenger car, and minivan. The given information tells us that P(SUV) $= .25$ and P(pickup) $= .15$. We are also told that P(passenger car) is five times P(minivan). take P(minivan) $= x$. Then we have:

Outcome	SUV	Pickup	Passenger Car	Minivan
Probability	.25	.15	$5x$	x

Since the sum of the probabilities must be 1, we get

$$.25 + .15 + 5x + x = 1$$
$$.40 + 6x = 1$$
$$6x = 1 - .40 = .60$$
$$x = .10$$

So, P(minivan) $= .10$, and P(passenger car) $= 5x = .50$. This gives the required distribution:

Outcome	SUV	Pickup	Passenger Car	Minivan
Probability	.25	.15	.50	.10

75. Start with

Outcome	1	2	3	4	5	6
Probability	0	x	x	x	x	0

$4x = 1$, so $x = 1/4 = .25$ giving

Outcome	1	2	3	4	5	6
Probability	0	.25	.25	.25	.25	0

77. Start with

Outcome	1	2	3	4	5	6
Probability	x	$2x$	$2x$	$2x$	$2x$	x

$10x = 1$, so $x = .1$, giving

Outcome	1	2	3	4	5	6
Probability	.1	.2	.2	.2	.2	.1

$P(\text{Odd}) = .1 + .2 + .2 = .5$

79. Let $x = P(\text{matching numbers})$. Then
$$P(\text{Mismatching numbers}) = 2x$$

There are 6 matching pairs and 30 mismatching ones. Since the probabilities add to 1,

$6x + 30(2x) = 1$

$66x = 1$, so $x = \dfrac{1}{66}$.

This gives:

$$P(1,1) = P(2,2) = \ldots = P(6,6) = \frac{1}{66}$$

$$P(1,2) = \ldots = P(6,5) = \frac{2}{66} = \frac{1}{33}$$

To obtain an odd sum, the pair must consist of an even and odd number, and there are nine such (mismatching) pairs:

$$P(\text{odd sum}) = 9 \times \frac{1}{33} = \frac{6}{11}$$

81. Take $P(2) = 15x$. This gives

Outcome	1	2	3	4	5	6
Probability	$5x$	$15x$	$5x$	$3x$	$5x$	$5x$

$38x = 1$, so $x = \dfrac{1}{38}$, giving

Outcome	1	2	3	4	5	6
Probability	$\dfrac{5}{38}$	$\dfrac{15}{38}$	$\dfrac{5}{38}$	$\dfrac{3}{38}$	$\dfrac{5}{38}$	$\dfrac{5}{38}$

$$P(\text{Odd}) = \frac{5}{38} + \frac{5}{38} + \frac{5}{38} = \frac{15}{38}$$

83. The estimated probability of an event E is defined to be the fraction of times E occurs.

85. Wrong. For a pair of fair dice, the theoretical probability of a pair of matching numbers is 1/6, as Ruth says. However, it is quite possible, although not very likely, that if you cast a pair of fair dice 20 times, you will never obtain a matching pair (in fact, there is approximately a 2.6% chance that this will happen). In general, a nontrivial claim about theoretical probability can never be absolutely validated or refuted experimentally. All we can say is that the evidence suggests that the dice are not fair.

87. For a (large) number of days, record the temperature prediction for the next day and then check the actual temperature the next day. Record whether the prediction was accurate (within, say, 2°F of the actual temperature). The fraction of times the prediction was accurate is the estimated probability.

89. He is wrong. It is possible to have a run of losses of any length. Each time he plays the game,. the chances of losing are the same, regardless of his history of wins or losses. Tony may have grounds to *suspect* that the game is rigged, but no proof.

7.3

1. $P(A) = .1$, $P(B) = .6$, $P(A \cap B) = .05$
$P(A \cup B) = P(A) + P(B) - P(A \cap B)$
$= .1 + .6 - .05 = .65$

3. $A \cap B = \varnothing$, $P(A) = .3$, $P(A \cup B) = .4$
$P(A \cup B) = P(A) + P(B)$ Mutually exclusive
$4 = .3 + P(B)$
$P(B) = ,4 - ,3 = ,1$

5. $A \cap B = \varnothing$, $P(A) = .3$, $P(B) = .4$
$P(A \cup B) = P(A) + P(B)$ Mutually exclusive
$= .3 + .4 = .7$

7. $P(A \cup B) = .9$, $P(B) = .6$, $P(A \cap B) = .1$
$P(A \cup B) = P(A) + P(B) - P(A \cap B)$
$.9 = P(A) + .6 - .1$
$P(A) = .9 - .6 + .1 = .4$

9. $P(A) = .75$
$P(A') = 1 - P(A) = 1 - .75 = .25$

11. $P(A) = .3$, $P(B) = .4$, $P(C) = .3$
Since A, B and C are mutually exclusive,
$P(A \cup B \cup C) = P(A) + P(B) + P(C)$
$= .3 + .4 + .3 = 1$

13. $P(A) = .3$, $P(B) = .4$
Since A, and B are mutually exclusive,
$P(A \cup B) = P(A) + P(B)$
$= .3 + .4 = .7$
$P((A \cup B)') = 1 - P(A \cup B)$
$= 1 - .7 = .3$

15. Since $A \cup B = S$,
$P(A \cup B) = 1$
Since $A \cap B = \varnothing$, A and B are mutually exclusive, so
$P(A \cup B) = P(A) + P(B)$

Substituting $P(A \cup B) = 1$ gives
$1 = P(A) + P(B)$

17. $P(A) = .2$, $P(B) = .1$; $P(A \cup B) = .4$
$P(A \cup B)$ is more than $P(A) + P(B)$. But, since $P(A \cup B) = P(A) + P(B) - P(A \cap B)$, $P(A \cup B)$ cannot be more than $P(A) + P(B)$. Therefore, the given information does not describe a probability distribution.

19. $P(A) = .2$, $P(B) = .4$; $P(A \cap B) = .2$
Since $P(A \cap B)$ is $\leq$ both $P(A)$ and $P(B)$, the given information is consistent with a probability distribution.

21. $P(A) = .1$, $P(B) = 0$; $P(A \cup B) = 0$
No; $P(A \cup B)$ should be $\geq P(A)$, since $A \cup B \supseteq A$

23.

Outcome	a	b	c	d	e
Probability	.1	.05	.6	.05	x

Since the probabilities of all the outcomes add to 1,
$.1 + .05 + .6 + .05 + x = 1$
$.8 + x = 1$
$x = .2$

(a) $P(\{a, c, e\}) = P(a) + P(c) + P(e)$
$= .1 + .6 + .2 = .9$

(b) $P(E \cup F) = P(\{a, b, c, e\})$
$= P(a) + P(b) + P(c) + P(e)$
$= .1 + .05 + .6 + .2 = .95$

(c) $P(E') = P(\{b, d\}) = P(b) + P(d)$
$= .05 + .05 = .1$

(d) $P(E \cap F) = P\{(c, e)\} = P(c) + P(e)$
  $= .6 + .2 = .8$

25. Take

D: I will meet a tall dark stranger; $P(D) = \dfrac{1}{3}$

T: I will travel; $P(T) = \dfrac{2}{3}$

$P(D \cap T) = \dfrac{1}{6}$

$P(D \cup T) = P(D) + P(T) - P(D \cap T)$
  $= \dfrac{1}{3} + \dfrac{2}{3} - \dfrac{1}{6} = \dfrac{5}{6}$

27. Take A: A randomly selected person polled ranked jobs or health care as the top domestic priority.
We are told that $P(A) = .61$, and we are asked to find $P(A')$.

$P(A') = 1 - P(A)$
  $= 1 - .61 = .39$

29. Take

U: A randomly the bag of cement was consumed in the US; $P(U) = .06$
C: It was consumed in China; $P(C) = .40$
The events U and C are mutually exclusive, so

$P(U \cup C) = P(U) + P(C)$
  $= .06 + .40 = .46$

We are asked to find

$P[(U \cup C)'] = 1 - P(U \cup C) = 1 - .46 = .54$

31. S = set of all applicants; $n(S) = 39{,}802$
E = set of admitted applicants; $n(E) = 9{,}534$

$P(E) = \dfrac{n(E)}{n(S)} = \dfrac{9{,}534}{39{,}802} \approx .24$

33. S = set of all applicants; $n(S) = 39{,}802$
E = set of admitted applicants who had a Math SAT below 400; $n(E) = 7$

$P(E) = \dfrac{n(E)}{n(S)} = \dfrac{7}{39{,}802} \approx .00018 \approx .00$

35. A: An applicant was admitted.

$P(A) = \dfrac{n(A)}{n(S)} = \dfrac{9534}{39{,}802} \approx .24$

$P(A') = 1 - P(A) = 1 - .24 = .76$

37. R: An applicant had a Math SAT in the range 500–599.
A: An applicant was admitted.

$P(R) = \dfrac{9813}{39{,}802} \approx .247$

$P(A) = \dfrac{9534}{39{,}802} \approx .240$

$P(R \cap A) = \dfrac{1124}{39{,}802} \approx .028$

$P(R \cup A) = P(R) + P(A) - P(R \cap A)$
  $= .247 + .240 - .028 \approx .46$

39. R: An applicant had a Math SAT in the range 500–599.
A: An applicant was admitted.
We are asked to find $P[(R \cup A)']$. In Exercise 37 we computed

$P(R \cup A) \approx .46$

41. S = set of all applicants who had a Math SAT below 400; $n(S) = 694$
E = set of admitted applicants who had a Math SAT below 400; $n(E) = 7$

$P(E) = \dfrac{n(E)}{n(S)} = \dfrac{7}{694} \approx .01$

43. S = set of all admitted applicants; $n(S) = 9534$
E = set of admitted applicants who had a math SAT of 700 or above; $n(E) = 5309$

$P(E) = \dfrac{n(E)}{n(S)} = \dfrac{5309}{9534} \approx .56$

Section 7.3

45. S = set of all rejected applicants; $n(S) = 30{,}268$
E = set of rejected applicants who had a math SAT below 600;

$n(E) = 687 + 3512 + 8689 = 12{,}888$

$P(E) = \dfrac{n(E)}{n(S)} = \dfrac{12{,}888}{30{,}268} \approx .43$

47. Use the following events:
E: Agree that it should be the government's responsibility to provide a decent standard of living for the elderly.
F: Agree that it would be a good idea to invest part of their Social Security taxes on their own.

$P(E\cup F) = P(E) + P(F) - P(E\cap F)$
$P(E\cup F) = .79 + .43 - P(E\cap F)$
$P(E\cup F) = 1.22 - P(E\cap F)$ ... (1)

Since $P(E\cup F)$ must be less than or equal to 1, $P(E\cap F)$ must be at least .22, so the smallest percentage of people that could have agreed with both statements is 43%.

For the second part of the question, we ask how large $P(E\cap F)$ can be for the formula (1) to make sense. The largest conceivable value for $P(E\cap F)$ is .43 (because $E\cap F$ is a subset of E and F and hence its probability cannot exceed either .79 or .43). Thus, the largest percentage of people that could have agreed with both statements is 43%.

49. Use the following events:
B: Failed it backwards; $P(B) = \dfrac{2}{3}$
D: Failed it sideways; $P(D) = \dfrac{3}{4}$
We are also told that $P(B\cap D) = \dfrac{5}{12}$

$P(B\cup D) = P(B) + P(D) - P(B\cap D)$
$= \dfrac{2}{3} + \dfrac{3}{4} - \dfrac{5}{12} = 1$

Therefore, all of them failed it either backwards or sideways, so all were disqualified.

51. Use the following events:
V: Vaccinated; $P(V) = .80$
F: Gets flu; $P(F) = .10$
We are also told that 2% of the vaccinated population gets the flu. Therefore, the percentage that are vaccinated and also get the flu is
$P(V\cap F) = .02P(V) = .02\times.80 = .016$
Now
$P(V\cup F) = P(V) + P(F) - P(V\cap F)$
$= .80 + .10 - .016 = .884$

53. The probability of the union of two events is the sum of the probabilities of the two events if <u>they are mutually exclusive</u>.

55. Wrong. For example, the theoretical probability of winning a state lotto is small but nonzero. However, the vast majority of people who play lotto day of their lives never win, no mater how frequently they play.

57. When $A\cap B = \emptyset$ we have
$P(A\cap B) = P(\emptyset) = 0$,
so $P(A\cup B) = P(A) + P(B) - P(A\cap B)$
$= P(A) + P(B) - 0$
$= P(A) + P(B)$.

59. Zero. According to the assumption, no matter how many thunderstorms occur, lightning cannot strike your favorite spot more than once, and so, after n trials the estimated probability will never exceed $1/n$, and so will approach zero as the number of trials gets large.

61. $P(A\cup B\cup C) = P(A) + P(B) + P(C) - P(A\cap B) - P(A\cap C) - P(B\cap C) + P(A\cap B\cap C)$

7.4

1. $n(S) = C(10, 5) = 252$

$n(E) = C(4, 4)C(6, 1) = 6$

4 red, 1 non-red

$$P(E) = \frac{n(E)}{n(S)} = \frac{6}{252} = \frac{1}{42}$$

3. $n(S) = C(10, 5) = 252$

E: At least 1 white marble

The complementary event is

F: No white ones

$n(F) = C(8, 5) = 56$

6 non-white

$$P(F) = \frac{n(F)}{n(S)} = \frac{56}{252} = \frac{2}{9}$$

Therefore,

$$P(E) = 1 - P(F) = 1 - \frac{2}{9} = \frac{7}{9}$$

5. $n(S) = C(10, 5) = 252$

$n(E) = C(4, 2)C(3, 1)C(2, 1)C(1, 1) = 36$

2 red, 1 green, 1 white, 1 purple

$$P(E) = \frac{n(E)}{n(S)} = \frac{36}{252} = \frac{1}{7}$$

7. $n(S) = C(10, 5) = 252$

$n(E) = C(7, 5) + C(3, 1)C(7, 4)$

5 non-green or 1 green, 4 non-green

$= 21 + 105 = 126$

$$P(E) = \frac{n(E)}{n(S)} = \frac{126}{252} = \frac{1}{2}$$

9. $n(S) = C(10, 5) = 252$

E: She does not have all the red ones

The complementary event is

F: She has all the red ones.

In Exercise 1, we computed this probability:

$$P(F) = \frac{1}{42}$$

Therefore,

$$P(E) = 1 - \frac{1}{42} = \frac{41}{42}$$

11. $n(S) = C(10, 2) = 45$

There are only 3 stocks listed with yields of 5% or more. Therefore,

$n(E) = C(3, 2) = 3$

$$P(E) = \frac{n(E)}{n(S)} = \frac{3}{45} =$$

13. $n(S) = C(10, 4) = 210$

$n(E) = C(1, 1) \, C(8, 3) = 56$

(Choose SBC and then 3 out of the 8 that remain when you exclude SBC and GE.)

$$P(E) = \frac{n(E)}{n(S)} = \frac{56}{210} = \frac{4}{15}$$

15. $n(S) = C(10, 2) = 45$

Ending up 200 shares of PFE means that PFE was one of the stocks you chose:

$n(E) = C(1, 1) \, C(9, 1) = 9$

(Choose PFE and then 1 out of the 9 that remain.)

$$P(E) = \frac{n(E)}{n(S)} = \frac{9}{45} = \frac{1}{5}$$

17. Number of possible completed tests:

$n(S) = 2^8 \times 5^5 \times 5!$

8 true/false, 5 multiple choice, 5 matching

Number of correct answers: $n(E) = 1$

$$P(E) = \frac{n(E)}{n(S)} = \frac{1}{2^8 \times 5^5 \times 5!}$$

19. $n(S) = C(52, 5) = 2,598,960$

$n(E) = 1,098,240$ (6.4 Exercise 47)

$$P(E) = \frac{n(E)}{n(S)} = \frac{1,098,240}{2,598,960} \approx .4226$$

21. $n(S) = C(52, 5) = 2,598,960$

$n(E) = 123,552$ (6.4 Exercise 49)

$$P(E) = \frac{123,552}{2,598,960} \approx .0475$$

23. $n(S) = C(52, 5) = 2,598,960$

Decision algorithm for computing the number of flushes:

First compute the number of hands in which all 5 cards are of the same suit:

Step 1: Select a suit: 4 choices

Step 2: Select 5 cards of that suit: $C(13, 5)$ choices

Total number of choices: $4 \times C(13, 5)$

Among these are hands the 5 cards are consecutive: A, 2, 3, 4, 5 up through 10, J, Q, K, A (straight flushes). The number of straight flushes is 4×10 (Step 1: Select a suit; Step 2: Select a starting card for the run). Excluding these gives

Total number of flushes:

$n(E) = 4 \times C(13, 5) - 4 \times 10 = 5108$

$$P(E) = \frac{n(E)}{n(S)} = \frac{5108}{2,598,960} \approx .0020$$

25. $n(S) = 27^{39}$

$n(E) = 1$ (The correct sequence)

$$P(E) = \frac{n(E)}{n(S)} = \frac{1}{27^{39}}$$

27. $n(S) = C(7, 4) = 35$

$n(E) = C(5, 4) = 5$

$$P(E) = \frac{n(E)}{n(S)} = \frac{5}{35} = \frac{1}{7}$$

29. $n(S) = C(50, 5) = 2,118,760$

Big Winner: $n(E) = 1$

$$P(E) = \frac{n(E)}{n(S)} = \frac{1}{2,118,760} \approx .000\,000\,472$$

Small-Fry Winner:

$n(F) = C(5, 4)C(45, 1) = 225$

　　4 winning numbers & 1 losing number

$$P(F) = \frac{n(F)}{n(S)} = \frac{225}{2,118,760} \approx .000\,106\,194$$

$P(E \cup F) = P(E) + P(F)$

　　　　　　　　　Mutually exclusive

$$= \frac{226}{2,118,760} \approx .000\,106\,666$$

31. $n(S) = C(700, 400)$

(a) $n(E) = C(100, 100)C(600, 300) = C(600, 300)$

　　　　100 managers, 300 non-managers

$$P(E) = \frac{n(E)}{n(S)} = \frac{C(600, 300)}{C(700, 400)}$$

(b) $n(E) = C(1, 1)C(699, 399) = C(699, 399)$

　　　You, 399 others

$$P(E) = \frac{n(E)}{n(S)} = \frac{C(699, 399)}{C(700, 400)}$$

33. $n(S) = 10 \times 10 \times 10 = 10^3$

$n(E) = 10 \times 9 \times 8 = P(10, 3) = 720$

$$P(E) = \frac{n(E)}{n(S)} = \frac{720}{1000} = .72$$

35. A perfect progression is a sequence of 8 "Won" scores in the range 0–7 that are all different. S is the set of all sequences of 8 digits in the range 0–7.

$n(S) = 8^8$; $n(E) = 8!$

$$P(E) = \frac{n(E)}{n(S)} = \frac{8!}{8^8}$$

37. Each of the two random moves from the starting position be go up, down, left, or right. Since this gives 4 choices per move, there are

$$n(S) = 4 \times 4 = 16$$

possible sequences of two moves. Only two of these sequences will get you to the Finish node: right+down and down+right. Therefore,

$$n(E) = 2$$
$$P(E) = \frac{n(E)}{n(S)} = \frac{2}{16} = \frac{1}{8}$$

39. The number of possible outcomes is

$$n(S) = 2 \times 2 \times 2 = 8 \qquad \text{Select a winner 3 times}$$

The only way North Carolina (NC) will beat Central Connecticut (CC) but lose to Virginia is if NC beats CC, Virginia beats Syracuse, and CC loses to Virginia. Therefore,

$$n(E) = 1$$
$$P(E) = \frac{n(E)}{n(S)} = \frac{1}{8}$$

41. The number of possible sequences of 4 digits in the range 0–9 is

$$n(S) = 10 \times 10 \times 10 \times 10 = 10,000$$

Number of correct codes: 1
Number of codes that are correct except for a single digit:

Select a slot for the incorrect digit: 4 choices
Select the incorrect digit: 9 choices
Fill in the remaining slots with the correct digits: 1 choice

Total number of incorrect codes: $4 \times 9 \times 1 = 36$

Therefore, $n(E) = 1 + 36 = 37$
$$P(E) = \frac{n(E)}{n(S)} = \frac{37}{10,000} = .0037$$

43. (a) Decision algorithm for forming a committee:

Select a chief investigator: $C(6, 1)$ choices

Select an assistant investigator: $C(6, 1)$ choices
Select 2 at-large investigators: $C(10, 2)$ choices
Select 5 ordinary members: $C(8, 5)$ choices
This gives
$$n(S) = C(6, 1)C(6, 1)C(10, 2)C(8, 5) = 90,720$$

(b) Decision algorithm to make Larry happy:

Alternative 1: Larry is chief and Otis is assistant:

Select Larry for chief and Otis for assistant: 1 choice
Select 2 at-large investigators: $C(10, 2)$ choices
Select 5 ordinary members: $C(8, 5)$ choices
This gives $C(10, 2)C(8, 5)$ choices for this alternative

Alternative 2: Larry is not on the committee:

Select a chief investigator: $C(5, 1)$ choices
Select an assistant investigator: $C(6, 1)$ choices
Select 2 at-large investigators: $C(9, 2)$ choices
Select 5 ordinary members: $C(7, 5)$ choices
This gives $C(5, 1)C(6, 1)C(9, 2)C(7, 5)$ choices for this alternative

Adding gives a total of
$$n(E)$$
$$= C(10,2)C(8,5) + C(5,1)C(6,1)C(9,2)C(7,5)$$
$$= 25,200$$

(c) $P(E) = \dfrac{n(E)}{n(S)} = \dfrac{25,200}{90,720} \approx .28$

45. The four outcomes listed are not equally likely; for example, (red, blue) can occur in four ways. The methods of this section yield a probability for (red, blue) of $C(2,2)/C(4,2) = 1/6$

47. No. If we do not pay attention to order, the probability is $C(5,2)/C(9,2) = 10/36 = 5/18$.

If we do pay attention to order, the probability is $P(5,2)/P(9,2) = 20/72 = 5/18$ again.

The difference between permutations and combinations cancels when we compute the probability.

7.5

1. $P(A \mid B) = \dfrac{P(A \cap B)}{P(B)} = \dfrac{.2}{.5} = .4$

3. $P(A \cap B) = P(A \mid B)P(B) = (.2)(.4) = .08$

5. $P(A \cap B) = P(A \mid B)P(B)$

$\quad .3 = .4P(sB)$

$\quad P(B) = \dfrac{.3}{.4} = .75$

7. If A and B are independent, then

$\quad P(A \cap B) = P(A)P(B) = (.5)(.4) = .2$

9. If A and B are independent, then

$\quad P(A \mid B) = P(A) = .5$

11.

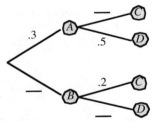

Since the probabilities leaving each branching point must add to 1, we can fill in the missing probabilities:

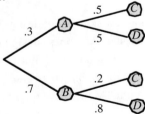

The probability of each outcome is obtained by multiplying the probabilities along the branches leading to the corresponding node:

$\quad P(A \cap C) = .3 \times .5 = .15$

$\quad P(A \cap D) = .3 \times .5 = .15$

$P(B \cap C) = .7 \times .2 = .14$

$P(B \cap D) = .7 \times .8 = .56$

13.

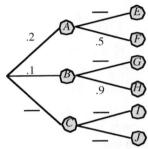

Since the probabilities leaving each branching point must add to 1, we can fill in some of the missing probabilities:

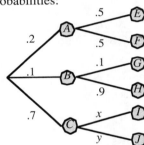

To get the remaining two, use the given information about the probabilities of the outcomes:

$\quad P(C \cap I) = .7 \times x = .14 \Rightarrow x = \dfrac{.14}{.7} = .2$

And so $y = 1 - .2 = .8$

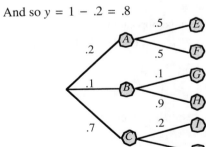

The probability of each outcome is obtained by multiplying the probabilities along the branches leading to the corresponding node:

$P(A \cap E) = .2 \times .5 = .10$

$P(A \cap F) = .2 \times .5 = .10$

$P(B \cap G) = .1 \times .1 = .01$

$P(B \cap H) = .1 \times .9 = .09$

$P(C \cap I) = .7 \times .2 = .14$

$P(C \cap J) = .7 \times .8 = .56$

15. A: The sum is 5.

B: The green one is not a 1.

$P(A \cap B) = P\{(3, 2), (2, 3), (1, 4)\} = \dfrac{3}{36} = \dfrac{1}{12}$

$P(B) = \dfrac{30}{36} = \dfrac{5}{6}$

$P(A|B) = \dfrac{P(A \cap B)}{P(B)} = \dfrac{1/12}{5/6} = \dfrac{1}{10}$

17. A: The red one is 5.

B: The sum is 6.

$P(A \cap B) = P\{(5, 1)\} = \dfrac{1}{36}$

$P(B) = P(\{1, 5), (2, 4), (3, 3), (4, 2), (5, 1)\}$

$\quad = \dfrac{5}{36}$

$P(A|B) = \dfrac{P(A \cap B)}{P(B)} = \dfrac{1/36}{5/36} = \dfrac{1}{5}$

19. A: The sum is 5.

B: The dice have opposite parity.

$P(A \cap B) = P\{(4, 1), (3, 2), (2, 3), (1, 4)\}$

$\quad = \dfrac{4}{36} = \dfrac{1}{9}$

$P(B) = \dfrac{18}{36} = \dfrac{1}{2}$

$P(A|B) = \dfrac{P(A \cap B)}{P(B)} = \dfrac{1/9}{1/2} = \dfrac{2}{9}$

21. A: She gets all 3 red ones.

B: She gets the fluorescent pink one.

$n(S) = C(10, 4) = 210$

$P(A \cap B) = \dfrac{C(1, 1)C(3, 3)}{210} = \dfrac{1}{210}$

$P(B) = \dfrac{C(1, 1)C(9, 3)}{210} = \dfrac{84}{210}$

$P(A|B) = \dfrac{P(A \cap B)}{P(B)} = \dfrac{1/210}{84/210} = \dfrac{1}{84}$

23. A: She gets no red ones.

B: She gets the fluorescent pink one.

$n(S) = C(10, 4) = 210$

$P(A \cap B) = \dfrac{C(1, 1)C(6, 3)}{210} = \dfrac{20}{210}$

$P(B) = \dfrac{C(1, 1)C(9, 3)}{210} = \dfrac{84}{210}$

$P(A|B) = \dfrac{P(A \cap B)}{P(B)} = \dfrac{20/210}{84/210} = \dfrac{5}{21}$

25. A: She gets one of each color other than fluorescent pink.

B: She gets at least one red one.

Notice that $A \cap B = A$

$n(S) = C(10, 4) = 210$

$P(A \cap B) = P(A)$

$\quad = \dfrac{C(3, 1)C(2, 1)C(2, 1)C(2, 1)}{210} = \dfrac{24}{210}$

$P(B) = 1 - \dfrac{C(7, 4)}{210} = 1 - \dfrac{35}{210} = \dfrac{175}{210}$

$P(A|B) = \dfrac{P(A \cap B)}{P(B)} = \dfrac{24/210}{175/210} = \dfrac{24}{175}$

27. A: Your new skateboard design is a success.

B: Your new skateboard design is a failure.

If your new skateboard design is a success, it cannot be a failure, and *vice-versa*. Therefore, A and B are mutually exclusive (choice B).

29. A: Your new skateboard design is a success.

B: Your competitor's new skateboard design is a failure.

Since The likelihood of A does effect the likelihood of B, the events are not independent. Since it is possible for A and B to occur together, they are not mutually exclusive. Therefore, the correct choice is (C).

31. A: The red die is 1, 2, or 3; $P(A) = \dfrac{1}{2}$

B: The green die is even; $P(B) = \dfrac{1}{2}$

$A \cap B$: The red die is 1, 2, or 3 and the green one is even; $P(A \cap B) = \dfrac{9}{36} = \dfrac{1}{4}$

$P(A)P(B) = \dfrac{1}{2} \times \dfrac{1}{2} = \dfrac{1}{4}$

$P(A \cap B) = \dfrac{1}{4}$

Since $P(A \cap B) = P(A)P(B)$, A and B are independent.

33. A: Exactly one die is 1; $P(A) = \dfrac{10}{36} = \dfrac{5}{18}$

B: The sum is even; $P(B) = \dfrac{1}{2}$

$A \cap B$: Exactly one die is 1 and the sum is even;

$P(A \cap B) = \dfrac{4}{36} = \dfrac{1}{9}$

$P(A)P(B) = \dfrac{5}{18} \times \dfrac{1}{2} = \dfrac{5}{36}$

$P(A \cap B) = \dfrac{1}{9}$

Since $P(A \cap B) \neq P(A)P(B)$, A and B are dependent.

35. A: Neither die is 1; $P(A) = \dfrac{25}{36}$

B: Exactly one die is 2; $P(B) = \dfrac{10}{36}$

$A \cap B$: Neither die is 1 and exactly one is 2;

$P(A \cap B) = \dfrac{8}{36}$

$P(A)P(B) = \dfrac{25}{36} \times \dfrac{10}{36} \approx .1929$

$P(A \cap B) = \dfrac{8}{36} = \dfrac{2}{9} \approx .2222$

Since $P(A \cap B) \neq P(A)P(B)$, A and B are dependent.

37. The outcome of each coin toss is independent of the others. Therefore,

$P(HTTHHHHTHHTT) = P(H)P(T)P(H) \ .. \ P(T)$

$= \dfrac{1}{2} \cdot \dfrac{1}{2} \cdot \ \cdots \ \cdot \dfrac{1}{2} = \dfrac{1}{2^{11}} = \dfrac{1}{2048}$

39. The two events we are interested in are:

B: A person in the US declared personal bankruptcy.

E: A person in the US recently experienced a "big three" event.

We are told that:

The probability that a person in the US would declare personal bankruptcy was .006. That is,

$P(B) = .006$.

The probability that a person in the US would declare personal bankruptcy *and* had recently experienced a "big three" event was .005. That is,

$P(E \cap B) = .005$

We are asked to find the probability that a person had recently experienced one of the "big three" events given that she had declared personal bankruptcy. That is, we are asked to find $P(E \mid B)$.

$P(E \mid B) = \dfrac{P(E \cap B)}{P(B)} = \dfrac{.005}{.006} \approx .8$

41. Take

A: A registered eBay user was an active user.

B: A registered eBay user had registered by the end of 2003.

$A \cap B$ is the event that a registered eBay user was an active user and had also registered by the end of 2003.

$P(A \cap B) = \dfrac{41}{136} \ ; P(B) = \dfrac{95}{136}$

$P(A \mid B) = \dfrac{P(A \cap B)}{P(B)} = \dfrac{41/136}{95.136} = \dfrac{41}{95} \approx .43$

Section 7.5

43. The two events are

A: Agreed that it should be the government's responsibility to provide a decent standard of living for the elderly.
$P(A) = .79$

B: Agreed that it would be a good idea to invest part of their Social Security taxes on their own. $P(B) = .43$

For independence,
$P(A \mid B) = P(A)P(B) = (.79)(.43) \approx .34$

45. Take

X: Used Brand X; $P(X) = .40$
G: Gave up doing laundry; $P(G) = .05$
$P(X \cap G) = .04$
$P(X)P(G) = .40 \times .05 = .02 \neq P(X \cap G)$

Therefore X and G are not independent.
$P(G \mid X) = \dfrac{P(G \cap X)}{P(X)} = \dfrac{.04}{.10} = .10,$

which is larger than $P(G)$. Therefore, a user of Brand X is more likely to give up doing laundry than a randomly chosen person.

47. Use the following events:

D: Involved in deadly accident.
T: Involved in tire-related accident.

We are told that
$P(D \cap T) = .000003$

We are also told that the probability that a tire-related accident would prove deadly was .02. Rephrasing this,

The probability that an accident is deadly, *given that* it is tire related, is .02.
$P(D \mid T) = .02$

We are asked to find $P(T)$. Start with the formula:
$P(D \mid T) = \dfrac{P(D \cap T)}{P(T)}$
$.02 = \dfrac{.000003}{P(T)}$
$.02P(T) = .000003$

$P(T) = \dfrac{.000003}{.02} = .00015$

49. $P(E \mid S) = \dfrac{P(E \cap S)}{P(S)} = \dfrac{.25}{.30} = \dfrac{5}{6}$
$or \, P(E \mid S) = \dfrac{fr(E \cap S)}{fr(S)} = \dfrac{25}{30} = \dfrac{5}{6}$

	N	E	Total
S	5	25	30
U	15	55	70
Total	20	80	100

51. $P(U \mid N) = \dfrac{P(U \cap N)}{P(N)} = \dfrac{.15}{.20} = .75$
$or \, P(U \mid N) = \dfrac{fr(U \cap N)}{fr(N)} = \dfrac{15}{20} = .75$

	N	E	Total
S	5	25	30
U	15	55	70
Total	20	80	100

53. $P(U \mid E) = \dfrac{P(U \cap E)}{P(E)} = \dfrac{.55}{.80} = \dfrac{11}{16}$
$or \, P(U \mid E) = \dfrac{fr(U \cap E)}{fr(E)} = \dfrac{55}{80} = \dfrac{11}{16}$

	N	E	Total
S	5	25	30
U	15	55	70
Total	20	80	100

55. P(An unsuccessful author is established)
$= P(E \mid U)$

$P(E \mid U) = \dfrac{P(E \cap U)}{P(U)} = \dfrac{.55}{.70} = \dfrac{11}{14}$
$or \, P(E \mid U) = \dfrac{fr(E \cap U)}{fr(U)} = \dfrac{55}{70} = \dfrac{11}{14}$

	N	E	Total
S	5	25	30
U	15	55	70
Total	20	80	100

57.

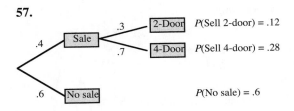

$P(\text{Sell 2-door}) = .12$

$P(\text{Sell 4-door}) = .28$

$P(\text{No sale}) = .6$

Note that the probability of each outcome was obtained by multiplying the probabilities of the corresponding edges.

59. Notice that the total number of vehicles add up to 100, so that the numbers give the probabilities for the fist branches of the tree:

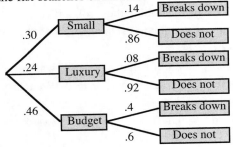

The probability of each outcome was obtained by multiplying the probabilities of the corresponding edges:

$P(\text{Small and breaks down})$
$= .30 \times .14 = .042$

$P(\text{Small and does not break down})$
$= .30 \times .86 = .258$

$P(\text{Luxury and breaks down})$
$= .24 \times .08 = .0192$

$P(\text{Luxury and does not break down})$
$= .24 \times .92 = .2208$

$P(\text{Budget and breaks down})$
$= .46 \times .4 = .184$

$P(\text{Budget and does not break down})$
$= .46 \times .6 = .276$

61.

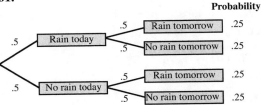

	Probability
Rain tomorrow	.25
No rain tomorrow	.25
Rain tomorrow	.25
No rain tomorrow	.25

(The probability of each outcome was obtained by multiplying the probabilities of the corresponding edges.)

From the tree,

$P(\text{No rain today or tomorrow}) = .25$

63. Use the following events:

A: Employed

B: Bachelor's degree or higher

$$P(A \mid B) = \frac{\text{fr}(A \cap B)}{\text{fr}(B)} = \frac{39.6}{51.8} \approx .76$$

65. Use the following events:

A: Bachelor's degree or higher

B: Employed

$$P(A \mid B) = \frac{\text{fr}(A \cap B)}{\text{fr}(B)} = \frac{39.6}{120.8} \approx .33$$

67. Rephrasing the question, we want to find the probability that someone was not in the labor force, given that they had not completed a Bachelor's degree or higher.

A: Not in labor force

B: Not completed a Bachelor's degree or higher

$$P(A \mid B) = \frac{\text{fr}(A \cap B)}{\text{fr}(B)} = \frac{61.5 - 11.2}{187.7 - 51.8} \approx .37$$

69. Rephrasing the question, we want to find the probability that someone was employed, given that he or she had completed a Bachelor's degree or higher and was in the labor force.

A: Employed

B: Bachelor's degree or higher and in the labor force

$$P(A \mid B) = \frac{\text{fr}(A \cap B)}{\text{fr}(B)} = \frac{39.6}{51.8\text{-}11.2} \approx .98$$

71. We compare

P(High school diploma only | Unemployed)

with

P(High school diploma only | Employed)

P(High school diploma only | Unemployed)

$$= \frac{1.9}{5.4} \approx .35$$

P(High school diploma only | Employed)

$$= \frac{36.2}{120.8} \approx .30$$

Your friend is right: The probability that an unemployed person has a high school diploma only is .35, while the corresponding figure for an employed person is .30.

73. $P(K|D) = 1.31P(K|D')$

75. (a) If the course has a positive effect on productivity, then the probability of a productivity increase is greater for a person who took the course. That is,

$P(I|T) > P(I)$

(b) If T and I are independent, than $P(I|T) = P(I)$, so that taking the course had no effect on productivity: The course was ineffective.

77. (a) Use the following events:

A: The person was an Internet user

B: The person's family income was at least $35,000

$$P(A|B) = \frac{P(A \cap B)}{P(B)} = \frac{.35}{.35 + .24} \approx .59$$

(b) We compare

P(Internet User | Income $\geq$ $35K)

with

P(Internet User | Income $<$ $35K)

By part (a),

P(Internet User | Income $<$ $35K) = .59

By a similar calculation

P(Internet User | Income $<$ $35K)

$$= \frac{.11}{.11 + .30} \approx .27$$

Therefore, a person was more likely to be an Internet user if his or her family income was $35,000 or more.

79. From the table, the probability that Jeep Wrangler was reported stolen was .0170. In other words,

The probability that a vehicle was reported stolen given that it was a Jeep Wrangler was .0170.

$P(R|J) = .0170$

81. From the table, the probability that Toyota Land Cruiser was reported stolen was .0143. In other words,

The probability that a vehicle was reported stolen given that it was a Toyota Land Cruiser was .0143.

This is exactly what Choice (D) says.

83. Consider the following events:

B: My BMW 3-series will be stolen;

$P(B) = .0077$

L: My Lexus GS300 will be stolen;

$P(L) = .0074$

(a) Since the events are independent,

$P(B \cap L) = P(B)P(L) = .0077 \times .0074 \approx .000057$

(b) $P(B \cup L) = P(B) + P(L) - P(B \cap L)$
$$= .0077 + .0074 - .000057$$
$$= .015043$$

85. Use the following events:

V: Tests positive

U: Use steroids

$P(U|V) = .90$, $P(U \cap V) = .10$

$$P(U|V) = \frac{P(U \cap V)}{P(V)}$$

$$.90 = \frac{.10}{P(V)} \Rightarrow .90P(V) = .10$$

$$P(V) = \frac{.10}{.90} \approx .11, \text{ or } 11\%$$

87. Use the following events:

A: Contaminated by salmonella

B: Contaminated by a strain of salmonella resistant to at least three antibiotics

We are told that

$P(A) = .20,$

and also that the probability that a salmonella-contaminated sample was contaminated by a strain resistant to at least three antibiotics was .53. Rewriting this gives:

The probability that a sample was contaminated by a strain resistant to at least three antibiotics *given that* it was contaminated by salmonella is .53.

$P(B|A) = .53$

We are asked to find $P(B \cap A)$ (which is actually the same as $P(B)$), so we use the formula:

$$P(B|A) = \frac{P(B \cap A)}{P(A)}$$

$$.53 = \frac{P(B \cap A)}{.20}$$

$$P(B \cap A) = .53 \times .20 = .106$$

89. Take

A: Contaminated by salmonella

$B1$: Contaminated by a strain of salmonella resistant to at least one antibiotic

$B3$: Contaminated by a strain of salmonella resistant to at least three antibiotics

We are told:

$P(A) = .20$, $P(B1|A) = .84$, $P(B3|A) = .53$

We are asked to find $P(B3|B1)$.

$$P(B3|B1) = \frac{P(B3 \cap B1)}{P(B1)} = \frac{P(B3)}{P(B1)} = \frac{P(B3 \cap A)}{P(B1 \cap A)}$$

because $B3 \cap B1 = B3 = B3 \cap A$, and $B1 = B1 \cap A$. But we have

$$P(B1|A) = \frac{P(B1 \cap A)}{P(A)} = .84$$

$$P(B3|A) = \frac{P(B3 \cap A)}{P(A)} = .53$$

Taking their ratio gives

$$\frac{P(B3 \cap A)}{P(B1 \cap A)} = \frac{.53}{.84} \approx .631$$

Therefore,

$$P(B3|B1) = \frac{P(B3 \cap A)}{P(B1 \cap A)} \approx .631$$

91. Answers will vary. Here is a simple one: E: the first toss is a head, F: the second toss is a head, G: the third toss is a head.

93. The probability you seek is $P(E|F)$, or should be. If, for example, you were going to place a wager on whether E occurs or not, it is crucial to know that the sample space has been reduced to F (you know that F did occur). If you base your wager on $P(E)$ rather than $P(E|F)$ you will misjudge your likelihood of winning.

95. If $A \subseteq B$ then $A \cap B = A$, so
$$P(A \cap B) = P(A)$$
and
$$P(A|B) = \frac{P(A \cap B)}{P(B)} = \frac{P(A)}{P(B)}$$

97. Your friend is correct. If A and B are mutually exclusive then

$P(A \cap B) = 0$.

On the other hand, if A and B are independent then

$P(A \cap B) = P(A)P(B)$.

Thus,

$P(A)P(B) = 0$.

If a product is 0 then one of the factors must be 0, so either $P(A) = 0$ or $P(B) = 0$. Thus, it cannot be true that A and B are mutually exclusive, have nonzero probabilities, and are independent all at the same time.

99. Suppose that A and B are independent, so that

$P(A \cap B) = P(A)P(B)$

Then,

$P(A' \cap B') = 1 - P(A \cup B)$

$= 1 - [P(A) + P(B) - P(A \cap B)]$

$= 1 - [P(A) + P(B) - P(A)P(B)]$

   [Since $P(A \cap B) = P(A)P(B)$]

$= (1 - P(A))(1 - P(B))$

$= P(A')P(B')$

Therefore,

$P(A' \cap B') = P(A')P(B')$

and so A' and B' are independent.

7.6

1. $P(A|B) = .8$, $P(B) = .2$, $P(A|B') = .3$

$P(B') = 1 - P(B) = 1 - .2 = .8$

$$P(B|A) = \frac{P(A|B)P(B)}{P(A|B)P(B) + P(A|B')P(B')}$$

$$= \frac{(.8)(.2)}{(.8)(.2) + (.3)(.8)} = \frac{.16}{.16 + .24} = .4$$

3. $P(X|Y) = .8$, $P(Y') = .3$, $P(X|Y') = .5$

$P(Y) = 1 - P(Y') = 1 - .3 = .7$,

$$P(Y|X) = \frac{P(X|Y)P(Y)}{P(X|Y)P(Y) + P(X|Y')P(Y')}$$

$$= \frac{(.8)(.7)}{(.8)(.7) + (.5)(.3)} = \frac{.56}{.56 + .15}$$

$$\approx 7887$$

5. $P(X|Y_1) = .4$, $P(X|Y_2) = .5$, $P(X|Y_3) = .6$,

$P(Y_1) = .8$, $P(Y_2) = .1$

$P(Y_3) = 1 - (.8 + .1) = .1$

$P(Y_1|X)$

$$= \frac{P(X|Y_1)P(Y_1)}{P(X|Y_1)P(Y_1) + P(X|Y_2)P(Y_2) + P(X|Y_3)P(Y_3)}$$

$$= \frac{(.4)(.8)}{(.4)(.8) + (.5)(.1) + (.6)(.1)}$$

$$= \frac{.32}{.32 + .05 + .06} \approx .7442$$

7. $P(X|Y_1) = .4$, $P(X|Y_2) = .5$, $P(X|Y_3) = .6$,

$P(Y_1) = .8$, $P(Y_2) = .1$

$P(Y_3) = 1 - (.8 + .1) = .1$

$P(Y_2|X)$

$$= \frac{P(X|Y_2)P(Y_2)}{P(X|Y_1)P(Y_1) + P(X|Y_2)P(Y_2) + P(X|Y_3)P(Y_3)}$$

$$= \frac{(.5)(.1)}{(.4)(.8) + (.5)(.1) + (.6)(.1)}$$

$$= \frac{.05}{.32 + .05 + .06} \approx .1163$$

9. Use the following events:

D: Decreased spending on music

I: Internet user

$P(D|I) = .11$, $P(I) = .40$, $P(D|I') = .2$

We are asked to compute $P(I|D)$

$$P(I|D) = \frac{P(D|I)P(I)}{P(D|I)P(I) + P(D|I')P(I')}$$

$$= \frac{(.11)(.4)}{(.11)(.4) + (.2)(.6)} = \frac{.044}{.044 + .12}$$

$$\approx .268, \text{ or } 26.8\%$$

11. Use the following events:

A: Snows in Greenland;

B: Glaciers grow;

$P(A) = \dfrac{1}{25} = .04$

$P(B|A) = .20$, $P(B|A') = .04$

We are asked to compute $P(A|B)$.

$$P(A|B) = \frac{P(B|A)P(A)}{P(B|A)P(A) + P(B|A')P(A')}$$

$$= \frac{(.2)(.04)}{(.2)(.04) + (.04)(.96)} = \frac{.008}{.008 + .0384}$$

$$\approx .1724$$

13. Use the following events

C: Driving a car

D: Driver dies (in a severe side-impact)

We are given $P(C) = .454$. Also, the information in the table tells us that $P(D \mid C) = 1$, $P(D \mid C')$ = .3. We are asked to find $P(C|D)$. Bayes' theorem states

$$P(C \mid D) = \frac{P(D|C)P(C)}{P(D|C)P(C) + P(D|C')P(C')} =$$

$$\frac{.454}{.454 + (.3)(1 - .454)} \approx .73$$

15. Use the following events:

A: Fit enough to play

B: Failed the fitness test and therefore dropped from the team

$P(B|A) = 5$, $P(B|A') = 1$

$P(A) = .45$, so $P(A') = .55$

We are asked to compute the probability that Mona was justifiably dropped.

$$P(A|B) = \frac{P(B|A)P(A)}{P(B|A)P(A) + P(B|A')P(A')}$$

$$= \frac{(.5)(.45)}{(.5)(.45) + (1)(.55)} = \frac{.225}{.225 + .55}$$

$$\approx .2903$$

This is the probability that she was unjustifiably dropped. Therefore, the probability that Mona was justifiably dropped is

$$P(A'|B) = 1 - P(A|B) \approx 1 - .2309 = .7097$$

17. Use the following events

L: Driving a light truck

V: Driving an SUV

C: Driving a car

D: Driver dies (in a severe side-impact)

We are given the following information:

$P(L) = .273$, $P(V) = .273$, $P(C) = .454$. Also, the information in the table tells us that

$P(D|T) = .210$, $P(D|V) = .271$, $P(D|C) = 1$.

We are asked to find the probability that a driver was driving an SUV, given that the driver died (in a severe side-impact), that is, $P(V|D)$. Bayes' theorem states

$$P(V|D)$$

$$= \frac{P(D|V)P(V)}{P(D|L)P(L) + P(D|V)P(V) + P(D|C)P(C)}$$

$$= \frac{(.371)(.273)}{(.210)(.273) + (.371)(.273) + (1)(.454)}$$

$$\approx .165$$

19. Use the following events:

A: Admitted into UCLA

C: California applicant

U: Applicant from another U.S. state

I: International applicant

$P(A|C) = .24$, $P(A|U) = .22$, $P(A|I) = .20$

$P(C) = .87$, $P(U) = .09$, $P(I) = .04$

We are asked to compute $P(C|A)$

$$P(C|A)$$

$$= \frac{P(A|C)P(C)}{P(A|C)P(C) + P(A|U)P(U) + P(A|I)P(I)}$$

$$= \frac{(.24)(.87)}{(.24)(.87) + (.22)(.09) + (.20)(.04)}$$

$$\approx .8825, \text{ or approximately } 88\%$$

21. Use the following events:

I: Used the Internet for e-mail

C: Caucasian

A: African American

H: Hispanic

R: Other

$P(I|C) = .86$, $P(I|A) = .77$, $P(I|H) = .77$,

$P(I|R) = .85$

$P(C) = .69$, $P(A) = .12$, $P(H) = .13$,

$P(R) = 1 - (.69 + .12 + .13) = .06$

We are asked to compute $P(H|I)$

$$P(H|I)$$

$$= \frac{P(I|H)P(H)}{P(I|C)P(C) + P(I|A)P(A) + P(I|H)P(H) + P(I|R)P(R)}$$

$$= \frac{(.77)(.13)}{(.86)(.69) + (.77)(.12) + (.77)(.13) + (.85)(.06)}$$

$$\approx .1196, \text{ or approximately } 12\%$$

23. Use the following events:

M: Married

L: Have pool

$P(M|L) = .86$, $P(L) = .15$

(a) Given that $P(M|L') = .90$, we are asked to find $P(L|M)$

$$P(L|M) = \frac{P(M|L)P(L)}{P(M|L)P(L) + P(M|L')P(L')}$$

$$= \frac{(.86)(.15)}{(.86)(.15) + (.9)(.85)} = \frac{.129}{.129 + .765}$$

$$\approx .1443, \text{ or } 14.43\%$$

(b) From part (a), 14.43% of married couples have pools. The percentage of single people with pools is obtained as follows: Take

S: Single

L: Have pool

$P(L) = .15$,

$P(S|L) = 1 - .86 = .14$

$P(S|L') = 1 - .90 = .10$ (From part (b))

$$P(L|S) = \frac{P(S|L)P(L)}{P(S|L)P(L) + P(S|L')P(L')}$$

$$= \frac{(.14)(.15)}{(.14)(.15) + (.1)(.85)} = \frac{.021}{.021 + .085}$$

$$\approx .1981, \text{ or } 19.81\%$$

Therefore, 19.81% of single homeowners have pools. Thus pool manufacturers should go after the single homeowners.

25. Use the following events:

A: Former student of Prof. A.

C: Earned C– or lower

All of Professor A's former students wound up with a C– or lower. In other words,

$P(C|A) = 1$

Two thirds of students not from Prof A's class got better than a C–. In other words,

$$P(C|A') = \frac{1}{3}$$

Three quarters of Professor F's class consisted of former students of Professor A, so

$P(A) = .75$

We are asked to find what percentage of the students who got C– or worse were former students of Prof. A: $P(A|C)$.

$$P(A|C) = \frac{P(C|A)P(A)}{P(C|A)P(A) + P(C|A')P(A')}$$

$$= \frac{(1)(.75)}{(1)(.75) + (1/3)(.25)} = \frac{.75}{.75 + 1/3}$$

$$= .9$$

Therefore, we estimate that 9 of the 10 students in the delegation were former students of Prof. A.

27. Use the following events:

H: Husband employed

W: wife employed

$P(H) = .95$, $P(W|H) = .71$

Since either the husband or wife in a couple with earnings had to be employed,

$P(W|H') = 1$

We are asked to find $P(H|W)$.

$$P(H|W) = \frac{P(W|H)P(H)}{P(W|H)P(H) + P(W|H')P(H')}$$

$$= \frac{(.71)(.95)}{(.71)(.95) + (1)(.05)} = \frac{.6745}{.6745 + .05}$$

$$\approx .9310$$

29. Use the following events:

A: Arrested by age 14

C: Become a chronic offender

$P(C|A) = 17.9P(C|A')$

$P(A) = .001$

$$P(A|C) = \frac{P(C|A)P(A)}{P(C|A)P(A) + P(C|A')P(A')}$$

$$= \frac{17.9P(C|A')(.001)}{17.9P(C|A')(.001) + P(C|A')(.999)}$$

Now cancel the $P(C|A')$ to get

$$\frac{(17.9)(.001)}{(17.9)(.001) + .999} = \frac{.0179}{.0179 + .999}$$

$$\approx .0176, \text{ or } 17.6\%$$

31. Use the following events:

D: has diabetes

A: very active

$P(D|A) = .5P(D|A')$

$$P(A) = \frac{1}{3}$$

$$P(A|D) = \frac{P(D|A)P(A)}{P(D|A)P(A) + P(D|A\,')P(A\,')}$$

$$= \frac{.5P(D|A')P(A)}{.5P(D|A)P(A) + P(D|A\,')P(A\,')}$$

Cancel the $P(D|A')$ to obtain

$$\frac{.5P(A)}{.5P(A) + P(A\,')}$$

$$= \frac{(.5)(1/3)}{(.5)(1/3) + 2/3} = \frac{1/6}{1/6 + 2/3} = .2$$

33. Use the following events:

K: The child was killed

D: The airbag deployed

$P(K|D) = 1.31P(K|D\,')$

$P(D) = .25$

We are asked to compute $P(D|K)$.

$$P(D|K) = \frac{P(K|D)P(D)}{P(K|D)P(D) + P(K|D\,')P(D\,')}$$

$$= \frac{1.31P(K|D\,')P(D)}{1.31P(K|D\,')P(D) + P(K|D\,')P(D\,')}$$

Cancel the terms $P(K|D')$ to obtain

$$\frac{1.31P(D)}{1.31P(D) + P(D\,')}$$

$$= \frac{(1.31)(.25)}{(1.31)(.25) + .75} = \frac{.3275}{.3275 + .75}$$

$$\approx .30$$

35. Show him an example like Example 1 of this section, where $P(T|A) = .95$ but $P(A|T) \approx .64$.

37. Suppose that the steroid test gives 10% false negatives and that only 0.1% of the tested population uses steroids. Then the probability that an athlete uses steroids, given that he or she has tested positive, is $\approx .083$.

39. Draw a tree in which the first branching shows which of R_1, R_2, or R_3 occurred, and the second branching shows which of T or $T\,'$ then occurred. There are three final outcomes in which T occurs:

$P(R_1 \cap T) = P(T \mid R_1)P(R_1)$

$P(R_2 \cap T) = P(T \mid R_2)P(R_2)$

$P(R_3 \cap T) = P(T \mid R_3)P(R_3)$

In only one of these, the first, does R_1 occur. Thus,

$$P(R_1 \mid T) = \frac{P(R_1 \cap T)}{P(T)} =$$

$$\frac{P(T|R_1)P(R_1)}{P(T|R_1)P(R_1) + P(T|R_2)P(R_2) + P(T|R_3)P(R_3)}$$

41. The reasoning is flawed. Let A be the event that a Democrat agrees with Safire's column, and let F and M be the events that a Democrat reader is female and male respectively. Then A. D. makes the following argument:

$P(M|A) = .9$, $P(F|A') = .9$.

Therefore,

$P(A|M) = .9$.

According to Bayes' Theorem we cannot conclude anything about $P(A|M)$ unless we know $P(A)$, the percentage of all Democrats who agreed with Safire's column. This was not given.

7.7

1. The transition matrix P is organized as follows:

$$P = \begin{bmatrix} 1\to1 & 1\to2 \\ 2\to1 & 2\to2 \end{bmatrix}$$

Therefore, the diagram

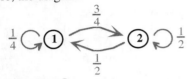

has $P = \begin{bmatrix} 1/4 & 3/4 \\ 1/2 & 1/2 \end{bmatrix}$

3. The transition matrix P is organized as follows:

$$P = \begin{bmatrix} 1\to1 & 1\to2 \\ 2\to1 & 2\to2 \end{bmatrix}$$

Therefore, the diagram

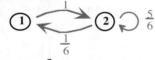

has $P = \begin{bmatrix} 0 & 1 \\ 1/6 & 5/6 \end{bmatrix}$

5. The transition matrix P is organized as follows:

$$P = \begin{bmatrix} 1\to1 & 1\to2 & 1\to3 \\ 2\to1 & 2\to2 & 2\to3 \\ 3\to1 & 3\to2 & 3\to3 \end{bmatrix}$$

Therefore, the diagram

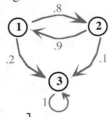

has $P = \begin{bmatrix} 0 & .8 & .2 \\ .9 & 0 & .1 \\ 0 & 0 & 1 \end{bmatrix}$

7. The transition matrix P is organized as follows:

$$P = \begin{bmatrix} 1\to1 & 1\to2 & 1\to3 \\ 2\to1 & 2\to2 & 2\to3 \\ 3\to1 & 3\to2 & 3\to3 \end{bmatrix}$$

Therefore, the diagram

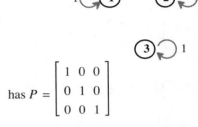

has $P = \begin{bmatrix} 1 & 0 & 0 \\ 0 & 1 & 0 \\ 0 & 0 & 1 \end{bmatrix}$

9. The diagram

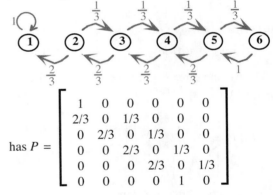

has $P = \begin{bmatrix} 1 & 0 & 0 & 0 & 0 & 0 \\ 2/3 & 0 & 1/3 & 0 & 0 & 0 \\ 0 & 2/3 & 0 & 1/3 & 0 & 0 \\ 0 & 0 & 2/3 & 0 & 1/3 & 0 \\ 0 & 0 & 0 & 2/3 & 0 & 1/3 \\ 0 & 0 & 0 & 0 & 1 & 0 \end{bmatrix}$

11. (a) $P^2 = \begin{bmatrix} .5 & .5 \\ 0 & 1 \end{bmatrix} \begin{bmatrix} .5 & .5 \\ 0 & 1 \end{bmatrix}$

$= \begin{bmatrix} .25 & .75 \\ 0 & 1 \end{bmatrix}$

(b) distribution after one step:

$$vP = [1 \ \ 0] \begin{bmatrix} .5 & .5 \\ 0 & 1 \end{bmatrix} = [.5 \ \ .5]$$

after two steps:

$$vP^2 = [.5 \ \ .5] \begin{bmatrix} .5 & .5 \\ 0 & 1 \end{bmatrix} = [.25 \ \ .75]$$

after three steps:

$$vP^3 = [.25 \quad .75] \begin{bmatrix} .5 & .5 \\ 0 & 1 \end{bmatrix} = [.125 \quad .875]$$

13. (a) $P^2 = \begin{bmatrix} .2 & .8 \\ .4 & .6 \end{bmatrix} \begin{bmatrix} .2 & .8 \\ .4 & .6 \end{bmatrix} = \begin{bmatrix} .36 & .64 \\ .32 & .68 \end{bmatrix}$

(b) distribution after one step:

$$vP = [.5 \quad .5] \begin{bmatrix} .2 & .8 \\ .4 & .6 \end{bmatrix} = [.3 \quad .7]$$

after two steps:

$$vP^2 = [.3 \quad .7] \begin{bmatrix} .2 & .8 \\ .4 & .6 \end{bmatrix} = [.34 \quad .66]$$

after three steps:

$$vP^3 = [.34 \quad .66] \begin{bmatrix} .2 & .8 \\ .4 & .6 \end{bmatrix} = [.332 \quad .668]$$

15. (a) $P^2 = \begin{bmatrix} 1/2 & 1/2 \\ 1 & 0 \end{bmatrix} \begin{bmatrix} 1/2 & 1/2 \\ 1 & 0 \end{bmatrix}$

$$= \begin{bmatrix} 3/4 & 1/4 \\ 1/2 & 1/2 \end{bmatrix}$$

(b) distribution after one step:

$$vP = [2/3 \quad 1/3] \begin{bmatrix} 1/2 & 1/2 \\ 1 & 0 \end{bmatrix} = [2/3 \quad 1/3]$$

after two and three steps, the result will remain the same: [2/3 1/3]

17. (a) $P^2 = \begin{bmatrix} 3/4 & 1/4 \\ 3/4 & 1/4 \end{bmatrix} \begin{bmatrix} 3/4 & 1/4 \\ 3/4 & 1/4 \end{bmatrix}$

$$= \begin{bmatrix} 3/4 & 1/4 \\ 3/4 & 1/4 \end{bmatrix}$$

(b) distribution after one step:

$$vP = [1/2 \quad 1/2] \begin{bmatrix} 3/4 & 1/4 \\ 3/4 & 1/4 \end{bmatrix} = [3/4 \quad 1/4]$$

after two steps:

$$vP^2 = [3/4 \quad 1/4] \begin{bmatrix} 3/4 & 1/4 \\ 3/4 & 1/4 \end{bmatrix} = [3/4 \quad 1/4]$$

after three steps, the result will remain the same: [3/4 1/4]

19. (a) $P^2 = \begin{bmatrix} .5 & .5 & 0 \\ 0 & 1 & 0 \\ 0 & .5 & .5 \end{bmatrix} \begin{bmatrix} .5 & .5 & 0 \\ 0 & 1 & 0 \\ 0 & .5 & .5 \end{bmatrix}$

$$= \begin{bmatrix} .25 & .75 & 0 \\ 0 & 1 & 0 \\ 0 & .75 & .25 \end{bmatrix}$$

(b) distribution after one step:

$$vP = [1 \quad 0 \quad 0] \begin{bmatrix} .5 & .5 & 0 \\ 0 & 1 & 0 \\ 0 & .5 & .5 \end{bmatrix} = [.5 \quad .5 \quad 0]$$

after two steps:

$$vP^2 = [.5 \quad .5 \quad 0] \begin{bmatrix} .5 & .5 & 0 \\ 0 & 1 & 0 \\ 0 & .5 & .5 \end{bmatrix}$$

$$= [.25 \quad .75 \quad 0]$$

after three steps:

$$vP^3 = [.25 \quad .75 \quad 0] \begin{bmatrix} .5 & .5 & 0 \\ 0 & 1 & 0 \\ 0 & .5 & .5 \end{bmatrix}$$

$$= [.125 \quad .875 \quad 0]$$

21. (a) $P^2 = \begin{bmatrix} 0 & 1 & 0 \\ 1/3 & 1/3 & 1/3 \\ 1 & 0 & 0 \end{bmatrix} \begin{bmatrix} 0 & 1 & 0 \\ 1/3 & 1/3 & 1/3 \\ 1 & 0 & 0 \end{bmatrix}$

$$= \begin{bmatrix} 1/3 & 1/3 & 1/3 \\ 4/9 & 4/9 & 1/9 \\ 0 & 1 & 0 \end{bmatrix}$$

(b) distribution after one step:

$$vP = [1/2 \quad 0 \quad 1/2] \begin{bmatrix} 0 & 1 & 0 \\ 1/3 & 1/3 & 1/3 \\ 1 & 0 & 0 \end{bmatrix}$$

$$= [1/2 \quad 1/2 \quad 0]$$

after two steps:

$$vP^2 = [1/2 \quad 1/2 \quad 0] \begin{bmatrix} 0 & 1 & 0 \\ 1/3 & 1/3 & 1/3 \\ 1 & 0 & 0 \end{bmatrix}$$

$$= [1/6 \quad 2/3 \quad 1/6]$$

after three steps:

$$vP^3 = [1/6 \quad 2/3 \quad 1/6] \begin{bmatrix} 0 & 1 & 0 \\ 1/3 & 1/3 & 1/3 \\ 1 & 0 & 0 \end{bmatrix}$$

$$= [7/18 \quad 7/18 \quad 2/9]$$

23. (a) $P^2 = \begin{bmatrix} .1 & .9 & 0 \\ 0 & 1 & 0 \\ 0 & .2 & .8 \end{bmatrix} \begin{bmatrix} .1 & .9 & 0 \\ 0 & 1 & 0 \\ 0 & .2 & .8 \end{bmatrix}$

$= \begin{bmatrix} .01 & .99 & 0 \\ 0 & 1 & 0 \\ 0 & .36 & .64 \end{bmatrix}$

(b) distribution after one step:

$vP = [.5 \ \ 0 \ \ .5] \begin{bmatrix} .1 & .9 & 0 \\ 0 & 1 & 0 \\ 0 & .2 & .8 \end{bmatrix} = [.05 \ \ .55 \ \ .4]$

after two steps:

$vP^2 = [.05 \ \ .55 \ \ .4] \begin{bmatrix} .1 & .9 & 0 \\ 0 & 1 & 0 \\ 0 & .2 & .8 \end{bmatrix}$

$= [.005 \ \ .675 \ \ .32]$

after three steps:

$vP^3 = [.005 \ \ .675 \ \ .32] \begin{bmatrix} .1 & .9 & 0 \\ 0 & 1 & 0 \\ 0 & .2 & .8 \end{bmatrix}$

$= [.0005 \ \ .7435 \ \ .256]$

25. To find the steady-state vector $v_\infty = [x \ \ y]$ we solve

$x + y = 1$

$[x \ \ y] \begin{bmatrix} 1/2 & 1/2 \\ 1 & 0 \end{bmatrix} = [x \ \ y]$

The above equations give:

$x + y = 1$

$x/2 + y = x$

$x/2 = y$

Rewriting in standard form and dropping the last equation (which is the same as the next-to-last) we get

$x + y = 1$

$-x/2 + y = 0$

Solving this system gives

$x = 2/3, \ y = 1/3$

So, $v_\infty = [2/3 \ \ 1/3]$

27. To find the steady-state vector $v_\infty = [x \ \ y]$ we solve

$x + y = 1$

$[x \ \ y] \begin{bmatrix} 1/3 & 2/3 \\ 1/2 & 1/2 \end{bmatrix} = [x \ \ y]$

The above equations give:

$x + y = 1$

$x/3 + y/2 = x$

$2x/3 + y/2 = y$

Rewriting in standard form and dropping the last equation (which is the same as the next-to-last) we get

$x + y = 1$

$-2x/3 + y/2 = 0$

Solving this system gives

$x = 3/7, \ y = 4/7$

So, $v_\infty = [3/7 \ \ 4/7]$

29. To find the steady-state vector $v_\infty = [x \ \ y]$ we solve

$x + y = 1$

$[x \ \ y] \begin{bmatrix} .1 & .9 \\ .6 & .4 \end{bmatrix} = [x \ \ y]$

The above equations give:

$x + y = 1$

$.1x + .6y = x$

$.9x + .4y = y$

Rewriting in standard form and dropping the last equation (which is the same as the next-to-last) we get

$x + y = 1$

$-.9x + .6y = 0$

Solving this system gives

$x = 2/5, \ y = 3/5$

So, $v_\infty = [2/5 \ \ 3/5]$

31. To find the steady-state vector $v_\infty = [x \ \ y \ \ z]$ we solve

$x + y + z = 1$

$[x \ \ y \ \ z] \begin{bmatrix} .5 & 0 & .5 \\ 1 & 0 & 0 \\ 0 & .5 & .5 \end{bmatrix} = [x \ \ y \ \ z]$

The above equations give:

$x + y + z = 1$
$.5x + y = x$
$.5z = y$
$.5x + .5z = z$

Rewriting in standard form we get

$x + y \quad + z = 1$
$-.5x + y \quad = 0$
$\quad -y + .5z = 0$
$.5x \quad - .5z = 0$

Solving this system gives

$x = 2/5, \ y = 1/5, \ z = 2/5$

So, $v_\infty = [2/5 \quad 1/5 \quad 2/5]$

33. To find the steady-state vector $v_\infty = [x \ y \ z]$ we solve

$x + y + z = 1$

$[x \ y \ z] \begin{bmatrix} 0 & 1 & 0 \\ 1/3 & 1/3 & 1/3 \\ 1 & 0 & 0 \end{bmatrix} = [x \ y \ z]$

The above equations give:

$x + y + z = 1$
$y/3 + z = x$
$x + y/3 = y$
$y/3 = z$

Rewriting in standard form we get

$x \quad + y \quad + z = 1$
$-x + \quad y/3 + z = 0$
$x - 2y/3 \quad = 0$
$\quad y/3 - z = 0$

Solving this system gives

$x = 1/3, \ y = 1/2, \ z = 1/6$

So, $v_\infty = [1/3 \quad 1/2 \quad 1/6]$

35. To find the steady-state vector $v_\infty = [x \ y \ z]$ we solve

$x + y + z = 1$

$[x \ y \ z] \begin{bmatrix} .1 & .9 & 0 \\ 0 & 1 & 0 \\ 0 & .2 & .8 \end{bmatrix} = [x \ y \ z]$

The above equations give:

$x + y + z = 1$
$.1x = x$
$.9x + y + .2z = y$
$.8z = z$

Rewriting in standard form we get

$x + y + z = 1$
$-.9x \quad = 0$
$.9x \quad + .2z = 0$
$\quad -.2z = 0$

Solving this system gives

$x = 0, \ y = 1, \ z = 0$

So, $v_\infty = [0 \quad 1 \quad 0]$

37. Take 1 = Sorey State, 2 = C&T

$P = \begin{bmatrix} 1/2 & 1/2 \\ 1/4 & 3/4 \end{bmatrix}$

We are asked to find the (1, 1) entry of the 2-step transition probability matrix.

$P^2 = \begin{bmatrix} 1/2 & 1/2 \\ 1/4 & 3/4 \end{bmatrix} \begin{bmatrix} 1/2 & 1/2 \\ 1/4 & 3/4 \end{bmatrix}$

$= \begin{bmatrix} 3/8 & 5/8 \\ 5/16 & 11/16 \end{bmatrix}$

The (1, 1) entry is $3/8 = .375$

39. (a) Take 1 = Not checked in, 2 = Checked in

$P = \begin{bmatrix} .4 & .6 \\ 0 & 1 \end{bmatrix}$

$P^2 = \begin{bmatrix} .4 & .6 \\ 0 & 1 \end{bmatrix} \begin{bmatrix} .4 & .6 \\ 0 & 1 \end{bmatrix} = \begin{bmatrix} .16 & .84 \\ 0 & 1 \end{bmatrix}$

$P^3 = \begin{bmatrix} .4 & .6 \\ 0 & 1 \end{bmatrix} \begin{bmatrix} .16 & .84 \\ 0 & 1 \end{bmatrix}$

$= \begin{bmatrix} .064 & .936 \\ 0 & 1 \end{bmatrix}$

(b) 1 hour: $P_{12} = .6$

2 hours: $(P^2)_{12} = .84$

3 hours: $(P^3)_{12} = .936$

(c) Eventually, all the roaches will have checked in.

41. Take 1 = High risk, 2 = Low risk

$$P = \begin{bmatrix} .50 & .50 \\ .10 & .90 \end{bmatrix}$$

To find the steady-state vector $[x \ y]$ we solve

$x + y = 1$

$$[x \ y] \begin{bmatrix} .50 & .50 \\ .10 & .90 \end{bmatrix} = [x \ y]$$

The above equations give:

$x + y = 1$

$.5x + .1y = x$

$.5x + .9y = y$

Rewriting in standard form and dropping the last equation, we get

$x + y = 1$

$-.5x + .1y = 0$

Solving this system gives

$x = 1/6, \ y = 5/6$

So, $[x \ y] = [1/6 \ 5/6]$

In the long-term, $1/6 \approx 16.67\%$ fall into the high-risk category and $5/65 \approx 83.33\%$ into the low risk category.

43. (a) Take 1 = User, 2 = Non-User

From Exercise 25 in Section 9.1,

$$P = \begin{bmatrix} 2/3 & 1/3 \\ 1/10 & 9/10 \end{bmatrix}$$

The 2-year transition matrix is

$$P^2 = \begin{bmatrix} 2/3 & 1/3 \\ 1/10 & 9/10 \end{bmatrix}\begin{bmatrix} 2/3 & 1/3 \\ 1/10 & 9/10 \end{bmatrix}$$

$$= \begin{bmatrix} 43/90 & 47/90 \\ 47/300 & 253/300 \end{bmatrix}$$

Non-user→User in 2 steps:

$(P^2)_{21} = 47/300 \approx .156667$

(b) To find the steady-state vector $[x \ y]$ we solve

$x + y = 1$

$$[x \ y] \begin{bmatrix} 2/3 & 1/3 \\ 1/10 & 9/10 \end{bmatrix} = [x \ y]$$

The above equations give:

$x + y = 1$

$(2/3)x + (1/10)y = x$

$(1/3)x + (9/10)y = y$

Rewriting in standard form and dropping the last equation, we get

$x + y = 1$

$-(1/3)x + (1/10)y = 0$

Solving this system gives

$x = 3/13, \ y = 10/13$

So, $[x \ y] = [3/13 \ 10/13]$

In the long-term, $3/13$ of the college instructors will be users of this book.

45. Take 1 = Paid up, 2 = 0–90 Days, 3 = Bad debt

$$P = \begin{bmatrix} .5 & .5 & 0 \\ .5 & .3 & .2 \\ 0 & .5 & .5 \end{bmatrix}$$

Steady-state vector:

$x + y + z = 1$

$$[x \ y \ z] \begin{bmatrix} .5 & .5 & 0 \\ .5 & .3 & .2 \\ 0 & .5 & .5 \end{bmatrix} = [x \ y \ z]$$

The above equations give:

$x + y + z = 1$

$.5x + .5y = x$

$.5x + .3y + .5z = y$

$.2y + .5z = z$

Rewriting in standard form and dropping the last equation, we get

$x + y + z = 1$

$-.5x + .5y = 0$

$.5x - .7y + .5z = 0$

Solving gives

$x = 5/12, \ y = 5/12, \ z = 1/6$

So, $[x \ y \ z] = [5/12 \ 5/12 \ 1/6]$

Answer:

5/12 or approximately 41.67% of the customers will be in the Paid up category, 5/12 or approximately 41.67% in the 0-90 days category, and 1/6 or approximately 16.67% in the bad debt category.

47. (a) Take 1 = Affluent, 2 = Middle class, 3 = Poor

$p_{11} = 1-.729 = .271, p_{12} = .729, p_{13} = 0$

$p_{21} = .075, p_{22} = 1-(.075+.085) = .84, p_{23} = .085$

$p_{31} = 0, p_{32} = .304, p_{33} = 1-.304 = .696$

$$P = \begin{bmatrix} .729 & .271 & 0 \\ .075 & .84 & .085 \\ 0 & .304 & .696 \end{bmatrix}$$

(b) The period 1980–2002 is a 22-year period, corresponding to two 11-year transition steps.

$$P^2 = \begin{bmatrix} .729 & .271 & 0 \\ .075 & .84 & .085 \\ 0 & .304 & .696 \end{bmatrix}\begin{bmatrix} .729 & .271 & 0 \\ .075 & .84 & .085 \\ 0 & .304 & .696 \end{bmatrix}$$

$$\approx \begin{bmatrix} .552 & .425 & .023 \\ .118 & .752 & .131 \\ .023 & .467 & .51 \end{bmatrix}$$

Affluent→Poor in 2 steps:

$(P^2)_{12} = .023$, or 2.3%

(c) Let us use technology to compute a high enough power of P directly so that the row are approximately the same:

$$P^\infty \approx P^{64} \approx \begin{bmatrix} .178 & .643 & .18 \\ .178 & .643 & .18 \\ .178 & .643 & .18 \end{bmatrix}$$

(We used the 64th power because powers of 2 are computed most efficiently in the online Matrix Algebra Tool.)

Answer:

Affluent: 17.8%; Middle class: 64.3%; Poor: 18.0%

49. Take 1 = Bottom 10%, 2 = 10–50%, 3 = 50–99%, 4 = Top 10%.

The transition matrix is given by the table:

$$P = \begin{bmatrix} .3 & .52 & .17 & .01 \\ .1 & .48 & .38 & .04 \\ .04 & .38 & .48 & .1 \\ .01 & .17 & .52 & .3 \end{bmatrix}$$

Let us use technology to compute a high enough power of P directly so that the row are approximately the same:

$$P^\infty \approx P^{64} \approx \begin{bmatrix} .0843 & .4157 & .4157 & .0843 \\ .0843 & .4157 & .4157 & .0843 \\ .0843 & .4157 & .4157 & .0843 \\ .0843 & .4157 & .4157 & .0843 \end{bmatrix}$$

(We used the 64th power because powers of 2 are computed most efficiently in the online Matrix Algebra Tool.)

Thus, the percentages in each category are given by [8.43 41.57 41.57 8.43].

51. (a) Number the states as follows: 1: Verizon, 2: Cingular, 3: AT&T, 4: Other. The figures next to the arrows show all the transition probabilities (obtained by dividing by 100) except the ones from each state to itself. To compute the missing probabilities, use the fact that the entries in each row add to 1. This leads to:

$$P = \begin{bmatrix} .981 & .005 & .005 & .009 \\ .01 & .972 & .006 & .012 \\ .01 & .006 & .973 & .011 \\ .008 & .006 & .005 & .981 \end{bmatrix}$$

(b) Each time-step is one quarter. At the *end* of the third quarter, the distribution is given as

$w = [0.297 \quad 0.193 \quad 0.181 \quad 0.329]$

We are asked to find the distribution v at the beginning of that quarter: one time-step earlier. We know that

Distribution after one quarter

$= vP = [0.297 \quad 0.193 \quad 0.181 \quad 0.329]$

To obtain v, multiply both sides on the right by P^{-1}.

$v = [0.297 \quad 0.193 \quad 0.181 \quad 0.329]P^{-1}$

On the Matrix Algebra Tool, use the format w*P^-1: We obtain (rounding to 3 decimal places:

$v = [0.296 \quad 0.194 \quad 0.182 \quad 0.328]$.

Thus, the market shares at the beginning of the quarter were:

Verizon: 29.6%, Cingular: 19.3%, AT&T: 18.1%, Other: 32.8%.

(c) The end of 2005 is 9 quarters from the end of the third quarter in 2003. Therefore, the distribution is predicted as

$wP^9 = [0.303 \quad 0.186 \quad 0.176 \quad 0.335]$

Verizon: 30.3%, Cingular: 18.6%, AT&T: 17.6%, Other: 33.5%. (Technology format: w*P^9) The biggest gainers are Verizon and Other, each gaining 0.6%

53. From the diagram, the transition matrix is

$$P = \begin{bmatrix} 1/2 & 1/2 & 0 & 0 & 0 \\ 1/2 & 0 & 1/2 & 0 & 0 \\ 0 & 1/2 & 0 & 1/2 & 0 \\ 0 & 0 & 1/2 & 0 & 1/2 \\ 0 & 0 & 0 & 1/2 & 1/2 \end{bmatrix}$$

For the steady-state vector, we solve:

$x + y + z + u + v = 1$

$(1/2)x + (1/2)y = x$

$(1/2)x + (1/2)z = y$

$(1/2)y + (1/2)u = z$

$(1/2)z + (1/2)v = u$

$(1/2)u + (1/2)v = v$

Rewriting in standard form and dropping the last equation, we get the system

$x + y + z + u + v = 1$

$-(1/2)x + (1/2)y = 0$

$(1/2)x - y + (1/2)z = 0$

$(1/2)y - z + (1/2)u = 0$

$(1/2)z - u + (1/2)v = 0$

Solving gives

$x = 1/5, y = 1/5, z = 1/5, u = 1/5, v = 1/5$

Hence the steady-state vector is

$[1/5 \quad 1/5 \quad 1/5 \quad 1.5]$.

The system spends an average of 1/5 of the time in each state.

55. Answers will vary.

57. There are two assumptions made by Markov systems that may not be true about the stock market: the assumption that the transition probabilities do not change over time, and the assumption that the transition probability depends only on the current state.

59. If q is a row of Q, then by assumption $qP = q$. Thus, when we multiply the rows of Q by P, nothing changes, and $QP = Q$.

61. At each step only 0.4 of the population in state 1 remains there, and nothing enters from any other state. Thus, when the first entry in the steady-state distribution vector is multiplied by 0.4 it must remain unchanged. The only number for which this true is 0.

63. An example is

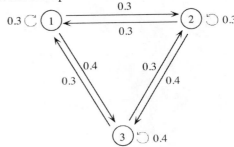

Its transition matrix is

$$P = \begin{bmatrix} .3 & .3 & .4 \\ .3 & .3 & .4 \\ .3 & .3 & .4 \end{bmatrix}$$

and is therefore already in a steady state!

65. If $vP = v$ and $wP = w$, then

$$\tfrac{1}{2}(v + w)\,P = \tfrac{1}{2}\,vP + \tfrac{1}{2}\,wP$$

$$= \tfrac{1}{2}\,v + \tfrac{1}{2}\,w = \tfrac{1}{2}(v + w)\,.$$

Further, if the entries of v and w add up to 1, then so do the entries of $(v + w)/2$.

Chapter 7 Review Exercises

1. $n(S) = 2 \times 2 \times 2 = 8$

$E = \{\text{HHT, HTH, HTT, THH, THT, TTH, TTT}\}$

$$P(E) = \frac{n(E)}{n(S)} = \frac{7}{8}$$

2. $n(S) = 2 \times 2 \times 2 \times 2 = 16$

$E = \{\text{HTTT, THTT, TTHT, TTTH, TTTT}\}$

$$P(E) = \frac{n(E)}{n(S)} = \frac{5}{16}$$

3. $n(S) = 6 \times 6 = 36$

$E = \{(1,6), (2,5), (3,4), (4,3), (5,2), (6,1)\}$

$$P(E) = \frac{n(E)}{n(S)} = \frac{6}{36} = \frac{1}{6}$$

4. $n(S) = 6^3 = 216$

$E = \{(1,1,3),\ (1,3,1),\ (3,1,1),\ (1,2,2),\ (2,1,2),\ (2,2,1)\}$

$$P(E) = \frac{n(E)}{n(S)} = \frac{6}{216} = \frac{1}{36}$$

5. $n(S) = 6$ and $E = \{2\}$. However, the outcomes are not equally likely, so we need the probability distribution. Start with

Outcome	1	2	3	4	5	6
Probability	$2x$	x	x	x	x	$2x$

Since the sum of the probabilities of the outcomes must be 1, we get

$$8x = 1, \text{ so } x = \frac{1}{8} = .125$$

This gives:

Outcome	1	2	3	4	5	6
Probability	.25	.125	.125	.125	.125	.25

$P(E) = P(\{2\}) = .125$ (or 1/8)

6. $n(S) = 2$ and $E = \{(1,6), (2,5), (3,4)\}$. However, the outcomes are not equally likely. To compute $P(E)$, we pretend that the dice were distinguishable (they dice don't behave any differently if they are different colors). For distinguishable dice,

$n(S) = 36,$

$E = \{(1,6), (2,5), (3,4), (4,3), (5,2), (6,1)\}$

$$P(E) = \frac{n(E)}{n(S)} = \frac{6}{36} = \frac{1}{6}$$

7. $P(2 \text{ heads}) = \dfrac{fr(2 \text{ heads})}{N} = \dfrac{12}{50}$

$P(\text{At least 1 tail}) = 1 - P(2 \text{ heads})$

$$= 1 - \frac{12}{50} = \frac{38}{50} = .76$$

8. The given information can be placed in a table:

	Internet	Other	Total
Increased			7
Decreased	2		
Total	8		10

We can now fill in the missing quantities (starting with the totals):

	Internet	Other	Total
Increased	6	1	7
Decreased	2	1	3
Total	8	2	10

E: Increased $\cup$ Internet

$$P(E) = \frac{fr(E)}{N} = \frac{9}{10} = .9$$

	Internet	Other	Total
Increased	6	1	7
Decreased	2	1	3
Total	8	2	10

$$\left\{ \begin{array}{cccccc} \mathbf{(1,1)} & \mathbf{(1,2)} & (1,3) & \mathbf{(1,4)} & (1,5) & \mathbf{(1,6)} \\ (2,1) & \mathbf{(2,2)} & (2,3) & (2,4) & (2,5) & (2,6) \\ (3,1) & \mathbf{(3,2)} & \mathbf{(3,3)} & \mathbf{(3,4)} & (3,5) & \mathbf{(3,6)} \\ (4,1) & (4,2) & (4,3) & \mathbf{(4,4)} & (4,5) & (4,6) \\ (5,1) & \mathbf{(5,2)} & (5,3) & \mathbf{(5,4)} & \mathbf{(5,5)} & \mathbf{(5,6)} \\ (6,1) & (6,2) & (6,3) & (6,4) & (6,5) & \mathbf{(6,6)} \end{array} \right\}$$

9. We use the following events for a randomly selected novel:

U: You have read the novel.

R: Roslyn has read the novel.

$$P(U \cup R) = P(U) + P(R) - P(U \cap R)$$
$$= \frac{150}{400} + \frac{200}{400} - \frac{50}{400} = \frac{300}{400} = .75$$

Therefore, the probability that a novel has been read by neither you nor your sister is

$$P[(U \cup R)'] = 1 - P(U \cup R) = 1 - .75 = .25$$

10. For the sum of the numbers to be even, either both numbers must be odd or both numbers must be even. This occurs 8 times out of the 10 (since only on 2 rolls is one number odd and the other even). Therefore,

$$P(E) = \frac{fr(E)}{N} = \frac{8}{10} = .8$$

11. Let us use the following events:

A: A student is in category A

B: A student is in category B

$$P(A \cup B) = 1, \quad P(A) = P(B) = \frac{24}{32} = .75$$

We are asked to find $P(A \cap B)$

$$P(A \cup B) = P(A) + P(B) - P(A \cap B)$$
$$1 = .75 + .75 - P(A \cap B)$$
$$P(A \cap B) = .75 + .75 - 1 = .5$$

12. $n(S) = 36$, and there are 15 losing combinations, shown in dark (the first number represents the green die):

$$P(\text{Lose}) = \frac{n(\text{Lose})}{n(S)} = \frac{15}{36}$$

Therefore,

$$P(\text{Win}) = 1 - P(\text{Lose}) = 1 - \frac{15}{36} = \frac{21}{36} = \frac{7}{12}$$

13. Use the following events:

A: Is a Model A

R: Is orange

$$P(A \cup R) = P(A) + P(R) - P(A \cap R)$$
$$= \frac{1}{3} + \frac{1}{5} - \frac{1}{15} = \frac{7}{15}$$

14. Use the following events:

T; tornadoes

M: monsoon

$$P(T \cup M) = P(T) + P(M) - P(T \cap M)$$
$$= .5 + .2 - .1 = .6$$

Therefore,

$$P(\text{Neither}) = P[(T \cup M)']$$
$$= 1 - P(T \cup M) = 1 - .6 = .4$$

15. $n(S) = C(12, 5) = 792$

$$P(E) = \frac{n(E)}{n(S)} = \frac{C(4,4)C(8,1)}{792} = \frac{8}{792}$$

16. $n(S) = C(12, 5) = 792$

$$P(E) = \frac{n(E)}{n(S)} = \frac{C(2,2)C(10,3)}{792} = \frac{120}{792}$$

17. $n(S) = C(12, 5) = 792$

$$P(E) = \frac{n(E)}{n(S)}$$

$$= \frac{C(4,1)C(2,1)C(1,1)C(3,1)C(2,1)}{792}$$

$$= \frac{48}{792}$$

18. $P(E') = P(\text{All are red}) = 0$

(There are only 4 red ones)

Therefore,

$$P(E) = 1 - P(E') = 1 - 0 = 1$$

19. $n(S) = C(12, 5) = 792$

Decision algorithm for constructing s set with at least 2 yellow

Alternative 1: 2 yellow: $C(3, 2)C(9, 3) = 252$ choices

Alternative 2: 3 yellow: $C(3, 3)C(9, 2) = 36$ choices

This gives a total of 252+36 = 288 choices.

$$P(E) = \frac{n(E)}{n(S)} = \frac{288}{792}$$

20. $n(S) = C(12, 5) = 792$

Decision algorithm for constructing s set with none yellow and at most one red:

Alternative 1: No yellow, no red: $C(5, 5) = 1$ choice

Alternative 2: No yellow, 1 red: $C(4, 1)C(5, 4) = 20$ choices

This gives a total of 1+20 = 21 choices.

$$P(E) = \frac{n(E)}{n(S)} = \frac{21}{792}$$

21. $n(S) = C(52, 5)$

$$P(E) = \frac{n(E)}{n(S)} = \frac{C(8, 5)}{C(52, 5)}$$

22. $n(S) = C(52, 5)$

$$P(E) = \frac{n(E)}{n(S)} = \frac{C(12, 5)}{C(52, 5)}$$

23. $n(S) = C(52, 5)$

$$P(E) = \frac{n(E)}{n(S)} = \frac{C(4, 3)C(1, 1)C(3, 1)}{C(52, 5)}$$

24. $n(S) = C(52, 5)$

Decision algorithm for constructing a prime full house:

Step 1: Select the denomination for the triple (2, 3, 5, 7, 11, 13): $C(6, 1)$ choices

Step 2: Select one of the remaining 5 denominations for the double: $C(5, 1)$ choices

Step 3: Select 3 cards of the chosen denomination for the triple: $C(4, 3)$ choices

Step 4: Select 2 cards of the chosen denomination for the double: $C(4, 2)$ choices

$$P(E) = \frac{n(E)}{n(S)}$$
$$= \frac{C(6, 1)C(5, 1)C(4, 3)C(4, 2)}{C(52, 5)}$$

25. $n(S) = C(52, 5)$

Decision algorithm for constructing a full house of commons:

Step 1: Select the denomination for the triple (2, 3, 4, 5, 6, 7, 8, 9, 10): $C(9, 1)$ choices

Step 2: Select one of the remaining 8 denominations for the double: $C(8, 1)$ choices

Step 3: Select 3 cards of the chosen denomination for the triple: $C(4, 3)$ choices

Step 4: Select 2 cards of the chosen denomination for the double: $C(4, 2)$ choices

$$P(E) = \frac{n(E)}{n(S)}$$
$$= \frac{C(9, 1)C(8, 1)C(4, 3)C(4, 2)}{C(52, 5)}$$

26. $n(S) = C(52, 5)$

Decision algorithm for constructing a black two pair:

Step 1: Select the two denominations for the pairs $C(12, 2)$ choices

Step 2: Choose 2 black cards from the highest-ranked denomination chosen in Step 1: $C(2, 2) = 1$ choice

Step 3: Choose 2 black cards from the other denomination chosen in Step 1: $C(2, 2) = 1$ choice

Step 4: Choose a denomination for the single: $C(13, 1)$ choices

Step 5: Choose a single black card of that denomination: $C(2, 1)$ choices

$$P(E) = \frac{n(E)}{n(S)} = \frac{C(12, 2)C(13, 1)C(2, 1)}{C(52, 5)}$$

27. A: The sum is 5.

B: The green one is not a 1 and the yellow one is 1.

$A \cap B$: The sum is 5, the green one is not 1 and the yellow one is 1.

$B = \{(2, 1), (3, 1), (4, 1), (5, 1), (6, 1)\}$

$A \cap B = \{((4, 1)\}$

$P(A \cap B) = \frac{1}{36}, P(B) = \frac{5}{36}$

$P(A|B) = \frac{P(A \cap B)}{P(B)} = \frac{1/36}{5/36} = \frac{1}{5}$

$P(A) = \frac{4}{36} = \frac{1}{9}$

Since $P(A|B) \neq P(A)$, the events A and B are dependent.

28. A: The sum is 6.

B: The green one is either 1 or 3 and the yellow one is 1.

$A \cap B$: The sum is 6, the green one is either 1 or 3, and the yellow one is 1.

$B = \{(1, 1), (3, 1)\}$

$A \cap B = \emptyset$

$P(A \cap B) = 0, \; P(B) = \frac{2}{36}$

$P(A|B) = \frac{P(A \cap B)}{P(B)} = \frac{0}{2/36} = 0$

$P(A) = \frac{4}{36} = \frac{1}{9}$

Since $P(A|B) \neq P(A)$, the events A and B are dependent.

29. A: The yellow one is 4.

B: The green one is 4.

$A \cap B$: Both the yellow and green dice are 4.

$P(A \cap B) = \frac{1}{36}, P(B) = \frac{1}{6}$

$P(A|B) = \frac{P(A \cap B)}{P(B)} = \frac{1/36}{1/6} = \frac{1}{6}$

$P(A) = \frac{1}{6} = P(A|B)$. Therefore, the events A and B are independent.

30. A: The yellow one is 5.

B: The sum is 6.

$A \cap B$: The yellow one is 5 and the sum is 6.

$A \cap B = \{(1, 5)\}$

$P(A \cap B) = \frac{1}{36}, P(B) = \frac{5}{36}$

$P(A|B) = \frac{P(A \cap B)}{P(B)} = \frac{1/36}{5/36} = \frac{1}{5}$

$P(A) = \frac{1}{6}$

Since $P(A|B) \neq P(A)$, the events A and B are dependent.

31. A: The dice have the same parity.

B: Both dice are odd.

$A \cap B$: The dice have the same parity and are both odd.

Note that $A \cap B = B$

$P(A \cap B) = \frac{9}{36}, P(B) = \frac{9}{36}$

$P(A|B) = \frac{P(A \cap B)}{P(B)} = \frac{9/36}{9/36} = 1$

$$P(A) = \frac{18}{36} = \frac{1}{2}$$

Since $P(A|B) \neq P(A)$, the events A and B are dependent.

32. A: The sum is 7.

B: The dice do not have the same parity.

$A \cap B$: The sum is 7 and the dice do not have the same parity.

Note that $A \cap B = A$

$$P(A \cap B) = \frac{6}{36}, \quad P(B) = \frac{18}{36}$$

$$P(A|B) = \frac{P(A \cap B)}{P(B)} = \frac{6/36}{18/36} = \frac{6}{18} = \frac{1}{3}$$

$$P(A) = \frac{6}{36} = \frac{1}{12}$$

Since $P(A|B) \neq P(A)$, the events A and B are dependent.

33. Take 1 = Brand A, 2 = Brand B.

$$P = \begin{bmatrix} 1/2 & 1/2 \\ 1/4 & 3/4 \end{bmatrix}$$

34. Take 1 = Brand A, 2 = Brand B.

$$P = \begin{bmatrix} 1/2 & 1/2 \\ 1/4 & 3/4 \end{bmatrix}$$

$$P^2 = \begin{bmatrix} 1/2 & 1/2 \\ 1/4 & 3/4 \end{bmatrix} \begin{bmatrix} 1/2 & 1/2 \\ 1/4 & 3/4 \end{bmatrix}$$

$$= \begin{bmatrix} 3/8 & 5/8 \\ 5/16 & 11/16 \end{bmatrix}$$

$$P^3 = \begin{bmatrix} 1/2 & 1/2 \\ 1/4 & 3/4 \end{bmatrix} \begin{bmatrix} 3/8 & 5/8 \\ 5/16 & 11/16 \end{bmatrix}$$

$$= \begin{bmatrix} 11/32 & 21/32 \\ 21/64 & 43/64 \end{bmatrix}$$

Prob. of Brand $\rightarrow$ Brand B in 3 years

$$= (P^3)_{12} = \frac{21}{32}$$

35. From Exercise 33,

$$P = \begin{bmatrix} 1/2 & 1/2 \\ 1/4 & 3/4 \end{bmatrix}$$

$$P^2 = \begin{bmatrix} 1/2 & 1/2 \\ 1/4 & 3/4 \end{bmatrix} \begin{bmatrix} 1/2 & 1/2 \\ 1/4 & 3/4 \end{bmatrix}$$

$$= \begin{bmatrix} 3/8 & 5/8 \\ 5/16 & 11/16 \end{bmatrix}$$

$$P^3 = \begin{bmatrix} 1/2 & 1/2 \\ 1/4 & 3/4 \end{bmatrix} \begin{bmatrix} 3/8 & 5/8 \\ 5/16 & 11/16 \end{bmatrix}$$

$$= \begin{bmatrix} 11/32 & 21/32 \\ 21/64 & 43/64 \end{bmatrix}$$

Take $v = [2/3 \quad 1/3]$

Distribution after 3 years is

$$vP^3 = [2/3 \quad 1/3] \begin{bmatrix} 11/32 & 21/32 \\ 21/64 & 43/64 \end{bmatrix}$$

$$= [65/192 \quad 127/192]$$

Brand A: $65/192 \approx .339$, Brand B: $127/192 \approx .661$

36. To find the steady-state vector $[x \quad y]$ we solve

$$x + y = 1$$

$$[x \quad y] \begin{bmatrix} 1/2 & 1/2 \\ 1/4 & 3/4 \end{bmatrix} = [x \quad y]$$

The above equations give:

$$x + y = 1$$

$$x/2 + y/4 = x$$

$$x/2 + 3y/4 = y$$

Rewriting in standard form and dropping the last equation (see the note in the textbook before Example 4) we get

$$x + y = 1$$

$$-x/2 + y/4 = 0$$

Solving this system gives

$$x = 1/3, \quad y = 2/3$$

So, $[x \quad y] = [1/3 \quad 2/3]$

Brand A: 1/3, Brand B: 2/3

Exercises 37–42

In Exercises 37–42 we take the first letter of each category shown in the table to stand for the corresponding event:

S: The book is Sci-Fi

W The book is stored in Washington,

and so on. Here is the table, with all figures shown in thousands:

	S	H	R	O	Total
W	10	12	12	30	64
C	8	12	6	16	42
T	15	15	20	44	94
Total	33	39	38	90	200

37. $P(S \cup T) = \dfrac{94 + 33 - 15}{200} = \dfrac{112}{200} = \dfrac{14}{25}$

	S	H	R	O	Total
W	10	12	12	30	64
C	8	12	6	16	42
T	15	15	20	44	94
Total	33	39	38	90	200

38. $P(S \cap T) = \dfrac{15}{200} = \dfrac{3}{40}$

	S	H	R	O	Total
W	10	12	12	30	64
C	8	12	6	16	42
T	15	15	20	44	94
Total	33	39	38	90	200

39. $P(S|T) = \dfrac{n(S \cap T)}{n(T)} = \dfrac{15}{94}$

	S	H	R	O	Total
W	10	12	12	30	64
C	8	12	6	16	42
T	15	15	20	44	94
Total	33	39	38	90	200

40. $P(T|S) = \dfrac{n(T \cap S)}{n(S)} = \dfrac{15}{33} = \dfrac{5}{11}$

	S	H	R	O	Total
W	10	12	12	30	64
C	8	12	6	16	42
T	15	15	20	44	94
Total	33	39	38	90	200

41. $P(T|S') = \dfrac{n(T \cap S')}{n(S')} = \dfrac{79}{167}$

	S	H	R	O	Total
W	10	12	12	30	64
C	8	12	6	16	42
T	15	15	20	44	94
Total	33	39	38	90	200

42. $P(T'|S) = \dfrac{n(T' \cap S)}{n(S)} = \dfrac{18}{33} = \dfrac{6}{11}$

	S	H	R	O	Total
W	10	12	12	30	64
C	8	12	6	16	42
T	15	15	20	44	94
Total	33	39	38	90	200

Exercises 43–47

In Exercises 43–47 we use the following events:

H: Visited OHaganBooks.com; $P(H) = .02$

C: Visited a competitor; $P(C) = .05$

43. $P(H') = 1 - P(H) = 1 - .02 = .98$, or 98%

44. Since H and C are independent,

$P(H \cap C) = P(H)P(C) = .02 \times .05 = .001$

Therefore,

$P(H \cup C) = P(H) + P(C) - P(H \cap C)$

$= .02 + .05 - .001 = .069$, or 6.9%

45. $P[(H \cup C)'] = 1 - P(H \cup C) = 1 - .069 = .931$

46. We are told that an online shopper visiting a competitor was more likely to visit OHaganBooks.com that a randomly selected online shopper. That is,

$P(H|C) > P(H)$

Multiplying both sides by $P(C)$ gives

$P(H|C)P(C) > P(H)P(C)$

That is,

$P(H \cap C) > P(H)P(C)$

Therefore, $P(H \cap C)$ is greater.

47. $P(H|C) = .25$

We are asked to compute $P(H \cap C')$.

Since we are given $P(H|C)$ we can also compute $P(H \cap C)$ using the multiplication principle:

$P(H \cap C) = P(H|C)P(C) = .25 \times .05 = .0375$

Now we can find $P(H \cap C')$:

$P(H \cap C') = P(H) - P(H \cap C)$
$= .02 - .0125 = .0075$, or 0.75%

Exercises 48–50

In Exercises 48–50 we use the following events:

    H: visited OHaganBooks.com

    B: purchased books

    $P(B|H) = .08$

    $P(H) = ,02$

    $P(B|H') = .005$

48. We compare $P(B|H)$ and $P(B|H')$. From the figures above,

    $P(B|H) = 16P(B|H')$

Therefore, online shoppers who visit the OHaganBooks.com web site are <u>16</u> times as likely to purchase books than shoppers who do not.

49. We are asked to find $P(B \cap H')$. Using the multiplication principle,

    $P(B \cap H') = P(B|H')P(H')$
               $= .005 \times (1 - .02) = .0049$

50. We are asked to compute $P(H|B)$. Using Bayes theorem,

$$P(H|B) = \frac{P(B|H)P(H)}{P(B|H)P(H) + P(B|H')P(H')}$$

$$= \frac{(.08)(.02)}{(.08)(.02) + (.005)(.98)} = \frac{.0016}{.0016 + .0049}$$

$$\approx .2462$$

Exercises 51–54

In Exercises 51–54 we number the states in the order shown in the table:

$$P = \begin{bmatrix} .8 & .1 & .1 \\ .4 & .6 & 0 \\ .2 & 0 & .8 \end{bmatrix}$$

51. Starting distribution (July 1):

    $v = [2000 \quad 4000 \quad 4000]$

Distribution after 1 month:

$$vP = [2000 \quad 4000 \quad 4000]\begin{bmatrix} .8 & .1 & .1 \\ .4 & .6 & 0 \\ .2 & 0 & .8 \end{bmatrix}$$

$$= [4000 \quad 2600 \quad 3400]$$

52. Starting distribution (July 1):

    $v = [2000 \quad 4000 \quad 4000]$

Distribution after 1 month:

$$vP = [2000 \quad 4000 \quad 4000]\begin{bmatrix} .8 & .1 & .1 \\ .4 & .6 & 0 \\ .2 & 0 & .8 \end{bmatrix}$$

$$= [4000 \quad 2600 \quad 3400]$$

Distribution after 2 months (End of August):

$$(vP)P = [4000 \quad 2600 \quad 3400]\begin{bmatrix} .8 & .1 & .1 \\ .4 & .6 & 0 \\ .2 & 0 & .8 \end{bmatrix}$$

$$= [4920 \quad 1960 \quad 3120]$$

53. Here are three factors that the Markov model does not take into account:

(1) It is possible for someone to be a customer at two different enterprises.

(2) Some customers may stop using all three of the companies.

(3) New customers can enter the field.

54. Steady-state vector:

$$x + y + z = 1$$

$$[x \quad y \quad z] \begin{bmatrix} .8 & .1 & .1 \\ .4 & .6 & 0 \\ .2 & 0 & .8 \end{bmatrix} = [x \quad y \quad z]$$

The above equations give:

$$x + y + z = 1$$
$$.8x + .4y + .2z = x$$
$$.1x + .6y = y$$
$$.1x + .8z = z$$

Rewriting in standard form and dropping the last equation, we get

$$x + y + z = 1$$
$$-.2x + .4y + .2z = 0$$
$$.1x - .4y = 0$$

Solving gives

$$x = 2/5, \ y = 1/5, \ z = 2/5$$

So, $[x \quad y \quad z] = [2/5 \quad 1/5 \quad 2/5]$

$$P^{\infty} = [4/7 \quad 1/7 \quad 2/7]$$

Answer:

OHaganBooks.com: 4/7, JungleBooks.com: 1/7, FarmerBooks.com: 2/7

Chapter 8

8.1

1. X = the sum of the numbers facing up when you roll two dice. The possible sums are 2, 3, ..., 12. Thus, X is finite and has the set $\{2, 3, ..., 12\}$ of possible values.

3. X is the profit, to the nearest dollar, earned in a year if you purchase one share of a stock selected at random. The possible values of X are 0, ±1, ±2, ... (negative numbers indicate a loss). This gives an infinite number of discrete possible values for X. Therefore, X is a discrete infinite random variable with the set of possible values

$\{0, 1, -1, 2, -2, ...\}$

5. X is the time the second hand of your watch reads in seconds. This can be any real number between 0 and 60. Therefore, X is a continuous random variable that can assume any value between 0 and 60.

7. The total number of goals that can be scored, up to a maximum of 10, is 1, 2, 3, ..., 9, or 10. Therefore, X is a finite random variable with values in the set

$\{0, 1, 2, ..., 10\}$.

9. The possible energies of an electron in a hydrogen atom are $k/1$, $k/4$, $k/9$, $k/16$, ... Thus, X is a discrete infinite random variable with set of possible values

$\{ k/1, k/4, k/9, k/16, ...\}$

11. (a) $S = \{HH, HT, TH, TT\}$

(b) X is the rule that assigns to each outcome the number of tails.

(c) Counting the number of tails in each outcome gives us the following values of X:

Outcome	HH	HT	TH	TT
Value of X	0	1	1	2

13. (a) $S = \{(1,1), (1,2), ..., (1,6), (2,1), (2,2), ..., (6,6)\}$

(b) X is the rule that assigns to each outcome the sum of the two numbers.

(c) Computing the sum of the numbers in each outcome gives us the following values of X:

Outcome	(1, 1)	(1, 2)	(1, 3)	(1, 4)
Value of X	2	3	4	5
Outcome	...	(6, 4)	(6, 5)	(6, 6)
Value of X	...	10	11	12

15. Take each outcome to be a pair of numbers (# of red marbles, # of green ones). Thus,

$S = \{(4,0), (3,1), (2,2)\}$

(The pairs (1, 3) and (0, 4) are impossible since there are only 2 green marbles.)

(b) X is the rule that assigns to each outcome the number of red marbles.

(c) Writing down the number of red marbles (the first coordinate) in each outcome gives the following values of X.

Outcome	(4, 0)	(3, 1)	(2, 2)
Value of X	4	3	2

17. (a) S = the set of students in the study group.

(b) X is the rule that assigns to each student his or her final exam score.

(c) The values of X, in the order given, are 89%, 85%, 95%, 63%, 92%, 80%.

19. (a) Assign letters to the missing values (we use the same letter for both since they are given to be equal):

x	2	4	6	8	10
$P(X = x)$	.1	.2	x	x	.1

Since the probabilities add to 1, we get

$.1 + .2 + 2x + .1 = 1$
$.4 + 2x = 1$
$2x = .6$
$x = .3$

The completed table is:

x	2	4	6	8	10
$P(X = x)$	.1	.2	.3	.3	.1

(b) $P(X \geq 6) = P(X = 6) + P(X = 8) + P(X = 10)$
$= .3 + .3 + .1 = .7$

21. Since the probability that any specific number faces up is 1/6, the probability distribution for X is given by the following table and histogram:

x	1	2	3	4	5	6
$P(X = x)$	$\frac{1}{6}$	$\frac{1}{6}$	$\frac{1}{6}$	$\frac{1}{6}$	$\frac{1}{6}$	$\frac{1}{6}$

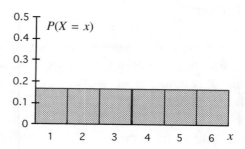

23. The number of heads showing when you toss 3 fair coins is 0, 1, 2, or 3. The corresponding probabilities are 1/8, 3/8, 3/8, and 1/8 (see Example 3). The corresponding values of X are the squares of 0, 1, 2, and 3; that is, 0, 1, 4, and 9. This gives us the following distribution and histogram:

x	0	1	4	9
$P(X = x)$	$\frac{1}{8}$	$\frac{3}{8}$	$\frac{3}{8}$	$\frac{1}{8}$

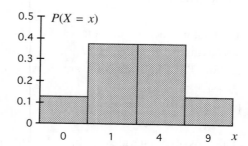

25. When two distinguishable dice are thrown, there are 36 possible outcomes. The values of X are the possible sums of the numbers facing up: 2, 3, ..., 11, 12. To calculate their probabilities, we can use

$P(X = 2) = P\{(1, 1)\} = \frac{1}{36}$

$P(X = 3) = P\{(1, 2), (2, 1)\} = \frac{2}{36}$

$P(X = 4) = P\{(1, 3), (2, 2), (3, 1)\} = \frac{3}{36}$

and so on. The completed probability distribution is as follows:

x	2	3	4	5	6	7
$P(X = x)$	$\frac{1}{36}$	$\frac{2}{36}$	$\frac{3}{36}$	$\frac{4}{36}$	$\frac{5}{36}$	$\frac{6}{36}$
x	8	9	10	11	12	
$P(X = x)$	$\frac{5}{36}$	$\frac{4}{36}$	$\frac{3}{36}$	$\frac{2}{36}$	$\frac{1}{36}$	

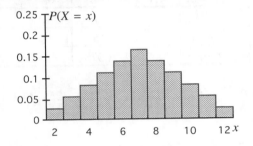

27. The possible values of X are 1, 2, ..., 6 (the larger of the two numbers facing up). We calculate their probabilities as follows:

$P(X = 1) = P\{(1, 1)\} = \frac{1}{36}$

$P(X = 2) = P\{(1, 2), (2, 2), (2, 1)\} = \frac{3}{36}$

$P(X = 3) = P\{(1, 3), (2, 3), (3, 3), (3, 2), (3, 1)\}$
$\quad = \frac{5}{36}$

and so on. The completed probability distribution is as follows:

x	1	2	3	4	5	6
$P(X = x)$	$\frac{1}{36}$	$\frac{3}{36}$	$\frac{5}{36}$	$\frac{7}{36}$	$\frac{9}{36}$	$\frac{11}{36}$

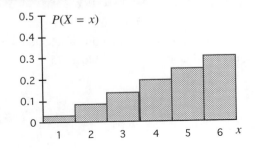

29. (a) The values of X are the possible tow ratings: 2000, 3000, 4000, 5000, 6000, 7000, 8000 (7000 is optional)

(b) The following table shows the frequency (number of models with each tow rating) and the resulting probabilities (divide each frequency by the sum of all the frequencies):

x	2000	3000	4000	5000
Freq	2	1	1	1
$P(X = x)$	.2	.1	.1	.1
x	6000	7000	8000	
Freq	2	0	3	
$P(X = x)$	.2	.0	.3	.2

(c) $P(X \le 5000) = .2 + .1 + .1 + .1 = .5$
(Add the probabilities in the first row of the above table.)

31. The associated random variable is shown on the x axis of the histogram: X = mold count on a given day.

To determine the probability distribution of X, use a frequency table. For the values of X, use the rounded midpoints of the measurement classes given:

$(0+1499)/2 \approx 750$

$(1500+2999)/2 \approx 2250$

$(3000+4499)/2 \approx 3750$

$(4500+5999)/2 \approx 5250$

To obtain the probabilities, divide each frequency by the sum of all the frequencies, 16:

x	750	2250	3750	5250
Freq	11	2	1	2
$P(X = x)$	$\frac{11}{16}$	$\frac{2}{16}$	$\frac{1}{16}$	$\frac{2}{16}$

33. (a) For the values of X, use the rounded midpoints of the measurement classes given:

$(19,999–0)/2 \approx 10,000$

$(39,999–20,000)/2 \approx 30,000$

$(59,999–40,000)/2 \approx 50,000$

$(79,999–60,000)/2 \approx 70,000$

$(99,999–80,000)/2 \approx 90,000$

To obtain the probabilities, divide each frequency by the sum of all the frequencies, 1000:

x	10,000	30,000	50,000	70,000	90,000
Freq	270	280	200	150	100
$P(X = x)$	.27	.28	.20	.15	.10

(b) Here is the histogram with the area corresponding to $X > 50,000$ shaded:

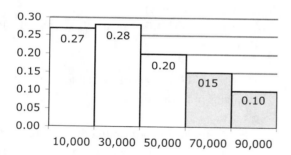

$P(X > 50,000) = ,15 + .10 = .25$

35. To obtain the frequency table, count how many of the scores fall in each measurement class:

Class	1.1–2.0	2.1–3.0	3.1–4.0
Freq	4	7	9

To obtain the probability distribution, divide each frequency by the sum of all the frequencies, 20:

x	1.5	2.5	3.5
$P(X = x)$	.20	.35	.45

37. First, construct the probability distribution:

x	0	1	2	3
Freq	140	350	450	650
$P(X = x)$	.07	.175	.225	.325
x	4	5	6	7
Freq	200	140	50	10
$P(X = x)$	.1	.06	.025	.005
x	8	9	10	
Freq	5	15	10	
$P(X = x)$	.0025	.0075	.005	

$P(X < 6) = .07+.175+.225+.325+.1+.06 = .955$

Therefore, 95.5% of cars are newer than 6 years old.

39. The frequency distribution for X is obtained from the first row (small cars) of the given table, and we can use that to compute the probability distribution (divide each frequency by the sum of the frequencies, 16):

x	3	2	1	0
Freq	1	11	2	2
$P(X = x)$	.0625	.6875	.125	.125

41. $P(X \geq 2) = P(X = 2) + P(X = 3)$
$= .6875 + .0625 = .75$

The probability that a randomly selected small car is rated Good or Acceptable is .75.

43. The following tables give the frequency distributions for Y and Z:

y	3	2	1	0
Freq	1	4	4	1
$P(Y = y)$	.1	.4	.4	.1

z	3	2	1	0
Freq	3	5	3	4
$P(Z = z)$	.2	.3333	.2	.3333

$$P(Y \geq 2) = P(Y = 2) + P(Y = 3)$$
$$= .4 + .1 = .5$$
$$P(Z \geq 2) = P(Z = 2) + P(Z = 3)$$
$$\approx .333 + .2 = .533$$

Since a crash rating of at least 2 indicates "acceptable" or "good," the data suggest that medium SUVs are safer than small SUVs in frontal crashes.

45. $P(X = 3) = \dfrac{1}{16} = .0625$

$P(Y = 3) = \dfrac{1}{10} = .1$

$P(Z = 3) = \dfrac{3}{15} = .2$

$P(U = 3) = \dfrac{3}{13} \approx .2308$

$P(V = 3) = \dfrac{3}{15} = .2$

$P(W = 3) = \dfrac{9}{19} \approx .4737$

The lowest is $P(Y = 3)$, for small cars.

47. From Exercise 41, the probability that a randomly selected small car will be rated at least 2 is

$P(X \geq 2) = .75$

From Exercise 43, the probability that a randomly selected small SUV will be rated at least 2 is

$P(Y \geq 2) = .5$

Since these two events are independent (both vehicles are selected at random), the probability that both will be rated at least 2 is

$P(X \geq 2) \times P(Y \geq 2) = .75 \times .5 = .375$

49. The sample space is the set of all sets of 4 tents selected from 7; $n(S) = C(7, 4) = 35$.

The possible values of X are 1, 2, 3, and 4. (X cannot equal 0, since that would require 4 green tents, and there are only 3.)

$$P(X = 1) = \frac{C(4, 1)C(3, 3)}{35} = \frac{4}{35}$$

$$P(X = 2) = \frac{C(4, 2)C(3, 2)}{35} = \frac{18}{35}$$

$$P(X = 3) = \frac{C(4, 3)C(3, 1)}{35} = \frac{12}{35}$$

$$P(X = 4) = \frac{C(4, 4)C(3, 0)}{35} = \frac{1}{35}$$

Probability distribution:

x	1	2	3	4
$P(X = x)$	$\dfrac{4}{35}$	$\dfrac{18}{35}$	$\dfrac{12}{35}$	$\dfrac{1}{35}$

$$P(X \geq 2) = 1 - P(X = 1)$$
$$= 1 - \frac{4}{35} = \frac{31}{35} \approx 0.886$$

51. Various answers possible

53. No; examples are infinite discrete random variables. For example, if X is the number of times you must toss a coin until heads comes up, then X is infinite but not continuous.

55. By measuring the values of X for a large number of outcomes, and then using the estimated probability (relative frequency).

57. Here are two examples:
(1) Let X be the number of times you have read a randomly selected book.
(2) Let X be the number of days a diligent student waits before beginning to study for an exam scheduled in 10 days' time.

59. Answers may vary. If we are interested in exact page-counts, then the number of possible values is very large and the values are (relatively speaking) close together, so using a continuous random variable might be advantageous. In general, the finer and more numerous the measurement classes, the more likely it becomes that a continuous random variable could be advantageous..

8.2

1. $n = 5, p = .1, q = .9$
$P(X = 2) = C(5, 2)(.1)^2(.9)^3 = .0729$

3. $n = 5, p = .1, q = .9$
$P(X = 0) = C(5, 0)(.1)^0(.9)^5 = .59049$

5. $n = 5, p = .1, q = .9$
$P(X = 5) = C(5, 5)(.1)^5(.9)^0 = .00001$

7. $n = 5, p = .1, q = .9$
$P(X \le 2) = P(X = 0) + P(X = 1) + P(X = 2)$
$P(X = 0) = C(5, 0)(.1)^0(.9)^5 = .59049$
$P(X = 1) = C(5, 1)(.1)^1(.9)^4 = .32805$
$P(X = 2) = C(5, 2)(.1)^2(.9)^3 = .0729$
Therefore,
$P(X \le 2) = .59049 + .32805 + .0729 = .99144$

9. $n = 5, p = .1, q = .9$
$P(X \ge 3) = P(X = 3) + P(X = 4) + P(X = 5)$
$P(X = 3) = C(5, 3)(.1)^3(.9)^2 = .0081$
$P(X = 4) = C(5, 4)(.1)^4(.9)^1 = .00045$
$P(X = 5) = C(5, 5)(.1)^5(.9)^0 = .00001$
Therefore,
$P(X \ge 3) = .0081 + .00045 + .00001 = .00856$

11. $n = 6, p = .4, q = 1-p = .6$
$P(X = 3) = C(6, 3)(.4)^3(.6)^3 = .27648$

13. $n = 6, p = .4, q = 1-p = .6$
$P(X \le 2) = P(X = 0) + P(X = 1) + P(X = 2)$
$P(X = 0) = C(6, 0)(.4)^0(.6)^6 = .046656$
$P(X = 1) = C(6, 1)(.4)^1(.6)^5 = .18662$
$P(X = 2) = C(6, 2)(.4)^2(.6)^4 = .31104$
Therefore,
$P(X \le 2) = .046656 + .18662 + .31104$
$= .54432$

15. $n = 6, p = .4, q = 1-p = .6$
$P(X \ge 5) = P(X = 5) + P(X = 6)$
$P(X = 5) = C(6, 5)(.4)^5(.6)^1 = .036864$
$P(X = 6) = C(6, 6)(.4)^6(.6)^0 = .004096$
Therefore,
$P(X \ge 5) = .036864 + .004096 = .04096$

17. $n = 6, p = .4, q = 1-p = .6$
$P(1 \le X \le 3) = P(X = 1) + P(X = 2) + P(X = 3)$
$P(X = 1) = C(6, 1)(.4)^1(.6)^5 = .18662$
$P(X = 2) = C(6, 2)(.4)^2(.6)^4 = .31104$
$P(X = 3) = C(6, 3)(.4)^3(.6)^3 = .27648$
Therefore,
$P(1 \le X \le 3) = .18662 + .31104 + .27648$
$= .77414$

19. $n = 5, p = \frac{1}{4}, q = \frac{3}{4}$

We used Excel to generate the distribution.
Format: BINOMDIST(x,n,p,0)

	H	I
1	x	P(X=x)
2	0	BINOMDIST(H2,5,0.25,0)
3	1	
4	2	
5	3	
6	4	
7	5	

TI-83/84: binompdf(5,0.25,x)

The resulting values are shown on the histogram:

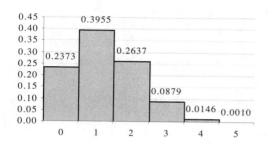

21. $n = 4$, $p = \frac{1}{3}$, $q = \frac{2}{3}$

We used Excel to generate the distribution.

Format: `BINOMDIST(x,n,p,0)`

	A	B
1	x	P(X=x)
2		0 BINOMDIST(A2,4,1/3,0)
3		1
4		2
5		3
6		4

TI-83/84: `binompdf(4,1/3,x)`

The resulting values are shown on the histogram, with the portion corresponding to $P(X \le 2)$ shaded.

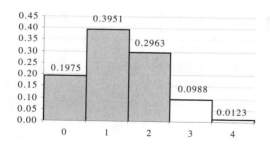

$P(X \le 2) \approx .1975 + .3951 + .2963 = 0.8889$

23. Take "success" = of pension age.

$n = 5$, $p = .25$, $q = 1 - .25 = .75$

$P(X = 2) = C(5, 2)(.25)^2(.75)^3 \approx .2637$

374

25. Take "success" = a stock market success.

$n = 10$, $p = .2$, $q = 1 - .2 = .8$

$P(X \ge 1) = 1 - P(X = 0)$

$P(X = 0) = C(10, 0)(.2)^0(.8)^{10} \approx .10737$

Therefore,

$P(X \ge 1) \approx 1 - 10737 = .8926$

27. Take "success" = selecting a male.

$n = 3$, $p = .5$, $q = 1 - .5 = .5$

$P(X \ge 1) = 1 - P(X = 0)$

$P(X = 0) = C(3, 0)(.5)^0(.5)^3 = .125$

Therefore,

$P(X \ge 1) = 1 - 125 = .875$

29. Take "success" = selecting a defective bag.

$n = 5$, $p = .1$, $q = 1 - .1 = .9$

(a) $P(X = 3) = C(5, 3)(.1)^3(.9)^2 = .0081$

(b) $P(X \ge 2) = 1 - P(X \le 1)$

$P(X = 0) = C(5, 0)(.1)^0(.9)^5 = .59049$

$P(X = 1) = C(5, 1)(.1)^1(.9)^4 = .32805$

Therefore,

$P(X \ge 2) = 1 - (.59049 + .32805) = .08146$

31. Take "success" = watching a rented video at least once.

$n = 10$, $p = .71$, $q = 1 - .71 = .29$

$P(X \ge 8) = P(X = 8) + P(X = 9) + P(X = 10)$

$P(X = 8) = C(10, 8)(.71)^8(.29)^2 \approx .244$

$P(X = 9) = C(10, 9)(.71)^9(.29)^1 \approx .133$

$P(X = 10) = C(10, 10)(.71)^{10}(.29)^0 \approx .033$

Therefore,

$P(X \ge 8) \approx .244 + .133 + .033$

$\approx .41$

33. (a) Take "success" = purchased by McDonalds.

$n = 10$, $p = .04$, $q = 1 - .04 = .96$

$P(X = 2) = C(10, 2)(.04)^2(.96)^8 \approx .0519$

(b) Following is the distribution generated by the online utility at

Chapter 8 → Binomial Distribution Utility:

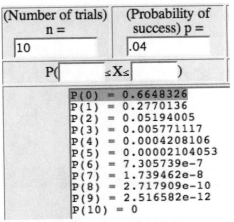

(Number of trials) n =	(Probability of success) p =
10	.04

P(≤ X ≤)

```
P(0)  = 0.6648326
P(1)  = 0.2770136
P(2)  = 0.05194005
P(3)  = 0.005771117
P(4)  = 0.0004208106
P(5)  = 0.00002104053
P(6)  = 7.305739e-7
P(7)  = 1.739462e-8
P(8)  = 2.717909e-10
P(9)  = 2.516582e-12
P(10) = 0
```

(c) The value of X with the largest probability is $X = 0$. So, the most likely number of pounds of beef purchased by McDonalds is 0.

35. Take "success" = computer malfunction.

$n = 3$, $p = .01$, $q = 1 - .01 = .99$

$P(X \geq 2) = P(X = 2) + P(X = 3)$

$P(X = 2) = C(3, 2)(.01)^2(.99)^1 \approx .000297$

$P(X = 3) = C(3, 3)(.01)^3(.99)^0 \approx .000001$

Therefore,

$P(X \geq 2) \approx .000297 + .000001 = .000298$

37. Take "success" = answering a question correctly.

$n = 100$, $p = .80$

We use technology to compute

$P(75 \leq X \leq 85) = P(X \leq 85) - P(X \leq 74)$

TI-83/84:

```
binomcdf(100,0.8,85)-
binomcdf(100,0.8,74)
```

Excel:

```
BINOMDIST(85,100,0.8,1)-
BINOMDIST(74,100,0.8,1)
```

Answer: $P(75 \leq X \leq 85) \approx .8321$

39. Take "success" = containing more than 10 grams of fat.

$n = 50$, $p = .43$

We use technology to generate the probability distribution:

Excel: `BINOMDIST(x,50,0.43,0)`

TI-83/84: `binompdf(50,0.43,x)`

	A	B
1	x	P(X=x)
2	0	6.21932E-13
3	1	2.34588E-11
4	2	4.33577E-10
5		
49		
50	48	1.01478E-15
51	49	3.12463E-17
52	50	4.71436E-19

(a) Since 43% of all the burgers contain more than 10 grams of fat, we can expect about

$0.43 \times 50 = 21$

of them to contain more than 10 grams of fat.

(b) The probability that k or more patties contain more than 10 grams of fat is

$P(X \geq k) = 1 - P(X \leq k-1)$

We want this to equal approximately .71:

$1 - P(X \leq k-1) \approx .71$

$P(X \leq k-1) \approx .29$

To answer this question, use the cumulative probability distribution:

Excel: `BINOMDIST(x,50,0.43,1)`

TI-83/84: `binomcdf(50,0.43,x)`

	A	B
1	k	P(X ≤ k)
2	0	6.21932E-13
3	1	2.40808E-11
4	2	4.57658E-10
5		
19		
20	18	0.19634337
21	19	0.285669736
22	20	0.390118898
23	21	0.502683159
24	22	0.61461907

From the table, $k-1 = 19$, so $k = 20$.

There is approximately a 71% chance that a batch of 50 ZeroFat patties contains <u>20</u> or more patties with at least 10 grams of fat.

(c) Graphs

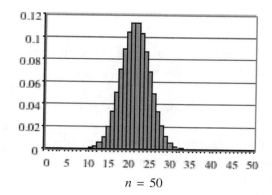

$n = 50$

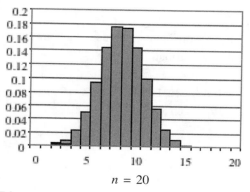

$n = 20$

The graph for $n = 50$ trials is more widely distributed than the graph for $n = 20$

41. Take "success" = bad bulb.

$p = .01$, $n = ?$

We want $P(X \geq 1) \geq .5$

$1 - P(X = 0) \geq .5$

$P(X = 0) \leq .5$

$C(n, 0)\,(.01)^0 (.99)^n \leq .5$

$1 \times 1 \times .9^n \leq .5$

$.99^n \leq .5$

Using technology, we compute the values of $.9^n$ for $n = 0, 1, 2, \ldots$:

n	$.9^n$
0	1
1	0.99
. . .	. . .
68	0.50488589
69	0.49983703
. . .	. . .

The probability first dips below .5 when $n = 69$ trials.

43. The estimated probability of an accident in a given mile is

$$p = \frac{\text{Number of accidents}}{\text{Number of miles}} = \frac{562}{100,000,000}$$

$$= 0.562 \times 10^{-5}$$

Since this is the estimated probability of a success, it is the probability that a male driver will have an accident in a one-mile trip.

45. Let X be the number of cases of mad cow disease found among the 243,000 tested. Then X is a binomial random variable with

$$p = \frac{5}{45,000,000} \approx 1.1111 \times 10^{-7}$$

and $n = 243,000$

We are asked to compute the probability of at least one "success": $P(X \geq 1)$.

$$P(X \geq 1) = 1 - P(X = 0)$$
$$= 1 - C(n, 0)p^0(1-p)^{n-0}$$
$$= 1 - (1)(1)(1-1.1111 \times 10^{-7})^{243,000}$$
$$= 1 - (.999\ 999\ 889)^{243,000}$$
$$\approx 1 - .9734 = .0266$$

Since there is only a 2.66% chance of detecting the disease in a given year, the government's claim seems dubious.

47. No; in the given scenario, the probability of success depends on the outcome of the previous shot. However, in a sequence of Bernoulli trials, the occurrence of one success does not effect the probability of success on the next attempt.

49. No; if life is a sequence of Bernoulli trials, then the occurrence of one misfortune ("success") does not affect the probability of a misfortune on the next trial. Hence, misfortunes may very well not "occur in threes."

51. The probability of selecting a red marble changes after each selection, as the number of marbles left in the bag decreases. This violates the requirement that, in a sequence of Bernoulli trials, the probability of "success" does not change.

8.3

1. $\bar{x} = \dfrac{-1 + 5 + 5 + 7 + 14}{5} = 6$

To compute the median, arrange the scores in order and take the (average of the) middle score(s).

−1, 5, **5**, 7, 14

Median = middle score = 5

Mode = most frequent score(s) = 5

3. $\bar{x} = \dfrac{2 + 5 + 6 + 7 - 1 - 1}{6} = 3$

To compute the median, arrange the scores in order and take the (average of the) middle score(s).

−1. −1, **2, 5**, 6, 7

Median = average of middle scores = $\dfrac{2+5}{2}$ = 3.5

Mode = most frequent score(s) = −1

5. In decimal notation, the given scores are:

0.5, 1.5, −4, 1.25

$\bar{x} = \dfrac{0.5 + 1.5 - 4 + 1.25}{4} = -0.1875$

To compute the median, arrange the scores in order and take the (average of the) middle score(s).

−4, **0.5, 1.25**, 1.5

Median = average of middle scores = $\dfrac{0.5 + 1.25}{2}$

= 0.875

Mode = most frequent score(s) = Every value

7. $\bar{x} = \dfrac{2.5 - 5.4 + 4.1 - 0.1 - 0.1}{5} = 0.2$

To compute the median, arrange the scores in order and take the (average of the) middle score(s).

−5.4, −0.1, **−0.1**, 2.5, 4.1

Median = middle score = −0.1

Mode = most frequent score(s) = −0.1

9. Answers may vary. One example is:

0, 0, 0, 0, 0, 6

$\bar{x} = 1$, Median = 0

Another example is:

0, 0, 0, 1, 2, 3

$\bar{x} = 1$, Median = 0

11. We use the tabular method described in Example 3 in the textbook:

x	0	1	2	3
$P(X = x)$	.5	.2	.2	.1
$xP(X = x)$	0	0.2	0.4	0.3

$E(X)$ = Sum of entries in the bottom row = 0.9

13. We first convert the fractions into decimals, and then use the tabular method described in Example 3 in the textbook:

x	10	20	30	40
$P(X = x)$	.3	.4	.2	.1
$xP(X = x)$	3	8	6	4

$E(X)$ = Sum of entries in the bottom row = 21

15. We use the tabular method described in Example 3 in the textbook:

x	−5	−1	0	2	5	10
$P(X=x)$	.2	.3	.2	.1	.2	.0
$xP(X=x)$	−1	−0.3	0	0.2	1	0

$E(X)$ = Sum of entries in the bottom row = − 0.1

17. The probability distribution of X is

x	1	2	3	4	5	6
$P(X=x)$	1/6	1/6	1/6	1/6	1/6	1/6

Using the tabular method described in Example 3, we get

x	1	2	3	4	5	6
$P(X=x)$	1/6	1/6	1/6	1/6	1/6	1/6
$x \cdot P(X=x)$	1/6	2/6	3/6	4/6	5/6	6/6

$E(X)$ = sum of numbers in bottom row = 3.5

19. X is a binomial random variable with $n = 2$ and $p = .5$. Therefore,

$$E(X) = np = 2(.5) = 1$$

21. We compute the probability distribution of X (see the solution to Exercise 27 in Section 8.1) and then use the tabular method of Example 3 to compute $E(X)$:

x	1	2	3	4	5	6
Freq	1	3	5	7	9	11
$P(X=x)$	$\frac{1}{36}$	$\frac{3}{36}$	$\frac{5}{36}$	$\frac{7}{36}$	$\frac{9}{36}$	$\frac{11}{36}$
$xP(X=x)$	$\frac{1}{36}$	$\frac{6}{36}$	$\frac{15}{36}$	$\frac{28}{36}$	$\frac{45}{36}$	$\frac{66}{36}$

$E(X)$ = Sum of entries in the bottom row = $\frac{161}{36} \approx$ 4.4722

23. Number of sets of 4 marbles = $C(6, 4) = 15$. If X is the number of red marbles, then the possible values of X are 2, 3, 4, and

$$P(X = x) = \frac{C(4, x)C(2, 4-x)}{15}$$

x	2	3	4
$P(X = x)$	$\frac{6}{15}$	$\frac{8}{15}$	$\frac{1}{15}$
$xP(X = x)$	$\frac{12}{15}$	$\frac{24}{15}$	$\frac{4}{15}$

$E(X)$ = Sum of entries in the bottom row = $\frac{40}{15}$

≈ 2.6667

25. X is a binomial random variable with $n = 20$ and $p = .1$. Therefore,

$$E(X) = np = 20(.1) = 2$$

27. The number of possible hands of 5 cards is $C(52, 5) = 2,598,960$. The possible values of X are 0, 1, 2, 3, 4, and

$$P(X = x) = \frac{C(4, x)C(48, 5-x)}{2,598,960}$$

x	0	1	2
$P(X=x)$	.6588	.2995	.0399
$xP(X=x)$	0.0000	0.2995	0.0799
x	3	4	
$P(X=x)$	.0017	.0000	
$xP(X=x)$	0.0052	0.0001	

$E(X)$ = Sum of entries $xP(X=x) \approx 0.3846$

29. The sum of the given tow ratings is

2000+2000+3000+4000+5000+6000
+6000+6000+8000+8000 = 50,000

Therefore,

$$\bar{x} = \frac{50,000}{10} = 5000$$

The two middle scores are 5000 and 6000. Therefore,

$$\text{Median} = \frac{5000 + 6000}{2} = 5500$$

There were as many light trucks with a tow rating of more than 5500 pounds as there were with tow ratings below that. (See the definition of the median.)

31. Arranging the scores in order gives:

423, 424, 425, 425, 425, 426, 427, 428, 428, 429

$$\bar{x} =$$
$$\frac{423+424+425+425+425+426+427+428+428+429}{10}$$
$$= \$426$$

The two middle scores are 425 and 426. Therefore,

$$\text{Median} = \frac{425+426}{2} = \$425.50$$

The mode is the most frequently occurring score: 425.

Over the 10-business day period sampled, the price of gold averaged $426 per ounce. It was above $425.50 as many times as it was below that, and stood at $425 per ounce for most of the days sampled.

33. (a) We use the tabular method described in Example 3 in the textbook:

x	1	2	3	4	5
$P(X=x)$	.01	.04	.04	.08	.10
$xP(X=x)$	0.01	0.08	0.12	0.32	0.5
x	6	7	8	9	10
$P(X=x)$	.15	.25	.20	.08	.05
$xP(X=x)$	0.9	1.75	1.6	0.72	0.5

$\mu = E(X) = $ sum of values $xP(X=x) = 6.5$

There were an average of 6.5 checkout lanes in a supermarket that was surveyed.

(b) $P(X < \mu) = P(X < 6.5)$
$$= .01 + .04 + .04 + .08 + .10 + .15$$
$$= .42$$

$$P(X > \mu) = P(X > 6.5)$$
$$= .25 + .20 + .08 + .05$$
$$= .58,$$

and is thus larger. Most supermarkets have more than the average number of checkout lanes.

35. Using the rounded midpoints of the given measurement classes, we get the following table. (The probabilities are obtained by dividing the given frequencies by their sum, 72 and then rounding to 2 decimal places.)

x	5	10	15
$P(X=x)$	.17	.33	.21
$xP(X=x)$	0.85	3.3	3.15
x	20	25	35
$P(X=x)$	.19	.03	.07
$xP(X=x)$	3.8	0.75	2.45

$E(X) = $ sum of values $xP(X=x) = 14.3$

Interpretation: The average age of a student in 1998 was 14.3.

37. The associated random variable is shown on the x axis of the histogram: $X = $ mold count on a given day. We use the tabular method described in Example 3 in the textbook:

x	750	2250	3750	5250
$P(X = x)$	0.6875	0.125	0.0625	0.125
$xP(X = x)$	515.625	281.25	234.375	656.25

Expected mold count = sum of entries in bottom row = 1687.5

39. Using the rounded midpoints of the measurement classes, we get the following table:

x	10,000	30,000	50,000	70,000	90,000
$P(X = x)$	.27	.28	.20	.15	.10
$x \cdot P(X = x)$	2700	8400	10000	10500	9000

The sum of the entries in the bottom row is 40,600, representing an average income of $40,600 ≈ $41,000

41. The probabilities in the tables below are obtained by dividing the given frequencies by the sum, 16 for X and 10 for Y:

x	3	2	1	0
$P(X = x)$	.0625	.6875	.125	.125
$xP(X = x)$	0.1875	1.375	0.125	0

$E(X)$ = sum of values in the bottom row = 1.6875

y	3	2	1	0
$P(Y = y)$	.1	.4	.4	.1
$yP(Y = y)$	0.3	0.8	0.4	0

$E(Y)$ = sum of values in the bottom row = 1.5

Since small cars (X) have a higher average rating, small cars performed better in frontal crashes.

43. Small cars:

x	3	2	1	0
$P(X = x)$	.0625	.6875	.125	.125
$xP(X = x)$	0.1875	1.375	0.125	0

$E(X) = 1.6875$

Midsize cars:

v	3	2	1	0
$P(V = v)$	.2	.333	0	.467
$vP(V = v)$	0.6	0.667	0	0

$E(V) ≈ 1.267$

Large cars:

w	3	2	1	0
$P(W = w)$	.474	.263	.158	.105
$wP(W = w)$	1.421	0.526	0.158	0

$E(W) ≈ 2.105$

Of the three large cars (W) performed best.

45. Either you lose $1 ($X = -1$) or win $1 ($X = 1$).

There are 20 losing numbers out of 38, so

$$P(X = -1) = \frac{20}{38}$$

There are 18 winning numbers out of 38, so

$$P(X = 1) = \frac{18}{38}$$

x	-1	1
$P(X = x)$	$\frac{20}{38}$	$\frac{18}{38}$
$xP(X = x)$	$-\frac{20}{38}$	$\frac{18}{38}$

$$E(X) = -\frac{20}{38} + \frac{18}{38} = -\frac{2}{38} ≈ -0.53$$

Expect to lose 53¢.

47. The given experiment consists of $n = 40$ Bernoulli trials with $p = .63$. Thus, the expected number of students that will shop at a mall during the next week is

$$\mu = np = 40 \times .63 = 25.2 \text{ students}$$

49. (a) The given experiment consists of $n = 20$ Bernoulli trials with $p = .10$. Therefore, the expected number of defective air bags is

$$\mu = np = 20 \times .10 = 2$$

(b) $p = .10$, $\mu = 12$ and n is unknown

$$\mu = np$$
$$12 = .10n$$
$$n = \frac{12}{.10} = 120 \text{ airbags}$$

51. The probability distribution for X = number of red tents is derived in the solution to Set 8.1, Exercise 47. Here, we add a new row to compute the expected value:

x	1	2	3	4
$P(X = x)$	$\frac{4}{35}$	$\frac{18}{35}$	$\frac{12}{35}$	$\frac{1}{35}$
$xP(X = x)$	$\frac{4}{35}$	$\frac{36}{35}$	$\frac{36}{35}$	$\frac{4}{35}$

$E(X)$ = Sum of bottom row entries

$$= \frac{80}{35} \approx 2.2857 \text{ tents}$$

53. Let X = rate of return of Fastforward Funds, and let Y = rate of return of SolidState Securities. The following worksheets show the computation of the expected values of X and Y:

	A	B	C
1	x	P(X = x)	xP(X = x)
2	-0.4	0.015	-0.006
3	-0.3	0.025	-0.0075
4	-0.2	0.043	-0.0086
5	-0.1	0.132	-0.0132
6	0	0.289	0
7	0.1	0.323	0.0323
8	0.2	0.111	0.0222
9	0.3	0.043	0.0129
10	0.4	0.019	0.0076
11		Total:	0.0397

	E	F	G
1	y	P(Y = y)	yP(Y = y)
2	-0.4	0.012	-0.0048
3	-0.3	0.023	-0.0069
4	-0.2	0.05	-0.01
5	-0.1	0.131	-0.0131
6	0	0.207	0
7	0.1	0.33	0.033
8	0.2	0.188	0.0376
9	0.3	0.043	0.0129
10	0.4	0.016	0.0064
11		Total:	0.0551

From the worksheets, we read off the following expected rates of return: FastForward: 3.97%; SolidState: 5.51%; SolidState gives the higher expected return.

55. If a driver wrecks a car, the net cost to the insurance company is

$$\$100,000 - \$5000 = \$95,000$$

so $X = -95,000$

If a driver does not wreck a car, the net profit for the insurance company is the premium: $\$5000$, so

$$X = 5000$$

To compute the probability distribution, use the following tree:

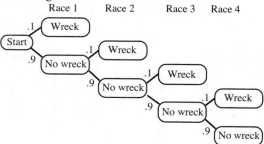

Probability of no wreck is $.9^4 = .6561$

Probability of a wreck is $1 - .9^4 = .3439$

x	−95,000	5000
$P(X = x)$	.3439	.6561
$xP(X = x)$	−32670.5	3280.5

$E(X)$ = sum of entries in bottom row = −29390

The company can expect to lose $29,390 per driver.

57. Since there are as many scores above the median as below the median, the correct choice is (A): The median and mean are equal.

59. He is wrong; for example, the collection 0, 0, 300 has mean 100 and median 0.

61. No. The expected number of times you will hit the dart-board is the average number of times you will hit the bull's eye per 50 shots; the average of a set of whole numbers need not be a whole number.

63. Wrong. It might be the case that only a small fraction of people in the class scored better than you but received a exceptionally high scores that raised the class average. Suppose, for instance, that there are 10 people in the class. Four received 100%, you received 80%, and the rest received 70%. Then the class average is 83%, 5 people have lower scores that you, but only four have higher scores.

65. No; the mean of a very large sample is only an *estimate* of the population mean. The means of larger and larger samples *approach* the population mean as the sample size increases.

67. Wrong. the statement attributed to President Bush asserts that the mean tax refund would be $1000, whereas the statements referred to as "The Truth" suggest that the *median* tax refund would be close to $100 [and that the 31st percentile would be zero].

69. Select a U.S. household at random, and let X be the income of that household. The expected value of X is then the population mean of all U.S. household incomes.

8.4

1. For ease of computation, we arrange the data in a table. The first column lists the values of X, giving us the mean $\bar{x}$. The second column lists the numbers $(x_i-\bar{x})$, and the third column lists their squares.

x	$(x-\bar{x})$	$(x-\bar{x})^2$
−1	−7	49
5	−1	1
5	−1	1
7	1	1
14	8	64
30	0	116

The bottom row shows the sums.

$$\bar{x} = \frac{30}{5} = 6 \qquad \sum_{i=1}^{n} (x_i-\bar{x})^2 = 116$$

$$s^2 = \frac{\sum_{i=1}^{n} (x_i-\bar{x})^2}{n-1} = \frac{116}{5-1} = 29$$

$$s = \sqrt{s^2} \approx 5.39$$

3.

x	$(x-\bar{x})$	$(x-\bar{x})^2$
2	−1	1
5	2	4
6	3	9
7	4	16
−1	−4	16
−1	−4	16
18	0	62

The bottom row shows the sums.

$$\bar{x} = \frac{18}{6} = 3 \qquad \sum_{i=1}^{n} (x_i-\bar{x})^2 = 62$$

$$s^2 = \frac{\sum_{i=1}^{n} (x_i-\bar{x})^2}{n-1} = \frac{62}{6-1} = 12.4$$

$$s = \sqrt{s^2} \approx 3.52$$

5. In the following table, we first converted all the fractions to decimals.

x	$(x-\bar{x})$	$(x-\bar{x})^2$
0.5	0.6875	0.4727
1.5	1.6875	2.8477
−4	−3.8125	14.5352
1.25	1.4375	2.0664
−0.75	0	19.9219

The bottom row shows the sums.

$$\bar{x} = \frac{-0.75}{4} = -0.1875$$

$$\sum_{i=1}^{n} (x_i-\bar{x})^2 \approx 19.9219$$

$$s^2 = \frac{\sum_{i=1}^{n} (x_i-\bar{x})^2}{n-1} \approx \frac{19.9219}{4-1} \approx 6.64$$

$$s = \sqrt{s^2} \approx 2.58$$

7.

x	$(x-\bar{x})$	$(x-\bar{x})^2$
2.5	2.3	5.29
−5.4	−5.6	31.36
4.1	3.9	15.21
−0.1	−0.3	0.09
−0.1	−0.3	0.09
1	0	52.04

The bottom row shows the sums.

$$\bar{x} = \frac{1}{5} = 0.2 \qquad \sum_{i=1}^{n} (x_i-\bar{x})^2 = 52.04$$

$$s^2 = \frac{\sum\limits_{i=1}^{n} (x_i - \bar{x})^2}{n-1} = \frac{52.04}{5-1} = 13.01$$

$$s = \sqrt{s^2} \approx 3.61$$

9. We use the tabular method of arranging the data described in Example 3:

x	0	1	2	3
$P(X = x)$	.5	.2	.2	.1
$xP(X = x)$	0	0.2	0.4	0.3

μ = Sum of entries in bottom row = 0.9

$x - \mu$	−0.9	0.1	1.1	2.1
$(x-\mu)^2$	0.81	0.01	1.21	4.41
$P(X=x)$ $\times(x-\mu)^2$	0.405	0.002	0.242	0.441

σ^2 = Sum of entries in bottom row = 1.09

$\sigma = \sqrt{\sigma^2} \approx 1.04$

11. We first convert all fractions into decimals, and then use the tabular method described in Example 3:

x	10	20	30	40
$P(X = x)$	.3	.4	.2	.1
$xP(X = x)$	3	8	6	4

μ = Sum of entries in bottom row = 2.8

$x - \mu$	−11	−1	9	19
$(x-\mu)^2$	121	1	81	361
$P(X=x)$ $\times(x-\mu)^2$	36.3	0.4	16.2	36.1

σ^2 = Sum of entries in bottom row = 89

$\sigma = \sqrt{\sigma^2} \approx 9.43$

13. We use the tabular method of arranging the data described in Example 3. (Note that, for ease of layout, we are arranging the data in columns.)

x	$P(X = x)$	$xP(X = x)$
−5	.2	−1
−1	.3	−0.3
0	.2	0
2	.1	0.2
5	.2	1
10	0	0

μ = Sum of entries in right-hand column = −0.1

$x - \mu$	$(x - \mu)^2$	$P(X = x)\times$ $(x - \mu)^2$
−4.9	24.01	4.802
−0.9	0.81	0.243
0.1	0.01	0.002
2.1	4.41	0.441
5.1	26.01	5.202
10.1	102.01	0

σ^2 = Sum of entries in right-hand column = 10.69

$\sigma = \sqrt{\sigma^2} \approx 3.27$

15. The probability distribution and expected value calculated in Exercise 17 of Section 8.3:

x	$P(X = x)$	$xP(X = x)$
1	1/6	1/6
2	1/6	1/3
3	1/6	1/2
4	1/6	2/3
5	1/6	5/6
6	1/6	1

μ = Sum of entries in right-hand column = 3.5

$x-\mu^1$	$(x-\mu)^2$	$P(X=x)\times$ $(x-\mu)^2$
−2.5	6.25	1.042
−1.5	2.25	0.375
−0.5	0.25	0.042
0.5	0.25	0.042
1.5	2.25	0.375
2.5	6.25	1.042

σ^2 = Sum of entries in right-hand column ≈ 2.918

$\sigma = \sqrt{\sigma^2} \approx 1.71$

17. X is a binomial random variable with $n = 2$ and $p = .5$.

$$\mu = E(X) = np = 2(.5) = 1$$
$$\sigma^2 = npq = 2(.5)(.5) = 0.5$$
$$\sigma = \sqrt{\sigma^2} \approx 0.71$$

19. The probability distribution was calculated in Exercise 27 of Section 8.1. To continue the calculation, we use decimal approximations of the fractions:

x	$P(X=x)$	$xP(X=x)$
1	.0278	0.0278
2	.0833	0.1667
3	.1389	0.4167
4	.1944	0.7778
5	.2500	1.2500
6	.3056	1.8333

μ = Sum of entries in right-hand column ≈ 4.47

$x-\mu$	$(x-\mu)^2$	$P(X=x)\times$ $(x-\mu)^2$
−3.4722	12.0563	0.3349
−2.4722	6.1119	0.5093
−1.4722	2.1674	0.3010
−0.4722	0.2230	0.0434
0.5278	0.2785	0.0696
1.5278	2.3341	0.7132

σ^2 = Sum of entries in right-hand column ≈ 1.9715

$\sigma = \sqrt{\sigma^2} \approx 1.40$

21. The probability distribution was calculated in Exercise 21 of Section 8.3. To continue the calculation, we use decimal approximations of the fractions:

x	$P(X=x)$	$xP(X=x)$
2	.40	0.8000
3	.5333	1.6000
4	.0667	0.2667

μ = Sum of entries in right-hand column ≈ 2.67

$x-\mu$	$(x-\mu)^2$	$P(X=x)\times$ $(x-\mu)^2$
−0.6667	0.4444	0.1778
0.3333	0.1111	0.0593
1.3333	1.7778	0.1185

σ^2 = Sum of entries in right-hand column ≈ 0.36

$\sigma = \sqrt{\sigma^2} \approx 0.60$

23. X is a binomial random variable with $n = 20$ and $p = .1$

$$\mu = np = 20(.1) = 2$$
$$\sigma^2 = npq = 20(.1)(.9) = 1.8$$
$$\sigma = \sqrt{\sigma^2} \approx 1.34$$

25. (a)

x	$(x-\bar{x})$	$(x-\bar{x})^2$
3	0	0
2	−1	1
0	−3	9
9	6	36
1	−2	4
15	0	50

(The bottom row shows the sums.)

$$\bar{x} = \frac{15}{5} = 3$$

$$\sum_{i=1}^{n} (x_i-\bar{x})^2 = 50$$

$$s^2 = \frac{\sum_{i=1}^{n} (x_i-\bar{x})^2}{n-1} \approx \frac{50}{5-1} \approx 12.5$$

$$s = \sqrt{s^2} \approx 3.54$$

(b) The empirical rule states that approximately 68% of the class will rank you between

$$\bar{x} - s = 3 - 3.54 = -0.54$$

and

$$\bar{x} + s = 3 + 3.54 = 6.54$$

That is, in the interval [0, 6.54] (We replaced the negative score by 0, since no rankings can be negative.

We must assume that the population distribution is bell-shaped and symmetric.

27. (a)

x	$(x-\bar{x})$	$(x-\bar{x})^2$
4.2	−0.8	0.6
4.7	−0.3	0.09
5.4	0.4	0.16
5.8	0.8	0.64
4.9	−0.1	0.01
25	0	1.54

(The bottom row shows the sums.)

$$\bar{x} = \frac{25}{5} = 5.0$$

$$\sum_{i=1}^{n} (x_i-\bar{x})^2 = 1.54$$

$$s^2 = \frac{\sum_{i=1}^{n} (x_i-\bar{x})^2}{n-1} \approx \frac{1.54}{5-1} \approx 0.385$$

$$s = \sqrt{s^2} \approx 0.6$$

(b) The empirical rule states that approximately 95% of the data will fall between

$$\bar{x} - 2s = 5.0 - 1.2 = 3.8$$

and

$$\bar{x} + 2s = 5.0 + 1.2 = 6.2$$

29. (a)

x	$(x-\bar{x})$	$(x-\bar{x})^2$
2000	−3000	9,000,000
2000	−3000	9,000,000
3000	−2000	4,000,000
4000	−1000	1,000,000
5000	0	0
6000	1000	1,000,000
6000	1000	1,000,000
6000	1000	1,000,000
8000	3000	9,000,000
8000	3000	9,000,000
50,000	0	44,000,000

(The bottom row shows the sums.)

$$\bar{x} = \frac{50,000}{10} = 5000$$

$$\sum_{i=1}^{n} (x_i-\bar{x})^2 = 44,000,000$$

$$s^2 = \frac{\sum\limits_{i=1}^{n} (x_i - \bar{x})^2}{n-1} \approx \frac{44{,}000{,}000}{10-1} \approx 4{,}888{,}888.9$$

$$s = \sqrt{s^2} \approx 2211$$

(b) The empirical rule states that approximately 68% of the data will fall between

$$\bar{x} - s = 5000 - 2211 = 2789$$

and

$$\bar{x} + s = 5000 + 2211 = 7211$$

To obtain the actual percentage of scores that fall within the range [2789. 7211] count how many of the tow ratings are in this range: 6 of them (highlighted)

2000, 2000, **3000, 4000, 5000, 6000, 6000, 6000**, 8000, 8000

Since 6 scores is 60% of the original 10, we conclude that 60% of scores in the sample that fall in the range [2789. 7211].

31. (a) Following is a worksheet computation of the variance:

	A	B
1	x	(x − mu)^2
2	17	0.5625
3	18	0.0625
4	17	0.5625
5	18	0.0625
6	21	10.5625
7	16	3.0625
8	21	10.5625
9	18	0.0625
10	16	3.0625
11	14	14.0625
12	15	7.5625
13	22	18.0625
14	17	0.5625
15	19	1.5625
16	17	0.5625
17	18	0.0625
18	mean (mu)	variance
19	17.75	4.733333

From the worksheet,

$$s^2 \approx 4.7333$$

$$s = \sqrt{s^2} \approx 2.1756 \approx 2.18$$

(b) Chebyshev's inequality predicts that at least 8/9 of the scores will fall between

$$\bar{x} - 3s \approx 17.75 - 3(2.1756) \approx 11.22$$

and

$$\bar{x} + 3s \approx 17.75 + 3(2.1756) \approx 24.28$$

That is, they will fall in the interval [11.22, 24.28].

(c) Every single score, or <u>100%</u> of the scores fall in the range [11.22, 24.28]. Since the empirical rule predicts that 99.7% of the scores will fall in the given range, making it a more accurate predictor.

33. We use the tabular method of arranging the data described in Example 3. (Note that, for ease of layout, we are arranging the data in columns.)

x	P(X = x)	xP(X = x)
0	0.4	0
1	0.1	0.1
2	0.2	0.4
3	0.2	0.6
4	0.1	0.4

μ = Sum of entries in right-hand column = 1.5

$x - \mu$	$(x - \mu)^2$	$P(X = x) \times (x - \mu)^2$
−1.5	2.25	0.9
−0.5	0.25	0.025
0.5	0.25	0.05
1.5	2.25	0.45
2.5	6.25	0.625

σ^2 = Sum of entries in right-hand column = 2.05
$\sigma = \sqrt{\sigma^2} \approx 1.43$

The range of X within 2 standard deviations of the mean is given by
Lower limit = $\mu - 2\sigma \approx 2.05 - 2(1.43) = -0.81$
Upper limit = $\mu + 2\sigma \approx 2.05 + 2(1.43) = 4.91$

All (100%) of the values of X are within the interval $[-0.81, 4.91]$. Therefore, 100% of malls have a number of movie theater screens within two standard deviations of μ.

35. The probabilities for the distribution are computed by dividing the frequencies by the total: 100.

x	$P(X = x)$	$xP(X = x)$
10	.27	2.7
30	.28	8.4
50	.20	10.0
70	.15	10.5
90	.10	9.0

μ = Sum of entries in right-hand column = 40.6, representing $40,600

$x - \mu$	$(x - \mu)^2$	$P(X = x) \times (x - \mu)^2$
−30.6	936.36	252.8172
−10.6	112.36	31.4608
9.4	88.36	17.672
29.4	864.36	129.654
49.4	2440.36	244.036

σ^2 = Sum of entries in right-hand column = 675.64
$\sigma = \sqrt{\sigma^2} \approx 26$, representing $26,000

The range of X within 1 standard deviation of the mean is given by
Lower limit = $\mu - \sigma \approx 40.6 - 26 \approx 14.6$
Upper limit = $\mu + \sigma \approx 40.6 + 26 \approx 66.6$
The difference is $66.6 - 14.6 = 52$ (2 standard deviations), representing an income gap of $52,000.

37. (a) The rounded midpoints of the measurement classes (ages) are:
$(15+24.9)/2 \approx 20$
$(25+54.9)/2 \approx 40$
$(55 + 64.9)/2 \approx 60$

In the following table, all intermediate calculations have been rounded to two decimal places.

X	20	40	60	**Total**
N	16,000	13,000	1600	30600
$P(X=x)$	.52	.42	.05	
$x \cdot P(X=x)$	10.4	16.8	3	30.2
$x - \mu$	-10.2	9.8	29.8	
$(x - \mu)^2$	104.04	96.04	888.04	
$P(X=x) \cdot (x - \mu)^2$	54.1	40.34	44.4	138.84

Expected value = μ = Sum of entries in 4th row ≈ 30.2 yrs old
Variance = σ^2 = Sum of entries in bottom row ≈ 138.84
St. Deviation = $\sigma = \sqrt{138.84} \approx 11.78$ years
(b) According to the empirical rule approximately 68% of all male Hispanic workers fall in the interval
$[\mu - \sigma, \mu + \sigma] = [30.2 - 11.78, 30.2 + 11.78]$
$= [18.42, 41.98] \approx [18, 42]$
So, approximately 68% of all male Hispanic workers are 18–42 years old

39. Since the probability distribution is highly skewed, we need to use Chebyshev's rule.

$\mu = 2$, $\sigma = 0.15$

Following is a representation of the mean ± several standard deviations (each division is one standard deviation in width):

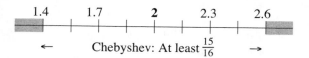

The range [1.4, 2.6] represents the interval $\mu \pm 4\sigma$, so by Chebyshev's rule, at least 15/16 of companies have a lifespan in this range. Therefore, *at most* 1/16 have a lifespan outside this range (the gray regions above). Since the distribution is not symmetric, we cannot conclude that half of the 1/16 is in the range on the right (companies at least as old as yours).

Therefore, all we can say is that at most 1/16, or 6.25% of all companies are in the range on the right (at least as old as yours).

41. Since the probability distribution is not known to be bell-shaped, we need to use Chebyshev's rule.

$\mu = 9$, $\sigma = 2$

Following is a representation of the mean ± several standard deviations (each division is one standard deviation in width):

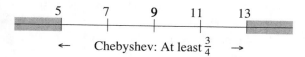

The range [5, 13] represents the interval $\mu \pm 2\sigma$, so by Chebyshev's rule, at least 3/4 of all Batmobiles have a lifespan in this range. Therefore, *at most* 1/4 have a lifespan outside this range (the gray regions

above). Since the distribution is symmetric, at most half of these: 1/8 of all Batmobiles, have life spans more than 13 years (the gray region on the right).

Therefore, there is *at most* (Choice B) a 12.5% chance that your new Batmobile will last 13 years or more.

43. (a) Take "success" to mean shopping at a mall. The given distribution is a binomial distribution with

$n = 40$, $p = .63$, $q = 1 - .63 = .37$

$\mu = np = 40 \times .63 = 25.2$ students

$\sigma = \sqrt{npq} = \sqrt{40 \times .63 \times .37} \approx 3.05$

(b) Since the binomial distribution is symmetric and bell-shaped, we can use the Empirical rule, which says there is a 95% chance that that between

$\mu - 2\sigma = 25.2 - 2(3.05) \approx 19$

and

$\mu + 2\sigma = 25.2 + 2(3.05) \approx 31$

teenagers in the sample will shop at a mall during the next week.

Therefore, there is a 5% chance of this not happening—that is, a 5% chance that either 19 or fewer students will shop at a mall, or that 31 or more will.

Since the distribution is symmetric, there is a 2.5% chance that <u>31</u> or more students in the group will shop at a mall during the next week.

45. (a) Take "success" to mean *not* having a checking account. The given distribution is a binomial distribution with

$n = 1000$, $p = 1 - .22 = .78$, $q = .22$

$\mu = np = 1000 \times .78 = 780$

$\sigma = \sqrt{npq} = \sqrt{1000 \times .78 \times .22} \approx 13.1$

(b) Since the binomial distribution is symmetric and bell-shaped, we can use the Empirical rule, which says there is a 95% chance that that between $\mu - 2\sigma$ and $\mu + 2\sigma$ teenagers in the sample will not have checking accounts. These numbers, are respectively,

$$\mu - 2\sigma \approx 780 - 2(13.1) \approx 754$$
$$\mu + 2\sigma \approx 780 + 2(13.1) \approx 806$$

47. (a) Following is a worksheet computation of the variance (the bottom row gives the column sums):

	A	B	C	D
1	x	P(x)	x*P(x)	(x-mu)^2*P(x)
2	1	0.01	0.01	0.3025
3	2	0.04	0.08	0.81
4	3	0.04	0.12	0.49
5	4	0.08	0.32	0.5
6	5	0.1	0.5	0.225
7	6	0.15	0.9	0.0375
8	7	0.25	1.75	0.0625
9	8	0.2	1.6	0.45
10	9	0.08	0.72	0.5
11	10	0.05	0.5	0.6125
12	55	1	6.5	3.99

From the worksheet,

$$\mu = 6.5$$
$$\sigma^2 = 3.99 \approx 4.0$$
$$\sigma = \sqrt{\sigma^2} \approx 2.0$$

(b) According to Chebyshev's inequality, at least 3/4 or 75% of all supermarkets will have between

$$\mu - 2\sigma \approx 6.5 - 2(2.0) = 2.5$$

and

$$\mu + 2\sigma \approx 6.5 + 2(2.0) = 10.5$$

checkout lanes.
The smallest (whole) number of checkout lanes in this range is 3.

49. The household income of a poor family in the U.S. is

$$38,000 - 1.3(21,000) = \$10,700 \text{ or less.}$$

51. The household income of a rich family in the U.S. is

$$38,000 + 1.3(21,000) = \$65,300 \text{ or more.}$$

53. Cutoffs for poor families (1.3 standard deviations):

U.S.: $38,000 - 1.3(21,000) = \$10,700$
Canada: $35,000 - 1.3(17,000) = \$12,900$
Switzerland: $39,000 - 1.3(16,000) = \$18,200$
Germany: $34,000 - 1.3(14,000) = \$15,800$
Sweden: $32,000 - 1.3(11,000) = \$17,700$

The U.S. has the poorest households.

55. The gap between rich and poor is measured by $2 \times 1.3 = 2.6$ standard deviations. Since the U.S. has the largest standard deviation listed, it has the largest gap between rich and poor.

57. An income of $17,000 is 1 standard deviation below the U.S. mean income. By the empirical rule, approximately 68% earned within 1 standard deviation, so that approximately 32% earn outside the 1-standard deviation interval. Half of that, approximately 16%, earn less.

59. By the empirical rule, approximately 99.7% earned within 3 standard deviations of the mean. This is the range

$$34,000 - 3(14,000) = -8000$$

to

$$34,000 + 3(14,000) = 76,000$$

Since income can't be negative, the answer is 0–$76,000.

61. Using technology, $\mu = 12.56\%$, $\sigma \approx 1.8885\%$.

63. The empirical rule predicts approximately 68%. The 1-standard deviation interval based on the calculations in Exercise 61 is

$\mu - \sigma = 12.56 - 1.888 = 10.672$

$\mu + \sigma = 12.56 + 1.888 = 14.448$

The scores in that range are shown in bold:

6, 9, 10, 10, 10, **11, 11, 11, 11, 11, 11, 11, 12, 12, 12, 12, 12, 12, 12, 12, 12, 12, 13, 13, 13, 13, 13, 13, 13, 13, 13, 13, 13, 13, 13, 14, 14, 14, 14, 14, 14, 14,** 15, 15, 15, 15, 16, 18

These are 39 states, representing 39/50 = 78% of all states. This differs substantially from the empirical rule prediction. One reason for the discrepancy is that the associated probability distribution is roughly bell-shaped but not symmetric:

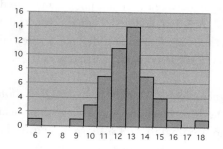

65. The 2-standard deviation interval based on the calculations in Exercise 61 is

$\mu - \sigma = 12.56 - 2(1.888) \approx 8.78$

$\mu + \sigma = 12.56 + 2(1.888) \approx 16.34$

The scores in that range are shown in bold:

6, **9, 10, 10, 10, 11, 11, 11, 11, 11, 11, 11, 12, 12, 12, 12, 12, 12, 12, 12, 12, 12, 13, 13, 13, 13, 13, 13, 13, 13, 13, 13, 13, 13, 13, 14, 14, 14, 14, 14, 14, 14, 15, 15, 15, 15, 16,** 18

These are 48 states, representing 48/50 = 96% of all states. Chebyshev's rule is valid, since it predicts that *at least* 75% of the scores are in this range.

67. (A) The graph shows standard deviations, not the actual power grid frequency, so we cannot conclude (A).

(B) The standard deviation indicates the variability in the power supply frequency. Since it was lower in mid-1999 than in 1995, this indicates greater stability in mid-1999 than in 1995, so the assertion is true.

(C) The standard deviation indicates the variability in the power supply frequency. Since it was higher in mid-2002 than in mid-1995, this indicates *less* stability in mid-2002, so we cannot conclude (C).

(D) The standard deviation was greatest in 2001–2002, indicating that the greatest fluctuations in the power grid frequency occurred during that period, so the assertion is true.

(E) Around January 1995, the standard deviation was closest to its average of 0.9, but was lower around January 1999, so the power grid was more stable around January 1999, than around January 1995. Thus, (E) is false.

69. The sample standard deviation is bigger; the formula for sample standard deviation involves division by the smaller term $n-1$ instead of n, which makes the resulting number larger.

71. The grades in the first class were clustered fairly close to 75. By Chebyshev's inequality, at least 88% of the class had grades in the range 60–90. On the other hand, the grades in the second class were widely dispersed. The second class had a much wider spread of ability than did the first class.

73. Since the standard deviation is 0, there is no variability at all; that is, the variable must be constant. Therefore, the variable must take on only the value 10, with probability 1.

75. Since there are 2 data points, the mean is midway between them, at a distance of $(y-x)/2$ from each. Summing the square of this distance twice gives

$$\Sigma(x_i - \mu)^2 = \frac{(y-x)^2}{4} + \frac{(y-x)^2}{4} = \frac{(y-x)^2}{2}$$

Dividing by 2 gives the variance:

$$\sigma^2 = \frac{(y-x)^2}{4}$$

Therefore, $\sigma = \frac{y-x}{2}$.

8.5

Note: Answers for Section 8.5 were computed using 4-digit tables, and may differ slightly from (more accurate) answers generated using technology.

1. We use the table.

$P(0 \leq Z \leq 0.5) = .1915$

0.4	0.1554	0.1591
0.5	0.1915	0.1950
0.6	0.2257	0.2291

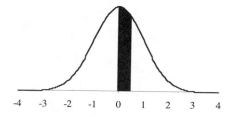

3. The table gives us

$P(0 \leq Z \leq 0.71) = .2611$

0.6	0.2257	0.2291
0.7	0.2580	0.2611
0.8	0.2881	0.2910

$P(-0.71 \leq Z \leq 0.71)$ is twice this area, as shown:

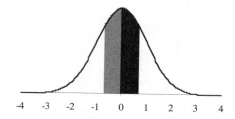

$P(-0.71 \leq Z \leq 0.71) = 2(.2611) = .5222$

5. The table gives us

$P(0 \leq Z \leq 1.34) = .4099$

1.2	0.3849	0.3869	0.3888	0.3907	0.3925
1.3	0.4032	0.4049	0.4066	0.4082	0.4099
1.4	0.4192	0.4207	0.4222	0.4236	0.4251

$P(0 \leq Z \leq 0.71) = .2611$

0.6	0.2257	0.2291	0.2324	0.2357	0.2389
0.7	0.2580	0.2611	0.2642	0.2673	0.2704
0.8	0.2881	0.2910	0.2939	0.2967	0.2995

$P(-0.71 \leq Z \leq 1.34)$ is obtained by adding these areas (see figure).

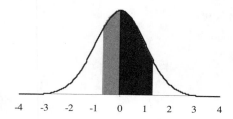

$P(-0.71 \leq Z \leq 1.34) = .4099 + .2611 = .6710$

7. From the table,

$P(0 \leq Z \leq 1.5) = .4332$

$P(0 \leq Z \leq 0.5) = .1915$

To obtain $P(0.5 \leq Z \leq 1.5)$, subtract the smaller from the larger (see figure).

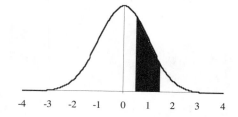

$P(0.5 \leq Z \leq 1.5) = .4332 - .1915 = .2417$

9. $\mu = 50$, $\sigma = 10$

Standardize the given problem:

$$P(a \le X \le b) = P\left(\frac{a-\mu}{\sigma} \le Z \le \frac{b-\mu}{\sigma}\right)$$

$$P(35 \le X \le 65) = P\left(\frac{35-50}{10} \le Z \le \frac{65-50}{10}\right)$$

$$= P(-1.5 \le Z \le 1.5)$$

$$= 2(.4332) = .8664$$

11. $\mu = 50$, $\sigma = 10$

Standardize the given problem:

$$P(a \le X \le b) = P\left(\frac{a-\mu}{\sigma} \le Z \le \frac{b-\mu}{\sigma}\right)$$

$$P(30 \le X \le 62) = P\left(\frac{30-50}{10} \le Z \le \frac{62-50}{10}\right)$$

$$= P(-2 \le Z \le 1.2)$$

$$= .3849 + .4772 = .8621$$

13. $\mu = 100$, $\sigma = 15$

Standardize the given problem:

$$P(a \le X \le b) = P\left(\frac{a-\mu}{\sigma} \le Z \le \frac{b-\mu}{\sigma}\right)$$

$$P(110 \le X \le 130) = P\left(\frac{110-100}{15} \le Z \le \frac{130-100}{15}\right)$$

$$\approx P(0.67 \le Z \le 2)$$

$$= .4772 - .2486 = .2286$$

15. The Z-value measures the number of standard deviations form the mean, Therefore, the given problem translates to

$$P(-0.5 \le Z \le 0.5) = 2(.1915) = .3830$$

17. This is the probability that Z is either $> \frac{2}{3}$ or $< -\frac{2}{3}$. The complement of this event is the event that

$$-\frac{2}{3} \le Z \le \frac{2}{3}$$

$$P(-0.67 \le Z \le 0.67) = 2(.2486) = .4972$$

Therefore, the desired probability is

$$1 - .4972 = .5028$$

19. $P(10 \le X \le 15) = P(9.5 \le Y \le 15.5)$

where Y has a mean of

$$\mu = np = 100 \times \frac{1}{6} \approx 16.6667$$

$$\sigma = \sqrt{npq} = \sqrt{100 \times \frac{1}{6} \times \frac{5}{6}} \approx 3.7268$$

We now standardize Y:

$$P(9.5 \le Y \le 15.5)$$

$$= P\left(\frac{9.5-16.6667}{3.7268} \le Z \le \frac{15.5-16.6667}{3.7268}\right)$$

$$\approx P(-1.92 \le Z \le -0.31)$$

$$= .4726 - .1217 = .3509 \approx .35$$

21. $P(X < 25) = P(0 \le X \le 24)$

$$= P(-0.5 \le Y \le 24.5)$$

where Y has a mean of

$$\mu = np = 200 \times \frac{1}{6} \approx 33.3333$$

$$\sigma = \sqrt{npq} = \sqrt{200 \times \frac{1}{6} \times \frac{5}{6}} \approx 5.2705$$

We now standardize Y:

$$P(-0.5 \le Y \le 24.5)$$

$$= P\left(\frac{-0.5-33.3333}{5.2705} \le Z \le \frac{24.5-33.3333}{5.2705}\right)$$

$$\approx P(-6.42 \le Z \le -1.68)$$

$$= .5000 - .4535 = .0465 \approx .05$$

23. $\mu = 500$, $\sigma = 100$

$$P(450 \le X \le 550)$$

$$= P\left(\frac{450-500}{100} \le Z \le \frac{550-500}{100}\right)$$

$$= P(-0.5 \le Z \le 0.5)$$

$$= .1915 + .1915 = .3830$$

25. $\mu = 500$, $\sigma = 100$

$$P(300 \le X \le 550)$$

$$= P\left(\frac{300-500}{100} \le Z \le \frac{550-500}{100}\right)$$

$$= P(-2 \le Z \le 0.5)$$

$$= .4772 + .1915 = .6687$$

27. $\mu = 100$, $\sigma = 16$

$P(110 \leq X \leq 140)$

$= P\left(\dfrac{110-100}{16} \leq Z \leq \dfrac{140-100}{16}\right)$

$\approx P(0.63 \leq Z \leq 2.5)$

$= .4938 - .2357 = .2581$,

approximately 26%

29. $\mu = 100$, $\sigma = 16$

$P(X \geq 120) = P\left(Z \geq \dfrac{120-100}{16}\right)$

$= P(Z \geq 1.25) = .5 - .3944 = .1056$

The total number of such people in the U.S. is
.1056×280,000,000 ≈ 29,600,000.

31. $\mu = 0.25$, $\sigma = 0.03$

$P(X \geq 0.4) = P\left(Z \geq \dfrac{0.4-0.25}{0.03}\right)$

$= P(Z \geq 5) = .5 - .500 = .0000$

The total number of such batters is therefore 0.

33. $\mu = 7.5$, $\sigma = 1$

$P(X \geq 9) = P\left(Z \geq \dfrac{9-7.5}{1}\right)$

$= P(Z \geq 1.5) = .5 - .4332 = .0668$

The total number of jars is therefore
.0668×100,000 ≈ 6680

35. $\mu = 38$, $\sigma = 21$ (in thousands of dollars)

$P(X \geq 50) = P\left(Z \geq \dfrac{50-38}{21}\right)$

$\approx P(Z \geq 0.57) = .5 - .2157 = .2843$;

approximately 28%

37. $\mu = 39$, $\sigma = 16$ (in thousands of dollars)

Very rich: $P(X \geq 100) = P\left(Z \geq \dfrac{100-39}{16}\right) \approx P(Z$

$\geq 3.81) = .5 - .5000 = 0$

Very poor: $P(X \leq 12) = P\left(Z \leq \dfrac{12-39}{16}\right)$

$\approx P(Z \leq -1.69) = .5 - .4545 = .0455$;

approximately 5%

39. U.S:

$\mu = 38$, $\sigma = 21$ (in thousands of dollars)

$P(X \leq 12) = P\left(Z \leq \dfrac{12-38}{21}\right)$

$\approx P(Z \leq -1.24) = .5 - .3925 = .1075$

Canada:

$\mu = 35$, $\sigma = 17$ (in thousands of dollars)

$P(X \leq 12) = P\left(Z \leq \dfrac{12-35}{17}\right)$

$\approx P(Z \leq -1.35) = .5 - .4115 = .0885$

The U.S. has a higher proportion of very poor
people.

41. Wechsler. Since this test has a smaller
standard deviation, a greater percentage of scores
fall within 20 points of the mean.

43. $\mu = 6$, $\sigma = 1$

The Z-value corresponding to $X = 1$ month is

$Z = \dfrac{1-6}{1} = -5$

This is surprising, because the time between
failures was more than 5 standard deviations
away from the mean, which happens with an
extremely small probability.

45. Task 1: $\mu = 11.4$, $\sigma = 5.0$

$P(X \geq 10) = P\left(Z \geq \dfrac{10-11.4}{5}\right)$

$\approx P(Z \geq -0.28) = .5 + .1103 = .6103$

47.

Task 1: By Exercise 41,

$P(X_1 \leq 10) = .6103$

Task 2: $\mu = 11.9$, $\sigma = 9$

$P(X_2 \geq 10) = P\left(Z \geq \dfrac{10-11.9}{9}\right)$

$\approx P(Z \geq -0.21) = .5 + .0832 = .5832$

Since the times taken to complete the tasks are independent,

$P(X_1 \leq 10 \text{ and } X_2 \leq 10)$

$= P(X_1 \leq 10) \times P(X_2 \leq 10)$

$= .6103 \times .5832 \approx .3559$

49. Tasks 1&2:

$\mu = 11.4 + 11.9 = 23.3$

$\sigma = \sqrt{5^2 + 9^2} \approx 10.2956$

$P(X \geq 20) = P\left(Z \geq \dfrac{20-23.3}{10.2956}\right)$

$\approx P(Z \geq -0.32) = .5 + .1255 = .6255$

51. $\mu = np = 800 \times .51 = 408$

$\sigma = \sqrt{npq} = \sqrt{800 \times .51 \times .49} \approx 14.139$

$\mu - 3\sigma \approx 366$

$\mu + 3\sigma \approx 450$

Since these are between 0 and $n = 800$, the normal approximation of the binomial distribution is valid.

$P(400 \leq X \leq 800)$

$= P(399.5 \leq Y \leq 800.5)$

$= P\left(\dfrac{399.5-408}{14.139} \leq Z \leq \dfrac{800.5-408}{14.139}\right)$

$\approx P(-0.6 \leq Z \leq 27.76) = .5 + .2257 = 0.7257$

53. $n = 100,000,000$, $p = .00000165$

$\mu = np = 165$

$\sigma = \sqrt{npq} = \sqrt{165 \times .99999835} \approx 12.845$

$\mu - 3\sigma \approx 126$

$\mu + 3\sigma \approx 204$

Since these are between 0 and $n = 100,000,000$, the normal approximation of the binomial distribution is valid.

$P(X < 180) = P(X \leq 179)$

$= P(Y \leq 179.5)$

$= P\left(Z \leq \dfrac{179.5-165}{12.845}\right)$

$\approx P(Z \leq 1.13)$

$= .3708 + .5 = .8708$

55. $n = 100,000,000$, $p = .00000165$

$\mu = np = 165$

$\sigma = \sqrt{npq} = \sqrt{165 \times .99999835} \approx 12.845$

$\mu - 3\sigma \approx 126$

$\mu + 3\sigma \approx 204$

Since these are between 0 and $n = 100,000,000$, the normal approximation of the binomial distribution is valid.

Suppose there are X crashes. Since 10 people buy insurance,

Payout = $10 \times 1,000,000$ per crash; that is,

$10,000,000X$

Premium = $10 \times 2 \times 100,000,000$

$= 2,000,000,000$

For break-even

$10,000,000X \quad = 2,000,000,000$

$X = 200$ flights

To lose money, $X > 200$

$P(X > 200) = P(X \geq 201)$

$= P(Y \geq 200.5)$

$= P\left(Z \geq \dfrac{200.5-165}{12.845}\right)$

$\approx P(Z \geq 2.76)$

$= .5 - .4971 = .0029$

57. Let "success" = preferring Goode.

$p = .9 \times .55 + .1 \times .45 = .54$

$n = 1000$

$\mu = np = 540$

$\sigma = \sqrt{npq} = \sqrt{1000 \times .54 \times .46} \approx 15.761$

$\mu - 3\sigma \approx 493$

$\mu + 3\sigma \approx 587$

Since these are between 0 and $n = 1000$, the normal approximation of the binomial distribution is valid.

$P(X > 520) = P(X \geq 521) = P(Y \geq 520.5)$

$= P\left(Z \geq \dfrac{520.5 - 540}{15.761}\right) \approx P(Z \geq -1.25)$

$= .5 + .3925 = .8925$

59. $\mu = 100$, σ unknown.

Let the Z-score of someone who just barely qualifies be k.

To be in the top 2%, $P(Z \geq k) = .02$

Therefore, $P(0 \leq z \leq k) = .5 - .02 = .48$

Looking in the table for the closest score to .48 (which is .4798) gives

$k \approx 2.05$

Since $Z = \dfrac{148 - 100}{\sigma}$, we have

$k = 2.05 = \dfrac{148 - 100}{\sigma} = \dfrac{48}{\sigma}$

$2.05\sigma = 48$, so $\sigma = \dfrac{48}{2.05} \approx 23.4$

Note: A more accurate guess at k would be midway between 2.05 and 2..06. Using $k = 2.055$ gives the same answer to one decimal place.

61. Since the empirical rule is based on the normal distribution (see the remarks before Example 3 in the textbook), the empirical rule gives the exact results when the distribution is exactly normal.

63. Neither. they are equal, since they differ by $P(X = a)$, which is zero for a continuous random variable.

65. The total area under the curve must be equal to 1, and the area is a rectangle with width $(b-a)$. Therefore, its height must be $\dfrac{1}{b-a}$.

67. A normal distribution with standard deviation 0.5, since it is narrower near the mean, but must enclose the same amount of area as the standard curve, and so it must be higher.

Chapter 8 Review Exercises

1. X = the number of boys. X is a binomial random variable with $p = .5$, $n = 2$

$$P(X = 0) = C(2, 0)(.5)^0(.5)^2 = \frac{1}{4}$$

$$P(X = 1) = C(2, 1)(.5)^1(.5)^1 = \frac{1}{2}$$

$$P(X = 0) = C(2, 0)(.5)^0(.5)^2 = \frac{1}{4}$$

Probability distribution:

x	0	1	2
$P(X = x)$	$\frac{1}{4}$	$\frac{1}{2}$	$\frac{1}{4}$

Histogram:

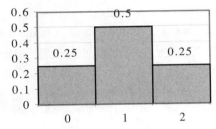

2. X = the sum of the two numbers when a four-sided dice is rolled twice

There are 4×4 = 16 possible outcomes

$X = 2$: Outcomes {(1, 1)}

$$P(X = 2) = \frac{1}{16}$$

$X = 3$: Outcomes {(1, 2), (2, 1)}

$$P(X = 3) = \frac{2}{16}$$

$X = 4$: Outcomes {(1, 3), (2, 2), (3, 1)}

$$P(X = 4) = \frac{3}{16}$$

$X = 5$: Outcomes {(1, 4), (2, 3), (3, 2), (4, 1)}

$$P(X = 5) = \frac{4}{16}$$

$X = 6$: Outcomes {(2, 4), (3, 3), (4, 2)}

$$P(X = 6) = \frac{3}{16}$$

$X = 7$: Outcomes {3, 4), (4, 3)}

$$P(X = 7) = \frac{2}{16}$$

$X = 8$: Outcomes {(4, 4)}

$$P(X = 2) = \frac{1}{16}$$

Probability distribution:

x	2	3	4	5	6	7	8
$P(X = x)$	$\frac{1}{16}$	$\frac{2}{16}$	$\frac{3}{16}$	$\frac{4}{16}$	$\frac{3}{16}$	$\frac{2}{16}$	$\frac{1}{16}$

Histogram:

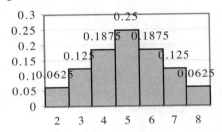

3. X = age of an *X Box* player

Probability distribution:

x	15	25	35	45
$P(X = x)$	.482	.386	.116	.016

Histogram:

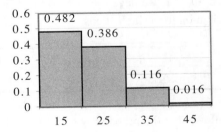

4. X = number of defective joysticks chosen when 3 are selected from a bin that contains 20 defective joysticks and 30 good ones

$$P(X = 0) = \frac{C(20,0)C(30,3)}{C(50,3)} \approx .2071$$

$$P(X = 1) = \frac{C(20,1)C(30,2)}{C(50,3)} \approx .4439$$

$$P(X = 2) = \frac{C(20,2)C(30,1)}{C(50,3)} \approx .2908$$

$$P(X = 3) = \frac{C(20,3)C(30,1)}{C(50,3)} \approx .0582$$

Histogram:

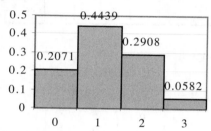

5. X = the number of ones that face up when 2 weighted dice are thrown

Probability distribution for each weighted dice: Take p to be the probability of a 2. The given information implies that the probability distribution for the die is

x	1	2	3	4	5	6
$P(X = x)$	$2p$	p	p	p	p	$2p$

Since $8p = 1$, we have $p = 1/8$, so the probability distribution for a single die is

x	1	2	3	4	5	6
$P(X = x)$	$\frac{1}{4}$	$\frac{1}{8}$	$\frac{1}{8}$	$\frac{1}{8}$	$\frac{1}{8}$	$\frac{1}{4}$

It follows that the probability of rolling a 1 is 1/4. Take this as the "success".

$$P(X = 0) = C(2, 0) \left(\frac{1}{4}\right)^0 \left(\frac{3}{4}\right)^2 = \frac{9}{16}$$

$$P(X = 1) = C(2, 1) \left(\frac{1}{4}\right)^1 \left(\frac{3}{4}\right)^1 = \frac{3}{4}$$

$$P(X = 2) = C(2, 2) \left(\frac{1}{4}\right)^2 \left(\frac{3}{4}\right)^0 = \frac{1}{16}$$

Probability distribution:

x	0	1	2
$P(X = x)$	$\frac{9}{16}$	$\frac{3}{4}$	$\frac{1}{16}$

Histogram:

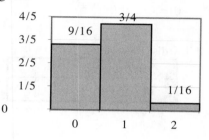

6. The first column lists the values of X, giving us the mean $\bar{x}$. The second column lists the numbers $(x_i - \bar{x})$, and the third column lists their squares.

x	$(x - \bar{x})$	$(x - \bar{x})^2$
-1	-3	9
2	0	0
0	-2	4
3	1	1
6	4	16
10	0	30

The bottom row shows the sums.

$$\bar{x} = \frac{10}{5} = 2 \qquad \sum_{i=1}^{n} (x_i - \bar{x})^2 = 30$$

$$s^2 = \frac{\sum_{i=1}^{n} (x_i - \bar{x})^2}{n-1} = \frac{30}{5-1} = 7.5$$

$$s = \sqrt{s^2} \approx 2.7386$$

For the median m, arrange the scores in order and select the middle score:

$-1, 0,$ **2**$, 3, 6$

$m = 2$

7. Two examples are:

0, 0, 0, 4

and

−1, −1, 1, 5

8. Since the standard deviation is 0, the scores must all be equal. Therefore, the only possibility is

2, 2, 2, 2, 2, 2

9. An example is −1, −1, −1, 1, 1, 1.

(Also see Exercise 68 in Section 8.4.)

Here is the calculation of the population standard deviation (the bottom row shows the sums):

x	$(x-\bar{x})$	$(x-\bar{x})^2$
−1	−1	1
−1	−1	1
−1	−1	1
1	1	1
1	1	1
1	1	1
0	0	6

$$\bar{x} = \frac{0}{6} = 0 \qquad \sum_{i=1}^{n} (x_i-\bar{x})^2 = 6$$

$$\sigma^2 = \frac{\sum_{i=1}^{n} (x_i-\bar{x})^2}{n} = \frac{6}{6} = 1$$

$$\sigma = \sqrt{\sigma^2} = 1$$

10. An example is −1, −1, 0, 1, 1.

Here is the calculation of the sample standard deviation (the bottom row shows the sums):

x	$(x-\bar{x})$	$(x-\bar{x})^2$
−1	−1	1
−1	−1	1
0	0	0
1	1	1
1	1	1
0	0	4

$$\bar{x} = \frac{0}{5} = 0 \qquad \sum_{i=1}^{n} (x_i-\bar{x})^2 = 4$$

$$s^2 = \frac{\sum_{i=1}^{n} (x_i-\bar{x})^2}{n-1} = \frac{4}{5-1} = 1$$

$$s = \sqrt{s^2} = 1$$

11–15. To construct the probability distribution for the weighted die, take p to be the probability of a 1. The given information implies that the probability distribution for the die is

x	1	2	3	4	5	6
$P(X = x)$	p	p	p	p	p	$2p$

Since $7p = 1$, we have $p = 1/7$, so the probability distribution for a single die is

x	1	2	3	4	5	6
$P(X = x)$	$\frac{1}{7}$	$\frac{1}{7}$	$\frac{1}{7}$	$\frac{1}{7}$	$\frac{1}{7}$	$\frac{2}{7}$

Throwing the die 4 times is a sequence of 4 Bernoulli trials with $p = 2/7$ and $q = 1-2/7 = 5/7$.

11. $P(X = 1) = C(4, 1) \left(\frac{2}{7}\right)^1 \left(\frac{5}{7}\right)^3 \approx .4165$

12. The probability that 6 comes up at most twice is

$P(X \le 2) = P(X = 0) + P(X = 1) + P(X = 2)$

$P(X = 0) = C(4, 0) \left(\frac{2}{7}\right)^0 \left(\frac{5}{7}\right)^4 \approx .2603$

$P(X = 1) = C(4, 1) \left(\frac{2}{7}\right)^1 \left(\frac{5}{7}\right)^3 \approx .4165$

$P(X = 2) = C(4, 2) \left(\frac{2}{7}\right)^2 \left(\frac{5}{7}\right)^2 \approx .2499$

Therefore,

$P(X \le 2) \approx .2603 + .4165 + .2499 = .9267$

13. The probability that X is at least 2 is

$P(X \ge 2) = 1 - P(X \le 1)$

$\approx 1 - (.2603 + .4165)$

(from part (b))

$= .3232$

14. $P(X > 3) = 1 - P(X \le 2)$

$\approx 1 - .9267 = .0067$

(using the answer from part (b))

15. $P(1 \le X \le 3) = P(X = 1) + P(X = 2) + P(X = 3)$
We computed $P(X = 1)$ and $P(X = 2)$ in part (b).

$P(X = 3) = C(4, 3) \left(\frac{2}{7}\right)^3 \left(\frac{5}{7}\right)^1 \approx .0666$

Therefore,

$P(1 \le X \le 3) = P(X = 1) + P(X = 2) + P(X = 3)$

$= .4165 + .2499 + .0666 = .7330$

16. One can approach this in two ways: (1) As a sequence of 3 Bernoulli trials with "success" = girl, $n = 3$ and $p = .5$ (2) As analogous to the 3-coin toss experiment, with H (heads) replaced by G (girl) and T (tails) replaced by B (boy). the problem then becomes identical to Example 3 in Section 8.1, and we get the following probability distribution:

x	0	1	2	3
$P(X = x)$	$\frac{1}{8}$	$\frac{3}{8}$	$\frac{3}{8}$	$\frac{1}{8}$

To compute the mean, we can either use

$\mu = np$ $3(.5) = 1.5$

or extend the above table to include a row for $xP(X = x)$ and then add.

$\sigma = \sqrt{npq} = \sqrt{3 \times .5 \times .5} \approx 0.8660$

To answer the last part,

$[\sigma - \mu, \, \sigma + \mu] = [0.634, \, 2.366]$

This interval does not include all 3.

$[\sigma - 2\mu, \, \sigma + 2\mu] = [-0.232, \, 3.232]$

This interval includes all the scores, so all values of X lie within 2 standard deviations of the expected value.

17. The frequencies add up to 16, so we obtain the probabilities by dividing the frequencies by 16:

x	$P(X = x)$	$xP(X = x)$
−3	.0625	−0.1875
−2	.125	−0.25
−1	.1875	−0.1875
0	.25	0
1	.1875	0.1875
2	.125	0.25
3	.0625	0.1875

$\mu =$ Sum of entries in right-hand column $= 0$.

$x - \mu$	$(x - \mu)^2$	$P(X = x) \times$ $(x - \mu)^2$
−3	9	0.5625
−2	4	0.5
−1	1	0.1875
0	0	0
1	1	0.1875
2	4	0.5
3	9	0.5625

$\sigma^2 = $ Sum of entries in right-hand column $= 2.5$

$\sigma = \sqrt{\sigma^2} \approx 1.5811$

For the last part, note that 14 of the 16 scores in the frequency table (exclude the first and last value) are in the interval $[-2, 2]$; that is, in the interval

$[\mu - 2, \mu + 2]$

because $\mu = 0$. The number of standard deviations that 2 represents is approximately

$\dfrac{2}{1.5811} \approx 1.3$ standard deviations

Therefore, 87.5% (or 14/16) of the time, X is within 1.3 standard deviations of the expected value.

18. By Chebyshev's rule, X is guaranteed to lie within k standard deviations with of μ with a probability of at least $1 - \dfrac{1}{k^2}$. Thus,

$1 - \dfrac{1}{k^2} = .90$

$\dfrac{1}{k^2} = 1 - .90 = .10$

$k^2 = \dfrac{1}{.10} = 10$

$k = \sqrt{10} \approx 3.162$

The associated interval is therefore

$[\mu - k\sigma, \mu + k\sigma]$

$\approx [100 - 3.162(16), 100 + 3.162(16)]$

$\approx [49.4, 150.6]$

19. Since X is bell-shaped and symmetric, the empirical rule applies so approximately 95% of samples of X are within the interval

$[\mu - 2\sigma, \mu + 2\sigma]$

$= [100 - 2(30), 100 + 2(30)] = [40, 160]$

20. From the table,

$P(0 \le Z \le 1.5) = .4332$

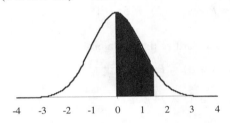

21. $P(Z \le -1.5) = .5 - .4322 = .0668$

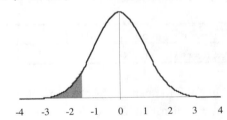

22. $P(|Z| \ge 2.1) = 2(.5 - .4821) = .0358$

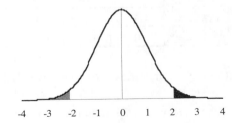

23. Standardize the given problem:

$$P(a \le X \le b) = P\left(\frac{a-\mu}{\sigma} \le Z \le \frac{b-\mu}{\sigma}\right)$$

$$P(80 \le X \le 120) = P\left(\frac{80-100}{16} \le Z \le \frac{120-100}{16}\right)$$

$$= P(-1.25 \le Z \le 1.25)$$

$$= 2(.3944) = .7888$$

24. Standardize the given problem:

$$P(X \le b) = P\left(Z \le \frac{b-\mu}{\sigma}\right)$$

$$P(X \le -1) = P\left(Z \le \frac{-1-0}{2}\right)$$

$$= P(Z \le -0.5) = .5 - .1915 = .3085$$

25. Standardize the given problem:

$$P(X \ge a) = P\left(Z \ge \frac{a-\mu}{\sigma}\right)$$

$$P(X \ge 1) = P\left(Z \ge \frac{1-(-1)}{0.5}\right)$$

$$= P(Z \ge 4) = .5 - .5000 = .0000$$

26. (a) The frequency distribution for the price is

x	5.50	10	12	15
fr(X=x)	1	2	3	4

Dividing the frequencies by the sum, 10 gives the probability distribution. We also add a row for the computation of $E(X)$:

x	5.50	10	12	15
$P(X = x)$	.1	.2	.3	.4
$xP(X = x)$	0	0.2	0.4	0.3

$E(X) = $ Sum of entries in bottom row = $12.15

(b) The frequency distribution for the weekly sales is

x	6200	3500	3000	1000
fr(X=x)	1	2	3	4

Dividing the frequencies by the sum, 10 gives the probability distribution. We also add a row for the computation of $E(X)$:

x	6200	3500	3000	1000
$P(X = x)$	.1	.2	.3	.4
$xP(X = x)$	620	700	900	400

$E(X) = $ Sum of entries in bottom row = 2620 copies

27. (a) Revenue = Price per copy sold × Number of copies sold. The values of the weekly revenue are obtained by multiplying the prices by the weekly sales:

x (Revenue)	34,100	35,000	36,000	15,000
$P(X = x)$	.1	.2	.3	.4
$xP(X = x)$	3410	7000	10,800	6000

$E(X) = $ Sum of entries in bottom row = $27,210

(b) False; let X = price and Y = weekly sales. Then weekly revenue = XY. However, $27,210 \ne 12.15 \times 2620$. In other words, $E(XY) \ne E(X)E(Y)$.

28. Take X to be the number of orders per million residents. For the values of X, use the rounded midpoints of the measurement classes given. To obtain the probabilities, divide each frequency by the sum of all the frequencies, 100:

x	2	4	6	8	10
Freq	25	35	15	15	10
$P(X = x)$	.25	.35	.15	.15	.10

To compute the expected value, add the $xP(X = x)$ row as usual:

x	2	4	6	8	10
$P(X = x)$	.25	.35	.15	.15	.10
$xP(X = x)$	0.5	1.4	0.9	1.2	1

μ = Sum of entries in bottom row = 5

$x - \mu$	−3	−1	1	3	5
$(x-\mu)^2$	9	1	1	9	25
$P(X=x)$ $\times (x-\mu)^2$	2.25	0.35	0.15	1.35	2.5

σ^2 = Sum of entries in bottom row = 6.6

$\sigma = \sqrt{\sigma^2} \approx 2.5690$

29. The empirical rule predicts that 68% of all orders will lie in the interval

$$[\mu - \sigma, \mu + \sigma] \approx [5 - 2.569, 5 + 2.569]$$
$$= [2.431, 7.569]$$

The empirical rule does not apply because the distribution is not symmetric. (However, it does give a fairly accurate prediction in this case.)

30. The original frequency distribution is

x	1-2.9	3-4.9	5-6.9	7-8.9	9-10.9
Freq	25	35	15	15	10

The shaded cells correspond to the cities from which you obtain between 3 and 8 orders per million residents. These cells account for a total of 35% + 15% = 50% of the cities. Therefore, *at least*

50% of the cities gave orders between 3 and 8 per million residents.

However, we know more: The additional 15 cities in the 7–8.9 category could conceivably all be ≤ 8, so we can conclude that *up to* 50% + 15% = 65% of the cities gave orders between 3 and 8 per million residents.

Conclusion: Between 50% and 65% of the cities gave orders between 3 and 8 per million residents [choice (A)].

31–35. Let X be the number of Mac OS users who will order books in the next hour, and let Y be the number of Windows users who will order books in the next hour. X and Y are both binomial random variables with the following properties:

X: $n = 10$, $p = .05$

Y: $n = 20$, $p = .10$

31. $P(Y = 3) = C(20, 3)(.10)^3(.90)^{17} \approx .190$

32. $P(Y \leq 3) = P(Y = 0) + P(Y = 1) + P(Y = 2) + P(Y = 3)$

$P(Y = 0) = C(20, 0)(.10)^0(.90)^{20} \approx .1216$
$P(Y = 1) = C(20, 1)(.10)^1(.90)^{19} \approx .2702$
$P(Y = 2) = C(20, 2)(.10)^2(.90)^{18} \approx .2852$
$P(Y = 3) = C(20, 3)(.10)^3(.90)^{17} \approx .1901$

Therefore, $P(Y \leq 3) \approx .1216 + .2702 + .2852 + .1901 \approx .867$

33. $P(X = 1) = C(10, 1)(.05)^1(.95)^9 \approx .3151$
$P(Y = 3) = C(20, 3)(.10)^3(.90)^{17} \approx .1901$

By independence,

$P(X = 1 \text{ and } Y = 3) = P(X = 1)P(Y = 3)$
$= .3151 \times .1901 \approx .060$

34. The event that a *Mac OS* visitor to the site orders books is independent of the event that a *Windows* visitor to the site orders books.

35. $E(X) = np = (10)(.05) = 0.5$

$E(Y) = np = (20)(.10) = 2$

The total number of expected orders is $0.5 + 2 = 2.5$

36. Skin cream: $\mu = 38$, $\sigma = 21$ (in thousands of dollars)

$P(X \geq 50) = P\left(Z \geq \dfrac{50-38}{21}\right) \approx P(Z \geq 0.571) =$

$.5 - .2157 \approx .284$

37. Hair products: $\mu = 34$, $\sigma = 14$ (in thousands of dollars)

$P(X \leq 50) = P\left(Z \leq \dfrac{50-34}{14}\right) \approx P(Z \leq 1.14) = .5 +$

$.3729 \approx .873$

38. Skin cream: $\mu = 38$, $\sigma = 21$ (in thousands of dollars)

$P(X \leq 12) = P\left(Z \leq \dfrac{12-38}{21}\right) \approx P(Z \leq -1.238)$

$= .5 - .3925 \approx .108$

Hair products: $\mu = 34$, $\sigma = 14$ (in thousands of dollars)

$P(X \leq 12) = P\left(Z \leq \dfrac{12-34}{14}\right) \approx P(Z \leq -1.57)$

$= .5 - .4418 \approx .058$

Conclusion: Skin cream is most likely to yield sales of less than $12,000 next month with a probability of approximately .108, compared with

hair products, with a probability of approximately .058.

39. People in the 3 Sigma Club have IQs with Z-values of at least 3.

$P(Z \geq 3) \approx .5 - .4987 = .0013$

Given a population of 280 million, this gives

$.0013 \times 280{,}000{,}000 = 364{,}000$

people. (A more accurate answer using technology is 378,000 people.)

40. To qualify for Mensa, your IQ must be at least 132.

$\mu = 100$, σ is unknown, $X = 132$

$Z = \dfrac{132 - 100}{\sigma} = \dfrac{32}{\sigma}$.

$P\left(Z \geq \dfrac{32}{\sigma}\right) = .02$

(Your IQ must be in the top 2%).

Therefore,

$P\left(0 \leq Z \leq \dfrac{32}{\sigma}\right) = .5 - .02 = .48$

Using the table backwards, we get

$Z \approx 2.06$

Therefore,

$\dfrac{32}{\sigma} \approx 2.06$

so $\sigma \approx \dfrac{32}{2.06} \approx 16$

41. Given $\mu = 100$, $\sigma = 16$, to get into the 3 Sigma club, your IQ must be at least 3 standard deviations above the mean; that is

$\mu + 3\sigma = 100 + 3(16) = 148$

Appendix A: Logic

1. False statement

3. Not a statement, since it is not a declarative sentence

5. True statement: You can add 1 to any number to get a larger one.

7. True (we hope!) statement

9. Not a statement, since it is self-referential.

11. Rephrase this as "Our Mayor is not trustworthy and he is not a good speller." In symbol form: $(\sim p)\wedge q$

13. Rephrase this as "Our Mayor is trustworthy and a patriot, and a good speller." In symbol form: $(p\wedge r)\wedge q$ or just $p\wedge q\wedge r$.

15. Rephrase this as "Our Mayor is trustworthy or our mayor is not trustworthy." In symbol form: $p\vee(\sim p)$

17. $p\wedge(\sim r)$: Willis is a good teacher and his students do not hate math.

19. $q\vee(\sim q)$: Either Carla is a good teacher, or she is not.

21. $r\wedge(\sim r)$: Willis' students both hate and do not hate math.

23. $\sim(q\vee s)$: It is not true that either Carla is a good teacher or her students hate math.

25. Polly sings well and Quentin writes well. This statement has the form $p\wedge q$. Since q is false, $p\wedge q$ is false as well by the truth table for conjunction $\wedge$. Answer: F

27. Polly sings poorly and Quentin writes well. This statement has the form $(\sim p)\wedge q$. Since q is false, $(\sim p)\wedge q$ is false as well by the truth table for conjunction $\wedge$. Answer: F

29. Either Polly sings well and Quentin writes poorly, or Rita is good at math. This statement has the form $[p\wedge(\sim q)]\vee r$. Since r is true, $[p\wedge(\sim q)]\vee r$ is true as well, regardless of the truth value of $[p\wedge(\sim q)]$. Answer: T

31. Either Polly sings well or Quentin writes well, or Rita is good at math. This statement has the form $(p\vee q)\vee r$. p is true, $p\vee q$ is true as well, and so $(p\vee q)\vee r$ must be true. Answer: T

33. "If 1=1, then 2=2." T
Solution:
The given statement has the form $p\to q$ where p: "$1 = 1$" is true and q: "$2 = 2$" is also true. Therefore the statement is true. (First row of the truth table for the conditional)

35. "If $1\neq 0$, then $2\neq 2$."
The given statement has the form $p\to q$ where p: "$1 \neq 0$" is true and q: "$2 \neq 2$" is false. Therefore the statement is false. (Second row of the truth table for the conditional)

37. "A sufficient condition for 1 to equal 2 is 1=3."
The given statement has the form $p\to q$ where p: "$1 = 3$" is false and q: "$1 = 2$" is false. Therefore

the statement is true. (Fourth row of the truth table for the conditional).

39. "1 = 0 is a necessary condition for 1 to equal 1."

The given statement has the form $p \rightarrow q$ where p: "1 = 1" is true and q: "1 = 0" is false. Therefore the statement is false. (Second row of the truth table for the conditional)

41. "If I pay homage to the great Den, then the sun will rise in the east."

The given statement has the form $p \rightarrow q$ where p: "I pay homage to the great Den" may be true or false and q: "The sun will rise in the east" is true. Therefore the statement is true. (First or third row of the truth table for the conditional)

43. "In order for the sun to rise in the east, it is necessary that it sets in the west."

The given statement has the form $p \rightarrow q$ where p: "The sun will rise in the east" is true, and q: "The sun will set in the west" is true. Therefore the statement is true. (First row of the truth table for the conditional)

45. "The sun rises in the west only if it sets in the west."

The given statement has the form $p \rightarrow q$ where p: "The sun will rise in the west" is false, and q: "The sun will set in the east" is false. Therefore the statement is true. (Fourth row of the truth table for the biconditional)

47. "In order for the sun to rise in the east, it is necessary and sufficient that it sets in the west."

The given statement has the form $p \rightarrow q$ where p: "The sun will rise in the eat" is true, and q: "The sun will set in the west" is true. Therefore the

statement is true. (First row of the truth table for the biconditional)

49.

p	q	$\sim q$	$p \wedge \sim q$
T	T	F	F
T	F	T	T
F	T	F	F
F	F	T	F

51.

p	$\sim p$	$\sim(\sim p)$	$\sim(\sim p) \vee p$
T	F	T	T
F	T	F	F

53.

p	q	$\sim p$	$\sim q$	$(\sim p) \wedge (\sim q)$
T	T	F	F	F
T	F	F	T	F
F	T	T	F	F
F	F	T	T	T

55.

p	q	r	$p \wedge q$	$(p \wedge q) \wedge r$
T	T	T	T	T
T	T	F	T	F
T	F	T	F	F
T	F	F	F	F
F	T	T	F	F
F	T	F	F	F
F	F	T	F	F
F	F	F	F	F

57.

p	q	r	$q \vee r$	$p \wedge (q \vee r)$
T	T	T	T	T
T	T	F	T	T
T	F	T	T	T
T	F	F	F	F
F	T	T	T	F
F	T	F	T	F
F	F	T	T	F
F	F	F	F	F

59.

p	q	q∨p	p→(q∨p)
T	T	T	T
T	F	T	T
F	T	T	T
F	F	F	T

61.

p	q	p∨q	p↔(p∨q)
T	T	T	T
T	F	T	T
F	T	T	F
F	F	F	T

63.

p	p∧p
T	T
F	F

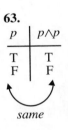

same

65.

p	q	p∨q	q∨p
T	T	T	T
T	F	T	T
F	T	T	T
F	F	F	F

same

67.

p	q	p∨q	~(p∨q)	~p	~q	(~p)∧(~q)
T	T	T	F	F	F	F
T	F	T	F	F	T	F
F	T	T	F	T	F	F
F	F	F	T	T	T	T

same

69.

p	q	r	p∧q	(p∧q)∧r	q∧r	p∧(q∧r)
T	T	T	T	T	T	T
T	T	F	T	F	F	F
T	F	T	F	F	F	F
T	F	F	F	F	F	F
F	T	T	F	F	T	F
F	T	F	F	F	F	F
F	F	T	F	F	F	F
F	F	F	F	F	F	F

same

71.

p	q	p→q	~p	~q	(~q)→(~p)
T	T	T	F	F	T
T	F	F	F	T	F
F	T	T	T	F	T
F	F	T	T	T	T

same

73. First construct the truth table:

p	~p	p∧~p
T	F	F
F	T	F

Since the last column contains only Fs, the given statement is a contradiction.

75. First construct the truth table:

p	q	p∨q	~(p∨q)	p∧~(p∨q)
T	T	T	F	F
T	F	T	F	F
F	T	T	F	F
F	F	F	T	F

Since the last column contains only Fs, the given statement is a contradiction.

77. First construct the truth table:

p	q	p∧q	~(p∧q)	p∨~(p∧q)
T	T	T	F	T
T	F	F	T	T
F	T	F	T	T
F	F	F	T	T

Since the last column contains only Ts, the given statement is a tautology.

79. The commutative law for disjunction can be stated as $A \lor B \equiv B \lor A$.

Applying it with $A = p$ and $B = {\sim}p$ gives $p \lor ({\sim}p) \equiv ({\sim}p) \lor p$.

Answer: $({\sim}p) \lor p$

81. One form of DeMorgan's Law can be stated as ${\sim}(A \land B) \equiv ({\sim}B) \lor ({\sim}A)$.

Applying it with $A = p$ and $B = {\sim}q$ gives ${\sim}(p \land ({\sim}q)) \equiv ({\sim}p) \lor {\sim}({\sim}q)$.

Answer: $({\sim}p) \lor {\sim}({\sim}q)$

83. One form of the Distributive Law can be stated as $A \lor (B \land C) \equiv (A \lor B) \land (A \lor C)$.

Applying it with $A = p$, $B = ({\sim}p)$ and $C = q$ gives $p \lor (({\sim}p) \land q) \equiv (p \lor ({\sim}p)) \land (p \lor q)$.

Answer: $(p \lor ({\sim}p)) \land (p \lor q)$.

85. Let p: "I am Julius Caesar," and q: "You are a fool."

The given statement is ${\sim}(p \land q)$.

By De Morgan's Law, it is equivalent to $({\sim}p) \lor ({\sim}q)$: "Either I am not Julius Caesar or you are not a fool."

87. Let p: "It is raining," q: "I have forgotten my umbrella," and r: "I have forgotten my hat."

The given statement is $(p \land q) \lor (p \land r)$.

By the Distributive Law (used in reverse), it is equivalent to $p \land (q \lor r)$: "It is raining and I have forgotten either my umbrella or my hat."

89. The given statement has the form $p \to q$ where p: "I am,", q: "I think."

The contrapositive of $p \to q$ is ${\sim}q \to {\sim}p$: "If I do not exist, then I do not think."

The converse of $p \to q$ is $q \to p$: "If I am, then I think."

91. The given statement has the form $p \to q$ where p: "You worship Den,", q: "You sacrifice beasts of burden." The choices for the answer are:

(A): ${\sim}q \to {\sim}p$

(B) $q \to p$

(C) ${\sim}p \to {\sim}q$

Only choice (A) is logically equivalent to $p \to q$. It is the contrapositive of the given statement.

93. $h \to t$

$\underline{\qquad h \qquad}$

$\therefore \ t$

Valid; Modus Ponens,

95. $r \to u$

$\underline{\qquad {\sim}r \qquad}$

$\therefore \ {\sim}u$

Invalid

97. $g \to m$

$\underline{\qquad {\sim}m \qquad}$

$\therefore \ {\sim}g$

Valid: Modus Tollens

99. $m \lor b$

$\underline{\qquad {\sim}b \qquad}$

$\therefore \ m$

Valid: Disjunctive Syllogism

101. $s \lor a$

$\underline{\qquad a \qquad}$

$\therefore \ {\sim}s$

Invalid

103. Let s: "John is a swan," and g: "John is green." The partial argument has the form

$$s \rightarrow g$$
$$\underline{s \qquad}$$
$$\therefore \quad \underline{\qquad}$$

Modus Ponens tells us to put g as the conclusion: "John is green."

105. Let s: "John is a swan," and g: "John is green." The partial argument has the form

$$s \rightarrow g$$
$$\underline{\sim g \qquad}$$
$$\therefore \quad \underline{\qquad}$$

Modus Tollens tells us to put $\sim s$ as the conclusion: "John is not a swan."

107. Let s: "John is a scholar," and g: "John is a gentleman." The partial argument has the form

$$s \vee g$$
$$\underline{\sim s \qquad}$$
$$\therefore \quad \underline{\qquad}$$

Disjunctive Syllogism tells us to put g as the conclusion: "He is a gentleman."

109. Their truth tables have the same truth values for corresponding values of the variables.

111. For the truth table of $A \vee B$ to be a contradiction, it must have Fs in the final column. For this to happen, the columns corresponding to A and to B must both be all Fs. In other words, both A and B must be contradictions.

113. Answers may vary. Let p: "You have smoker's cough," and q: "You smoke."

115. Let p: "It is summer in New York," and q: "It is summer in Seattle."